Proceedings of the Xth International Scientific Congress in fur animal production

Proceedings of the Xth International Scientific Congress in fur animal production

Scientifur volume 36 (3/4)

Edited by:
P.F. Larsen
S.H. Møller
T. Clausen
A.S. Hammer
T.M. Lássen
V.H. Nielsen
A.H. Tauson
L.L. Jeppesen
S.W. Hansen
J. Elnif
J. Malmkvist

Wageningen Academic
Publishers

ISBN: 978-90-8686-205-4
e-ISBN: 978-90-8686-760-8
DOI: 10.3921/978-90-8686-760-8

First published, 2012

Table of contents

Preface

Dear reader,

Scientifur Volume 36, No. 3/4 is the proceedings from the IFASA 2012 congress held in Copenhagen, Denmark in 21-24 August 2012 and hosted by Kopenhagen Fur. As editors we were pleased to have received 88 abstracts for IFASA 2012 congress. We were even more thrilled that the majority of scientists wanted to give an oral presentation of their work to their colleagues, even though this challenged the group of editors with the difficult task of disappointing some. We have therefore made an effort to give ample time and good location for the many poster presentations, in order to make room for thorough reading and discussion of results presented in this way. These proceedings contain the revised papers within the topics: Nutrition, feeding and management, Health and disease, Breeding, genetics and reproduction, Behaviour and welfare, and the theme on 'WelFur for mink and foxes'. Finally, some contributions in the form of short communications are presented in Part 6. The reviews have focused on making results clear and easy to understand. The content is the responsibility of the author and has not been part of Scientifur's paper review process.

We would like to thank all of the people who have made the Xth International Scientific Congress in Fur Animal Production possible in Copenhagen, Denmark. We hope you will enjoy the congress as well as the proceedings!

The editors
Peter Foged Larsen
Steen H. Møller
Tove Clausen
Anne Sofie Hammer
Tor Mikael Lássen
Vivi Hunnicke Nielsen
Anne-Helene Tauson
Leif Lau Jeppesen
Steffen W. Hansen
Jan Elnif
Jens Malmkvist

Part 1. Nutrition, feeding and management

Protein digestibility of some traditional and new feed ingredients for mink

Ø. Ahlstrøm[1], M. Tjernsbekk[1] and A-H. Tauson[1,2]
[1]Department of Animal and Aquacultural Sciences, Norwegian University of Life Sciences, Ås, Norway; [2]Department of Basic Animal and Veterinary Sciences, Faculty of Life Sciences, University of Copenhagen, Denmark; oystein.ahlstrom@umb.no

Abstract

The study comprised 19 protein sources, four meals made from salmon by-products, three enzyme-hydrolyzed protein concentrated products made from salmon by-products, two fish meals, three krill meals, two poultry meals, one lamb meal, raw poultry by-products, raw cod scraps, raw eggs and cooked eggs. Protein digestibility was measured in adult male mink. Protein digestibility was significantly different among the sources, it was highest for one of the hydrolyzed salmon protein concentrates, 91.4% (SD 0.5), and lowest for the lamb meal, 64.9% (SD 1.2). Within the different categories of protein meals (salmon by-products meals, the hydrolyzed by-products, krill meals) there were also significantly differences in protein digestibility. Some of these products revealed similar chemical content and protein digestibility as high quality fish meal, but some did not. Raw egg white has been shown to reduce protein digestibility in young dogs. No difference in protein digestibility between raw and cooked eggs was observed in mink. Generally, the raw protein sources revealed digestibility coefficients between 85 and 90%, while most of the meals revealed values between 75 and 85%. The study showed that animal derived meals most often have protein digestibility around 80%, but variation can be large and digestibility studies are needed to have exact information on each protein ingredient.

Keywords: protein sources, salmon by-products, krill meal, raw egg

Introduction

Numerous raw materials with various origin, conservation and storage are involved in fur animal feeding. Animal by-product ingredients have been the basis for the feed production because of the high value nutritional value and palatability for foxes and mink. These ingredients include fresh, frozen stored, acid preserved and dry meals made of raw materials of animal origin. Some of these raw materials have no other alternative use but destruction. The flexibility in choice of ingredients is high in fur animal production because the commercial product is none-food and concerns regarding product quality as in human food productions are not considered. Moreover, wet feed production allows for utilization of both wet and dry ingredients and the rapid turnover of the feed opens for short time changes in the ingredient composition if necessary. The feed industry will always search for new and cheaper feed raw materials and the market price will be a main decisive factor for which raw materials to be used. Nutritional quality is the important factor to take into consideration before new ingredients are incorporated into the feed. Protein supply, which is a main nutritional focus in fur animal production, is dependent on amino acid level, amino acid composition as well as digestibility. Digestibility values will directly affect the amino acids and energy supplied by the protein source. Protein digestibility is therefore an important factor in the evaluation of protein quality of feed ingredients.

Our study aimed at making screening of protein content and protein digestibility in some traditional and some new potential ingredients in fur animal feed. The inclusion of the traditional ingredients in study was done to make good a set of comparisons with the new ingredients.

Material and methods

Animals and diets

The digestibility study comprised 19 protein ingredients; four meals from salmon by-products (Marine Harvest, Bergen Norway), three enzyme hydrolyzed protein concentrate meals made of salmon by-products (Marine Bioproducts, Storbø, Norway), two fish meals (Nordsildmel, Bergen, Norway), three krill meals (Aker Biomarine, Oslo, Norway), one lamb meal (Norsk Protein, Ingeberg, Norway), two poultry meals (Meal A, low ash, Gepro, Diepholz, Germany, Meal B, Norsk Protein, Ingeberg, Norway), raw poultry by-products (deboned chicken meat, Prior, Rakkestad, Norway), raw cod scraps (Fryserienes Fôromsetning, Tromsø, Norway) and raw eggs and cooked eggs. The protein ingredients were added as the only protein source in the experimental diets. The other ingredients were soybean oil, precooked corn starch, cellulose powder, vitamin/mineral mixture and water. Protein, fat and carbohydrate accounted for approximately 32, 35 and 33% of dietary dry matter content in the experimental diets, respectively.

A group of 24 adult males of the black genotype (>6 months) comprised the experimental animals at the laboratory. Body weights (BW) of the males were between 2.0 and 2.5 kg. Four males were given each of diets and digestibility was calculated as the mean of the four replicates. Daily feed allowances were planned to approximately meet the maintenance metabolizable energy (ME) requirement. Daily rations were approximately 150 g per day corresponding to 60-65 g dry matter. The mink was kept in standard cages equipped for quantitative collection of faeces and separation of urine. The animals were fed once daily and had free access to drinking water. The experiment lasted for seven days of which the three first was an adaptation period. During the four last days the feed intake were measured precisely and faeces were collected daily. Faeces were kept frozen stored during the experiment and freeze dried before analysis.

Apparent protein digestibility was calculated quantitatively using this formula:

$$\text{Digestibility (\%)} = ((\text{Protein consumed} - \text{Protein excreted in faeces})/(\text{Protein consumed}) \times 100$$

Chemical analyses

Samples of protein sources and faeces were analysed for dry matter (by heating at 105 °C for 16-18 h), ash (by combustion at 550 °C to constant weight), crude protein as nitrogen×6.25 (by the semi-micro-Kjeldahl method, Kjeltec-Auto System, Tecator, Sweden), fat (by diethylether extraction in a Fosstec analyzer (Tecator, Sweden) after HCl-hydrolysis.

Statistics

Differences in protein digestibility were tested by analyses of variance (GLM procedure by SAS) and difference between sources were tested with Students T test. Significant differences between products were set at $P<0.05$.

Results and discussion

All the digestibility experiments were carried out without any problems and the feed intakes were close to 100% for all diets. Thus, palatability of all the protein sources was satisfactory. The standard deviations in protein digestibility were minor among animals given each diet, mainly equal or less than 1.0 (Table 1).

Chemical content and digestibility

The four salmon by-products meals was produced from partly viscera and partly scraps. According to the producer the mixture of viscera and scraps were random and not identical

*Table 1. Chemical content (g/kg), apparent crude protein (N*6.25) digestibility (%). Standard deviation in parentheses.*

Product	Dry matter	Ash	Protein	Fat	Protein digestibility[1]
Salmon meal A	900	82	528	237	81.8 (0.6) f
Salmon meal B	918	118	644	100	74.7 (0.8) g
Salmon meal C	929	126	576	182	80.0 (1.4) f
Salmon meal D	923	115	622	161	84.4 (0.6) e
Hydrolyzate A	928	115	646	161	84.7 (1.2) de
Hydrolyzate B	597	64	519	12	91.4 (0.5) a
Hydrolyzate C	934	108	823	6	89.9 (0.9) ab
Norse LT 94	928	130	699	92	87.9 (0.5) bc
Norseamink	908	145	695	76	81.5 (0.6) f
Krill meal A	900	82	528	237	81.7 (1.0) f
Krill meal B	918	118	644	100	85.1 (0.6) de
Krill meal C	943	117	663	80	76.1 (1.0) g
Poultry meal A	938	110	658	139	80.9 (1.7) f
Poultry meal B	952	127	601	167	67.4 (1.8) i
Lamb meal	933	205	563	120	64.9 (1.2) h
Raw cod scraps	279	58	202	8	85.2 (0.4) de
Raw poultry by.	403	21	153	205	88.2 (0.9) bc
Raw eggs	352	11	100	77	86.8 (0.6) cde
Raw boiled eggs	362	11	101	81	87.4 (0.5) cd

[1] Different letters with protein digestibility values indicate significant difference ($P<0.05$).

among the batches of meal. Viscera contain more fat and less protein and ash compared with scraps. The chemical content analyses among the four meals confirmed that mixture of the two ingredients varied. Protein content varied from 528 to 622 g/kg and fat content from 100 to 237 g/kg and ash from 82 to 126 g/kg (Table 1). Compared with the guaranteed chemical content of Norse LT 94 and Norseamink qualities of fish meal, the protein levels of the salmon meals were lower. The minimum guaranteed protein content of Norse LT 94 and Norseamink is 680 g/kg and the typical content is even higher, 710 g/kg. Regarding the fat and ash content, Norse LT 94 and Norseamink have maximum guaranteed fat content of 115 g and 130 g/ kg, respectively, and a maximum ash content of 140 g/kg. Only Salmon meal B obtained such low fat content, while all four salmon meals had lower ash content than 140 g/kg. The protein digestibility varied among the salmon meals (Table 1). High ash content is known to be an indicator of low protein digestibility because of increased proportion of protein from structural proteins such as collagen from connective tissue and bone (Skrede, 1979a). No such relation was observed with the four salmon meals, but the ash content was generally low for all the salmon meals in the present study compared with other type by-product meals made of cod scraps (Ahlstrøm *et al.*, 2004). The variation in chemical content and protein digestibility for the salmon meals suggests that the quality of these meals was not as high as for Norse LT 94, but comparable to that of Norseamink.

The enzyme-hydrolyzed protein concentrated products were made of similar by-products from salmon as the meals described above. Two of them were meals and one was a liquid, hydrolyzed protein B. (Table 1). The hydrolyzed proteins are mainly used in pet-foods to provide higher protein digestibility in special purpose foods. The chemical compositions for the three products were in accordance with the product declarations given by the producer. Previous tests carried out by the producer have showed protein digestibility values of more than 90% in mink and roosters. However, our study revealed somewhat lower digestibility values, but still the highest among the products in the present study.

Norse LT 94 and Norseamink are produced from raw, whole fish that is cooked, pressed, dried and milled. The raw material is gently dried to avoid any heat damage that may reduce protein digestibility. Norse LT 94 have a minimum biological protein digestibility of 90% (true digestibility in mink), while the Norseamink quality meals have somewhat lower digestibility (Ahlstrøm *et al.*, 2004). The batches of Norse LT 94 and Norseamink meal that were included in the present experiment revealed declared values for chemical content and protein digestibility, except for slightly higher ash content than maximum level for the Norseamink fishmeal. The protein digestibility value given for the Norse LT fishmeal is the apparent value (Table 1). If true digestibility is estimated by applying endogenous protein factors determined by Skrede (1979b), digestibility will be higher than 90% in accordance with the declaration.

Utilization of Antartic krill (*Euphasia superba*) has increased during the last decade because higher demand for marine protein and fat sources. The three krill meals included in our study was produced at sea on the vessel. Two of the meals were defatted (B,C) and therefore contained more protein and less fat than the crude meal (A). Krill meal A revealed fairly similar chemical content as found in a recent published study by Øvrum Hansen *et al.* (2010). Protein digestibility varied significantly among the meals in our study, from 76.1 to 85.1%. The reason for this variation is not clear, but it could be caused by different content of krill shells that have been shown to reduce protein digestibility in Atlantic salmon (Øvrum Hansen *et al.*, 2010). When comparing the krill meals with the other protein sources in this study, they had comparable chemical content and protein digestibility as the salmon meals.

The chemical composition of the poultry meals A and B was similar, but the protein digestibility was much lower for poultry meal B. The most likely reason for the difference was the temperature during heat treatment. Poultry meal B was heat treated at 133°C, while the drying temperature for poultry meal A was not known, but it was probably considerably lower than that of poultry meal B. Similar to poultry meal B, the lamb meal had gone through the same harsh heat treatment and came out with even lower protein digestibility. The chemical content and protein digestibility of poultry meal B and the lamb meal is typical for meat-and bone meal (Ahlstrøm *et al.*, 2004; Cramer *et al.*, 2007). These two products revealed the poorest protein digestibility in the present study. Moreover, the amino acid composition and digestibility (not presented) revealed that these ingredients cannot be applied as the only protein sources in fur animal feed.

The chemical composition and digestibility of the raw cod scraps were typical and similar to results from other reports (Skrede, 1979; Ahlstrøm *et al.*, 2004). The low fat content and high protein content make this ingredient very useful for regulation of protein level of fur animal feed. In contrast, the raw poultry by-products contained more fat than protein and is therefore primarily a fat source. Both these ingredients showed very high protein digestibility in comparison with the different types of meals.

Protein digestibility of raw and cooked eggs was carried because previous studies in young dogs have indicated poor protein digestibility of raw eggs because of an anti-trypsin factors present in raw egg white, but not in cooked (Lineweaver *et al.*, 1947). Since the use of raw eggs in mink nutrition only has been an issue due to the content of avidin that can destroy biotin, we wanted to examine whether protein digestibility also was affected. The results showed no difference in protein digestibility between raw and cooked eggs in mink.

Conclusions

Protein sources such as salmon meals, enzyme-hydrolyzed protein concentrates and krill meals reveal high protein digestibility in mink. The study confirmed that most animal derived meals have protein digestibility around 80%, but variation can be large and digestibility studies are needed to have exact information on each protein ingredient.

References

Ahlstrøm Ø., Alden E., Børsting C.F., Dahlman T., Elnif J., Hansen N.E., Mäkelä J., Pölönen I. and Skrede A., 2004. Handbok for fôrmidler til pelsdyr, NJF-utredning/rapport nr. 502, Nordic Association of Agricultural Scientists, Subsection for fur animals, Ås, Norway, 74 p.

Cramer, K.R., Greenwood, M.W., Moritz, J.S., Beyer, R.S. and Parsons, C.M. 2007. Protein quality of various raw and rendered by-products meals commonly incorporatedinto companion animal diets. J Anim Sci 85: 3285-3293.

European Community Directive, 1998. Commision directive 98/64/EC. European Commission, Brussels.

Linewater, H. and Murray, C.W., 1947. Identification of the trypsin inhibitor of egg white with ovomucid. J. Biol. Chem 171: 565-581.

Skrede, A., 1979a. Utilization of fish and animal by-products in mink nutrition. V. Content and digestibility of amino acids in cod (*Gadus morrhua*) byproducts. Acta Agric. Scand. 29, 353-362.

Skrede, A., 1979b. Utilization of fish and animal by-products in mink nutrition. IV. Fecal excretion and digestibility of nitrogen and amino acids by mink fed cod (*Gadus morrhua*) fillet or meat and bone meal.

Øvrum Hansen, J., Penn, M., Øverland, M., Shearer, K.D., Krogdahl, Å., Mydland, L.T. and Storebakken, T. 2010. High inclusion of partially deshelled and whole krill meals in diets for Atlantic salmon (*Salmo salar*). Aquaculture 310: 164-172.

Effects of low-protein, *DL*-methionine and lysine-supplemented diets on growth performance of blue foxes (*Alopex lagopus*) during the growing-furring period

H.H. Zhang, F.H. Yang, Q.K. Jiang, Z.G. Yue, X.M. Xing, W.L. Sun, G.Y. Li
Institute of Special Animal and Plant Sciences, Chinese Academy of Agricultural Sciences, Changchun 130112, P.R. China; zhh83@126.com

Abstract

An experiment was carried out to examine the effects of low-protein diets supplemented with different levels of *DL*-methionine (Met) and Lysine (Lys) on growth performance of growing-furring blue foxes in order to find the optimal dietary supplementation levels of Met and Lys. For two protein levels, conventional 27% (P27) and low 19% (P19) on dry matter basis, respectively, and the low-protein diets were supplemented with Met (0.3%, 0.5%, 0.7%) and Lys (0.4%, 0.6%, 0.8%). An entirely random experimental design was adopted with two factors (3×3) and totally 10 groups (P27, L1M1, L1M2, L1M3, L2M1, L2M2, L2M3, L3M1, L3M2 and L3M3). From mid-September to pelting, based on the average daily gain, daily N retention, N retention ratio and the performance of blue foxes in different groups, 0.6% Met supplementation in low-protein diet was optimum; based on the daily N retention, N biological value and the quality of the fur, 0.3% and 0.5% Lys supplementation were optimum; based on the N apparent digestibility and daily N output, 0.3% Lys supplementation was optimum. In this experiment, the performance of blue foxes in L1M2 0.3% Lys×0.6% Met group was better than that in the other groups, which indicates that low-protein diets supplemented with *DL*-methionine and lysine for blue foxes can be beneficial to reduce feed expenses and nitrogen emission to the environment.

Keywords: low-protein; growth performance; fur characteristic; blue fox (*Alopex lagopus*)

Introduction

Blue fox (*Alopex lagopus*) is by nature a carnivorous species with a high protein requirement, and diets used in commercial blue fox production characteristically contain high levels of animal protein. Feed constitutes the greatest individual item of cost, amounting to 50% of the total production costs in fox farming (Hernesniemi, 2000). As protein is the most expensive dietary nutrient, any reduction in protein level contributes to a saving in production costs, and moreover, to a reduction in nitrogen emission. In the light of findings in growing-furring silver foxes, Harris *et al.* (1951) concluded that 28% of protein in dry matter (DM) was sufficient for growth as that in higher protein levels. A reduction in protein levels, from 45% to 22% of metabolic energy (ME), was reported not to affect body weight. However, a protein level below 28%-30% of ME significantly reduced body length at pelting (Rimeslåtten, 1976). The established blue fox nutrition requirements do not contain any data on amino acids. In mink, another carnivorous fur-bearing animal, researches on dietary protein and amino acids are intensive (Kerminen-Hakkio *et al.*, 2000). These studies displayed that methionine (Met) was the first limiting amino acid for hair growth and thus for fur quality, and Lys was the second limiting amino acid. There were some experiments which confirmed that animals fed diets of low-protein with some essential amino acid supplementations had no significant differences in their growth performance compared with those fed conventional diets (Kerr *et al.*, 1995). Earlier feeding experiments showed that it was possible to decrease the recommended protein level during the growing-furring season of blue foxes (Dahlman *et al.*, 2002; Koskinen *et al.*, 2005). The crucial point of this study is to evaluate the growth performance, N balance and fur characteristics of blue foxes fed low-protein diets supplemented with Met and Lys during the growing-furring period, and to quantify the optimal supplementation levels of amino acids in the low-protein diets.

Materials and methods

The experiment was carried out at the Fur Animals Experiment Station of Institute of Special Animals and Plants Sciences of the Chinese Academy of Agricultural Sciences (44° N, 126° E) in Northeast China during the period from September 27 to December 13.

Diet

Foxes were fed two protein levels: conventional 27% (Control) and low 19% (Basal experimental diet) on dry matter basis (Table 1), and the low-protein diets were supplemented with Met (M1 0.3%, M2 0.5%, M3 0.7%) and Lys (L1 0.4%, L2 0.6%, L3 0.8%).

Animals and experimental design

One hundred and twenty male blue foxes were selected with similar weight (5.34±0.57 kg) from different litters in an attempt to minimize genetic variation. The animals were allotted to 10 treatments of 12 animals each and fed twice a day at 08:00 and 16:00. Water and experimental diets were provided for *ad libitum* consumption throughout the trial. Before the formal experiment, the animals were accustomed to the experimental feed for two weeks. The N-balance experiments were carried out with eight 21-week old male blue foxes from each treatment group. The feces and urine collection period lasted for 3 days.

Chemical analyses

The chemical compositions of feed and feces were analyzed by standard methods. Dry matter (DM), ash, crude protein (CP), calcium, and phosphorus contents were analyzed according to AOAC procedures. The calculation of ME content and the proportional composition of ME were based on the digestibility coefficients achieved and the following ME values: protein 18.8 MJ/kg, fat 39.8 MJ/kg and carbohydrate 17.6 MJ/kg.

Calculations and statistical analyses

Statistical analyses of experimental data were performed with the general linear model (GLM) procedure of the SAS statistical package.

Table 1. Composition nutrient levels of the diets (DM basis) %.

Ingredients	Control diet	Basal experimental diet
Extruded soybean	18.0	20.0
Extruded corn	34.0	42.0
Soybean meal	9.0	5.0
Bone and meat meal	10.0	6.0
Corn germ meal	7.5	10.0
Fish meal	16.0	12.0
Lysine	0.4	
DL-methionine	0.6	
Soy oil	3.5	4.0
Premix[1]	1.0	1.0
Total	100.0	100.0
Crude protein	27.12	19.01
Fat	10.22	10.59
Carbohydrates	45.65	52.28
ME/(MJ/kg)	17.20	16.99
Lys	1.38	1.22
Met	0.63	0.59
Cys	0.43	0.40

[1] The premix provided the following per kg of diet: Ca(as $CaHPO_4{\cdot}2H_2O$), 6.4mg; P(as $Ca(H_2PO_4)_2{\cdot}H_2O$), 4.4mg; Mg(MgO), 1.6mg; Na (NaCl), 24mg;Fe (as $FeSO_4{\cdot}H_2O$), 16 mg; Cu (as $CuSO_4{\cdot}5H_2O$), 4.0 mg; Zn (as $ZnSO_4{\cdot}H_2O$), 10 mg; Mn (as $MnSO_4{\cdot}H_2O$), 12mg; VA, 300 IU; VB_1, 0.15 mg; VB_2, 0.40 mg; VB_6, 0.30 mg; folic acid, 0.30 mg; nicotinic acid, 1.60 mg, *D*-pantothenic acid, 1.3 mg.

Results

Feed intake and growth performance

The ADFI values were affected by Lys levels ($P<0.01$), and were higher in L1 and L2 than in L3. Final body weight, average daily gain (ADG), body length and dry fur length were similar among the groups ($P>0.05$, Table 2). Different Met and Lys levels did not affect the ADG ($P>0.05$), L1M2 was the highest one. There was no Met effect or Met-Lys interaction on the ADFI and ADG ($P>0.05$).

N-balance

Daily N intake, urine nitrogen (UN) and N retention were significantly affected by dietary protein levels with amino acid supplements ($P<0.01$, Table 3). N intake was affected by different levels of Lys ($P<0.01$), and L2 group was the highest one. Fecal nitrogen (FN) of the control group was higher than most low-protein groups, except for L2M2, L2M3 and L3M1 ($P<0.05$). Different

Table 2. Effects of low-protein, Lys and Met-supplemented diets on performance of blue foxes.

Treatment[1]		Final weight (kg)	ADG (g/d)	ADFI (g/d)	Body length (cm)	Dry fur length (cm)
Control		8.69±0.67	55.67±6.71	334.88±5.37^{CD}	67.12±2.70	107.14±2.97
L1M1		8.57±0.44	53.61±5.33	350.48±29.04^{BC}	64.88±1.81	105.29±2.75
L1M2		8.79±0.84	57.51±5.27	364.93±14.17^{AB}	68.44±2.72	106.67±3.20
L1M3		8.71±0.81	56.90±8.21	355.73±17.33^{BC}	67.38±2.50	105.83±2.04
L2M1		8.77±0.49	57.17±7.62	350.92±14.01^{BC}	67.25±2.43	107.43±4.03
L2M2		8.65±0.83	55.17±4.93	382.02±19.68^{A}	67.81±2.70	105.86±4.10
L2M3		8.43±0.46	51.33±4.12	359.76±16.02^{B}	68.62±2.67	105.57±2.30
L3M1		8.52±0.84	53.00±4.74	335.80±20.31^{CD}	67.62±1.41	102.14±4.85
L3M2		8.64±0.48	54.83±5.17	327.11±13.80^{D}	67.00±1.51	106.57±2.82
L3M3		8.44±0.49	51.71±10.80	323.15±24.49^{D}	66.38±2.00	104.43±2.37
L1		8.69±0.69	55.78±6.27	356.99±20.18^{A}	67.90±2.34	105.90±2.66
L2		8.61±0.59	54.56±5.55	356.31±16.57^{A}	67.90±2.60	106.29±3.47
L3		8.53±0.60	53.11±6.90	328.83±19.46^{B}	67.00±1.64	104.38±3.34
M1		8.69±0.59	55.83±5.89	341.20±20.12	67.75±1.88	104.95±3.87
M2		8.62±0.71	54.67±5.12	351.00±15.83	67.58±2.31	106.35±3.37
M3		8.53±0.58	52.94±7.71	346.21±19.21	67.46±2.39	105.25±2.23
P-value	C&T	0.1524	0.3959	0.0001	0.0001	0.8981
	Lys	0.8341	0.5886	0.0004	0.0010	0.7133
	Met	0.6165	0.8514	0.2125	0.2685	0.9031
	Lys×Met	0.2613	0.8433	0.2303	0.3311	0.5597

[1] The values are the mean±SD, and each of the estimated requirement levels is based on 8 blue foxes. C&T means the control group compared with the other groups. In the same column of the same stage, values with different capital letter superscripts mean significant difference ($P<0.01$).

levels of Met and the interaction between Lys and Met had no effects on N intake, N retention, FN and UN of different groups ($P>0.05$). Different Lys levels had significant effects on UN of each group ($P<0.05$), and L1 and L2 groups were the highest and the lowest, respectively. N retention was the highest in the M2 level.

Discussion

Feed intake and growth performance

Feed intake can be affected by many factors, including animal growth period, feed energy, the balance of amino acids and other nutrients, palatability, ion concentrations in intestines, and feedstuff digestibility and so on. Feed intake of some animals would be reduced if protein in diets was absent (Han and Kong, 2007), and this may induce the synthesis of digestive enzyme being decreased and the body protein being broken down. Chickens and rats can slightly improve their

Table 3. Effects of low-protein, Lys and Met-supplemented diets on N-balance of blue foxes.

Treatment[1]		N intake (g/d)	FN (g/d)	UN (g/d)	N retention (g/d)	N apparent digestibility%
Control		13.55±0.22^{A}	4.68±0.40^{abc}	5.16±0.59^{A}	3.74±0.24^{A}	67.31±3.28^{A}
L1M1		11.16±0.89^{BC}	4.07±0.89^{c}	3.75±0.65^{BC}	3.34±0.37^{A}	63.64±6.28^{AB}
L1M2		10.88±0.49^{CD}	4.20±0.85^{c}	3.58±0.28^{BCD}	3.10±0.71^{AB}	61.44±7.31^{BC}
L1M3		10.69±0.52^{CDE}	4.22±0.66^{c}	3.46±0.75^{BCD}	3.19±0.52^{A}	60.58±5.39^{BCD}
L2M1		11.13±0.44^{BC}	4.38±0.41^{bc}	3.82±0.28^{B}	3.24±0.52^{A}	60.73±2.36^{BCD}
L2M2		11.87±0.59^{B}	5.19±0.65^{a}	3.41±0.59^{BCD}	3.35±0.66^{A}	56.29±2.58^{CDE}
L2M3		11.32±0.50^{BC}	5.09±0.81^{ab}	3.46±0.67^{BCD}	3.08±0.90^{AB}	55.12±6.32^{DE}
L3M1		10.30±0.45^{DE}	4.78±0.53^{abc}	3.30±0.31^{BCD}	2.22±0.66^{C}	53.61±4.44^{E}
L3M2		10.21±0.49^{DE}	4.16±0.38^{c}	2.99±0.83^{D}	2.98±1.28^{ABC}	57.70±2.10^{CDE}
L3M3		9.96±0.81^{E}	4.47±0.46^{abc}	3.05±0.67^{CD}	2.35±0.50^{BC}	55.07±2.75^{DE}
L1		10.91±0.63^{A}	4.16±0.80^{A}	3.58±0.56^{a}	3.21±0.53^{A}	61.40±6.32^{A}
L2		11.45±0.51^{A}	4.91±0.62^{B}	3.46±0.51^{ab}	3.22±0.69^{A}	56.92±3.75^{B}
L3		10.16±0.58^{B}	4.47±0.45^{AB}	3.1±0.60^{b}	2.50±0.81^{B}	55.48±3.09^{B}
M1		10.88±0.59	4.40±0.61	3.53±0.41	2.95±0.51	59.09±4.36
M2		11.02±0.52	4.53±0.62	3.32±0.56	3.15±0.88	58.51±3.99
M3		10.68±0.61	4.60±0.64	3.31±0.69	2.90±0.64	56.56±4.82
P-value	C & T	0.0001	0.0357	0.0001	0.0043	0.8318
	Lys	0.0001	0.0014	0.0308	0.0024	0.2880
	Met	0.2569	0.6453	0.4615	0.5277	0.9035
	Lys×Met	0.3375	0.0802	0.9639	0.4467	0.4931

[1] The values are the mean±SD, and each of the estimated requirement levels is based on 8 blue foxes. C & T means the control group compared with the other groups. In the same column of the same stage, values with different small letter superscripts mean significant difference (P<0.05), with different capital letter superscripts mean significant difference (P<0.01).

intake to compensate the shortage of amino acids when fed the diets with inadequate amino acids (Yan, 2009). The ADFI of this study was affected by different diets and may be influenced more by different dietary Lys levels. Our results suggested that supplemental Met could improve the biological value of low protein diets for blue foxes, which shows good agreement with Dahlman's results (Dahlman *et al.*, 2002). Our results were also consistent with the results reported in swine and poultry that free amino acids are used as efficiently as protein-bound ones, and it is therefore possible to obtain the same performance with low-protein, amino acid-supplemented diets as with high-protein controls (Gruber *et al.*, 2000; Roth *et al.*, 2001).

The development of hair and the quality of fur were crucially affected by dietary factors in later growing-furring period of mink in recent studies (Rasmussen and Borsting, 2001). In our study, there were no significant differences observed in body length and dry fur length. Body length of L1M2 and L2M2 (68.44 and 68.62 cm, respectively) was higher compared with other groups. In an earlier study of blue foxes, low-protein diets added with Met, Lys, or both, did not affect fur quality compared with high-protein control group (Työppönen *et al.*, 1987). These results were consistent with previous studies that Met supplementation improved the overall fur quality of blue foxes fed the lowest protein diet (Dahlman *et al.*, 2002), and brought them up to the level of animals in the highest protein diet group.

N-balance

N intake was decreased with dietary protein level being reduced. It may be related to the increased dietary content of extruded corn. The results were consistent with earlier findings (Jongbloed and Lenis, 1992; Zhang *et al.*, 2008). N excretion in urine declined significantly ($P<0.01$) when the protein level in the diet was lowered from 27% to 19% (Table 3). The results were consistent with the earlier findings that N excretion declined noticeably along with a reduction in dietary protein in pigs (Valaja *et al.*, 1993). N retention and N retention ratio were not significantly different among most groups. Our results showed very good agreement with the view that the lower the protein level in the diet, the better the utilization of N and the smaller the proportion excreted. A previous study showed that in 21-week-old blue foxes, N retention did not differ in relation to dietary protein level and even 15% of protein was sufficient to satisfy the animals' requirement (Dahlman *et al.*, 2002b). The protein requirement of blue foxes at this age is likely to be determined mainly by hair growth, suggested by studies of Blomstedt (1998) and also by recent studies in mink (Rasmussen *et al.*, 2001). Our results showed that when the dietary protein level decreased from 27% to 19% with Met and Lys supplements, the performance of blue foxes was not affected, but the N excretion was decreased. This discovery was confirmed by the present research on blue foxes which showed that there was a 36.8% reduction in N excretion when the protein level declined from about 320 to 250 g/kg (DM basis) (Dahlman *et al.*, 2002).

Fur parameters

The development of hair and the quality of fur were crucially affected by dietary factors in later growing-furring period of mink in recent studies (Rasmussen and Børsting, 2001). In our study, there were no significant differences observed in body length and dry fur length. Body length of L1M2 and L2M2 (68.44 and 68.62 cm, respectively) was higher compared with other groups. In an earlier study of blue foxes, low-protein diets added with Met, Lys, or both, did not affect fur quality compared with high-protein control group (Työppönen *et al.*, 1987). These results were consistent with previous studies that Met supplementation improved the overall fur quality of blue foxes fed the lowest protein diet (Dahlman *et al.*, 2002), and brought them up to the level of animals in the highest protein diet group.

Conclusions

Reducing the percentage of CP in dietary DM by 1 unit, on average, led to a 4.1 percentage unit decline in the daily amount of N excreted in urine. In absolute amounts, an approximately 1.72 g decline in N excretion per blue fox per day can be achieved by reducing the dietary protein level from about 27% to 19% of DM. The optimal dietary Lys and Met levels to attain the best growth performance and fur characters were 1,035 and 2,070 mg/d, respectively, during growing-furring period.

References

Blomstedt, L., 1998. Pelage cycle in blue fox (*Alopex lagopus*): a comparison between animals born early and late in the season. Acta Agr. Scand. 48: 122-128.

Dahlman, T., Kiiskinen, T., Mäkelä, J., Niemelä, P., Syrjälä-Qvist, L., Valaja, J. and Jalava, T., 2002. Digestibility and nitrogen utilisation of diets containing protein at different levels and supplemented with DL-methionine, L-methionine and L-lysine in blue fox (*Alopex lagopus*). Animal Feed Science and Technology 98: 219-235.

Gruber, K., Roth, F.X. and Kirchgessner, M., 2000. Effect of partial dietary amino acid deductions on growth rate and nitrogen balance in growing chicks. Archiv für Geflügelkunde 64: 244-250.

Han, H.J. and Kong, F.C., 2007. The study of animal intake adjustment. Private Scientific and Technological 3: 88.

Harris, L.E., Bassett, C.F., Llewellyn, L.M. and Loosli, J.K., 1951. The protein requirement of growing foxes. J. Nutr. 43: 167-180.

Hernesniemi, T., 2000. Ketun ravinnontarve jarehut. [The fox's nutrient requirements and feeds.] In: Hernesniemi, T. (ed.) Ketunkasvatuksen taito. Finnish Fur Breeders Association, Gummerus Kirjapaino, Jyväskylä, Finland.

Jongbloed, A.W. and Lenis, N.P., 1992. Alteration of nutrition as a means to reduce environmental pollution by pigs. Lives.Prod. Sci. 31: 75-94.

Kerminen-Hakkio, M., Dahlman, T., Niemelä, P., Jalava, T., Rekilä, T. and Syrjälä-Qvist, L., 2000. Effect of dietary protein level and quality on growth rate and fur parameters in mink. Proc. VII Int. Sci. Congr. Fur Anim. Prod. 24: 7-12.

Kerr, B.J., Mckeith, F.K., Easter, R.A., 1995. Effect of performance and carcass characteristics to finisher pigs fed reduced crude protein, amino acid-supplemented diets. J. Anim. Sci. 73: 433-440.

Koskinen, N., Dahman, T., Pölönen, I., Valaja, J. and Rekilä, T., 2005. Physiological bases for increasing the productivity of mammals introduced in zooculture seminar. Petroleum Science and Technology 2005: 107-108.

Rasmussen, P.V. and Børsting, C.F., 2001. Effects of variations in dietary protein levels on hair growth and pelt quality in mink (*Mustela vison*). Canadian Journal of Animal Science 80: 633-642.

Roth, F.X., Gruber, K. and Kirchgessner, M., 2001. The ideal dietary amino acid pattern for broiler-chicks of age 7 to 28 days. Archives of für Geflügelkunde 65: 199-206.

Rimeslåtten, H., 1976. Experiments in feeding different levels of protein, fat and carbohydrates to blue foxes. 1st Int. Sci. Congr. Fur Anim. Prod. 27-29 April 1976.

Työppönen, J., Berg, H. and Valtonen, M., 1987. Effects of dietary supplement of methionine and lysine on parameters and fur quality in blue fox during low-protein feeding. Agricultual Science Finland 59: 355-360.

Valaja, J., Alaviuhkola, T. and Suomi, K., 1993. Reducing crude protein content with supplementation of synthetic lysine and threonine in barley – rapeseed meal – pea diets for growing pigs. Agricultral Science Finland 2: 117-123.

Yan, X.H., 2009. Factor analysis of feed quality affect feed intake. Available at: http://siliao.d288.com/xmsl/756178088.html.

Zhang, H.H., Li, G.Y., Yang, F.H., Liu, B.Y., Chang, Z.J. and Zhong, W., 2008. Effect of Diets with Low Protein Supplemented with Lys and Met on the Performance and Digestive Metabolism of Growing Blue Fox. Chinese. J. Anim. Nutr. 20(6): 724-730.

Effects of dietary protein level on growth, health and physiological parameters in growing-furring mink

B.M. Damgaard[1], *P.F. Larsen*[2] *and T.N. Clausen*[2]
[1]*Aarhus University, Department of Animal Science, Blichers Allé 20, 8830 Tjele, Denmark;*
[2]*Kopenhagen Research, Agro Food Park 15, 8200 Aarhus N, Denmark;*
birthem.damgaard@agrsci.dk

Abstract

The aim of the study was to investigate the effects of the dietary protein level and the feeding strategy on growth, health and physiological blood and liver parameters in growing-furring male mink. Effects of dietary protein levels ranging from 22% of metabolizable energy (MEp) to 32% of MEp were included in the study. The dietary protein levels were unchanged during the experimental period or reduced from a higher to a lower level in August or September. The results indicated negative effects on health and signs on immunosuppression in mink fed low dietary protein levels throughout the growing-furring period. The liver content of fat and fatty acids was higher in mink fed low dietary protein levels than in mink fed high dietary protein levels. The mortality rate was influenced of the dietary protein level in October.

Keywords: fatty liver, mortality rate, erythrocytes, leucocytes

Introduction

The mink is a carnivorous animal and, consequently, its natural diet contains a high proportion of energy from protein, a variable amount from fat, and a low amount from carbohydrate. Until recently, protein has contributed with 30% to 35% of metabolizable energy (ME) in Danish mink feed in the growing-furring period.

Since protein constitutes the most expensive part of the mink feed, possibilities to decrease the level of dietary protein in order to provide inexpensive feed and still support animal performance and health have, therefore, been the subject of several investigations for several years (e.g. Skrede, 1978; Børsting and Clausen, 1996; Sandbøl *et al.*, 2004).

Furthermore, there is a general wish to minimize the emission of nitrogen via urine and faeces to the environment and the Danish mink farmers are required to reduce the emission of nitrogen from the mink production with 15% before 2015.

Hepatic fatty infiltration or hepatic lipidosis is a common pathological finding in carnivores such as the mink, where fat accumulation in the liver often occurs due to metabolic or nutritional causes (Rouvinen-Watt *et al.*, 2010). In mink, hepatic fatty infiltration may be caused by several factors, including amino acid or fatty acid imbalances, excess dietary carbohydrate intake, cholin and vitamin B deficiency, poor feed quality, restricted or food deprivation (Damgaard *et al.*, 1994; Mustonen *et al.*, 2005). In earlier investigations a deficient protein supply caused hepatic fatty infiltration, and the degree of hepatic fatty infiltration increased during the growth period (Damgaard *et al.*, 1998). Furthermore, the primary cause of death at low protein feeding could be related to fatty infiltration of the liver. Damgaard *et al.* (1998) found that a dietary supplement of essential amino acids to a low protein diet did not improve the health status of the mink, but had a positive effect on growth performance compared with un-supplemented low protein diets. The plasma activity of alanine-aminotransferase (ALAT) was found to be a useful tool in the estimation of hepatic fatty infiltration, and clinical blood parameters were helpful in determining the level of protein necessary to sustain animal health and support normal growth performance (Damgaard *et al.*, 1998).

It can be hypothesized that the dietary protein content can be reduced in relation to the usual protein level in Danish mink feed and still support performance and health. Furthermore, that the protein requirement in mink kits is decreasing during the growing-furring period. The aim of the present investigation was to investigate the effects of the dietary protein level and the feeding strategy on growth, health and physiological blood and liver parameters in growing-furring male mink. Growth response and effects on pelt quality of the experimental diets are reported in a companion paper (Clausen *et al.*, 2012).

Material and methods

Animals

The experiment was carried out at Kopenhagen Farm in Holstebro, Denmark. A total of 1,380 mink kits of the colour type Scanbrown were included in the experiment. The experimental period covered the period from primo July to pelting in mid November. The animals were weighed at the start of the experiment in July, in August and September and at pelting in November. Dead animals were recorded and subjected to autopsy with particular focus on macroscopic signs of fat infiltration in the liver.

The animals were housed in pairs (male and female) in standard mink cages (L:90 cm, W: 30 cm, H: 45 cm) with access to a wooden nest box (L:28 cm, W: 30 cm, H 24 cm) embedded and covered with straw.

Diets

The mink kits were randomly distributed on 5 experimental groups which were allocated different dietary protein contents and fed according to different feeding strategies: (1) 32% metabolizable energy from protein (MEp) during the whole experimental period (32MEp); (2) 32% MEp in July/August and 25% MEp from September (32/25MEp); (3) 28% MEp in July and 24% MEp from August (28/24MEp); (4) 22% MEp during the whole experimental period (22MEp); (5) 32% MEp in July and 22% MEp from August (32/22MEp). The composition of the diets is shown in Table 1. The mink kits were fed *ad libitum* and had free access to drinking water.

Table 1. Dry matter and ash content, calculated content of metabolizable energy (ME) and distribution of ME on protein, fat and carbohydrates in the experimental diets.

	32MEp		32/25MEp		28/24MEp		22MEp	32/22MEp	
	July-Sept.	Oct.-Nov.	July-Sept.	Oct.-Nov.	July	Aug.-Nov.	July-Nov.	July	Aug.-Nov.
Dry matter (DM), %	41.8	40.7	42.2	41.7	40.9	42.4	42.9	41.8	42.9
Ash, % of DM	2.3	2.4	2.3	1.6	2.3	2.0	1.9	2.3	1.9
ME, MJ/ kg	8.3	7.7	8.2	8.1	8.0	8.1	8.6	8.2	8.6
ME, MJ/kg DM	19.9	19.0	19.4	19.5	19.7	19.1	20.1	19.6	20.0
Distribution of ME[1] on:									
Protein, %	30	31	30	25	28	24	21	30	21
Fat, %	54	50	52	53	54	52	56	52	56
Carbohydrates, %	16	19	18	22	18	24	23	18	23

[1] The contents of ME and the quantitative proportional composition of ME were calculated on the basis of the analyzed amounts of nutrients, estimated digestibility coefficients of the nutrients, and the following values of ME in kJ per g: protein 18.8; fat 39.8; carbohydrate 17.6 (Hansen *et al.*, 1991).

Blood samples

Blood samples were collected from 20 male kits per experimental group in early October and at pelting in November. The blood samples were collected between 08:00 and 12:00 after a fasting period of 16 hours. The blood samples were taken by venipuncture of *vena cephalica antibrachii* in October and by heart-puncture at pelting. Whole blood for cell counting was collected into K-EDTA and stored at room temperature until analysis within 4 h as described by Damgaard *et al.* (2012).

Whole blood for clinical-chemical analysis was collected in plastic tubes added Na-heparin, stored on ice and centrifuged within 4 h at 2,000×*g* for 20 min at 4 °C. The plasma was separated from the blood cells and stored at -20 °C until clinical-chemical analysis according to the methods described by Damgaard *et al.* (2012).

Liver tissue samples

After blood sampling at pelting the animals were killed by an overdose pentobarbital sodium. The liver was immediately excised, emptied for blood by slight pressure between paper towels and weighed. Liver tissue samples for chemical analysis were taken, immediately frozen in liquid nitrogen and stored at -80 °C until analyzed according to the methods described by Damgaard *et al.* (2012).

Statistical methods

All variables were subjected to analysis of variance using the Restricted Maximum Likelihood method in the mixed model procedure with multiple error terms in the statistical package, SAS (SAS Institute Inc., 1996). The model used for blood parameters included experimental group (32MEp, 32/25MEp, 28/24MEp, 22MEp, 32/22MEp) and month (October, November), and interaction between these effects as general fixed effects. Mink kits were treated as random factor. The residual variance was assumed to be different for each month. This was modelled using repeated statement of PROC mixed with month as a GROUP factor. Due to not totally balanced data the number of degrees of freedom (df) was estimated by using Kenward-Roger's (KR) approximation. When necessary, log transformations were performed to normalize data distribution. The model used for liver parameters included only experimental group as general fixed effect.

All analyses were performed as two-tailed tests. Results of the analysis performed by the mixed model procedure are given in least squares means±SEM. Significance is reported for $P<0.05$.

Results

There were no differences in body weight at pelting among the experimental groups (Table 2). Furthermore, there were no effects of the experimental diets and the feeding strategies on the liver weight and on the liver weight relatively to body weight (Table 2). The liver content of triacylglycerol (TAG) ($P=0.01$) and non-esterified fatty acids (NEFA) ($P=0.01$) were higher for the 22MEp group than for the groups fed a higher dietary protein content throughout the experimental period (Table 2). The liver content of phospholipids showed tendency to differ among the experimental groups ($P=0.07$). The liver content of glucose (79.0±4.5 µmol/g) and glycogen (43.0±8.1 µmol/g) were not different among the experimental groups.

Table 2. Body weight (BW), liver weight, liver weight relative to BW and liver content of TAG, NEFA and phospholipids in mink males in the experimental groups. Values are least squares means and SEM.

Parameter	32MEp	32/25MEp	28/24MEp	22MEp	32/22MEp	*P*-value[1] (effect of group)
BW, g	3121±89	3190±92	3085±87	3114±92	3058±92	0.88
Liver, g	71.0±2.5	73.1±2.5	73.3±2.4	69.9±2.5	67.1±2.5	0.39
Liver, % af BW	2.28±0.072	2.31±0.074	2.40±0.070	2.25±0.074	2.20±0.074	0.36
TAG, µmol/g	115±12 b	134±12 ab	111±12 b	167±12 a	143±12 ab	0.01
NEFA, µEq/g	294±27 b	342±28 ab	285±26 b	411±28 a	357±28 ab	0.01
Phospholip., µmol/g	21.4±0.51	20.5±0.52	19.9±0.50	21.0±0.52	21.8±0.52	0.07

[1] Within rows, values with different letters are significant different ($P<0.05$).

The mortality rate was low in all groups (0.5% to 2.7%). The frequency of dead animals with macroscopic fatty liver had a tendency to be higher in the experimental groups fed less than 25% of energy from protein in October than in groups fed 25% or more of energy from protein in October (0.4% vs. 1.4%).

The blood content of hemoglobin was highest for the 32MEp group, lowest for the 32/22MEp group and in between for the other groups (P=0.03) (Table 3). The number of erythrocytes had a tendency to differ among the experimental groups (P=0.08). The number of leucocytes was lowest for the 22MEp group, highest for the 28/24MEp group and in between for the other groups (P=0.05). Similarly, the number of lymphocytes (1.2±0.15 10^9/L vs. 1.7±0.14 10^9/L) and monocytes (0.36±0.04 10^9/L vs. 0.54±0.04 10^9/L) were lowest for the 22MEp group and highest for the 28/24MEp group (P=0.02 and P=0.008, respectively).

The plasma concentration of protein was highest for the two groups fed the highest dietary content of protein and lower for the other groups (P=0.02) (Table 3). The plasma content of

Table 3. Hemoblobin content and number of erythrocytes and leucocytes in blood, and plasma concentration of protein, urea and ALAT in mink males in the experimental groups. Values are least squares means and SEM.

Parameter	32MEp	32/25MEp	28/24MEp	22MEp	32/22MEp	P-value[1] (effect of group)
Hemoglobin, mmol/l	a	ab	ab	ab	b	0.03
October	11.8 (0.12)	11.7 (0.12)	11.5 (0.11)	11.6 (0.12)	11.4 (0.12)	
Pelting	11.2 (0.15)	10.7 (0.16)	10.8 (0.14)	10.9 (0.16)	10.6 (0.16)	
Erythrocytes, 10^{12}/l						0.08
October	9.0 (0.11)	9.0 (0.11)	8.9 (0.11)	9.3 (0.11)	9.0 (0.12)	
Pelting	8.5 (0.14)	8.1 (0.14)	8.1 (0.13)	8.5 (0.14)	8.2 (0.15)	
Leucocytes, 10^9/l	ab	ab	a	b	ab	0.05
October	7.7 (0.60)	8.2 (0.61)	8.6 (0.57)	6.4 (0.60)	7.0 (0.63)	
Pelting	3.8 (0.37)	4.2 (0.38)	4.7 (0.35)	3.9 (0.38)	4.3 (0.39)	
Protein, g/l	a	a	ab	ab	b	0.02
October	63.0 (0.87)	62.9 (0.89)	60.9 (0.83)	60.8 (0.87)	60.0 (0.89)	
Pelting	59.3 (0.87)	59.0 (0.90)	59.8 (0.83)	57.4 (0.90)	57.1 (0.92)	
Urea, mmol/l						
October	6.1 (0.22) a	5.6 (0.22) a	5.9 (0.21) a	4.5 (0.22) b	4.4 (0.22) b	<0.001
Pelting	4.4 (0.18)	3.9 (0.18)	4.3 (0.17)	3.9 (0.18)	4.1 (0.19)	0.20
ALAT, U/l						
October	183 (33) a	122 (22) ab	131 (22) ab	89 (16) b	106 (20) ab	0.04
Pelting	171 (21)	200 (25)	157 (18)	169 (21)	188 (24)	0.60

[1] For significant group × month interaction the P-values both in October and at pelting are shown (interactions not shown in the table). Within rows, values with different letters are significant different (P<0.05).

urea was in October highest for the three groups fed the highest dietary content of protein and lower for the two other groups (*P*<0.001), but at pelting there were no differences between the groups. The plasma content of the enzyme ALAT was in October highest for the 32MEp group and lowest for the 22MEp group (*P*=0.04) whereas there were no differences between the groups at Pelting (Table 3).

Discussion

The effects of the experimental diets on growth performance were very limited as there were no differences among the groups in body weight and liver weight at pelting. In earlier experiments we found that the growth performance was depressed at 20% MEp in two of three experiments whereas 25% MEp had no effect on the growth performance (Damgaard *et al.*, 2008). The liver content of TAG and NEFA was increased in the 22MEp group compared to the other groups although the liver weight and the liver weight relative to body weight were not different for the other groups. Clausen (2007) and Damgaard *et al.* (1998) found that mink with increased fat content in the liver had hypertrophic liver and increased liver weight compared to mink with lower liver fat content, these results are contradictory to the present results. The plasma content of the enzyme ALAT has been shown to be a useful biomarker for fat infiltration in the liver (Damgaard *et al.*, 1994, 1998). However, in the present investigation the plasma ALAT content in October was higher in groups with low liver fat content and lower in groups with high liver fat content. The reason for these results is unclear, but indicates that the high liver fat content was not related to leakage of ALAT from the liver cells.

In general, the mortality rate was low. However, there was tendency to higher frequency of dead animals with microscopic fat infiltration in the liver in groups fed less than 25% MEp in October. Therefore, 25% MEp seems to secure normal fat metabolism of the liver in October. In the present investigation, both experimental groups that were offered at least 25% MEp in October were offered 32% MEp from July to September. Based on these results it is not possible to estimate the requirements for protein in the early growth phases. Skrede (1978) found that 31% MEp supported normal growth and viability of mink kits. In experiments during three consecutive years, Damgaard *et al.* (1998) found that 30% MEp throughout the growing-furring period ensured a low mortality rate.

The blood concentration of protein and urea reflected the dietary content of protein and can be assumed to be indicators of the metabolism of amino acids.

The hemoglobin content in the blood was highest for the 32MEp group and lowest for the 32/22MEp group indicating that high dietary protein content had a positive effect on the general health status of the animals. Rouvinen-Watt *et al.* (2010) suggested that decline in percentage of blood lymphocytes at food deprivation of mink in 7 days was sign of immunosuppression. Restricted feeding of mink females during the winter time had a negative response on the total number of leucocytes and the number of neutrophiles, lymphocytes and monocytes indicating signs of immunosuppression (Damgaard *et al.*, 2012). In the present experiment, the reduced total number of leucocytes, number of lymphocytes and monocytes in mink fed a low dietary protein content compared to mink fed higher dietary contents of protein throughout the growing period may reflect immunosuppression. The signs of immunosuppression may influence the mink's sensitivity to infections and the progression on diseases of the mink.

In conclusion, tendencies to increased frequency of dead mink with fatty liver were shown in mink fed less than 25MEp in October. The liver content of fat and fatty acids was higher in mink fed low dietary protein levels than in mink fed high dietary protein levels. Signs on immunosuppression and impaired health status were found in mink fed low dietary protein content compared to mink fed high dietary protein content. The blood concentration of protein and urea reflected the dietary content of protein.

Acknowlegements

This work has been supported by Kopenhagen Fur and Faculty of Science and Technology, Aarhus University. Gratitude is extended to I.M Jepsen and L.L.G. Jeppesen for data collection and analysis of blood and liver and to Kopenhagen Farm for carrying out the practical work associated with the presented research.

References

Børsting, C.F. and Clausen, T.N., 1996. Requirements of essential amino acids for mink in the growing-furring period. VIth International Scientific Congress on Fur Animal Production, Poland. Anim. Prod. Rev., Appl. Sci. Rep. 28: 15-24.

Clausen, T.N., 2007. Fasting mink kits fed different amounts of protein. Annual Report 2006, Danish Fur Breeders Research Center, Holstebro, Denmark, pp. 151-154 (In Danish).

Clausen, T.N., Lassén, T.M. and Larsen, P.F., 2012. Effect of reduced protein in the growth and furring period in mink. In: Proceedings of the Xth International Scientific Congress in fur animal production, Larsen, P.F. *et al.* (eds.) Wageningen Academic Publishers, the Netherlands, pp. 47-53.

Damgaard, B.M., Clausen, T.N. and Henriksen, P., 1994. Effect of protein and fat content in feed on plasma alanine aminotransferase and hepatic liver infiltration in mink. J. Vet. Med. A. 41: 620-629.

Damgaard, B.M., Clausen, T.N. and Dietz, H.H., 1998. Effect of dietary protein level on growth performance, mortality rate and clinical blood parameters in mink (*Mustela vison*). Acta Agric. Scand., Sect. A, Animal Sci. 48: 38-48.

Damgaard, B.M., Dalgaard, T.S., Larsen, T., Hedemann, M.S. and Hansen, S.W., 2012. The effects of feed restriction on physical activity, body weight, physiology, haematology and immunology in female mink. Res. Vet. Sci. in press. DOI: http://dx.doi.org/10.1016/j.rvsc.2011.10.014.

Hansen, N.E., Finne, L., Skrede, A. and Tauson, A.-H., 1991. Energiforsyningen hos mink og ræv, NJF utredning/rapport No. 63, DSR Forlag, Copenhagen, 59 pp. (in Danish).

Mustonen, A.-M., Pyykönen, T., Paakkonen, T., Ryökkynen, A., Asikainen, J., Aho, J., Mononen, J. and Nieminen, P., 2005. Adaptations to fasting in the American mink (*Mustela vison*): carbohydrate and lipid metabolism. Comp. Biochem. Physiol. 140A: 195-202.

Rouvinen-Watt, K., Mustonen, A.-M., Conway, R., Pal, C., Harris, L., Saarela, S., Strandberg, U. and Nieminen, P., 2010. Rapid development of fasting-induced hepatic lipidosis in the American mink (*Mustela vison*): Effects of food deprivation and re-alimentation on body fat depots, tissue fatty acid profiles, hematology and endocrinology. Lipids 45: 111-128.

Sandbøl, P., Clausen, T.N. and Hejlesen, C., 2004. Ideal protein for mink (*Mustela vison*) in the growing and furring period. VIII Int. Sci. Cong. Fur Animal Prod. 2004, Scientifur 28 (3): 120-128 and Erratum.
SAS Institute Inc., 1996. SAS® Systems for Mixed Models. SAS Institute Inc., Cary, NC, USA.
Skrede, A., 1978. Utilization of fish and animal byproducts in mink nutrition. I. Effect of source and level of protein on nitrogen balance, postweaning growth and characteristics of winter fur quality. Acta Agric. Scand. 28: 105-129.

Foetal life protein deprivation of mink dams affects the growth and metabolism of the F_2-generation

C.F. Matthiesen and A.-H. Tauson
Faculty of Health and Medical Sciences, University of Copenhagen. Groennegaardsvej 3, 1870 Frederiksberg C, Denmark: cmt@life.ku.dk

Abstract

Malnutrition during foetal life might induce structural and metabolic changes in animals and humans. Studies in mink have shown that maternal (F_0-generation) protein-malnutrition leads to lower birth weight of the F_1, but higher birth weight of the F_2-generation and leads to changes in some key hepatic enzymes in the F_1 and F_2-generation foetal tissue. Our objectives were to study a group of seventy seven F_1-generation mink dams during their first gestation and lactation in order to investigate how the level of protein provision to the their mothers during foetal life (low – L_1 or control – C_1) and their own gestational exposure to either low (L_2) or adequate control (C_2) protein provision affected reproductive performance, preweaning kit growth and quantitative metabolism traits Furthermore, the post weaning growth, and the response to either low or adequate protein diet at 10, 26, and 50 weeks of age was studied in 16 of their male kits. The quantitative metabolism traits of the dams were mainly affected by the gestation diet at not foetal life treatment. Despite this, birth weight, pre-weaning and post-weaning growth of the males were mainly affected by interactions between foetal life protein provision of both the F_1 and F_2 generation.

Keywords: protein provision, metabolic programming, energy metabolism, post-weaning growth, early life nutrition

Introduction

Poor foetal environment such as maternal malnutrition during gestation might induce structural and functional metabolic changes in the offspring, primarily through adaptations aimed towards maximizing the utilization of those nutrients that are available so as to optimize foetal development and survival (Hales and Barker, 1992). These adaptations are associated with the period during development in which the malnutrition occurs, i.e. whether it occurs during certain sensitive periods of foetal life. Indeed, it is known that for offspring exposed to a postnatal environment different from that predicted in foetal life, the resulting mismatch can put them at increased risk of developing metabolic and cardiovascular diseases later in life. Of considerable interest, however, is the finding that modification and consequences of a period of malnutrition during pregnancy are not necessarily limited to first-generation offspring (i.e. the F_1 generation) but can be transgenerational (Pinheiro *et al.*, 2008).

More specifically, protein restriction occurring during pregnancy has been found to reduce birth weight, affect glucose homeostasis, elevate blood pressure, cause structural changes in the liver of adult rat offspring (Desai *et al.*, 1996; Ozanne and Hales, 1999), and to reduce the islet and β-cell mass in the endocrine pancreas leading to reduced insulin secretion (Garofano *et al.*, 1997). The outcome of protein restriction during foetal life depends on time of exposure during the pregnancy but also of the diets used and both contribute to the differences between the different animal studies. Previous findings in mink have shown the foetal life protein provision affects the birth weight of both the F_1-generation (Matthiesen *et al.*, 2010a; Vesterdorf *et al.*, 2012) and F_2-generation offspring (Matthiesen *et al.*, 2010b). Furthermore, the gene expression of key hepatic enzymes important for the glucose (Matthiesen *et al.*, 2010a,b) and fat (Matthiesen *et al.*, 2010c) metabolism was significantly affected by foetal life protein provision. Recent proteome studies have confirmed that proteins in these pathways also are regulated by foetal life protein provision (Guerreiro, personal communication).

Our objectives were to study a group of F_1-generation mink dams during gestation and lactation, regarding reproductive performance and pre-weaning kit growth, and then to study their male offspring (F_2-generation) during their first 50 weeks of life. We investigated how the level of protein provision to the dams during foetal life and the level of protein provision during their own first gestation affected reproductive performance and pre-weaning kit growth. Furthermore, we investigated how foetal life protein provision of their male offspring affected postweaning growth and metabolism until 50 weeks of age.

Material and methods

The experimental procedures followed the guidelines of the European Convention for the Protection of Vertebrate Animals used for Experimental and Other Scientific Purposes as well as Danish national legislation (Anonymous, 1986).

Animals and housing

Seventy seven one-year-old dams (F_1-generation) exposed to either low (L) or adequate (Control – C) protein provision the last 21±3 days of foetal life and fed a conventional wet farm feed comprising an adequate amount of protein from birth onwards were used in present study. The yearling dams were fed either a control diet (31% of ME from protein) containing an adequate amount of protein or a low protein diet (19% of ME from protein) the last 17±1 days of their own gestation in a 2×2 design which resulted in four groups according to their treatment: $Control_1$-Control (C_1C, n=20), $Control_1$-Low (C_1L, n=20) Low_1-Control (L_1C, n=19) and Low_1-Low (L_1L, n=18) where the letters denote the dietary treatment and the numbers denote if the diet was applied during foetal life of the F_1 or F_2 generation offspring. The dams were used to study the reproductive performance and to record kit body weight at birth and body weight subsequently every 7 days until they reached an age of 8 weeks. Twenty four dams, six from each treatment were used to study the quantitative metabolism traits once during gestation and twice in lactation (week 2 and 4). Sixteen males of the F_2-generation offspring, four from each treatment (C_1C_2, C_1L_2, L_1C_2, L_1L_2, the second number denote that the diet was applied during foetal life of the F_2-generation) were used to study the post-weaning growth and metabolism. They were measured in balance and respiration experiments starting at 10, 26 and 50 weeks of age, respectively. Each balance period was performed so that the animals were fed both the adequate and the low protein diets in alternating weeks, with a one week interval for adaptation to new diet in between measurements. The animals were housed under Danish mink farm conditions at the experimental farm Rørrendegård, Taastrup. During the balance and respiration experiments the animals were transferred from the experimental farm to an intensive care unit, where they were placed in metabolic cages, and were maintained under natural daylight conditions (55 °N 12 °E) at 15-17 °C.

Diets

The animals were fed *ad libitum* once a day and had free access to drinking water. Dams and male offspring were fed either a low-protein diet (L, 19:48:33% of ME from protein, fat and carbohydrates) or an adequate-protein control diet (C, 31:53:16% of ME from protein, fat and carbohydrates) during gestation and the balance periods during post-weaning growth. The sixteen males were fed conventional wet farm feed between the balance periods. The diets consisted of fresh chicken, potato mash powder, oatmeal, fishmeal, sugarbeet pulp, maize starch (gelatinised), maize oil, vitamin and mineral mixture and water. The diets were prepared on one occasion, apportioned by weight in plastic bags, and frozen immediately. The ingredients and chemical composition of the two diets are given in Table 1.

Balance and respiration experiments

The balance experiments consisted of a three-day adaptation period and a four-day collection and recording period. Feed residues, faeces, urine, and citric acid rinse (see below) were quantitatively collected, weighed, and recorded daily and stored at -18 °C until the end of each balance period. Urine was collected in 10 ml of 5% sulphuric acid. After each collection of faeces and urine, collection screens and funnels were rinsed in 1% citric acid to minimize urinary nitrogen losses. Collection and recording was performed every morning between 9:00 and 11:00 h. Both dams and kits were weighed at the start and end of each collection period during lactation and otherwise once a week. The animals were measured in respiration chambers for 22 h once each balance period; during lactation, the dam and kits were measured together. The respiration experiments were performed by indirect calorimetry in an open-air circulation unit (see Chwalibog *et al.*, 2004).

Statistical analysis

The data were analysed for statistical significance using the mixed procedure in Statistical Analysis Software (SAS Institute, Cary, NC, USA, version 9.2). Foetal life protein provision, gestation diet and post weaning diet were fixed effects in the model with the interactions between them. Results are presented as least square means. Pairwise comparisons of least square means were performed using the PDIFF option. Differences were denoted as significant, if $P<0.05$, and as a tendency, if $P<0.10$.

Results and discussion

The reproductive performance of the F_1-generation dams was not significantly affected by foetal life treatment or gestation diet (Table 2). However, birth weight of their offspring (F_2-

Table 1. Dietary composition (g/kg) and results of chemical analyses of the experimental diets.

	Control diet – adequate protein	Low protein diet
Dietary composition (g/kg)		
Fresh chicken[1]	600	400
Potato mash powder	40	40
Steam rolled oats	40	105
Fishmeal	150	50
Sugar beet pulp	30	20
Corn starch, gelatinized	40	95
Corn oil	60	60
Vitamin and mineral mixture[2]	3	3
Water	37	227
Chemical composition		
Dry matter (DM), g/kg	413	406
Ash, g/kg DM	91	55
Crude protein, g/kg DM	421	259
Fat, g/kg DM	257	249
Carbohydrate, g/kg DM	229	438
Gross energy, MJ/kg DM	23.77	23.15
Metabolisable energy (ME), MJ/kg DM	18.74	19.42
Protein:fat:carbohydrate ratio (% of ME)	31:53:16	19:48:33

[1] Chicken prepared for human consumption, i.e. without heads, feet, feathers, or entrails.

[2] Per kg of supplement: retinol 2,800,500 IU, cholecalciferol 280,000 IU, α-tocopherol acetate 24,020 IU, thiamine 10 g, riboflavin 4.8 g, nicotiamide 80 g, pyridoxine 3.2 g, D-pantothenic acid 3.2 mg, biotin 80 mg, folic acid 240 mg, choline chloride 60 g, cyanocobalamin 16 mg, para-aminobenzoic acid 800 mg, betaine 33.6 g, Fe (FeSO4) 19,712 mg, Cu (CuSO4) 1,025 mg, Zn (ZnO) 12,560 mg, and Mn (MnO) 6,238 mg (Trow Nutrition, Denmark A/S, Vejen, Denmark).

generation) was significantly affected by both the foetal life treatment of their dams (P=0.01) but also by their own foetal life protein provision (P<0.001). This corresponds to previous findings where F_1-generations dams exposed to low protein provision during foetal life gave birth to F_2-generation offspring with significantly higher birth weights (12.2 g vs. 11.1 g, P=0.003) than controls (Matthiesen *et al.*, 2010b). The ME intake was not different between treatment groups but heat production (HE) was affected by gestation diet (P<0.001) and interactions between foetal life protein provision and gestation diet (P=0.01) resulting in L_1C_2 having the highest HE followed by C_1C_2 and C_1L_2 and L_1L_2 with the lowest HE whereas retained energy (RE) was affected by interactions between the foetal life protein provision and gestation diet (P=0.04) of the F_1 dams. The effect of foetal life treatment on HE and RE among yearling females in their own first gestation and lactation does, however, not correspond with previous findings, where HE was unaffected by foetal life protein provision (Matthiesen *et al.*, 2010b). The RN was not was not affected by neither foetal life treatment nor the gestation dietary treatment but significantly

Table 2. The reproductive performance and quantitative metabolism traits of the F_1-generation dams exposed to either low (L_1) or adequate (C_1) protein provision the last 21±3 days of foetal life and fed a low protein (L) or an adequate protein control (C) diet the last 17±1 days of their own gestation.

	Treatment				*P*-value		
	C_1C	C_1L	L_1C	L_1L	F	G	F×G
Barren, %	15.0	0	5.3	5.6	NS	NS	NS
Litter size	8.1	7.5	7.7	7.2	NS	NS	NS
Still born	1.1	1.6	1.5	1.7	NS	NS	NS
Birth weight, g	9.75^a	8.34^b	10.00^{ac}	8.92^d	0.01	<0.001	NS
					Gestation	**Week 2**	**Week 4**
ME kJ×kg$^{-0.75}$	920	920	930	970	555^A	937^B	1,314^C
HE kJ×kg$^{-0.75}$	664^a	635^{abd}	728^c	602^d	657	660	655
RE kJ×kg$^{-0.75}$	255	283	202^a	368^b	-100^A	276^B	658^C
RN g×kg$^{-0.75}$	0.70	0.66	0.65	0.71	0.26^A	0.83^B	0.95^{BC}

Note: F = foetal life treatment; G = gestation diet; NS = not significant. Mean values within a row with unlike lower-case superscript letters (a,b,c,d) were significantly different within treatment combination. Mean values within a row with unlike upper-case superscript letters (A,B,C) were significantly different within gestation/lactation. Values for second and forth week of lactation includes the milk production. HE = heat production, ME = metabolizable energy, RE = retained energy, RN = retained nitrogen.

different within stage of gestation/lactation which is similar to previous findings (Matthiesen *et al.*, 2010b).

The body weight of the F_2-generation male offspring is given in Table 3. The preweaning body weight of the male offspring was significantly affected by their foetal life treatment (P<0.01) and by interactions (P<0.01) between foetal life protein provision of their mothers and by their own foetal life protein provision.

The quantitative metabolism traits of the F_2-generation males were not affected by the foetal life treatment of their mothers or by their own foetal life treatment except for RE where a tendency of foetal life protein provision (P=0.08) and interactions between their mothers' foetal life treatment and their own foetal life treatment affected the energy retention significantly (P=0.04; Table 4). The substrate oxidation of protein and fat was significantly affected by the actual diet given to the males (P<0.001) and by the interactions between both their mothers' and their own foetal life treatment. This corresponds with previous studies of F_1-generation males exposed to low protein provision during foetal life which have shown that the protein oxidation from 7 to 10 weeks of age was significantly lower in male kits which had been exposed to foetal life protein deprivation (P=0.006) than in controls (Matthiesen *et al.*, 2012).

Table 3. The body weight of the F_2-generation offspring from 1 to 8 weeks of age born by dams exposed to either low (L_1) or adequate (C_1) protein provision the last 21±3 days of foetal life and exposed to either low (L_2) or adequate (C_2) protein provision the last 17±1 days of foetal life.

Treatment	Weeks of age						
	1	2	3	4	5	6	8
C_1C_2	33.4	75.3	131.1	187.8	264.1	421.0	797.4
C_1L_2	35.2	72.0	133.4	181.8	260.9	439.5	816.1
L_1C_2	36.4	72.2	130.7	187.5	281.3	443.1	818.9
L_1L_2	33.4	70.4	135.9	196.8	263.7	418.1	811.5

Table 4. Quantitative metabolism traits for F_2-generation male offspring born by dams exposed to either low (L_1) or adequate (C_1) protein provision the last 21±3 days of foetal life and exposed to either low (L_2) or adequate (C_2) protein provision the last 17±1 days of foetal life and fed either low or adequate protein diet at 10, 26 and 50 weeks of age.

	Treatment				*P*-value				
	C_1C_2	C_1L_2	L_1C_2	L_1L_2	W	D	M	F	M×F
ME kJ×kg$^{-0.75}$	575^{a}	553^{ab}	431^{b}	572^{a}	0.01	NS	NS	NS	0.08
HE kJ×kg$^{-0.75}$	633^{a}	627^{a}	620^{ab}	599^{b}	<0.001	NS	0.04	NS	NS
RE kJ×kg$^{-0.75}$	-58	-73	-188^{a}	-10^{b}	<0.001	NS	NS	0.08	0.04
RN g×kg$^{-0.75}$	-0.14	0	-0.12	-0.15	<0.001	0.1	NS	NS	NS
Substrate oxidation given as % of HE									
OXP	27^{a}	22^{ab}	20^{b}	27^{a}	NS	<0.001	NS	NS	0.004
OXF	56^{a}	67^{b}	69^{b}	58^{a}	0.01	NS	NS	NS	0.005
OXCHO	17	11	11	15	<0.001	<0.001	NS	NS	0.06
Body weight, g					<0.001	NS	NS	NS	<0.001
10 weeks	1,455	1,335	1,424	1,304					
26 weeks	3,275	2,785^{a}	3,166	3,479^{b}					
50 weeks	3,189^{a}	2,541^{b}	2,668^{bc}	3,083^{ac}					

Note: W = week 10, 26 or 50; D = diet – low or control diet with adequate protein content; M = their mothers' foetal life treatment; F = their own foetal life treatment; NS = not significant. Mean values within a row with unlike lower-case superscript letters (a,b,c,d) were significantly different within treatment combination. HE = heat production, ME = metabolizable energy, RE = retained energy, RN = retained nitrogen, OXP = oxidation of protein, OXF = oxidation of fat, OXCHO = oxidation of carbohydrates.

The body weight of the males was significantly affected by interactions between foetal life protein provision of their mothers and by their own foetal life protein provision resulting in significantly different body weights at 50 weeks of age resulting in C_1C_2 and L_1L_2 treated males with the highest body weights and C_1L_2 and L_1C_2 with the lowest body weights ($P<0.05$).

In conclusion, the pre-weaning and post-weaning growth of male offspring was affected by both their mothers' and their own level of protein provision during foetal life. The quantitative metabolism traits were mainly affected by the age of the animals except for the substrate oxidation where interactions between both their mothers' and their own level of protein provision during foetal life appeared. It can be concluded that protein provision during gestation not only affects the present generation, but also growth and metabolism of subsequent generations. Further studies needs to clarify if other metabolic traits and pathways are affected.

References

Anonymous, 1986. European convention for protection of vertebrate animals used for experimental and other scientific purposes. Strasburg, Council of Europe. European Treaty Series No. 123.

Chwalibog, A., Tauson, A.H. and Thorbek, G., 2004. Energy metabolism and substrate oxidation in pigs during feeding, starvation and re-feeding. Journal of Animal Physiology and Animal Nutrition 88: 101-112.

Desai, M., Crowther, N.J., Lucas, A. and Hales, C.N., 1996. Organ-selective growth in the offspring of protein-restricted mothers. British Journal of Nutrition 76: 591-603.

Garofano, A., Czernichow, P. and Breant, B., 1997. *In utero* undernutrition impairs rat beta-cell development. Diabetologia 40: 1231-1234.

Hales, C.N. and Barker, D.J.P., 1992. Type 2 (non-insulin-dependent) diabetes mellitus: the thrifty phenotype hypothesis. Diabetologia 35: 595-601.

Matthiesen, C.F., Blache, D., Thomsen, P.D., Hansen, N.E. and Tauson, A.-H., 2010a. Effect of late gestation low protein supply to mink (*Mustela vison*) dams on reproductive performance and metabolism of dam and offspring. Archives of Animal Nutrition 64: 56-76.

Matthiesen, C.F., Blache, D., Thomsen, P.D. and Tauson, A.-H., 2010b. Feeding mink (*Neovison vison*) a protein-restricted diet during pregnancy induces higher birth weight and altered hepatic gene expression in the F_2 offspring. British Journal of Nutrition 104: 544-553.

Ozanne, S.E. and Hales, C.N., 1999. The long-term consequences of intra-uterine protein malnutrition for glucose metabolism. Proceedings of the Nutrion Society 58: 615-619.

Pinheiro, A.R., Salvucci, I.D.M., Aguila, M.B. and MandarimDeLacerda, C.A., 2008. Protein restriction during gestation and/or lactation causes adverse transgenerational effects on biometry and glucose metabolism in F_1 and F_2 progenies of rats. Clinical Science 114: 381-392.

Vesterdorf, K., Harrison, A., Matthiesen, C.F. and Tauson, A.-H., 2012.. Effects of protein restriction in utero on the metabolism of mink dams (*Neovison vison*) and on mink kit survival as well as on postnatal growth. Open Journal of Animal Sciences 2: 19-31.

Effect of reduced protein in the growth and furring period in 2010 and 2011 in mink

T.N. Clausen[1], T.M. Lassén[2] and P.F. Larsen[1]
[1]Kopenhagen Research, Agro Food Park 15, 8200 Aarhus N, Denmark; [2]Kopenhagen Consulting, Agro Food Park 15, 8200 Aarhus N, Denmark; pfl@kopenhagenfur.com

Abstract

Effects of reduced protein content and content of some essential amino acids on skin length and skin quality was studied in the growth and furring period in Black and Brown/Glow mink in 2010 and 2011. In 2010 five groups of each 276 male and female Brown/Glow mink kits were used. In 2011 five groups of each 155 male and female Black mink and three groups of each 155 Brown/Glow male and female mink were used. The mink were feed diets containing varied protein and amino acids from mid-July to pelting. Results showed that reduced protein content down to 21.5% of the metabolizable energy (MEp) from July to pelting or from mid-August to pelting and a changed amino acid profile cannot be used, since the skin length is shorter, the pelt quality is poorer and there is an increased risk of fatty liver. 28 MEp from mid-July to mid-August and thereafter 24 MEp resulted in the longest skins but a little reduced pelt quality in both mink types, most likely due to low protein content in July in the period where plant for wool and hairs are formed, since the skins were generally flatter and less heavy/voluminous.

Keywords: mink, low protein diet, amino acids, skin quality, skin length

Introduction

Danish mink farmers have to reduce the content of protein in the feed by 15% before 2015 (compared to the protein level in 2009) in order to reduce the emission of ammonia from the mink farms to the environment (Clausen *et al.*, 2012a). Thus some studies have already shown that it is possible to lower the protein content in some periods (Clausen *et al.*, 2012b). To obtain more knowledge on the period from July to pelting a series of studies was initiated in order to detect the periods/windows where it would be safe to lower the protein content without jeopardizing the health of the animals, the reproduction or the skin quality/skin length.

Most experiments in the growing season, to reduce the protein content, have been conducted on Brown/Glow mink. Black mink has been used to study the need for some important amino acids (phe and tyr) for coat color (Clausen and Larsen, 2012; Clausen and Sandbøl, 2011, Clausen and Sandbøl, 2010) and to determine the ratio among met and cys (Sandbøl and Clausen, 2005). Thus it is necessary further to examine protein requirements in July and August, and clarify whether the same protein requirements apply to Black mink as for Brown/Glow mink.

The aim of this paper is to combine studies of reduced protein levels and some essential amino acids in the growth and furring period from mid-July to pelting in Black and Brown/Glow mink in 2010 and 2011.

Materials and methods

In 2010 there were 5 groups of Brown/Glow mink consisting of each 276 males and females. In 2011 there were 5 groups of Black mink consisting of 150 to 155 males and females, and 3 groups of Brown/Glow mink consisting of 130 to 135 males and females. The setup is shown in Table 1.

Mink kits were placed into trial groups in late June. First week of July feed production initiated and there was a gradual transition from feed kitchen feed (starting level 40:45:15) to experimental feed during the first two weeks of July. The 15th of July there were 32 MEp in control groups in Black and Brown/Glow mink, 30 MEp in 2 Black groups and 28 MEp in the remaining groups, except 2 groups with 30 MEp and one 21.5 MEp in the Brown/Glow mink in 2010. Control groups continued at 32 MEp until pelting. The 10th of August 2011the protein content for one Brown/Glow group and two Black groups were reduced to 24 MEp, and from the 15th of September 2011 one Brown/Glow group and two Black groups were reduced to 24 MEp. The 10th of August 2010 the protein content was lowered to 24 MEp and 21.5 MEp.

The norm of grams of digestible amino acids / 100 kcal was followed in July (Sandbøl and Lassen, 2006) and the norm for met and phe were followed throughout the whole period. The requirement for his (0.11), leu (0.41), ile (0.23) and val (0.22) was decreased slightly from 10/8 in the low protein groups.

Table 1. Experimental setup for Black and Brown/Glow mink in the growing season 2010 and 2011. The proportion of metabolizable energy from protein, fat and carbohydrate. 2011: Group 71-75 are Black mink (B), group 21-23 are Brown/Glow mink (BG) and 2010: 21-23 are Brown/Glow mink (BG).

Groups	Start level	Gradual transition 1/7-15/7	15/7	10/8	15/9-pelting
71 B Control	40:45:15	40 → 32:53:15	32	32:50:18	32:50:18
72 B		40 → 30:55:15	30	28:54:18	24:55:21
73 B		40 → 30:55:15	30	24:58:18	24:55:21
74 B		40 → 28:57:15	28	28:54:18	24:55:21
75 B		40 → 28:57:15	28	24:58:18	24:55:21
21 BG Control		40 → 32:53:15	32	32:50:18	32:50:18
22 BG		40 → 28:57:15	28	24:58:18	24:55:21
23 BG		40 → 28:57:15	28	28:54:18	24:55:21
31 BG Control		40 → 32:53:15	32	32:53:15	32:50:18
32 BG	40:45:15	40 → 30:52:18	30	30:52:18	25:54:21
33 BG		40 → 28:54:18	28	24:54:22	24:52:24
34 BG		40 → 21.5:57.5:21	21.5	21.5:57.5:21	21.5:57.5:21
35 BG		40 → 30:52:18	30	21.5:57.5:21	21.5:57.5:21

Male kits were weighed at the start of the experiment (6/7), 8/8, 14/9 and at pelting. The skins were length measured and quality sorted on Kopenhagen Farm and sent for sorting at Kopenhagen Fur.

Feed plans are available upon request from the corresponding author. The raw materials were analyzed before the start of experiment, feed composition for the experimental groups was checked by analysis of protein, fat, dry matter and ash at regular intervals throughout the period. Pooled samples of the feed for relevant sub-periods were analyzed for amino acid content.

The statistical calculations were performed using the statistical program SAS (SAS Institute Inc., 1996). The procedures GLM (SS4), LSMEANS / PDIFF were used with 5% significance level for calculations of the weight and length. Relevant covariates were included in the cases they were significant. For fur quality, colour and clarity using a non-parametric test (GENMOD). Silkiness, wool quality and metallic were analyzed with PROBIT procedure or χ^2.

Results and discussion

Analyses of the feed chemical composition indicated in general a good agreement with the plan. Amino acid analyses varied in proportion to the level numbers (data not shown), but in most cases followed the plan strictly. Some amino acids were below the norm, although most were within the measurement uncertainty of 10%.

At the start of the experiment there was no difference in the weights among groups (Table 2), but in august, for both mink types, the heaviest mink were observed in the groups that had received low protein diet and thus the highest energy content. This difference persisted for Brown/Glow mink groups until pelting in both 2010 and 2011. For the Black mink, there was no significant difference in weights at pelting. Skin lengths showed the same as the weights (Table 2), namely that the longest skins were observed in the group fed low protein diet. In previous studies it has been demonstrated that 32 MEp in July gives the best growth in July, but that low protein content in July and August does not reduce the skin length at pelting (Clausen *et al.*, 2008). In the growing period in 2010 the skins of groups fed a diet containing 28 MEp in July to mid-August, and then 24 MEp until pelting, were of at least the same size as the control group (32 MEp) (Clausen *et al.*, 2012b). Therefore there is nothing that suggests a negative effect on the size, either in Black or Brown/Glow mink, if we lower the protein content in mid-July to 28 MEp.

Independently of the protein content was lowered to 24 MEp in August or September skin quality in the Black mink was slightly reduced compared to the control group (Table 3). No difference was observed in colour, silkiness or wool, however there were the fewest flat skins in the control group. The quality of Brown/Glow mink group was not significantly different in 2011, the slightly lower quality in the low-protein groups are mainly a result of the skins being a little longer than in the control group (Table 3). Whether the protein was lowered to 24 MEp in August or September seems to have had the same influence on the size and quality of the skins. The wool of Brown/Glow mink skins was affected by protein level, there were more heavy/voluminous and fewer flat skins in the control group rather than the low-protein groups, especially when protein was lowered to 24 MEp already in august (group 22). The effect of low protein in the early growth period is most likely due to disturbance in the formation of the under-wool and the hair fibre

Table 2. Weight for male kits in grams (2011 and 2010).[1]

2011	**06 July**	**08 August**	**14 September**	**Pelting**
71 B Control	959 (136)	1,794 (167) b	2,474 (250) b	2,902 (308)
72 B	960 (150)	1,804 (198) b	2,510 (264) ab	2,973 (330)
73 B	954 (150)	1,798 (179) b	2,542 (273) a	2,952 (308)
74 B	952 (120)	1,869 (161) a	2,527 (249) a	2,978 (295)
75 B	952 (130)	1,863 (176) a	2,565 (319) a	2,972 (412)
	NS	<0.0001	0.03	NS
2011				
21 BG Control	1,065 (145)	1,945 (192) b	2,925 (315) b	3,496 (413) b
22 BG	1,065 (124)	2,005 (223) a	2,982 (339) ab	3,597 (441) a
23 BG	1,065 (121)	2,026 (169) a	3,026 (282) a	3,634 (403) a
	NS	0.004	0.03	0.03
2010	**08 July**	**10 August**	**27 September**	**Pelting**
31 BG Control	1,188 (129)	1,890 (192) ab	2,800 (275) b	3,168 (322) a
32 BG	1,189 (143)	1,842 (207) c	2,731 (328) c	3,051 (344) b
33 BG	1,200 (125)	1,898 (193) a	2,868 (289) a	3,150 (376) a
34 BG	1,196 (130)	1,865 (180) bc	2,824 (298) ab	3,146 (359) a
35 BG	1,198 (144)	1,862 (187) c	2,754 /319) c	3,037 (380) b
	NS	0.002	<0.0001	<0.0001

[1] Numbers in parentheses are the standard deviation; NS show that there is no difference between the groups, different letters in a column shows that there are differences among different groups.

thickness (Rasmussen and Børsting, 1997). In 2010 there was a significant difference in skin quality in the Brown/Glow mink in the low protein groups.

A slightly reduced skin quality and more flat and less heavy/voluminous furs at 28 MEp from July, and 24 MEp from August or September, corresponds among experiments in 2010 and 2011. Thus, whether the altered amino acid requirement at 24 MEp may have had an impact on the experiment will be examined in future studies with increased addition of essential amino acids in the low protein groups in the growth period in 2012.

Generally the health status on the farm was good (Table 4). In the Brown/Glow mink control group six mink died in 2011, half of these from bladder/kidney infections. Totally only three minks died from enlarged fatty liver, all in the groups fed 24 MEp from August. In 2010 more mink died but few in the control group and fewest from enlarged fatty liver. It is well known that a low protein diet during the growing season increases the risk of enlarged fatty liver (Clausen *et al.*, 2012a; Clausen, 2007; Clausen and Sandbøl, 2005; Damgaard *et al.*, 1998a,b, 1994, 2012).

Table 3. Skin length and quality parameters for male mink (2011 and 2010).[1]

2011	**Length cm**	**Quality** [2]	**Colour** [3]	**Metallic %**	**Silkiness %**	**Wool, %**		
						flat	**normal**	**heavy/ voluminous**
71 B Control	88.7 (4.0)	7.2 (2.2) a	2.9 (1.0)	1.6	7.0	2.3	85.9	11.7
72 B	89.2 (4.4)	6.3 (2.4) b	2.9 (0.9)	2.3	8.6	7.8	82.0	10.2
73 B	89.3 (4.6)	6.3 (2.4) b	3.0 (0.8)	6.8	5.3	8.3	85.0	6.8
74 B	89.6 (4.1)	6.5 (2.3) ab	3.1 (0.9)	5.3	6.0	4.5	86.8	9.0
75 B	89.3 (5.4)	6.5 (2.7) b	3.1 (1.0)	5.5	12.6	11.0	75.6	13.4
P-value	NS	0.03	NS	NS	NS	NS		
2011				Clarity [4]				
21 BG Control	94.7 (4.8) b	7.1 (2.4)	2.9 (1.0)	3.0 (1.0)	46.7	4.2	61.7	34.2
22 BG	96.1 (4.2) a	6.4 (2.3)	3.2 (0.9)	3.1 (0.9)	40.2	13.4	69.3	17.3
23 BG	96.1 (4.7) a	6.5 (2.3)	3.0 (1.0)	3.0 (1.0)	45.8	3.8	77.9	18.3
P-value	0.02	NS	NS	NS	NS	0.005		
2010								
31 BG Control	91.9 (3.9) ab	6.9 (2.3) a	2.9 (0.9)	3.2 (1.0) a	19.3	5.0	78.2	16.8
32 BG	90.6 (4.3) c	7.0 (2.4) ab	3.0 (0.9)	3.0 (0.9) a	21.4	3.0	81.6	15.4
33 BG	92.2 (4.4) a	6.4 (2.5) bc	3.0 (1.0)	3.1 (1.0) a	20.8	11.4	80.5	8.1
34 BG	91.3 (4.1) bc	6.1 (2.5) cd	3.1 (0.9)	2.8 (1.0) b	22.3	10.9	80.4	8.7
35 BG	90.7 (4.3) c	6.0 (2.4) d	3.0 (1.0)	3.0 (1.0) a	20.0	9.6	86.5	3.9
P-value	<0.0001	<0.0001	NS	0.002	NS		<0.0001	

[1] Figures in parentheses are the standard deviation; NS show that there is no difference between the groups, different letters in a column shows that there are differences between groups.

[2] Quality scale 1-12, 12 is best.

[3] Colour scale 1-5, 5 is darkest.

[4] Clarity scale 1-5, 5 is most reddish.

Conclusion

The mink were feed diets containing varied protein and amino acids from mid-July to pelting. Results showed that lowering protein content down to 21.5% of metabolizable energy (MEp) from July to pelting or from mid-August to pelting and a changed amino acid profile cannot be used, since the skin length is shorter, the pelt quality is poorer and there is an increased risk of fatty liver.

In august, for both mink types, the heaviest mink were in the groups that were fed the low protein diet (28 MEp) from mid-July rather than 32 MEp. This difference continued for the Brown/Glow mink group to pelting, where the skins were also longest in the low-protein groups. For the Black mink this difference in fur length was not significant.

Table 4. Registration of mortality in the experiment (2011 and 2010).

2011	Mortality	
	Total %	**Enlarged fatty liver %**
71 B Control	1.6	0
72 B	0.7	0
73 B	1.0	0.3
74 B	1.0	0
75 B	1.6	0.3
21 BG Control	2.3	0
22 BG	0.8	0.4
23 BG	0.7	0
2010		
31 BG Control	0.5	0.4
32 BG	2.2	0.5
33 BG	2.4	1.1
34 BG	2.4	1.4
35 BG	2.7	1.4

Skin quality in the Black group was slightly reduced compared to the control group, independently of whether the protein content was lowered to 24 MEp in August or September. In the Brown/Glow mink group the quality was not significantly different, but wool quality was affected by protein levels, indicated by more heavy and fewer flat skins in the control group compared to the low-protein groups.

References

Clausen, T.N. and Larsen, P.F., 2012. Continued investigations of the importance of phenylalanine and tyrosine for growth and pelt colour in black mink. Annual Report 2011, Kopenhagen Research, Denmark, pp. 64-69.

Clausen, T.N., Damgaard, B.M., Harslund, J.L.F. and Hammer, A.S., 2012a. Reduced protein in the growing-furring period: Blood and organ investigations. Annual Report 2011. Kopenhagen Research, Denmark.

Clausen, T.N., Lassén, T.M. and Larsen, P.F., 2012b. Reduced protein in the growing and furring period. Annual Report 2011. Kopenhagen Research, Denmark.

Clausen, T.N. and Sandbøl, P., 2010. Investigation on the importance of the aminoacids phenylalanine (phe) and tyrosine (tyr), and the minerals iron (Fe), cupper (Cu) and zink (Zn) on pelt colour in black and brown mink. Annual Report 2009, 57-65, Danish Fur Breeders Research Center, Holstebro, Denmark.

Clausen, T.N. and Sandbøl, P., 2011. Continued investigations on the importance of the amino acids phenylalanine (phe) and tyrosine (tyr), and the minerals iron (Fe) and cobber (Cu) for pelt colour in black mink. Annual Report 2010, 67-71, Danish Fur Breeders Research Center, Holstebro, Denmark.

Clausen, T.N., 2007. Fasting mink kits fed different amounts of protein. Annual Report 2006, Danish Fur Breeders Research Center, Holstebro, Denmark, pp. 151-154.

Damgaard, B.M., Larsen, P.F. and Clausen, T.N. 2012. Effects of dietary protein level on growth, health and physiological parameters in growing-furring mink. IFASA2012 Proceedings.

Clausen, T.N. and Sandbøl, P., 2005. Fasting of mink kits feed different feed rations and its effect on liver fat content, plasma metabolites and enzymes. NJF seminar nr 337, Uppsala, Oct. 5-7, pp. 5.

Clausen, T.N. and Sandbøl, P., 2005. Correlation between liver fat and dry matter. Danish Fur Breeders Research Center, Herningvej 112 C, Tvis, DK-7500 Holstebro, Denmark.

Clausen, T.N. and Sandbøl, P., 2007. Feed with different energy distribution to mink kits from July to mid-September. Danish Fur Breeders Research Center (PFC), Holstebro, Denmark.

Damgaard, B.M., Clausen, T.N. and Dietz, H.H., 1998a. Effect of dietary protein level on growth performance, mortality rate and clinical blood parameters in mink (*Mustela vison*). Acta Agric. Scand., Sect. A, Animal Sci. 48: 38-48.

Damgaard, B.M., Clausen, T.N. and Børsting, C.F., 1998b. Effect of dietary supplement of essential amino acids on mortality rate, liver traits and blood parameters in mink (*Mustela vison*) fed low-protein diets. Acta Agric. Scand., Sect. A, Animal Sci. 48: 175-183.

Damgaard, B.M., Clausen, T.N. and Henriksen, P., 1994. Effect of protein and fat content in feed on plasma alanine aminotransferase and hepatic liver infiltration in mink. J. Vet. Med. A. 41: 620-629.

Sandbøl, P. and Lassén, T.M., 2006. Amino acid requirements during different stages of mink production (In Danish: Minkens krav til aminosyrer under forskellige produktionsfaser). National Advisory Service Denmark (Fur Animals).

Rasmussen, P. V., Børsting, C. F., 1997 Fur development and quality in pastel mink fed at varying protein levels during the growth period. Danish Fur Breeders Research Centre, pp. 181-189.

SAS Institute Inc., 1996. SAS® Systems for Mixed Models. SAS Institute Inc., Cary, NC, USA, 633 pp.

Validation of the ^{13}C-bicarbonate tracer technique for estimation of CO_2 production and energy expenditure in mink (*Neovison vison*)

C. Larsson and A.-H. Tauson
Faculty of Health and Medical Sciences, University of Copenhagen, Grønnegårdsvej 3, 1870 Frederiksberg C, Denmark; aht@life.ku.dk

Abstract

The objective was to validate the ^{13}C-bicarbonate tracer technique (^{13}CT) for estimation of the CO_2 production (RCO_2) and energy expenditure (EE) against 'the golden standard' indirect calorimetry (IC), in mink in the fed state. Six male mink were measured in a Latin-square design, simultaneously by means of ^{13}CT and IC. A dose of $NaH^{13}CO_3$ was administrated orally, by subcutaneous or intra muscular injection. A zero breath sample was collected before the isotope administration. After dose administration breath samples were collected at frequent intervals for 6 h, and simultaneously the animals were measured by IC for 22 h. There was excellent agreement between results from ^{13}CT and IC, and when using the measured factor for recovery of ^{13}C in breath and measured RQ values, no significant differences between administration methods in estimates of RCO_2 and EE were found. The ^{13}CT was concluded to be a useful method for estimating EE, regardless of administration method, but accuracy is dependent on the estimates used for RF and RQ.

Keywords: stable isotopes, administration method, CO_2 production, energy expenditure

Introduction

Determination of energy expenditure (EE) of an individual makes it possible to obtain estimates of its energy requirements. Energy requirements of animals and humans are complex and change depending upon sex, physiological status such as body weight, life stage, physical activity, and a number of various environmental factors, e.g. ambient temperature. Hence, measurements of EE in an individual are important prerequisites for accurate estimates of the energy requirement and to provide guidelines for an appropriate nutrition, especially during demanding situations such as fasting, illness or hard physical work (Junghans *et al.*, 2008).

Indirect calorimetry (IC) based on respiratory exchange measurements is 'the golden standard' for obtaining correct estimates of EE in human subjects and animals. The disadvantage of this method is that the subjects have to be confined to respiratory chambers. Alternatively, the ^{13}C-bicarbonate tracer technique (^{13}CT) can be used for indirect determination of the carbon dioxide production (RCO_2) and estimation of EE (Elia, 1991; Junghans *et al.*, 1997). This stable isotope technique may, contrary to IC, be used in completely free-living conditions with minimal interference of the subjects' natural life style (Lachica and Aguilera, 2003, Prieto *et al.*, 2001; Speakman and Thomson, 1997). The ^{13}CT method is based on the ^{13}C kinetics in breath CO_2 after administration of labelled sodium bicarbonate ($NaH^{13}CO_3$). After administration of the isotope, the ratio between the ^{13}C and ^{12}C in samples of expired air can be used for estimation of RCO_2.

The body size of the subject correlates with the turnover time of the body bicarbonate pool. Thus, in very small animals the measurements only take minutes, whereas in larger subjects it can take several hours (Junghans and Chwalibog, 2001; Elia, 1991; Elia *et al.*, 1988). In most studies, the isotope has been administrated by either a continuous intra venous infusion (Pouteau *et al.*, 2002; Prieto *et al.*, 2001), or a single bolus infusion (Roecker *et al.*, 2001; Junghans *et al.*, 1997). However, there are further studies that indicate that also oral administration of $NaH^{13}CO_3$ may provide correct results (Armon *et al.*, 1990; Junghans *et al.*, 2008), which means that the ^{13}CT may be completely non-invasive.

The aim of the present study was to validate the ^{13}CT technique for estimation of RCO_2 and EE in mink in the fed state against IC by using three different administration methods: subcutaneous injection (SC), intra muscular injection (IM) and oral (O) administration. Our hypothesis was that the ^{13}CT is a useful method for providing information on RCO_2 and EE, and that results are similar among administration methods and not different from IC.

Materials and methods

Animals and diet

Six adult male mink of the standard brown genotype with a live weight of approximately 2.0 kg were used. During the experiment, the mink were housed individually in metabolic cages in an intensive care unit. The mink were transferred from the experimental farm to the intensive care unit for adaptation three days before the experiment started. The daily feed allowance was 400 g of Whiskas® Original (Mars Inc., Denmark), which is experienced to be a palatable diet. The feed supply was chosen so as to cover their daily maintenance energy requirements (MEm) + 10% (Hansen *et al.*, 1991). There was free access to drinking water all time.

General experimental design

Effects of administration methods of $NaH^{13}CO_3$ were evaluated in three periods in a Latin-square design by combining ^{13}CT (6 h) and IC (22 h) in an open-air circulation system. Each period lasted for three days when the mink were placed in respiration chambers for 22 hours at a time for measurements by IC, while each ^{13}CT administration method was evaluated during the first 6 hours of each 22 h measurement. The animals were administrated a dose of $NaH^{13}CO_3$ (98 atom % ^{13}C, Campro Scientific, Berlin, Germany), 5 mg/kg body weight by IM or SC injection, and by O administration. Before injection, the $NaH^{13}CO_3$ was dissolved in sterile isotonic saline (5 mg/ml), and sterile filtered in to a 2.5 ml syringe through a 0.22 μm filter (Cameo25Ss). For O administration, the dose was incorporated in a small amount of feed given after an over-night fast.

Data collection

The mink were weighed and separately placed in the two independently working respiration chambers (volume 760 l). For ^{13}CT, exhaled breath was drawn from the chambers into breath bags (volume ≈1 l) by a laboratory pump (Inotech Vaccum Pump, Wohlen, Switzerland), to get a baseline value. Fifteen minutes before the start of the IC measurements the mink were quickly taken out and given the dose of $NaH^{13}CO_3$ as a single pulse SC or IM injection and then immediately put back into the chambers again. When the O administration was given, the isotope was given approximately 5 minutes earlier to give time to eat. After $NaH^{13}CO_3$ administration, the mink had free access to feed. Immediately after isotope administration exhaled breath was drawn from the chambers into breath bags every 1 minute for 10 minutes (10 samples/mink). Thereafter, expired air was automatically drawn from the chambers and the ratio between ^{13}C and ^{12}C was measured online by an IRIS infrared isotope analyzer (Wagner Analysentechnik, Bremen, Germany) with a collection interval of 5 minutes for 6 hours. The expired air collected in breath bags was analyzed for $^{13}C/^{12}C$ ratio by IRIS when online measurements were finished. The IC measurements of O_2 consumption (VO_2) and CO_2 production (VCO_2) were performed as described in detail by Chwalibog *et al.*, (2004). When the IC experiment was ended the mink were taken out, brought back to the laboratory and weighed.

The diet was analyzed for dry matter (DM), ash, N, fat and gross energy (GE) according to standard methods. During IC, urine was collected in 5 ml of 10% sulphuric acid, and analysed for nitrogen (N), and any feed residues were recorded.

Calculations

Isotopic enrichment was reported in $\partial^{13}C$ (in ‰) against a Pee Dee Belemnite (PDB) standard. Data were converted into atom percentage (AP; in mg/kg), and isotopic enrichment of expired CO_2 was expressed in terms of atom percentage excess (APE; in mg/kg), which was calculated as follows:

$$APE = AP_t - AP_0$$

where AP_t is the atom percentage of the sample at the time (t (min)) after the tracer administration, and AP_0 is the atom percentage before administration of the isotope.

The $^{13}CO_2$ excretion rate was calculated by multiplying the value for isotopic enrichment of CO_2 with the value for RCO_2 (in mmol/min) determined by IC and dividing the product by 100. The fractional ^{13}C label recovery was estimated by dividing the value for the $^{13}CO_2$ excretion rate by the value for the tracer dose (mol ^{13}C) administrated, and multiplying the quotient by 100. The RCO_2 production was calculated from the ^{13}C data by means of the following equation:

$$RCO_2 = (D / AUC) \times RF \qquad (A)$$

where D is the tracer dose (mol ^{13}C) administrated, AUC is the area under the $^{13}CO_2$ enrichment-time curve (mg/kg × min), and RF is the fractional ^{13}C label recovery in breath (Elia, 1991). To express CO_2 production in litres per day (l/d), the following equation was used:

$$RCO_2\ (l/d) = ((D / AUC) \times 60 \times 24) / (22.4 \times RF) \qquad (B)$$

where 22.4 is the volume (l) per mole gas produced. In this study, RCO_2 was calculated by using the estimated fractional ^{13}C label recovery factor (RF) for each measurement (Equation A). In addition, RCO_2 was estimated by using a RF of 1.00, which agreed with the average value from the estimated fractional ^{13}C label recoveries (Equation B). For calculations of AUC the software GraphPad Prism 4 was used, where the value of AUC was determined by using the kinetics of ^{13}C enrichments. The time course of the ^{13}C enrichment y(t) of breath CO_2 after oral administration of a single bolus dose of $NaH^{13}CO_3$ was adequately approximated by the following function:

$$y(t) = A_1\, e^{-a1 \cdot t} + A_2\, e^{-a2 \cdot t} + A_3\, e^{-a3 \cdot t}$$

where a_1, a_2 and $a_3 > 0$. The area under the ^{13}C enrichment-time curve (mg/kg × min) was obtained by integration of y(t) from t = 0 to ∞, which yields (Junghans *et al.*, 2008):

$$AUC = A_1/a_1 + A_2/a_2 + A_3/a_3$$

Energy expenditure (kJ/d) was calculated from the estimated RCO_2 expressed in l/d, and the respiratory quotient (RQ), obtained by IC, using the following equation (Equation 1) (Junghans *et al.*, 2008):

$$EE\ (kJ/d) = 4.96 \times RCO_2 + 16.07 \times (RCO_2\ /\ RQ) \quad (1)$$

Additionally, EE was calculated in accordance to the equation of Brouwer (1965) (Equation 2):

$$EE\ (kJ/d) = 16.18 \times (RCO_2\ /\ RQ) + 5.02 \times RCO_2 - 5.99 \times UN \quad (2)$$

where RCO_2 is the estimated CO_2 production (L/d), RQ is the respiratory quotient obtained by IC and UN is the urinary nitrogen (g). Whole body EE (kJ/d) measured by IC was calculated from the gas-exchange monitoring (VO_2 and VCO_2, in L/d) using Equation 1 and Equation 2. The metabolizable energy (ME) content of the diet was calculated according to the predictive step-wise equation for determination of ME in cat foods suggested by the NRC (2006). The food quotient (FQ) of the diet was calculated as described by Black *et al.* (1986):

$$FQ = (p \times 0.81) + (f \times 0.71) + (c \times 1.00)$$

where p, f and c are the percentages by which ME from respectively protein, fat and carbohydrates contribute to the whole ME intake, and the values 0.81, 0.71 and 1.00 are the standard values for RQ when protein, fat and carbohydrates are being oxidised, respectively.

Statistical analyses

Statistical analyses were carried out according to the GLM procedure of SAS (Statistical Analysis Systems Institute Inc., 1990). The dependent variables CO_2 production (Equation A, Equation B and VCO_2), EE (Equation 1, Equation 2) and the value for RF, were analyzed by using the following model: $Y_{ij} = \mu + a_i + e_{ij}$; where Y_{ij} = the *ij*th observation, μ = general mean, a_i = fixed effect of method (i = SC, IM, O, IC) and e_{ij} = random error. Results are presented as least squares means (LSmeans) and the root mean square error (RMSE) is given as a measure of variance. Statistical significance was accepted if the null hypothesis was rejected with $P<0.05$. Pearson correlation coefficients between traits were calculated by the CORR procedure of SAS® (Statistical Analysis Systems Institute Inc., 1990).

Results and discussion

The objective with the present study was to investigate the hypothesis that the ^{13}CT based on breath ^{13}C recovery can be used for estimation of RCO_2 and EE, regardless of method for tracer administration. If the tracer is given orally and breath samples are collected, the method would be completely non-invasive and thus very valuable for studies within nutrition research in humans as well as animals.

The fractional ^{13}C label recovery in breath CO_2 (RF) differed significantly when using O administration compared with SC and IM administration methods (P=0.002). In the present study, where the ^{13}CT was combined with IC, the exact recovery of ^{13}C could be determined, and a label recovery of 52% was found for O tracer administration, compared to 123% and 112% using IM or SC, respectively (Table 1). Previous studies have found RF values ranging from 50-00% (Downey *et al.*, 1986; Junghans and Chwalibog, 2001; Poteau *et al.*, 2002). Several factors may explain these discrepancies, such as measurements in humans or different species of animals, route of ^{13}C administration, age, nutritional or metabolic status, etc. In a study by Junghans *et al.* (2008), where the ^{13}C dose was orally given to humans, the found RF ranged from 59.6% to 71.6%, and another similar study found mean RF values of 73% and 71% in children and adults, respectively (Armon *et al.*, 1990). The low RF found after oral administration in the present study may indicate that the ^{13}C was lost via a route other than the pulmonary route. Generally, incomplete ^{13}C label recovery in breath CO_2 after administration of $NaH^{13}CO_3$ is caused by reasons such as fixation of ^{13}C in body pools and constituents with a slow turnover (e.g. bone, skeletal tissue, body fat, etc.) and to a lesser part, excretion via urine and faeces (Elia *et al.*, 1988; Roecker *et al.*, 2001). Moreover, it is possible that the ^{13}C may be released via the mouth without crossing the body bicarbonate pool by the reaction of $NaH^{13}CO_3$ with hydrochloride acid of the stomach (i.e. by eructation) (Armon *et al.*, 1990; Elia *et al.*, 1988; Junghans *et al.*, 2008). During exercise (or during pathological conditions where the CO_2 pool is increased (e.g. respiratory failure), the slower distribution processes of CO_2 to deeper compartments seem to be reduced in favour of CO_2 expiration, and the recovery of labelled bicarbonate increases to be almost 100%. The increase may also be explained with marked changes in the bicarbonate pool due to increased substrate oxidation and mobilization of CO_2 from the body stores. These changes occur mainly in the central bicarbonate pool, which at rest is the pool of entry and removal of the label. Also, at higher exercise intensity, an additional amount of CO_2 is exhaled via bicarbonate buffering of lactic acid (Armon *et al.*, 1990; Roecker *et al.*, 2001). However, the fact that IM and SC injection in this study showed a ^{13}C recovery over 100% may be explained with the use of an averaged CO_2 production over 24 hours, although the breath sampling for ^{13}C analysis only lasted for 6 hours, and the time for respiration measurements by IC was 22 hours.

The results of estimated RCO_2 using the ^{13}CT showed excellent correlation with the results from IC (correlation coefficient = 0.979), when using the actual RF of each measurement (Equation

Table 1. Results of the estimated fractional ^{13}C label recovery in breath (RF), CO_2 production (calculated according to Equation A or Equation B), CO_2 production measured by indirect calorimetry (VCO_2), and energy expenditure (calculated according to Equation 1 or Equation 2) and respiratory quotients (RQ), after an IM or SC injected or orally administrated dose of ^{13}C-bicarbonate, and the effects of administration methods.

	Administration method[1]			**RMSE**[2]	***P*-value; effect of administration method**
	IM	**SC**	**O**		
RF	1.23[a]	1.12[a]	0.52[b]	0.3	0.002
CO_2 production					
Eq. A	41.6	40.8	42.0	5.6	0.94
Eq. B	36.7[a]	37.6[a]	107[b]	38.0	<0.01
VCO_2	41.0	40.2	40.2	4.9	0.95
Energy expenditure					
Eq. 1A	653	612	637	90.2	0.73
Eq. 1IC	644	604	612	84.9	0.70
Eq. 1B	566[a]	563[a]	1613[b]	559.5	<0.01
Eq. 2A	644	607	629	90.8	0.78
Eq. 2IC	634	634	603	85.3	0.75
Eq. 2B	556[a]	558[a]	1612[b]	563.7	<0.01
RQ	0.73	0.71	0.72	0.04	0.78

[1] Values that share no common superscript differ significantly $P<0.01$.

[2] RMSE = root mean square error.

A). There were no significant differences in RCO_2 between the methods of administration (i.e. IM, SC or O), neither by using the ^{13}CT ($P=0.94$) nor by indirect calorimetry ($P=0.95$) (Table 1). The RCO_2 was also estimated by using a RF of 1.00 (*Eq.B*), which was the average measured ^{13}C label RF in this study. However, using this factor gave a significant difference between O administration and the other two methods ($P<0.01$) (Table 1). Also, when using this average value for RF, the correlation between the ^{13}CT and IC was only 0.206, and did not give satisfying estimates for RCO_2.

In this study, two equations were used for estimation of EE. The reason for that was to investigate if the equation corrected for estimated UN excretion (Equation 1), gave the same results as the equation of Brouwer (1965), where the measured UN value is included in calculations (Equation 2). The correlation coefficient between the two equations was 0.998, which means that sampling of urine may be excluded and still get accurate results. This is of great importance when using the ^{13}CT in free-living animals where sampling of urine may be problematic.

When using the estimated RCO_2 based on the actual RF (Equation A), the ^{13}CT technique gave slightly higher values than IC measurements, regardless of equation used for EE estimation.

However, again there was an excellent correlation between the two techniques with correlation coefficients within 0.965-0.971. When using the CO_2 production estimated by means of Equation B (i.e. RF = 1.00), the oral administration method differed significantly ($P<0.01$) regardless of use of Equation 1 or Equation 2. The EE estimates were markedly higher (1,613 and 1,612 kJ/kg$^{0.75}$/day) than EE estimated by using the actual RF value (637 and 629 kJ/kg$^{0.75}$/day), or from IC measurements (612 and 603 kJ/kg$^{0.75}$/day, respectively).

Our results suggest that the RF value has decisive influence on the accuracy of the results when EE is estimated by the ^{13}CT. Therefore, it is necessary to assess an appropriate value for the RF in every conceivable situation, i.e. when using different administration methods, measurements in different species of animals, during different physiological situations, such as in fed or fasted state, or during of situations of illness, since different values of RF may be required. Another important factor when using this technique is the assessment of the respiratory quotient (RQ). In this study, RQ was measured by indirect calorimetry, and used in the equations for estimation of EE. However, since the ^{13}CT method only estimates CO_2 production, an appropriate estimate of RQ is required. As an alternative to use the actual RQ, it may be assumed that the food quotient (FQ) from the diet represents the value of RQ (Black *et al.*, 1986). In this study, however, the actual RQ for each measurement was markedly lower (0.72), than the calculated value for FQ (0.81). Junghans *et al.* (2008) reported that to avoid much too large misestimating of EE, the daily average RQ should be estimated better than 10%. In fact, by using the FQ value of 0.81, would have led to too low estimates of EE in this study (Table 2).

Table 2. Mean values of estimated EE (kJ/kg$^{0.75}$/d) calculated according to Equation 1 using the actual value of RQ (measured by indirect calorimetry), or by using the food quotient (FQ = 0.81).

Administration method	EE (kJ/kg$^{0.75}$/d), RQ = actual value	EE (kJ/kg$^{0.75}$/d), RQ = FQ (0.81)
IM	653±115	601±108
SC	601±98	562±99
O	639±77	584±83

Conclusion

In conclusion the ^{13}CT seems to be a useful method for estimating EE, regardless of administration method. Both equations used for EE estimation gave similar and correct estimates. However, the accuracy of the EE estimates is dependent on the values used for RF and the estimates for RQ, which need to be assessed in every measurement situation.

Acknowledgements

The authors wish to express their sincere gratitude to Merethe Stubgaard and Ebba de Neergaard Harrison for skilful performance of chemical analyses, Abdalla Ali for operation of the respiration and IRIS units, Boye K. Pedersen and Kim Dinesen for professional handling of the mink, and Dr. Peter Junghans for assistance with the calculations.

References

Armon, Y., Cooper, D.M., Springer, C., Barstow, T.J., Rahimizadeh, H., Landaw, E. and Epstein, S., 1990. Oral [^{13}C]bicarbonate measurement of CO_2 stores and dynamics in children and adults. Journal of Applied Physiology 69: 1754-1760.

Black, A.E., Prentice, A. M. and Coward W.A., 1986. Use of food quotient to predict respiratory quotients for the doubly labelled water method of measuring energy expenditure. Human Nutrition: Clinical Nutrition 40c: 381-391.

Brouwer, E., 1965. Report of sub-commitee on constants and factors. In: Blaxter, K.L. (ed.) Energy Metabolism. Proceedings of the 3rd symposium on energy metabolism. European Association of Animal Production, Publication. No. 11. Academic Press, London, UK, pp. 441-443.

Chwalibog A., Tauson A-H. and Thorbek, G., 2004. Energy metabolism and substrate oxidation in pigs during feeding, starvation and re-feeding. Journal of Animal Physiology and Animal Nutrition 88: 101-112.

Downey, R.S., Mellone, A. and Matthews, D.E., 1986. Effect of tracer infusion site on measurement of bicarbonate-carbon dioxide metabolism in dogs. Journal of Applied Physiology 60: 1248-1253.

Elia, M., 1991. Estimation of short-term energy expenditure by the labeled bicarbonate method. In: Whitehead R.G. and Prentice A. (eds.). New techniques in nutritional research. Academic Press, New York, NY, USA, pp. 207-227.

Elia, M., Fuller, N. and Murgatroyd, P., 1988. The potential use of the labelled bicarbonate method for estimating energy expenditure in man. Proceedings of the Nutrition Society 47: 247-258.

Hansen, N.E., Finne, L., Skrede, A. and Tauson, A-H., 1991. Energiforsyningen hos mink og rev. NJF-Utredning/Rapport, No.63, DSR Forlag, Lanbohøjskolen, Copenhagen, 59 pp.

Junghans, P., Derno, M., Gehre, M., Höfling, R., Kowski, P., Strauch, G., Jentsch, W., Voigt, J. and Hennig, U., 1997. Calorimetric validation of ^{13}C bicarbonate and doubly labelled water method for determining the energy expenditure in goats. Zeitschrift für Ernährungswissenschaft 36: 268-272.

Junghans, P. and Chwalibog, A., 2001. Complementary application of stable isotope techniques and indirect calorimetry for determining energy expenditure. In: Chwalibog, A. and Jakobsen, K. (eds.) Energy metabolism in animals, Proceedings of the 15th symposium on energy metabolism in animals. EAAP publication No.103. Wageningen Academic Publishers, the Netherlands, pp. 425-428.

Junghans, P., Jentsch, W. and Derno, M., 2008. Non-invasive ^{13}C bicarbonate tracer technique for measuring energy expenditure in men – a pilot study. E-SPEN, the European e-journal of Clinical Nutrition and Metabolism 3: e46.

Lachica, M. and Aguilera, J.F., 2003. Estimation of energy needs of the free-ranging goat with particular reference to the assessement of its energy expenditure by the ^{13}C-bicarbonate method. Small Ruminant Research 49: 303-318.

National Research Council (NRC), 2006. Nutrient requirements of dogs and cats. The National Academies Press. Washington, DC, USA.

Pouteau, E., Mariot, S.M., Martin, L.J., Dumon, H.J., Mabon, F.J., Krempf, M.A., Robins, R.J., Darmaun, D.H., Naulet N.A. and Nguyen, P.G., 2002. Rate of carbon dioxide production and energy expenditure in fed and food-deprived adult dogs determined by indirect calorimetry and isotopic methods. American Journal of Veterinary Research 63: 111-118.

Prieto, C., Lachica, M. and Aguilera, J.F., 2001. The ^{13}C-bicarbonate method: its suitability for estimating the energy expenditure in grazing goats. Livestock Production Science 69: 207-215.

Roecker, K., Landlaw, E., Striegel, H., Mayer, F. and Dickhuth, H.-H., 2001. First-pass effect of an intravenous bolus of [^{13}C]bicarbonate displayed breath-by-breath. Journal of Applied Physiology 90: 2181-2187.

Speakman, J.R and Thomson, S.C., 1997. Validation of the labeled bicarbonate technique for measurements of short-term energy expenditure in the mouse. Zeitschrift für Ernährungswissenschaft 36: 273-277.

Body condition and reproduction success in farmed mink males (*Neovison vison*)

B.K. Hansen[1], K. Gautason[1,2], M. Sønderup[1] and H. Bækgaard[1]
[1]Kopenhagen Fur, Kopenhagen Consulting, Agro Food Park 15, 8200 Aarhus N, Denmark; [2]Agricultural University of Iceland, Hvanneyri, 311, Iceland; bkh@kopenhagenfur.com

Abstract

Correct managing of dam body condition during the winter period results in increased number of live born kits. So far, male body condition has not been related to male reproduction success. We therefore wanted to study whether the male body condition, visually graded using a 1 to 9 point scale, has influence on re-production results. Body condition of 6500 mink males of the brown colour type from 12 farms were scored in December, January, February and March in the winter period 2011/2012. The data will be used to seek answers for the following questions: (1) Has male body condition, or changes in body condition at specific point in time influence on mating willingness (mating willingness defined as the number of successful mating's), percentage of barren dams and litter size? (2) Have changes in male body condition in specific periods influence on number of mating's, percentage of barren dams and litter size? (3) What would be the optimal body condition development for mink males from December to *medio* March in relation to reproduction results?

Keywords: mink production, obese, subjective score

Introduction

Correct managing of mink dam body condition during the winter period combined with correct flushing results in an increased number of live born kits. Flushing procedure is defined as moderate restricted feeding in a short period followed by feeding *ad libitum* in 4-5 days (Tauson, 1985, 1988) conducted just before mating. On commercial farms the number of live kits depends on body condition of the dam (Baekgaard *et al.*, 2008). Systematic grading of body condition, based on experience from practice and results from a large experiment conducted on commercial farms, is a useful tool on Danish mink farms (Sønderup *et al.*, 2008).

Even though management of body condition and the flushing procedure have been done correctly, the reproduction results occasionally are lower than expected. We therefore wanted to study whether male body condition has influence on reproduction results. So far, visually graded male body condition (Rouvinen-Watt and Amstrong, 2002, Clausen, 2005, Sønderup *et al.*, 2008), has not been related to male reproduction success. To conduct such study a large number of males is needed and therefore it is possible only on commercial farms.

The aim of this project is to study body condition and changes in male body condition in the winter period and relate it to number of mating's, barren percentage and litter size. The preliminary results of body condition of males from 10 farms are presented here.

Material and method

Farms

The project was conducted on around 5,800 males of the brown/glow colour types from 8 farms. One of the 8 farms had two separate farm units with breeding stock from a joint pedigree, thus a total of 9 farms was included. The number of males per recording time differs as farmers sold and bought new animals. Only males with information of the pedigree are included in the investigation (Table 1). The participating farms represent farms with varying average litter size. The farms were located in different parts of Denmark, so each round of scoring took one week. The animals had an individual id.

Body condition

Clausen (2005) has shown that there is a correlation between body condition score and an empirical measurement of the fat percentage of the body. Thus, all males were subjectively graded for body condition altogether 4 times in the period from December to March. Grading times were: *primo* December, *primo* January, mid February and in the mating season in March. Scoring of body condition was based on a method developed by Rouvinen-Watt and Armstrong (2002) with adjustments made to fit our circumstances (homogeneity of males in January and on certain farms throughout the grading period). In December the traditional 5-point scale was used and in January and onwards a 9-point scale was used (Table 1).

Using the 9 point scale:
- Score 1 Very Thin animal. The mink has an emaciated appearance with decreased muscle mass. The animal has a thin neck and a clearly V-shaped body. There is no body fat and the stomach is sunk in. Shoulder and hip bones can be seen and the ribs are easily felt.
- Score 2. Thin. The mink has a thin neck and a V-shaped waist line. There is no subcutaneous body fat layer. The hip bones can be seen and the shoulder and the ribs can be easily felt.
- Score 3. Thin. Like score 2 but with a bit more subcutaneous body fat layer. The shape of hip bones can be seen and shoulder and the ribs can be easily felt.
- Score 4. Ideal. The mink has a slender neck and a straight body shape. There is a slight amount of subcutaneous body fat. The shoulder and hip bones and the ribs can be easily felt.
- Score 5. Ideal. Like score 4 but with more subcutaneous body fat.
- Score 6. Heavy. The mink has a thicker neck and a pear-shaped body. The ribs are difficult to feel. The shoulder and hip bones are covered by a moderate fat layer. An abdominal fat pad is present.

Table 1. Subgrouping of males from January and onwards to a 9-point scale.

9-point scale	1	2	3	4	5	6	7	8	9
9-point scale	1	2-1	2-2	3-1	3-2	4-1	4-2	5-1	5-2
5-point scale	1	2		3		4		5	

- Score 7 Heavy. Like score 6, but with more abdominal fat pad.
- Score 8 Obese. The mink has a thick neck with a slight brisket and a full body shape. The ribs are very difficult to feel. The shoulder and hip bones are covered by a moderate to thick fat layer. A fat pad is present in the abdomen and results in folds in the skin. Fat deposits can be seen in the limbs and the face.
- Score 9 Obese. Like score 8 but with more fat deposits in the limbs and the face.

Body weight at live grading shows the maximum size of the male. Even though the dam body condition is important for the litter size no dam body condition was recorded because we assumed that each male mates dams of different body condition.

Statistical analyses

The frequency of body condition and the change in body condition is illustrated in general and on each farm. After the farmers have send in the final registrations the effect of body condition or changes in body condition on willingness to mate defined as number of mating's (first, second and total mating's), percentage of barren dams and litter size is estimated using model (1). The DMU package (Madsen and Jensen, 2008) will be used to estimate the (co) variance components, using an AI-REML algorithm (Jensen *et al.,* 1997). The effect of relatives is included in the model to get a clear effect of body condition.

The model for the analysis will most likely be:

$$y = X\beta + Za + e \qquad (1)$$

Where, y is the vector of observations on the traits; number of mating's, barren percentage and litter size, in each analysis, X, and Z are matrices relating fixed and random effects to observations. β is the vector of fixed effects, a is the vector of random direct additive genetic effects (direct additive effects), is the vector of random residual. In the assumptions, A is the numerator relationship matrix and I's are identity matrixes of proper order.

Random effects are assumed to follow a normal distribution, $a \sim N(0, A\sigma^2_a)$ and $e \sim N(0, I\sigma^2_e)$. Fixed effects include always farm and the regression of the measured trait on litter size. The effect of body condition will be included.

Results

Means and standard deviations of body conditions (9 point scale) are shown in Table 2 below. The level of uniformity in the animal's body condition is apparent from the standard deviation. In January and February the body condition is low.

An average of score 4.5 on the 9 point scale corresponds to the medium body condition. At the grading time in the month of December the major part of the farms had trimmed the males since live grading in November. So the scored body condition may have been higher before grading in December, but when investigations are done on commercial farms, scoring in December was

Tabel 2. No of animals, mean (± sd) of body condition (9 point scale) at different grading time

Time of grading	Animals	Mean	±sd
December	5,403	5.02	±1.58
January	5,679	2.39	±0.79
February	5,861	2.68	±0.93
March	4,612	4.61	±1.00

the compromise to do. From December to January the average body condition goes down more than 2.5 classes. In average males stay on that level in the month of February, while they gain 2 classes in average from February to March. Figure 1 shows the distribution of body condition of males during the period from December to March.

From December to January 35% of males lost two classes while 50% lost three or four classes and around 10% lost more than four classes. From January to February a third part of males lost one to two classes, while 65% gained one to two classes. From February to March 40% of males gained two classes and 30% gained four classes (Figure 2).

In December there was a large variation between farms, while they were more even in the month's of January and February (Figure 3).

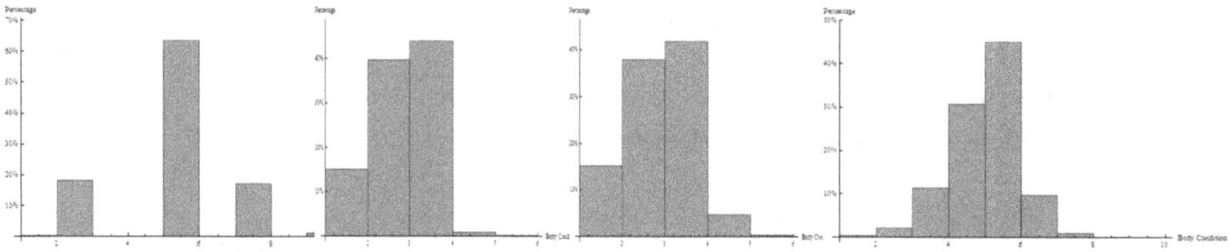

Figure 1. Histogram showing body condition in December (left), January, February and March. Body condition on the horizontal axis and percentage in each class on the vertical axis.

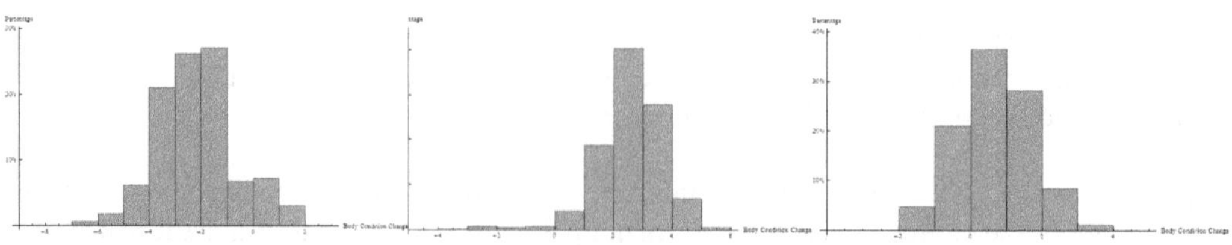

Figure 2. Change in body condition from December to January, January to February and from February to March. Change in body condition on the horizontal axis and percentage in each class on the vertical axis.

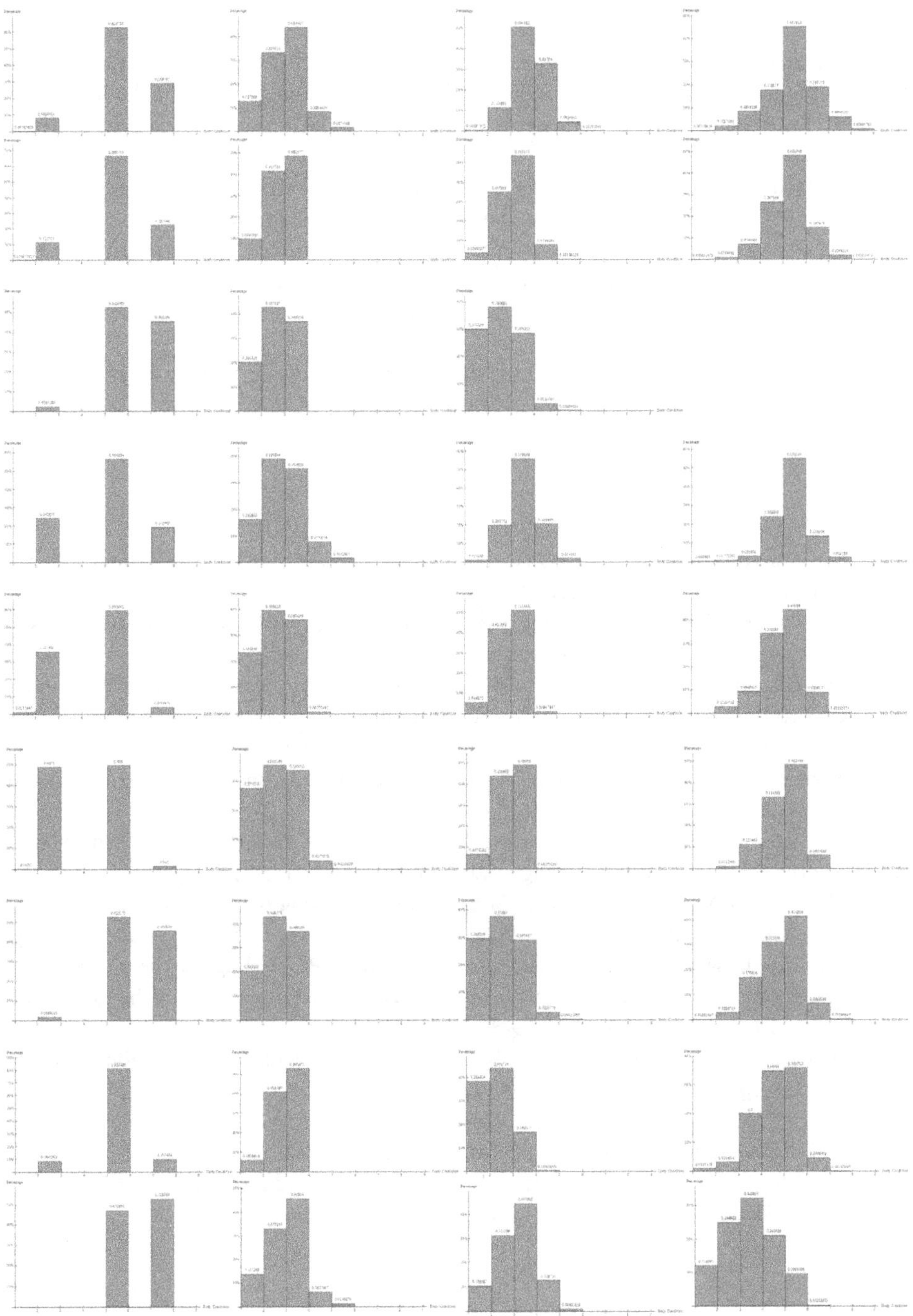

Figure 3. Body condition in December (left), January, February and March on nine farms (rows). Body condition on the horizontal axis and percentage in each class on the vertical axis.

Discussion

Mink have been bred for increased body size for decades. This entails better feed conversion rates along with a bigger genetic potential for size and therefore both a benefit for the farmer (bigger skins) and a challenge, the biggest mink is usually at the greatest risk of being obese. It has been shown in humans that body fat around the abdominal region can have deleterious effects on semen quality (e.g. Hammoud *et al.*, 2008). As the males that are selected for breeding are chosen in middle to late November, when the males are at their absolute maximum body weight, the farmer is faced with a dilemma. He has to trim the obese males into a slim condition in January to ensure a healthy development of semen, but then build up the condition of males again in February in order to have males with energy mate almost non-stop for two weeks in March.

The consequence of the body condition and of the changes of body condition to mating willingness and reproduction success will be presented at the meeting.

References

Baekgaard, H., Larsen, P.F. and Sønderup, M., 2008. Female body condition and early kit mortality: A description from practice. IX International Scientific Congress in Fur Animals, Halifax, Nova Scotia, Canada, August 19-23. 2008. Scientifur 32 (4) pp. 18-12.

Clausen, T.N, 2005. Huldvurdering af mink. Faglig Årsberetning 2005, Pelsdyrerhvervets Forsøgs- og Forsknings Center, Holstebro, Denmark. pp. 175-181.

Hammoud, A., Gibson, M., Peterson, M., Meikle, W., and Carrel, D. (2008). Impact of male obesity on infertility: a critical review of the current literature. Fertility and Sterility, 90(4): 897-904.

Jensen, J., Mäntysaari, E.A., Madsen, P. and Thompson, R., 1997. Residual Maximum Likelihood Estimation of (Co) Variance Components in Multivariate Mixed Linear Models using Average Information. Journal Indian Soc. Ag. Statistics 49: 215-236.

Madsen, P. and Jensen, J., 2008. A Users guide to DMU-package, A Package for Analyzing Multivariate Mixed models. Available at: http://dmu.agrsci.dk/dmuv6_guide-R4-6-7.pdf.

Rouvinen-Watt, K. and Armstrong, D., 2002. Body condition scoring of mink using a five-point scale. Appendix A. In: Hynes, A.M., Rouvinen-Watt, K. and Armstrong, D. Body condition and glycemic control in mink females during reproduction and lactation. Proceedings of the VIII International Scientific Congress in Fur Animal Production, De Ruwenberg, 's-Hertogenbosch, the Netherlands, 15-18 September 2004. Scientifur 28 (3) pp. 79-86.

Sønderup, M., Baekgaard, H. and Clausen, J., 2007. Recommendation with regard to feeding and control of body condition from pelting 2008 to birth 2008. Note to farmers.

Tauson, A.-H., 1985. Flushing of mink. Effects of level of flushing on reproductive performance, ovulation rate and plasma progesterone levels in mink. Acta Agric. Scand. 35: 295-309.

Tauson, A.-H., 1988. Flushing of mink. Effects of level of preceding feed restriction and length of flushing period on reproductive performance. Animal Reproduction Science 17: 243-250.

Maternal preconceptional nutrition leads to variable fat deposition of adult offspring mice (C57BL/6JBom)

E.L. Krogh Mortensen[1], T. Wang[1], H. Malte[1], D. Raubenheimer[2] and D. Mayntz[3,4]
[1]Department of Zoophysiology, University of Aarhus, C.F. Møllers Allé 3, 8000 Aarhus C, Denmark; [2]Institute of Natural Sciences, Massey University, Private Bag 102904, North Shore, 0745 Albany, New Zealand; [3]Department of Ecology and Genetics, University of Aarhus, Ny Munkegade 116, 8000 Aarhus C, Denmark; [4]Department of Genetics and Biotechnology, Research Centre Foulum, University of Aarhus, Blichers Allé 20, 8830 Tjele, Denmark; emo@kopenhagenfur.com

Abstract

Maternal nutrition during pregnancy or lactation may affect the chance of offspring becoming obese as adults, but little is known about the possible role of maternal nutrition before conception. Here we investigate how variable protein and carbohydrate content of the diet consumed before pregnancy affects fat deposition of offspring mice. Eight week old female mice (C57BL/6JBom) were fed isocaloric low protein (8.4% protein; LP), standard protein (21.5% protein; ST) or high protein (44.2% protein; HP) diets. After eight weeks of feeding, females were mated and fed a standard laboratory chow diet (22.5% protein) throughout periods of mating, gestation, lactation and weaning. Offspring mice were fed the same standard diet until 46 days of age. Then offspring were killed and measures of dissected fat deposits were taken. Fat deposition of the offspring was significantly affected by preconceptional maternal nutrition and the effects differed among sexes. Male offspring deposited most fat when mothers were fed LP diet, whereas female offspring deposited most fat when mothers were fed the ST diet. Our study demonstrates that preconceptional nutrition can have important influence on fat deposition in offspring in mice.

Keywords: protein; maternal diet; fat deposition

Introduction

Early nutritional programming is a concept in which differences in nutritional experience during critical periods of the early life stages program later development of metabolism and different health problems (Lucas, 1998; Patel and Srinivasan, 2002). Evidence of metabolic programming stems from both animal studies and epidemiological studies on humans (Taylor and Poston, 2007; Robinson, 2007). For example, increased preference for high fat foods was developed *in utero* as an effect of protein poor fetal diet in rodents (Ozanne *et al.*, 2004; Bellinger *et al.*, 2004).

Focus has recently been placed on the epigenetic determinants of obesity, especially through the maternal diet during pregnancy and breast feeding (Baker, 2004, Demmelmair and Rosen, 2006). Several studies have shown that early nutrition can cause subsequent health problems in the offspring, and that both pre- and postnatal diet treatments can program later development, metabolism and health. For example, low protein intake of pregnant mothers elevated blood pressure in offspring rats (Langley-Evans *et al.*, 1999), and increased later weight gain in mice offspring due to an elevated intake of high fat foods (Ozanne *et al.*, 2004). In rats (Lucas *et al.*,

1996), it was demonstrated that both pre- and postnatal stages are sensitive periods for nutritional programming, and dietary protein during pregnancy or lactation affected plasma concentrations of cholesterol and triacylglycerols in the offspring.

Although a link between prenatal nutrition and offspring obesity has been established over the past decade, less is known about the consequences of maternal diet and maternal body composition before pregnancy. In rats, blood glucose level of offspring may be affected by maternal body composition before pregnancy and a maternal low protein diet may increase blood glucose of the offspring, when they reach adulthood (Joshi *et al.*, 2003a). Thus, it seems that maternal body composition at the time of conception may affect metabolism and the obesity risks of the offspring. Considering that adult health might be programmed early in life, it is interesting to study the nutritional basis for effects of maternal diet on the offspring. Previous studies have investigated how a variety of nutritional compounds including fish oils (Hartvigsen *et al.*, 2003; Joshi *et al.*, 2003b), vitamins (Javaid *et al.*, 2006; Watson and McDonald, 2007), minerals (Hostetler *et al.*, 2003) and macronutrients (Saldana *et al.*, 2004; Moore and Davies, 2005) may affect offspring phenotype, but most studies have concerned protein composition in the mother's diet during pregnancy or lactation and obesity risk of offspring (Taylor and Poston, 2007; Bellinger *et al.*, 2004; Gorski *et al.*, 2006; Howie *et al.*, 2009).

In this study, we investigate how variation in protein to carbohydrate balance of maternal food before mating affects fat deposition of adult offspring mice.

Materials and methods

Housing and general procedures

Housing and general procedures are described in Mortensen *et al.* (2010).

Experimental diets

Three types of custom made experimental diets and a standard laboratory chow diet (dry, in pellet form) were purchased from Brogaarden /Lynge, Denmark) and manufactured by Altromin GmbH Company (Lage, Germany). The experimental diets were isocaloric and similar in composition, except for a variation in the ratio of protein to carbohydrate (Table 1). Breeding mice/rat diet Altromin 1314 was used as the standard laboratory chow diet. Before the mice were purchased, they were fed with a NIH-31 diet, with 18% protein, 4% fat, 5% fibre and 8% ash content.

Data analysis

To analyse the variable total body fat deposits, an analysis of covariance model was used with maternal diet and mouse gender as factors. When testing the mass of total fat deposits, we used mouse wet mass minus the removed fat deposits as the covariate. Throughout all these analyses, litter size was added as a covariate in the initial model and removed again if nonsignificant, which was always the case. If we found significant interactions between the maternal diet factor and gender, we conducted a new test on each of the genders, separately. Homogeneity of variances

Table 1. Composition of three types of experimental diets and one standard laboratory chow diet (g/100 g dry mass). The experimental diets, low protein (LP), standard protein (ST) and high protein (HP) varied in protein and carbohydrate concentrations, where other ingredients were kept similar. Main ingredients: casein, cornstarch, wheat starch, sucrose, sunflower oil, cellulose powder, a mineral mix (K_2CO_3, Coal.Chalk, NaCl, $MgCl_2$, K_2HPO_4, Fe_2O_3, $FeSO_4$, $MnSO_4$, NaF, KI, Na_2SeO_3, Na_2MoO_4, $CoCl_2$, $ZnCO_3$, $CuSO_4$) and a vitamin mixture (C 1000, Altromin).

Diets	LP	ST	HP	STlab
Crude protein %	8.4	21.5	44.2	22.5
Carbohydrates %	60.0	47.2	26.3	50.5
Crude fat %	7.2	7.2	7.1	5.0
Crude fibre %	5.0	5.1	5.4	4.5
Ash %	5.0	5.1	5.1	6.5
Energy (kcal/g)	3.3	3.3	3.5	3.0

was tested using Levene's test (P>0.05). If this assumption was violated, we tried using one of the following transformations log(x), $\sqrt{x}$, $1/\sqrt{x}$, x^2 or $1/x^2$. If variances were still unequal, we tested the two genders individually. Detailed differences of various fat deposits were tested using analysis of variance (ANOVA). Differences between feeding treatments in each variable were analyzed using Student's t-test. All statistics were done using JMP 7 (SAS Institute Inc., Cary, USA).

Results

Females feeding on the different diets did not consume different amount of energy (ANOVA, $F_{2,23}$ = 0.09, P=0.92) or mass (ANOVA, $F_{2,23}$ = 1.36, P=0.28) over the first 3 weeks of the preconceptional period. However, there was a large difference in the amount of consumed carbohydrates (g/day±SE.: LP=1.52±0.05, standard protein (ST)=1.21±0.03, high protein (HP)=0.63±0.03; ANOVA, $F_{2,23}$=120.7, P<0.0001) and protein (g/day±SE: LP=0.21±0.007, ST=0.55±0.01, HP=1.06±0.04; ANOVA, $F_{2,23}$=506.2, P<0.0001).

Although maternal diet did not affect body length (ANOVA, $F_{2,68}$=0.21, P=0.81) or body wet mass (ANOVA, $F_{2,68}$=0.73, P=0.49) of male offspring (Figure 1), we found significant effects on length (ANOVA, $F_{2,51}$=3.50, P=0.04) and body wet mass (ANOVA, $F_{2,51}$=3.87, P=0.03) of female offspring. Females of mothers fed LP diets were significantly shorter and lighter than those whose mothers had been fed ST diet or HP diet (Figure 1a,b).

Fat deposits

The total mass of fat deposits (subcutaneous, genital and kidney) varied positively with the covariate total fat-free mass of the mice (Table 2). We found a significant interaction between gender and maternal diet on total body fat (Table 2), which reflects that female offspring of mothers fed the ST diet had the largest fat deposits while male offspring had the most fat

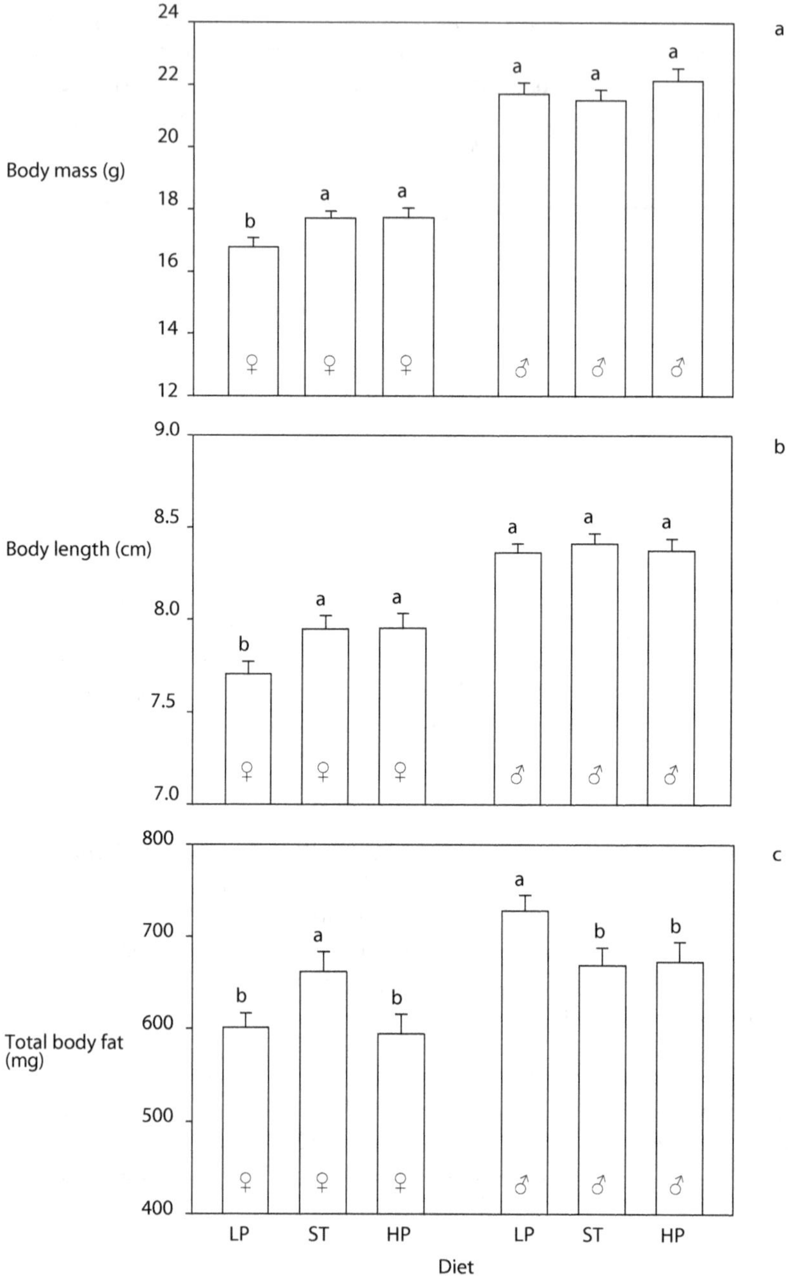

Figure 1. Effects of maternal preconceptional diet on (a) body mass, (b) total body fat and (c) body length of offspring at 46 days of age (mean±SE). Maternal experimental diets were given eight weeks before mating and a standard laboratory diet were fed to females and offspring during mating, lactation, weaning and to offspring during their entire lives. Letters indicate differences across maternal feeding treatments within each gender using Student's t*-test.*

Table 2. ANCOVA on total body fat, in offspring with maternal diet and mouse gender as factors. Covariates were included to adjust variables for total size or mass of the mice. In all analysis, litter size was used as a covariate in the initial model but it was never significant and was therefore removed in the final model.

Source of variation	DF	F	*P*-value
Total fat analysis			
Gender	1	0.32	0.57
Maternal diet	2	2.90	0.06
Gender × maternal diet	2	3.75	0.03
Wet mass minus total fat	1	15.45	<0.0001
Total fat – males			
Maternal diet	2	5.14	0.008
Wet mass minus total fat	1	29.96	<0.0001
Total fat – females			
Maternal diet	2	3.87	0.03
Wet mass minus total fat	1	0.62	0.43

when maternal diet was LP diet (Figure 1c). The separate analysis on each gender confirmed a significant effect of maternal diet on both males and females (Table 2). As this model adjusted for total fat-free mass, the result indicates that it is the relative amount of fat and not total body mass of the mice that was affected by maternal nutrition. This effect is also illustrated by the fact that the groups having significantly highest fat deposits (Figure 1c) did not have a significantly higher body mass (Figure 1a).

Focusing on the separate sections of fat deposits, females had more subcutaneous fat and kidney fat when mothers were fed the ST diet compared to LP and HP diets. No significant difference in genital fat was found (Table 3). The largest effect of maternal nutrition in males was found in

Table 3. Effects of maternal nutrition on fat deposits (mg). Maternal diets were low protein (LP), standard protein (ST) and high protein (HP) and they were fed to females for eight weeks until conception.

Gender	Female			Male		
Maternal diet	**LP**	**ST**	**HP**	**LP**	**ST**	**HP**
Subcutaneous fat mass	319.6±8.6[a]	360.3±15.7[b]	317.9±11.5[a]	329.5±7.9[a]	296.7±7.4[b]	296.2±7.7[b]
Genital fat mass	174.3±5.6	187.5±6.2	175.9±8.2	275.1±6.8	254.1±9.7	257.6±10.5
Kidney fat mass	107.4±3.2[ab]	114.4±2.7[a]	100.8±4.0[b]	123.2±3.2	118.2±3.8	119.1±4.5

[a,b] Letters indicate differences across maternal feeding treatments within each gender using Student's *t*-test. The test was only made if the initial ANOVA test found an overall significant effect of maternal nutrition.

subcutaneous fat mass, while differences in genital and kidney fat deposits were non-significant (Table 3).

Discussion

Diet composition during pregnancy or lactation is known to alter the risk of obesity in adulthood (Howie *et al.*, 2009). However, this study is, to our knowledge, the first experimental demonstration that diet composition supplied to mice before conception will lead to variable fat deposition in adult offspring.

To test for differences in palatability between the diets, we compared food intakes over the first 3 weeks of the experiment. Our results showed that there were no differences in food or energy intake over this period but large differences in protein and carbohydrate intake. This does not, however, rule out the possibility that there were differences in energy intake over the remaining 5 weeks of the experimental period. In an experiment using the same experimental feed but different strain (NMRI), sex and age group of mice, it was found that both energy consumption and final body composition differed (Sørensen *et al.*, 2008). Thus, it is possible that the females in this study had different energy consumption and/or different body composition when conception occurred and such variation might have contributed to the effect on the offspring. An important role of dietary protein density in the diets of mice (Sørensen *et al.*, 2008), as well as humans (Simpson *et al.*, 2003) and non-human primates (Felton *et al.*, 2009), is that it can influence health through its 'leverage' on calorie intake (Simpson and Raubenheimer, 2005). Although further work is needed to disentangle the effects of dietary protein content on energy intake *per se* vs. the balance of ingested macronutrients, our experiments have shown that the balance of macronutrients in foods of preconceptional females has effects on the developmental trajectories of their offspring.

The period of maximum susceptibility for imprinting of future metabolic physiology of offspring is controversial, but there is good evidence of an effect of maternal nutrition during the suckling period. For instance, rat offspring of lean prenatal dams became obese and insulin resistant when raised postnatally by obese dams (Gorski *et al.*, 2006). Other studies have found that vital effects of protein imbalance occur during different periods of pregnancy in animals (Howie *et al.*, 2009), including humans (Villar *et al.*, 1992; Mardonese-Santander *et al.*, 1998). Our study on mice indicates that protein composition in the maternal diet before conception may also be important in determining fat deposition of the offspring. In rats, variable preconceptional diet affected sizes of heart, brain and kidneys in the offspring, but not offspring body mass (Joshi *et al.*, 2003a). We similarly did not find differences in overall body mass, but fat deposition was significantly affected. Hence, it is possible that the rats also had variable fat deposition in response to preconceptional maternal diets (Joshi *et al.*, 2003a), although this was not reported.

In a previous study, in which young mice of an outbred strain were fed the same diet, males on the LP diet had relatively more body fat after 4.5 weeks of feeding than males on other diets (Sørensen *et al.*, 2008). We fed the dams for 8 weeks on the different diets, so body composition of the females by the time of conception may have been different, and we cannot, therefore, discern direct short-term nutritional effects from longer-term nutritional effects that are mediated by the body composition of the mother. There is some evidence that body composition in itself may mediate effects on offspring. In sheep, for example, offspring from mothers with low body condition score showed mild glucose intolerance compared with offspring from mothers with a normal body condition score (Cripps *et al.*. 2008). This indicates that the variable body fat of the female sheep, results in different glucose intolerance of offspring.

Fat deposition occurs when caloric intake exceeds expenditure. It is, therefore, an interesting question why male mice in this experiment deposited more fat when dams were fed LP diet. Although we did not measure food consumption in the offspring, other studies indicate that variable fat deposition may be due to differences in feeding behaviour (Sørensen *et al.*, 2008). The hypothalamic pathways responsible for appetite regulation develop already *in utero* (Bouret, 2009, Grove *et al.*, 2003, 2005) and continue to develop during lactation in rats (Grove *et al.*, 2003; Rinaman, 2003). An alternative explanation for the susceptibility of male offspring to maternal LP diet may be that energy expenditure is reduced compared to females, either by reduced basal metabolic rate or lower activity levels (Vickers *et al.*, 2003). Although we have no immediate explanation for the gender differences in our study, such interactions between offspring gender and maternal diets are not uncommon and, given the marked biological differences male and female mammals, not unexpected. For example, pregnant rats exposed to a maternal LP diet throughout gestation had female offspring with higher tendency to become obese than controls, but no effect was observed in male offspring (Bellinger *et al.*, 2004).

We only tested one strain of mice, but there may well be variation in how different strains respond to preconceptional maternal nutrition. Pre-pregnant nutrition may be a factor that can induce different susceptibility to overweight and obesity in adulthood. This is important from a management perspective, because unlike genetic effects diet can readily be altered.

References

Baker, J.L., 2004. Maternal prepregnant body mass index, duration of breastfeeding, and timing of complementary food introduction are associated with infant weight gain. American journal of Clinical Nutrition 80: 1579-1588.

Bellinger, L., Lilley, C. and Langley-Evans, S.C., 2004. Prenatal exposure to a maternal low-protein diet programmes a preference for high-fat foods in the young adult rat. British journal of Nutrition 92: 513-520.

Bouret, S.G., 2009. Early Life Origins of Obesity: Role of Hypothalamic Programming. Journal of Pedriatric Gastroenterology and Nutrition 48: S31-S38. Supplement 1.

Cripps, R.L. Green, L.R., Thompson, J., Martin-Gronert, M.S., Monk, M., Sheldon, I.M., Hanson, M.A., Hales, C.N. and Ozanne, S.E., 2008. The effect of maternal body condition score before and during pregnancy on the glucose tolerance of adult sheep offspring. Reproductive Sciences 15: 448-456.

Demmelmair, H., Von Rosen, J. and Koletzko, B., 2006. Long-term consequences of early nutrition. Early Human Development 82: 567-574.

Felton, A.M., Felton, A., Raubenheimer, D., Simpson, S.J., Foley, W.J., Wood, J.T., Wallis, I.R. and Lindenmayer, D.B., 2009. Protein content of diets dictates the daily energy intake of a free-ranging primate. Behavioral Ecology 20: 685-690.

Gorski, J.N., Dunn-Meynell, A.A., Hartman, T.G. and Levin, B.E., 2006. Postnatal environment overrides genetic and prenatal factors influencing offspring obesity and insulin resistance. American Journal of Physiology 291: R768-R778.

Grove, K.L., Allen, S., Grayson, B.E. and Smith, MS., 2003. Postnatal development of the hypothalamic neuropeptide Y system. Neuroscience 116: 393-406.

Grove, K.L., Grayson, B.E., Glavas, M.M., Xiao, X.Q. and Smith, M.S., 2005. Development of metabolic systems. Physiological Behaviour 86: 646-660.

Hartvigsen, S.M., Mu, H. and Høy, C.-E., 2003. 2003. Influence of maternal dietary n-3 fatty acids on breast milk and liver lipids of rat dams and offspring – a preliminary study. Nutrition Research 23: 747-760.

Hostetler, C.E., Kincaid, R.L. and Mirando, M.A., 2003. The role of essential trace elements in embryonic and fetal development in livestock. Veterinary Journal 166: 125-139.

Howie, G.J., Sloboda, D.M., Kamal, T. and Vickers, M.H., 2009. Maternal nutritional history predicts obesity in adult offspring independent of postnatal diet. Journal of Physiology 587: 905-915.

Javaid, M.K., Crozier, S.R., Harvey, N.C., Gale, C.R., Dennison, E.M., Boucher, B.J., Arden, N.K., Godfrey, K.M. and Cooper, C., 2006. Maternal vitamin D status during pregnancy and childhood bone mass at age 9 years: a longitudinal study. Lancet 367: 36-43.

Joshi, S., Garole, V., Daware, M., Girigosavi, S. and Rao, S., 2003. Maternal protein restriction before pregnancy affects vital organs of offspring in Wistar rats. Metabolism 52: 13-18.

Joshi, S., Rao, S., Golwilkar, A., Patwardhan, M. and Bhonde, R., 2003. Fish oil supplementation of rats during pregnancy reduces adult disease risks in their offspring. Journal of Nutrition 133: 3170-3174.

Langley-Evans, S.C., Welham, S.J.M. and Jackson, A.A., 1999. Fetal exposure to a maternal low protein diet impairs nephrogenesis and promotes hypertension in the rat. Life Sciences 64: 965-974.

Lucas, A., 1998. Programming by early nutrition: An experimental approach. Journal of Nutrition 128: 401S-406S.

Lucas, A., Baker, B.A., Desai, M. and Hales, C.N., 1996. Nutrition in pregnant or lactating rats programs lipid metabolism in the offspring. British Journal of Nutrition 76: 605-612.

Mardonese-Santander, F., Salazar, G., Rosso, P. and Villarroel, L., 1998. Maternal body composition near term and Birth weight. Obstetrics and Gynecology 91: 873-877.

Moore, V.M. and Davies, M.J., 2005. Diet during pregnancy, neonatal outcomes and later health. Reproduction Fertility and Development 17: 341-348.

Mortensen, E.L., Wang, T., Malte, H., Raubenheimer, D. and Mayntz, D., 2010. Maternal preconceptional nutrition leads to variable fat deposition and gut dimensions of adult offspring mice (C57BL/6JBom). International Journal of Obesity 1-7.

Ozanne, S.E., Lewis, R., Jennings, B.J. and Hales, C.N., 2004. Early programming of weight gain in mice prevents the induction of obesity by a highly palatable diet. Clinical Science 106: 141-145.

Patel, M.S. and Srinivasan M., 2002. Metabolic programming: Causes and consequences. Journal of Biological Chemistry 277: 1629-1632.

Rinaman, L., 2003. Postnatal development of hypothalamic inputs to the dorsal vagal complex in rats. Physiological Behaviour 79: 65-70.

Robinson, S., 2007. Metabolic programming during pregnancy: Epidemiological studies in humans. Genes and Nutrition 2: 31-32.

Saldana, T.M., Siega-Riz, A.M., Adair, L.S., 2004. Effect of macronutrient intake on the development of glucose intolerance during pregnancy. American Journal of Clinical Nutrition 79: 479-486.

Simpson, S.J. and Raubenheimer, D., 2005. Obesity: the protein leverage hypothesis. Obesity Reviews 6: 133-142.

Simpson, S.J., Batley, R. and Raubenheimer, D., 2003. Geometric analysis of macronutrient intake in humans: the power of protein? Appetite 41: 123-140.

Sørensen, A., Mayntz, D., Raubenheimer, D. and Simpson, S.J., 2008. Protein-leverage in mice: The geometry of macronutrient balancing and consequences for fat deposition. Obesity 16: 566-571.

Taylor, P.D. and Poston, L., 2007. Developmental programming of obesity in mammals. Experimental Physiology 92: 287-298.

Vickers, M.H., Breier, B.H., McCarthy, D. and Gluckman, P.D., 2003. Sedentary behavior during postnatal life is determined by the prenatal environment and exacerbated by postnatal hypercaloric nutrition. American Journal of Physiology 285: R271-R273.

Villar, J., Cogswell, M., Kestler, E., Castillo, P., Menendez, R. and Repke, J.T., 1992. Effect of fat and fat-free mass deposition during pregnancy on birth-weight. American Journal of Obstetrics and Gynecology 167: 1344-1352.

Watson, P.E. and McDonald, B.W., 2007. Seasonal variation of nutrient intake in pregnancy: effects on infant measures and possible influence on diseases related to season of birth. European Journal of Clinical Nutrition 61: 1271-1280.

Influence of nest box environment on kit survival

V.H. Lund[1] *and J. Malmkvist*[2]
[1]*Kopenhagen Counselling, Agro Food Park 15, 8200 Aarhus N;* [2]*University of Aarhus, Animal Science, Blichers alle 20, 8830 Tjele, Denmark; vlu@kopenhagenfur.com*

Abstract

Hypothermia is considered as one of the major risk factors increasing mortality in the kits born. However, very little is known about the preferred nesting materials for nest building and the link between nest building and kit survival in mink. To investigate these questions, we used 105 farmed mink females divided into 4 groups, non-pregnant females (NON), pregnant females with access to one resource of nest building material (RES-1), pregnant females with access to three resources (RES-3), and pregnant females with access to one resource but which were moved into a climate-controlled facility (CLIMA). CLIMA was video recoded in order to describe the time course of nest building prior to parturition. The quality of the nest was scored once a week for five weeks prior to delivery using nest scores of 1-5 (1: no nest, 5: full nest with cover). Kits were counted on day 1 and on day 7 after delivery. We discriminated between stillborn and live-born kits dying early using autopsy. The access to more than one resource (RES-3) of nest building material had a profound effect on kit mortality in the first week of life compared to access to straw only ($\chi^2(1) = 13.7$, $P<0.001$). In RES-1 ~20% of the kits died from day 1-7, and only ~5% in RES-3. The risk of dying was approx. 4 times higher for a kit live-born into the one resource environment. RES-3 females were building better nests and stayed in the nest box longer around parturition than RES-1, which could explain the higher mortality in this group. In conclusion, access to different nesting materials reduced the risk of dying in farm mink in relation with factors as e.g. nest quality and maternal behaviour. A decrease in kit mortality from 20% to 5% would give the Danish mink production a profound number of Danish kroner extra per year.

Introduction

Studies on the behaviour of female farmed mink around and after parturition have been studied when speaking about welfare, kit mortality, and maternal care (Baker *et al.*, 2008; Castella and Malmkvist, 2008). What happens during gestation e.g. maternal nest building behaviour can also have an important impact on the further life of the kit but this subject has not been studied in mink. In this study we were interested in describing the onset of nest building and to identify if mink has a structured maternal nest building behaviour.

The aim of this study was to determine if there was any specific pattern in the nest building behaviour in farmed mink during gestation. However the results on a structured nest building behaviour were not conclusive. Instead the results showed a distinct link between the number of material given and kit mortality. This paper will focus on this link.

Materials and methods

Details on the materials and methods of the study are described in the unpublished master thesis (Lund, 2009).

Results

Parturition and kit survival

The results of the variables number of live-born kits, the percentage of stillborn kits and the percentage of dead kits from day 0-7 are summarised in Table 1. There was no difference between treatments at parturition in the number of live-born or stillborn kits (RES-1 vs. RES-3). I found in average 8.0 (2.85) kits born/dam, out of these approximately 12% were stillborn. Dams with access to three types of material had significantly lower kit mortality the first seven days of life compared to dams that had access to straw alone. Using a binomial logistic regression we estimated the probability of dying to be 19.7% (11.5-31.6%, 95% confidence interval) in treatment RES-1 and 4.9% (1.5-15.1%, 95% confidence interval) in treatment RES-3. Consequently, the average litter size on Day 7 was 11.3% higher in RES-3 (6.4 vs. 5.8 kits in RES-1).

The number of still-born kits did not differ between the treatments RES-1 and CLIMA ($F_{1,53}$=0.12, P=0.73). The dams in CLIMA had in average 7.7 (3.02) kits out of these were approximately 13% still-born. The binomial logistic regression showed that RES-1 and CLIMA significantly differed regarding dead kits on day 7 ($\chi^2_{(1)}$ = 6.11, P=0.013). The probability of dying in CLIMA was estimated to 10.2% (5.1-9.3%, 95% confidence interval). CLIMA had half the risk of dying compared to RES-1.

In-nest climate

The temperature and relative humidity (RH) are illustrated graphically in Figure 1. There was no difference between RES-1 and RES-3 on either temperature or RH. Even though there is not a statistical difference between RES-1 and RES-3 up to parturition Figure 1 shows RES-3 to have a slightly higher in-nest temperature and lower RH than RES-1. NON had as expected a lower

Table 1. Results as mean (S.E.) for litters with one (RES-1) and three (RES-3) types of nest building materials

	RES-1	RES-3	Test statistics	*P*-value
No. of live-born kits	6.7 (0.61)	6.8 (0.47)	$F_{1,37} = 0.01$	0.94
Proportion of stillborn[b] kits (%)	12.8 (5.07)	11.8 (3.85)	$F_{1,37} = 0.02$	0.88
Mortality (%)				
Days 0-7	22.0 (28.3)	4.7 (8.4)	$\chi^2_1 = 13.7$	P<0.001[a]

[a] Significant difference between treatments.

[b] Stillborn: negative lung flotation test.

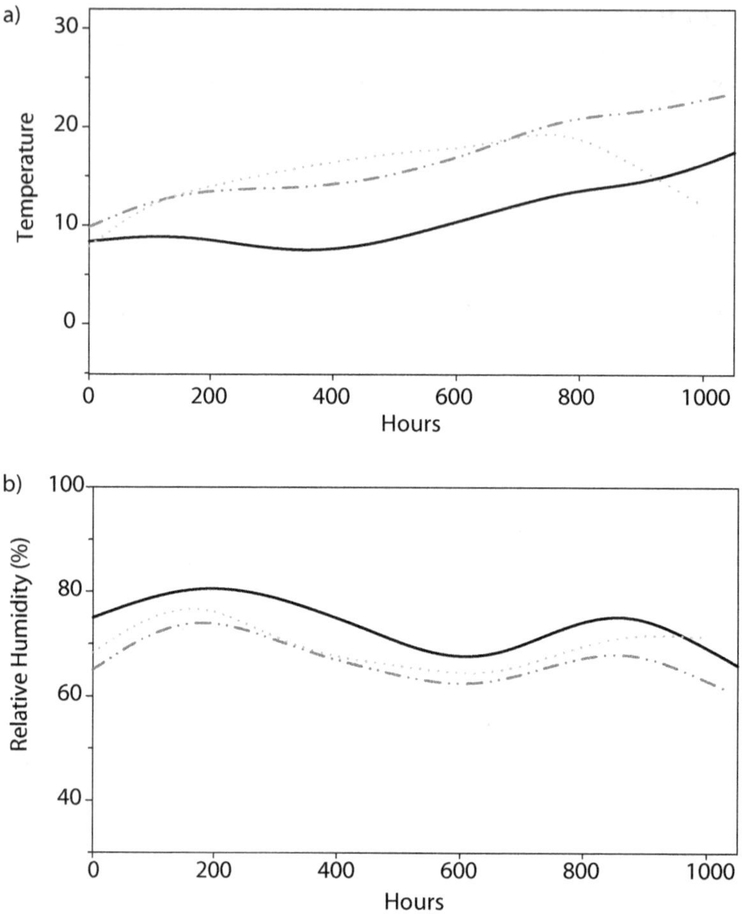

Figure 1. Time development in temperature and relative humidity (RH) from March 26 to May 8, (a) temperature, and (b) % RH. Black solid line, group NON; light grey dotted line, group RES-1; dark grey, dash-dot-dot-line, RES-3.

temperature and a higher RH than the other groups. No formal statistical analysis was performed on these data, but they are illustrated using Splines in SAS software.

Discussion

The neonate mink kit is born immature and maternal care from the dam is crucial for its survival. Hypothermia and starvation are the main causes of early kit mortality in farmed mink (Martino and Villar, 1990), therefore a good and warm nest is important when the dam is not present in the nest. We used the nest score in the current study as a measurement for nest quality and thereby also an indicator of maternal care (unpublished data). We found that mink in group RES-3 had a significantly higher mean nest score over the gestation period, and also after parturition, than in group RES-1. Furthermore RES-3 dams had a significantly higher nest weight. This could suggest that RES-3 mink were better mothers in beforehand or that the three materials given either stimulated the dam's nest building urge as seen in sows (Damm *et al.* 2003) or simply were better materials compared to straw. The latter may be more likely, since the probability of only RES-3 to be good mothers by chance is unlikely. Furthermore, since the animals used had not been giving birth before, the experience of this should not have had an impact on nest building or maternal care. Since RES-1 and RES-3 have been held under identical conditions before the experiment, they should all have the same quality for maternal care.

One may argue that also novelty (differ between the treatment RES-1 and RES-3, since all animals had previous experience with barley straw (given to RES-1) during their life (mandatory in the keeping of mink in Denmark (Anonymous, 2006)), but never experienced wood wool or rabbit wool (used as additional resources in RES-3) before this experiment) could have an effect on the females' behaviour towards the materials given. The experience of something new, the rabbit wool and wood wool, might have made the dams more interested in the materials, and therefore they took it into their nests. However, there would be no reason for choosing a material just because of the novelty value but rather because it might improve the nest and thereby decrease offspring mortality rate. There might also be a physical quality of the new material which straw lacks. In disagreement with this novelty hypothesis Malmkvist *et al.* (2003) demonstrated the reactions towards familiar and novel food in farmed mink. Mink from both lines (confident, fearful) showed an increased latency to approach the novel food source compared to the familiar food. The study showed that farmed mink may have evolved to be more 'picky' in their choice of food compared to mink in nature which have a broad food spectrum (Dunstone, 1993). However, foraging behaviour and nest building behaviour may not be directly comparable due to differences in the level of importance in e.g. survival (food keeps the dam alive, if she dies her kits would die too). In addition we noticed that the females from RES-3 treatment almost instantly after distribution pulled the materials inside their nests. If captive mink reacts more hesitating in situations where it has to explore something new our explanation of the novelty value in the new materials may not hold true but due to our observations neither can it be rejected.

The kit mortality in farmed mink is around 10-30% from day 0 until weaning at 7-8 weeks (Martino and Villar, 1990; Schneider and Hunter, 1993). This means that the mortality in RES-1 in this study did not differ from earlier studies. However, we did not expect to find such a large difference between RES-1 and RES-3 regarding kit mortality. The behaviour of the dam around parturition had a great impact on kit survival. Malmkvist *et al.* (2007) found that females with high kit mortality (MORT) circled less around the kits than females with low kit mortality (SURV) and suggested that the difference in behaviour in the two groups might have an impact on maternal care and kit survival. In our study a few days after parturition RES-1 dams spend more time in the wire cage (~60%) than with their kits in the nest box (unpublished data). This behaviour is probably the reason why RES-1 had a greater kit mortality rate from day 1-7 than RES-3, which spent ~75% of the time in the nest box. The lower nest score in RES-1, and thereby probably a poorer nest, could explain the increased kit mortality as well. The sudden change in RES-1 females' behaviour can also be seen in the temperature and RH changes. With the mother away from the nest, the kits can not maintain their own body temperature which increases the risk of dying of hypothermia and starvation (Martino and villar 1990), and this is what could have happened in this case. Malmkvist and Palme (2008) found that dams deprived straw as nest building material had significantly higher kit mortality than the other treatments with either an artificial nest and/or straw. One conclusion is that access to nest building material, and hence performing nest building behaviour, seems to be important for kit survival. However, the two studies are not directly comparable, since our results were obtained on dams having access to nest building materials and the females from the other study were completely deprived nest building material except from wood shavings. In addition Malmkvist and Palme (2008) distributed 70-80 g. of straw each day to some of their experimental groups. In my study the dams were only offered 30 g of straw three times a week. If this amount had not been sufficient for the dam it could trigger frustration as seen in sows deprived of nest building (Jensen and Toates, 1997) and this could also explain why RES-1 suddenly spend more time in the cage (decrease in in-nest temperature). Stress can arise as a consequence of adverse external stimulation (expressed as e.g. frustration) and as a consequence of internal causal factors driving the animal to attempt to carry out species-specific behaviour, such as nest building, in case of no or insufficient feedback (Jensen and Toates, 1997). Even though the weight of nest building materials was equal in both groups (RES-1: ~30 g.; RES-3: ~27 g. (total of all three materials) there was still a difference in the final nest weight. Another reason for the increased kit mortality could be a better climate inside the nest box, e.g. with more ideal humidity and temperatures, reducing the risk of kits dying of hypothermia. Since RES-3 had the highest mean nest score it would be reasonable to believe that the temperature would be higher and RH lower than in the poorer nests of RES-1. Due to this we expected to find a better climate in the RES-3 nest. RES-3 did have a lower RH and higher in-nest temperature compared to RES-1, though not significant. The results suggest that access to three kinds of materials reduced kit mortality, apparently acting together with effects on maternal care. Malmkvist and Palme (2008) saw an improved maternal behaviour in dams with access to both an artificial nest and straw compared to only straw or artificial nest. The dams had a shorter latency in the kit retrieval test which could suggest that an optimized nest also affected the maternal behaviour after parturition. It might be the same effect we saw with the groups RES-1 and RES-3; a poor nest makes poorer mothers thereby increased kit mortality. Another surprising finding was that CLIMA kits also survived better than RES-1 kits. CLIMA dams were disturbed more often than the other groups, and we expected that they had been affected negatively by this disturbance (Champagne and Meaney, 2009). Since the mink kits are solely dependant on the dam, a mother with less maternal care could be responsible of increased kit mortality. Stress in the gestational period increased anxiety-like behaviour in rats which induced smaller litter size and elevated mortality rate (Baker *et al.*, 2008). However, CLIMA dams were held under controlled conditions in the climate controlled facility. Constant temperature and humidity and not having to deal with shifting weather conditions may also have had a positive influence on kit mortality.

Conclusion

Providing pregnant mink with several types of nesting materials in the preparturient period had a marked effect on kit mortality, in comparison with pregnant mink having access to only one resource (straw as included in the Danish legislation for keeping of mink). The main effects of distributing other materials than straw alone could be less frustration in the females, better maternal care, and maybe a better in-nest climate seen as improved nests with higher nest scores.

This study did not answer many of the questions about the onset of nest building in farmed mink, however it has brought nest building behaviour up for discussion. If the onset of nest building behaviour in farmed mink can be determined, it will help the farmer in choosing the right time for distributing nest building material. More attention needs to be made on nest building behaviour in order to understand which factors play a role in maternal behaviour and thus kit mortality.

In addition if kit mortality can be minimised as seen from our results then Danish mink production has a great profit to gain.

Acknowledgements

I thank Birthe Houbak, Betty Skou and Eric L Decker, Aarhus University, Dept. Animal Health, Welfare and Nutrition and the farm staff of the research farm for the participation in the data collection and in set-up of equipment, and Jens Malmkvist, Aarhus University, Dept. Animal Health, Welfare and Nutrition, and Mark Bayley, Aarhus University, Dept. of Biological Sciences for manuscript improvements. This study was funded by Danish Fur Breeders Association.

References

Anonymous, 2006. Pelsdyrbekendtgørelsen nr. 1734 af 22. december 2006 om beskyttelse af pelsdyr. Justitsministeriet.

Baker, S., Chebli, M., Rees, S., LeMarec, N., Godbout, R. and Bielajew, C., 2008. Effects of gestational stress: 1. Evaluation of maternal and juvenile offspring behavior. Brain Research 1213: 98-110.

Castella, A. and Malmkvist, J., 2008. The effect of heat incubators on chilled mink kits. Applied Animal Behaviour Science 113: 265-269.

Champagne, F.A. and Meaney, M.J., 2006. Stress During Gestation Alters Postpartum Maternal Care and Development of the Offspring in a Rodent Model. Biological Psychiatry 59: 1227-1235.

Damm, B.I., Pedersen, L.J., Marchant-Forde, J.N. and Gilbert, C.L., 2003. Does feed-back from a nest affect periparturient behaviour, heart rate and circulatory cortisol and oxytocin in gilts. Applied Animal Behaviour Science 83: 55-76.

Dunstone, N., 1993. The Mink. T & A D Poyser Ltd., London, UK.

Jensen, P., Toates, F.M., 1997. Stress as a state of motivational systems. Applied Animal Behaviour Science 53: 145-156.

Lund, V.H., 2009. Nest building in farmed female mink (*Mustela vison*). Master thesis, Aarhus University, Denmark.

Malmkvist, J., Herskin, M.S. and Christensen, J.W., 2003. Behavioural responses of farmed mink towards familiar and novel food. Behavioural Processes 61: 123-130.

Malmkvist, J., Gade, M. and Damm, B.I., 2007. Parturient behaviour in farmed mink (*Mustela vison*) in relation to early kit mortality. Applied Animal Behaviour Science 107: 120-132.

Malmkvist, J. and Palme, R., 2008. Periparturient nest building: implications for parturition, kit survival, maternal stress and behaviour in farmed mink (*Mustela vison*). Applied Animal Behaviour Science 114: 270-283.

Martino, P.E. and Villar, J.A., 1990. A survey on perinatal mortality in young mink. Veterinary Research Communications 14: 199-205.

Schneider, R.R. and Hunter, D.B., 1993. Mortality in mink kits from birth to weaning. Canadian Veterinary Journal 34: 159-163.

The influence of succinic acid on the quality of fur-bearing animals skin

A.E. Kokorina, O.Yu. Bespyatykh and T.V. Tebenkova
Russian Research Institute of Game Management and Fur Farming, 79, Engels st., 610000 Kirov, Russia; b_oleg@mail.ru

Abstract

The introduction of succinic acid to the ration of juvenile foxes and minks allows for better skin quality, however the presence of it in the diet of Arctic foxes produces a negative effect on the skin. The optimum dose for juvenile fur-bearing animals (fox and mink) is 5 mg/kg LW. The succinic acid is being added to the fodder during the growth period for the first 10 days of every month with a 2-days break in the middle.

Keywords: fur-bearing animals, fox, mink, Arctic fox, skin quality, succinic acid.

Introduction

Biologically active substances are used to increase the productivity of fur-bearing animals. Efficient and environmentally safe substances are organic acids, especially succinic acid which has numerous properties (Kondrashova, 1976). In literature you may find the description of the application of succinic acid in order to increase the quality of skin of minks and foxes (Tutunnik and others, 1999; Blochin and others, 2006; Cherkashina, 2006; Bespyatykh, 2010). However, different authors recommend different optimum doses. There is no data on the influence of large doses of the preparation on fur-bearing animals. Moreover, there is no data on the influence of succinic acid on the quality of Arctic fox skin. The aim of this research was to study the influence of different doses of succinic acid on the quality of fur-bearing animals skin.

Materials and methods

The experiments were conducted on the red fox, sapphire mink and Arctic fox from 'Vyatka' Fur farm' Ltd and 'Nolinskoe Fur Farm' Ltd. (Kirov region) in 2006-2009. The groups of juvenile 2-months old animals were formed in compliance with the group-analogue principle. We added 5, 10, 15, 50 and 100 mg/kg LW of succinic acid to the ration of the animals of 2, 3, 4, 5 and 6 groups (correspondingly). The preparation was applied for the first 10 days of every month with two days break in the middle. The animals of the first (control) group did not recieve the preparation. The animals were kept in common conditions of cage keeping. After the winter skin was formed (in November) the skin quality was estimated in accordance with the Union State Standard (GOST). On the basis of these results we estimated an economic effect of the application of different succinic acid doses. Statistical processing of the results was done with 'Biostat' software.

Results and discussion

Succinic acid affects the forming of the winter skin of fur-bearing animals (Table 1). The largest skin with significant decrease of the rating (total evaluation of fell quallity parameters including the size and the presence of defects) were marked for red fox reciening the high dose of succinic acid, and we stated a negative economic effects in this case.

Table 1. The quality of fur-bearing animals skin under the influence of different doses of succinic acid.

Groups of animals		Skin area dm²	Rating of skin by quality %	Economic effect rub/animal
Red fox				
Males	1 (control, n=20)	21.3	110.12	
	2 (n=20)	21.8	116.70	238.83
	3 (n=20)	21.2	112.56	47.01
	4 (n=20)	21.2	113.29	71.47
	5 (n=20)	21.6	101.50	-203.03
	6 (n=20)	21.9	108.29	-21.73
Females	1 (control, n=10)	19.8	98.76	
	2 (n=10)	19.7	103.25	99.15
	3 (n=10)	19.9	101.65	82.17
	4 (n=10)	20.2	107.18	269.92
	5 (n=10)	19.6	97.50	-107.93
	6 (n=10)	20.5	96.82	-35.07
Mink				
Males	1 (control, n=22)	7.8	106.19	
	2 (n=22)	8.6	125.47	379.65
	3 (n=22)	8.5	121.29	304.66
	4 (n=22)	8.4	118.44	244.22
Females	1(control, n=20)	6.5	74.63	
	2 (n=20)	6.7	79.20	62.09
	3 (n=20)	6.5	72.72	-28.53
	4 (n=20)	6.6	78.10	37.13
Arctic fox				
Males	1 (control, n=15)	22.3	128.27	
	2 (n=15)	22.2	124.73	-91.41
	3 (n=15)	22.1	123.07	-140.57
	4 (n=15)	22.2	127.40	-32.14
Females	1 (control, n=12)	22.1	127.80	
	2 (n=12)	21.7	125.13	-109.06
	3 (n=12)	21.8	123.42	-133.82
	4 (n=12)	21.6	126.75	-86.58

The largest mink skin were marked in the group recieving the minimum dose of succinic acid, the smallest – in the control group. Moreover the largest skin were also characterized by higher quality, and as a result, we found a significant economic effect.

There were no differences in the case of Arctic fox for both skin size and quality.

Thus the introduction of succinic acid to the ration of foxes and minks allows to better skin quality, but the introduction of the preparation to the ration of the Arctic fox makes no difference. High doses of the preparation increase the size of the skin at the expense of its quality which is economically inefficient. Minimum dose of succinic acid (5 mg/kg LW) is optimal for foxes and minks is the. Our data coincide with the studies by Blochin *et al.* (2007), Bespyatykh (2010), but the doses are lower than used by Tyutyunik *et al.* (1999) for minks (30-50 mg/kg LW) and by Cherkashina (2006) for hypotrophic cubs foxes. But it is important to consider that hypotrophic cubs foxes were kept in severe conditions of Yakutia, where animals spend more energy on their vital activity and productivity than in the European part of Russia.

The lack of promotional influence of succinic acid on the quality of Arctic fox skin is probably caused by high homeostasis of the organism, which secure the adaptation of animal to the severe conditions of polar climate. Compared to fox, Arctic fox has higher overspecialized adaptation, i.e. less plasticity to changes of life conditions (Sergina and others, 2009).

Conclusion

Succinic acid promotes the increase of skin quality of foxes and minks and insignificant decline of Arctic fox skin quality. High doses of the preparation allow to get the largest skin at the expense of decreased quality. The optimum dose of succinic acid in the ration of foxes and minks is 5 mg/kg LW. The succinic acid is applied during the first 10 days of every month with a two days break in the middle.

References

Bespyatykh, O.Yu., 2010. The application of succinic acid to increase the economically valuable characteristics of foxes. Fur farming 1: 8-10. (in Russian).

Blochin, G.I., Blochina, T.V. and Selyukova, E.N., 2007. Succinic acid and reproductive quality of mink females. Agrarian science 4: 21-22. (in Russian).

Cherkashina, A.G., 2006. The use of adaptogens in fur farming of Yakutia. Agrarian science 7: 20-21. (in Russian).

Kondrashova, M.N., 1989. Structural-kinetic organization of carboxylic acid cycle in the period of active functioning of mitochondrion. Biophysics 34 (3): 450-457. (in Russian).

Sergina, S.N., Ilina, T.N., Ilyuha, V.A., Fatysheva, M.V. and Podlepina, L.G., 2009. The features of functioning of antioxidant system of predatory mammals under the influence of sodium selenite. Farming biology 6: 66-72. (in Russian).

Tyutyunik, L.A., Kozhevnikova, L.K., Meldo, Kh.I., Kondrashova, M.N., Badovskaya, L.A. and Unzhakov, A.P., 1999. Optimization of physiological conditions and reproductive quality of minks by succinic acid. Farming biology 4: 51-56. (in Russian).

Impact of metal amino acid complexes on arctic fox reproductive performance

D.J. Tomlinson[1], *N. Koskinen*[2] *and T.L. Ward*[1]
[1]*Zinpro Corporation, 10400 Viking Drive, Suite 240, Eden Prairie MN 55344, USA;* [2] *MTT Animal Production Research, Tervamäentie 179, 05840 Hyvinkää, Finland; dtomlinson@zinpro.com*

Abstract

Three hundred thirty one arctic blue, polar, half polar or shadow fox vixen one to six years of age were utilized to investigate effects of zinc, manganese and iron supplementation on whelping rate, morbidity and pup mortality. Treatments consisted of a control diet with no supplemental trace minerals, or or a control diet with additional 100 mg/kg Zn, 25 mg/kg Mn and 40 mg/kg Fe (CTM – complexed trace mineral diet). Treatment diets were individually fed to vixen and offered *ad libitum* one week prior to artificial insemination through weaning. As-fed intake averaged approximately 550 g/hd/d. Diets contained 30 and 105 mg/kg Zn, 9 and 27 mg/kg Mn, 90 and 130 mg/kg Fe for control and CTM treatments, DM basis. Vixen were bred artificially to one of 47 sires with 8 siring more than 10 vixen each. Results were analyzed using Proc Glimmix of SAS for categorical data. Eighty percent of vixen were primiparous with remaining vixen equally distributed 2 to 6 gestations. There was no significant age by treatment interaction effects. Each year of vixen age translated into 0.64 more pups per litter (P=0.025). Overall, 26% of vixen remained empty with a probability of 0.33 for control and 0.20 for CTM (P=0.007). The probability of at least one live pup at 2 d was Control = 0.55 vs. CTM = 0.68 (P=0.02). Analysis of sire and dam variation in litter size at 2 d (dead+alive) showed no significance. There were no significant treatment differences in number of pups alive at 2 d, 7 d or 49 d. The trend was more pups alive at 49 d for CTM (4.3 pups) vs. control (3.6 pups) (P=0.12). Results indicate supplementation with CTM helped reduced number of empty vixen and tended to increase live pups at weaning.

Keywords: fox, trace minerals, reproduction

Introduction

Trace mineral nutrition is an important component in production of fur bearing animals, as maintenance of skin and hair is essential to production of a high quality pelt. Trace minerals are also important components in immune and reproductive functions. Deficiencies may arise when food intake is depressed, diets are poorly formulated or stored, or the animal is unable to digest, absorb or utilize nutrients as a result of disease or antagonists. Many nutrient deficiencies result in skin disorders, most producing a range of clinical signs. Supplementation with highly bioavailable organic forms of iron, zinc, manganese, copper, and selenium have improved skin integrity, rate of healing, immune function, reproductive performance and production of higher quality pelts.

Justification

Interest has arisen in the fur bearing animal industry for use of organic trace minerals as a means of improving health and reproductive performance while reducing morbidity and positively impacting pelt quality traits such as haircoat quality, density and size. Few research studies have been found evaluating organic versus inorganic trace mineral sources in fur bearing animals, specifically fox. Previous work with trace mineral complexes of zinc and iron demonstrated that supplementation of Arctic Blue Fox diets with either (1) 80 mg/kg Zn, (2) 80 mg/kg Zn and 40 mg/kg Fe, or (3) 40 mg/kg $ZnSO_4$ and 40 mg/kg Fe resulted in performance improvements in growth, fur quality and pelt size while also enhancing immune function (Zang *et al.*, 2009). However, furriers are also interested in reproductive improvements resulting in larger litters and healthier pups. In other canine, Kuhlman and Rompala (1998) reported increased reproductive performance (more pups whelped) by adding chelated zinc, manganese, and copper to the diet of bitches during gestation and lactation (P=0.05). Similarly, response to supplementation with CTM in other monogastric species has resulted in more eggs produced, chicks or piglets born and improved health benefits of the offspring.

Therefore, the objective of this study was to investigate the effects of supplementing fox vixen with CTM beginning prior to breeding and continuing through weaning to evaluate potential effects on reproductive performance and production of offspring.

Experimental design

This study was conducted on a commercial furrier operation in Rahja, Finland. Three hundred thirty one healthy arctic blue, polar, half polar or shadow fox vixen between one and six years of age and of mixed parity were utilized in this study. Foxes were managed in individual cages with normal disinfection and epidemic prevention management. Cages were located in a larger housing complex consisting of wooden structures with corrugated steel roofing according to common farming practices meeting standards of MTT Agrifood Research, Finland.

The study was designed as a randomized block design with two treatments. Fox vixen (experimental units) were fed individually in their cages with all vixen within house fed the same treatment. There were two houses per treatment (replicates) with all animals within house receiving the assigned treatment. Vixen (within house) were blocked by potential whelping date to insure similar seasonal effects between treatments.

Treatments

The control (basal) diet was formulated and produced by Kalajoen Jäähdyttämö Oy feed kitchen of Kalajoki, Finland. The diet consisted of water, chicken slaughter by-product (cooked), pre-cooked and dehydrated barley/oats, boiled barley and oats, protein concentrate feedex 65, herring mäskiraes feed and cod, vitamin premix, inorganic/organic iron, vitamin E (liquid), and potassium (specific ingredient amounts are not listed to protect formulation integrity at the feed kitchen). The diet was delivered fresh to the farm daily.

CTM consisting of the control diet supplemented with 100 mg/kg Zn from Availa®Zn; 20 mg/kg Mn from Availa®Mn; and 40 mg/kg Fe from Availa®Fe (Zinpro Corp., MN, USA). The CTM treatment supplement was mixed at 45.31 grams of Availa®Mins (consisting of the above mentioned metal complexes from Zinpro. Corp) per 100 kg of basal diet feed. Mixing ratio was based on an assumed feeding rate of 550 grams per head per day in a batch size of sufficient quantity to feed approximately 170 mature vixen. Minerals were mixed homogenously in a stationary horizontal paddle mixer by the producer then dispatched into the feed cart and distributed to the test animals daily. The control group received the same ration without supplementation.

Diet samples of 100 g were pulled 2×/wk, placed in sealed plastic storage containers, and held in frozen storage until being sent to MTT Agrifoods. Daily samples were combined into weekly composites for analysis.

Prior to initiation of the study, all animals received the control diet that contained vitamin but no trace mineral supplementation, except the iron. Iron was supplemented at 54 mg/ml four d/wk as iron sulfate and three d/wk at 80 mg/ml from Biofer-Kolin (combination of iron sulfate and iron citrate). Diets were administered to four groups of vixen (each treatment went to animals in two houses) from April 27th (start date) until July 20th (end date). Fox vixen were fed 2 times/day, with *ad libitum* access to feed and water.

Performance measures

The following variables were recorded or available for each vixen: age, breed, parity, vaccinations and treatments (given only to sick animals upon recommendation of a veterinarian), insemination dates, sire identification, whelping dates, empty (breeding failure or abortion), litter size, litter weight at weaning (20 pups selected at random from each house and weighed at 7 weeks of age, i.e. 40 per treatment), and any events indicating morbidity or sickness in the pups or vixen. Results were analysed using Proc Glimmix of SAS for categorical data as a randomized complete block design.

Results and discussion

Of the 331 one to six year old arctic blue, polar, half polar or shadow fox vixen, 328 completed the study. Vixen were 80.5% first parity, 4.6% second, 5.5% third, 4.9% fourth and 4.65% fifth. Overall, vixen health was good with no reports of infectious disease or off-feed conditions made by the site management team. As specified, both control and treatment animals were individually fed for *ad libitum* intake on their cage attached feed tray. Based on daily deliveries and free choice distribution among study subjects, as-fed intake averaged approximately 550 g/hd/d. The original plan was to have vixen on study at least four weeks prior to breeding. However, due to unforeseen challenges with product delivery, volcanic ash clouds limiting air travel and movement of goods in the northern hemisphere, treatment diets were offered beginning only one week prior to artificial insemination through pup weaning. Analysed dietary mineral concentrations were 30 and 105 mg/kg Zn, 9 and 27 mg/kg Mn, 90 and 130 mg/kg Fe for control and CTM treatments respectively, DM basis (Table 1). Treatment Zn and Mn levels were slightly lower then predicted but were well within the normal laboratory variance of 20% for trace mineral analysis. Treatment

Table 1. Analyzed mineral composition of Finnish fox diets, DM basis.

Mineral	Control diet	CTM[1]
Calcium, %	9.58	6.13
Phosphorus, %	5.32	3.92
Potassium, %	2.11	2.62
Magnesium, %	0.50	0.42
Zinc, mg/kg	29.5	105.3
Manganese, mg/kg	9.1	26.5
Copper, mg/kg	2.9	3.6
Iron, mg/kg	91.0	129.5

[1] CTM = Control diet supplemented with 100 mg/kg Zn, 20 mg/kg Mn, and 40 mg/kg Fe.

diets should have contained sufficient Zn, Mn and Fe concentrations to meet or exceed the requirements for gestating and lactating vixen.

Vixen were bred artificially to one of 47 sires with 8 siring more than 10 vixen each. There was no significant age by treatment interaction effects. However, each year of vixen age translated into 0.64 more pups per litter (P=0.025). Overall, 26% of vixen remained empty with a probability of 0.33 for control and 0.20 for CTM (P=0.007). The probability of at least one live pup at 2 d was Control = 0.55 vs. CTM = 0.68 (P=0.02). Analysis of sire and dam variation in litter size at 2 d (dead+alive) showed no significance. There were no significant treatment differences in number of pups alive at 2 d, 7 d or 49 d or in pup growth rate. The trend was more pups alive at 49 d for CTM (4.3 pups) vs. control (3.6 pups) (P=0.12). Results indicate supplementation with CTM helped reduced number of empty vixen and tended to increase live pups at weaning.

References

Kuhlman, G. and Rompala, R.E., 1998. The influence of dietary sources of zinc, copper and manganese on canine reproductive performance and hair mineral content. J. Nutr. 128: 2603.

Zhang, Y., Wei, H., Tomlinson, D.J. and Ward, T.L., 2009. Effects of zinc amino acid complex and iron amino acid complex on performance, health and pelt quality of weanling blue fox (*Alopex lagopus*). J. Anim. Sci. 82, E. Suppl. 2: 318.

Yttrium oxide as an inert marker for nutrient digestibility in ferrets (*Mustela putoris furo)*, mink (*Neovison vison*) and cats (*Felis catus*)

L. Sundling[1], Ø. Ahlstrøm[2] and A.-H. Tauson[1,2]
[1]Department of Basic Animal and Veterinary Sciences, Faculty of Health and Medical Sciences, University of Copenhagen, Grønnegårdsvej 3, 1870 Frederiksberg C, Denmark; [2]Department of Animal and Aquacultural Sciences, Norwegian University of Life Sciences, Box 5003, 1432 Ås, Norway; aht@life.ku.dk

Abstract

Apparent total tract digestibility of major nutrients and energy was determined in a comparative study with mink (n=6), ferrets (n=6) and cats (n=2) which were fed three different diets. Apparent digestibility was determined by two methods, namely quantitative faeces collection and use of yttrium oxide as an inert dietary marker. Results achieved with the two methods agreed very well, and there were no significant differences in nutrient and energy digestibility when results generated with the two methods were compared. The recovery of the marker was close to 100%, and the correlation coefficients for AD measured by the two different methods were in most cases higher than 0.9 and highly significant. It was concluded that yttrium oxide is suitable for use as inert marker in digestibility studies with carnivorous mammals.

Keywords: carnivore, quantitative collection, recovery rate

Introduction

Determination of digestibility in monogastric species is often performed by quantitative faeces collection. However, for achieving reliable and reproducible results, very careful measurements of feed intake and faeces output are necessary, and especially when measuring small animal species any loss of material may perturb results significantly. Therefore, the experimental animals usually are kept in metabolic cages, an environment the animals must be habituated to in order not to disturb digestive processes. Therefore, there is a need for alternative methods that allow for reliable measurements on animals which are not confined in metabolic cages or even living in their home environment. Such an alternative is the use of inert markers which have the advantage that quantitative faeces collection is not necessary. However, inert markers must fulfill several criteria among which can be mentioned that they must be harmless for the animals consuming them and for staff handling them, that they are completely indigestible, and that they do not interact or interfere with the digestive processes. Furthermore, the marker must leave the body at a similar rate as the digesta passes the tract, i.e. it must have a standardized recovery rate. In addition, markers must be homogenously incorporated into the feed, and they ought to be simple to analyse (e.g. Austreng *et al.*, 2000; Sales and Janssens, 2003; Vhile *et al.*, 2007). Hitherto, chromic oxide (Cr_2O_3) has been the most commonly used marker in studies with monogastrics, but it does not fulfill all criteria for an inert marker and moreover, it may be toxic and allergenic to humans (Sales and Janssens, 2003), it needs to be incorporated at relative high concentrations

(5-10 g/kg) in diets for achieving reproducible analytical results, and the analytical procedure requires use of strong oxidizing agents (Austreng *et al.*, 2000).

Yttrium oxide (Y_2O_3) has been used as inert marker in studies with salmonid fish with satisfactory results, and it has been demonstrated that it fulfills the criteria for suitable inert markers better than chromic oxide. Another advantage is that it can be included in diets in very small concentrations (100 mg/kg) (Austreng *et al.*, 2000). Yttrium oxide has recently been introduced as an inert marker in digestibility studies with mammals. A single study reports comparative results on quantitative collection, chromic oxide and yttrium oxide for measurement of apparent digestibility in three carnivorous mammal species, i.e. dog, blue fox and mink. Although some differences were found between methods it was concluded that yttrium oxide provided precise measures of apparent total tract digestibility in the investigated species (Vhile *et al.*, 2007).

The objective of the present study was to evaluate the accuracy of Y_2O_3 as inert marker for digestibility in ferrets, mink and cats by comparing results with Y_2O_3 as a marker with those from quantitative faeces collection.

Materials and methods

Animals

The experiment was designed as a Latin square with six 2-year old male mink (2.31±0.67 kg body mass), and six 2-year old male ferrets (1.73±0.25 kg body mass) and six cats. However, only two cats (one 1-year old male and one 2-year old female, both neutered; 4.8±0.4 kg body mass) accepted the diets and completed the experiment. All animals were in good health and not under medical treatment. No clinical disease or mortality occurred during the experiment. The mink and ferrets were kept individually in metabolic cages. The animals adapted well to the cages, they did not show any stress behaviour and they accepted food from the first day. The temperature in the laboratory during the trial was between 6 and 10 °C. The cats were privately owned, both were domestic shorthaired cats, and they remained in their respective homes during the trial. The room temperatures were not measured but estimated to be about 20 °C.

Diets

Three diets were used, one standard mink feed, produced by Sjællands Pelsdyrfoder Amba (Højby, Denmark). A ferret feed composed with the mink feed as a basis (60% mink feed was added 22% whole chicken, 10% liver, 5% fish meal and 3% maize oil), and a commercial wet cat feed was bought from a grocery store (Whiskas® Original Beef, Mars Inc., Denmark). The chemical composition of the three diets appears from Table 1. According to the label the ingredient composition of the cat food was: meat and meat by-products, grain/barley, protein extracts of vegetable origin, minerals. Yttrium oxide (99.99%, Sigma Aldrich, Steinheim, Germany, mol weight 225.81) was added to the compounded, wet diets to give a concentration of 0.05%.

All the feed was prepared on a single occasion prior to the experimental start and then stored deep-frozen in daily potions. The different diets were mixed in a paddle mixer (Bear Varimixer

Table 1. Analysed chemical composition of the diets.

Diet fraction	Diet		
	mink feed	ferret feed	cat feed
Dry matter (DM), %	32.4	36.7	17.8
In % of DM			
Organic matter	86.5	89.9	85.1
Crude protein	44.8	50.8	56.3
Crude carbohydrate	21.5	10.3	8.7
Crude fibre	3.27	2.00	1.57
Crude fat	20.3	28.7	20.1
Ash	13.5	10.1	14.9
Gross energy, MJ/kg DM	21.94	24.73	21.95
Yttrium, g/kg DM	1.06	1.17	2.15

40 L, Wodschow and Co, Denmark) to a homogenous paste. The frozen diets were taken out of the freezer the day before feeding and thawed overnight in the laboratory (6 to 10 °C). The cats did not accept the wet feed, and therefore the feed was freeze-dried (Heto FD 8 freeze dryer) to approximately 94% dry matter before fed to the cats. The daily feed supply was calculated to give about 800 kJ metabolisable energy (ME) to the ferrets and 1,200 kJ ME to the mink. However, the ration was increased by about 30% during the trial to get as close as possible to *ad libitum* feeding. The cats were fed based on the owners' evaluation of the respective cat's appetite.

Experimental design

The study with yttrium oxide as inert marker was combined with a digestibility study using quantitative collection of faeces. Therefore, the ferrets and mink were housed in metabolism cages during the trial period to ensure total collection of faeces. The cat faeces were collected from their litter boxes, and any litter material clinging to the faeces was carefully removed. The mink and ferrets had a three days adaption period and the cats had a two days adaption period to each new diet, and the change between diets was abrupt. After the adaption period a collection period of four days started. The collection period was directly followed by a new adaption period until all three diets had been used. The collection period for the cats was three or four days due to feed shortage. The mink and ferrets had *ad libitum* access to drinking water through individual bottles connected to drinking nipples. They were fed once daily in bowls. The cats had access to water at all times in bowls and were fed multiple times during the day.

Data collection, chemical analyses and calculations

The animals were weighed at the start and end of each collection period. The feed was weighed and faeces and feed refusals were collected and weighed each morning between 09:00 and 12:00

h during the collection periods. The collected faeces were stored deep-frozen. After the collection periods the faeces was thawed and mixed and samples were taken before the faeces were freeze-dried for further analyses (Heto FD 8 freeze dryer). Material for yttrium analyses was taken from freeze dried diets and faeces. Chemical analyses were carried out on samples of the three diets and on faeces from all animals from all three collection periods. The feed refusals were not analyzed because of minor quantities. Wet samples were analyzed for dry matter (DM) and nitrogen (N). Gross energy (GE), crude ash, crude fat, crude fibre (CF) and DM analyses were carried out on freeze-dried samples, according to standard methods. All analyses were performed in duplicate except fat analyses of the diets that were carried out in triplicate. The crude carbohydrate (CHO) content was calculated as the difference between dry matter and sum of CP, crude fat and ash. Organic matter (OM) was calculated as the difference between dry matter and ash. Yttrium in freeze dried feed and faeces was analysed by inductivity coupled plasma mass spectrometry (ICP-AES analysis, Perkin Elmer Optia 3000DV; Perkin Elmer, Wellesley, MA, USA) at 371 nm as described by Vhile *et al.* (2007). The apparent total tract digestibility (AD) of nutrients when quantitative faeces collection was performed was calculated as AD % = ((ingested nutrient – nutrient in faeces)/(ingested nutrient))×100. Digestible energy (DE) was calculated as GE – energy in faeces (FE). When Y_2O_3 was used AD was calculated as $((a_1/a_2-b_1/b_2)/(a_1/a_2))\times100$ where a_1 is the concentration of nutrient in the diet, a_2 is the concentration of marker in the diet, b_1 is the concentration of nutrient in the diet and b_2 is the concentration of marker in faeces.

Statistical analyses

Analyses of variance were carried out according to the GLM procedure of SAS® (Statistical Analysis Systems Institute, 1990). Data from all species and all diets were analysed with method for determination of apparent digestibility (AD) as fixed effect. Results are presented as least squares means (LSmeans) and the root mean square error (RMSE) is used as a measure of variance. Correlation coefficients were calculated by use of the procedure CORR in SAS. Differences were denoted as significant if $P<0.05$ and a tendency if $P<0.10$.

Results and discussion

Dietary yttrium concentrations agreed well between the mink and ferret diets whereas the cat diet had a higher concentration on a DM basis. This can at least be partly explained by the cat diet's lower DM content compared to the mink and ferret diets, a difference that was unknown when the diets were prepared, and therefore the same amount of marker was added to all wet diets.

The agreement between AD values found by use of the two methods was mostly very good, and there were no significant differences between methods in the AD values found. Quantitative collection gave slightly higher values than Y_2O_3, but only for carbohydrates was the difference such that there was a tendency for a difference ($P=0.08$) (Table 2).

The good agreement between measurement methods was also reflected in high and significant correlation coefficients, most of which were higher than 0.90. Only for carbohydrates the correlation coefficient was relatively low (Table 3). Carbohydrates generally made up a small part

Table 2. Apparent total tract digestibility (%) of major nutrients and energy determined either by quantitative faeces collection or with yttrium oxide as an inert marker and using mink, ferrets and cats as experimental animals, and three different diets.

	Method		***P*-value; effect of method**
	Quantitative collection	**Y_2O_3 as inert marker**	
Organic matter	74.0	72.6	0.31
Crude protein	77.5	76.3	0.37
Fat	88.8	88.2	0.84
Carbohydrates	35.6	31.2	0.08
Energy	77.1	75.8	0.47

Table 3. Coefficients of correlation (r), and their level of significance (P), between apparent total tract digestibility of major nutrients and energy determined by quantitative faeces collection or by use of yttrium oxide as an inert marker in mink, ferrets and cats fed three different diets.

	Coefficient of correlation (r)	***P*-value**
Organic matter	0.92	<0.001
Crude protein	0.96	<0.001
Fat	0.99	<0.001
Carbohydrate	0.35	0.02
Energy	0.95	<0.001

of all diets used here, and dietary carbohydrate content was determined by difference. Therefore, it is fair to assume that values determined for this fraction are the least accurate.

The mean recovery of the marker was 96.0%, a value that supported the accuracy of the method. However, there was a tendency for a diet effect (P=0.08) with the lowest recovery (88.6%) being recorded for the cat diet, whereas the recoveries for the mink (96.3%) and ferret (102.8%) diets were similar. Furthermore, there was a significant interaction between diet and species (P=0.01) with cats having the lowest recovery when on the cat diet (70.9%), but the highest when on the ferret diet (123.6%). These findings concur with our previous view that the cat digestibility results were less reliable than those generated with mink and ferrets (Sundling *et al.*, 2012). The present study confirms that this was true regardless of method used for determination of AD. Our results also corroborate those of Vhile *et al.* (2007) who found slightly higher values for AD when using quantitative collection than when using Y_2O_3, but in general a very close agreement between methods, both when evaluating the results on different species (mink, blue fox and dog), or using different main protein sources. Vhile *et al.* (2007) also used chromic oxide, and among the two markers the results found with Y_2O_3 agreed more closely with quantitative collection results than those found with chromic oxide.

Conclusion

In conclusion, in agreement with results in the literature, this study confirms that yttrium oxide can be used as an inert dietary marker in studies with carnivorous mammal species. It has the advantage of being non-hazardous for animals and humans, it can be mixed into the diet in very small amounts and it produces reproducible results. Use of yttrium oxide as a marker will allow for measuring digestibility in an accurate way without quantitative faeces collection, and hence allow for measurements in animals which are not confined in metabolism cages.

References

Austreng, E., Storebakken, T., Thomassen, M.S., Refstie, S. and Thomassen, Y., 2000. Evaluation of selected trivalent metal oxides as inert markers used to estimate apparent digestibility in salmonids. Aquaculture 188: 65-78.

Sales, J. and Janssens, G.P.J., 2003. Acid-insoluble ash as a marker in digestibility studies: a review. Journal of Animal and Feed Sciences 12: 383-401.

SAS Institute Inc., 1990. SAS/STAT® User's Guide, Version 6, 4th edition, Volume 2. SAS Institute Inc., Cary, NC, USA.

Sundling, L., Ahlstrøm, Ø. and Tauson, A-H., 2012. Comparative digestibility of nutrient and energy in ferrets (*Mustela putorius furo*), mink (*Neovison vison*) and cats (*Felis catus*). In: Proceedings of the Xth International Scientific Congress in fur animal production, Larsen, P.F. *et al.* (eds.) Wageningen Academic Publishers, Wageningen, the Netherlands, pp. 112-120.

Vhile, S.G., Skrede, A., Ahlstrøm, Ø. and Hove, K. 2007. Yttrium oxide (Y_2O_3) as an inert marker in digestibility studies with dogs, blue foxes and mink fed diets containing different protein sources. Journal of Animal Physiology and Animal Nutrition 91: 381-389.

High storage stability of amino acids in mink feed

M. Schulin-Zeuthen and P.F. Larsen
Kopenhagen Research, Agro Food Park 15,8200 Aarhus N, Denmark

Abstract

Stability of 18 amino acids in common feed kitchen feed was examined during a period of three days and at three storage temperatures: 5, 20 and 30 °C. An extra 20% (round 1) or 50% (round 2) of each amino acid were added to the feed. None of the amino acid concentrations were affected when feed were stored at 5 °C. At 20 or 30 °C, contents of glycine, ornithine, hydroxyproline, alanine, tyrosine, isoleucine, proline, ammonia and total bacteria count increased in comparison to initial values. Furthermore, contents of arginine decreased when feed was stored at 20 or 30 °C.

Keywords: storage temperature, storage time, synthetic amino acids

Introduction

Application of synthetic amino acids in mink diets can optimise the dietary amino acid composition to meet the requirement of the mink and facilitate the potential use of feed ingredients with suboptimal amino acid contents. In mink feeding, especially methionine is used for optimising the diets (Lassén, 2007). However, investigations of liquid pig diets (Pedersen and Jensen, 2005) have shown that up till 90% of the synthetic amino acids lysine, methionine, threonine and tryptophan disappeared within 8 hours when added to a diet with 1 or 2 g / kg. In the moist mink diets, loss and degradation of added amino acids could potentially be a problem. The aim of this investigation was to examine the stability of amino acid contents in an ordinary feed kitchen diet during a period of three days and stored at three different temperatures.

Materials and methods

A suitable sample was taken from the feed kitchen feed storage on the research farm. A smaller sample was immediately taken and analysed for total bacterial contents at 21 °C, amount of sulphite-reducing bacteria at 21 °C and analysed for amino acid contents.

A mix of 18 amino acids was added to the sample. The amount added was calculated from planned dietary contents of each amino acid. Initially, 20% of the original contents of each amino acid were added. In the second investigation, 50% were added to the diet.

The samples were stored at 5, 20 and 25 °C for three days. In the second investigation, the samples were stored at 5, 20 and 35 °C for three days. A subsample was taken from each sample in 24 hours intervals (Table 1). The subsamples were analysed for amino acids, total bacteria count at 21 °C and amount of sulphite-reducing bacteria at 21 °C.

Table 1. Storage temperatures and sample schedules.

		0 hours	24 hours	48 hours	72 hours
	Control	+			
20/50% AA added	5 °C		+	+	+
20/50% AA added	20 °C	+	+	+	+
20/50% AA added	25/35 °C		+	+	+

Indexed results of dietary amino acids were calculated as:

Index = 100% × (AA/Control)

where AA = concentration of the individual amino acid, and
Control = concentration of the individual amino acid at initiation of trial.

Results and discussion

None of the amino acid concentrations were affected by storage time at 5 °C. Contents of tryptophan, cystine, methionine, histidine, threonine, phenylalanine, serine, lysine, leucine, valine, glutamic acid, glycine and aspartic acid were not affected by temperature or time (see Appendix 1). As an example, results for methionine are shown in Figure 1. The results are in agreement with both Pedersen and Jensen (2005) and Rasmussen (2011), where contents of methionine disappeared minimally in liquid pig feed and in a dry mineral-vitamin supplement.

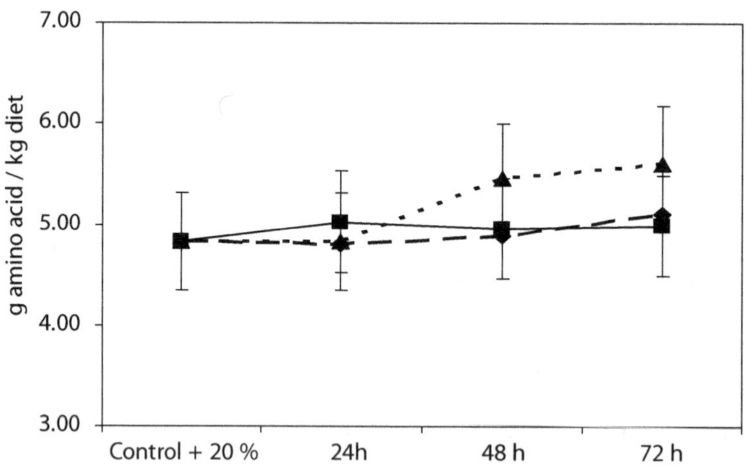

Figure 1. Contents of methionine in feed samples added with 20% synthetic amino acid and stored at 5 °C (–), 20 °C (---) and 30 °C (···). Error lines indicate 10% of individual values.

The results for lysine and threonine were in agreement with Rasmussen (2011), but disagreed with Pedersen and Jensen (2005), where concentrations of lysine, threonine and tryptophan dropped in the liquid pig feed. Contrary to our results, Rasmussen (2011) found that tryptophan and valine concentrations were reduced.

Some discrepancies between development of amino acid concentration in mink feed, liquid pig feed and a mineral supplement are to be expected since chemical conditions and microbial composition are different in the three unlike environments.

There was some effect of temperature and storage time on contents of glycine, arginine, ornithine, hydroxyproline, alanine, tyrosine, isoleucine and proline. Alanine contents (Figure 2) increased when feed were stored at 35 °C for 48 hours or more. This increase was the most notable of all. Arginine contents (Figure 3) was lowered after 24 hours when stored at either 20 or 35 °C, but did not change further during storage. Further, arginine was the only amino acid, where a decrease was observed.

Two products, ornithine and ammonia, were increased when storage temperature was above 5° C, presumable due to bacterial metabolism (Figure 4). This corresponds to an increasing amount of total bacteria count observed at the two highest temperatures (Figure 5). As pointed out by Pedersen and Jensen (2005), one has to distinguish between amino acids directly incorporated into bacteria and amino acids metabolised by bacteria. In the first case, the amino acid potentially still is available for the animal, in the last the amino acids disappears either by transformation or by degradation.

In conclusion, no amino acid concentrations were affected when stored at 5 °C. At 20 °C and 30 °C, dietary contents of glycine, ornithine, hydroxyproline, alanine, tyrosine, isoleucine and proline increased and contents of arginine decreased. Contents of ammonia and total count of bacteria were higher in feed stored at 20 °C and 30 °C.

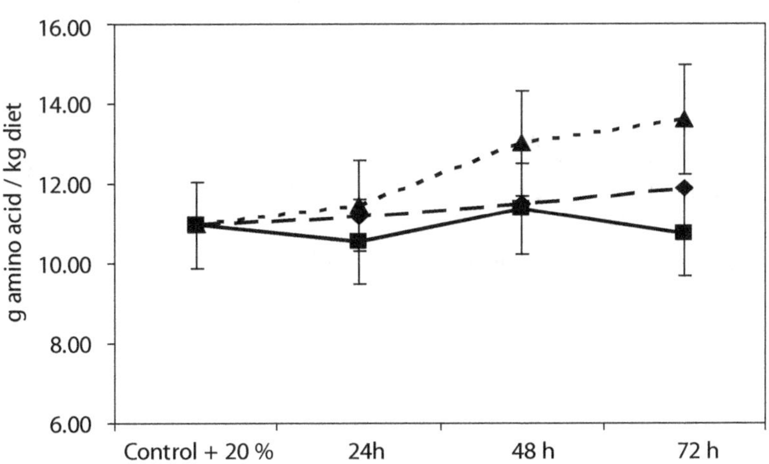

Figure 2. Contents of alanine in feed samples added with 20% synthetic amino acid and stored at 5 °C (–), 20 °C (---) and 30 °C (···). Error lines indicate 10% of individual values.

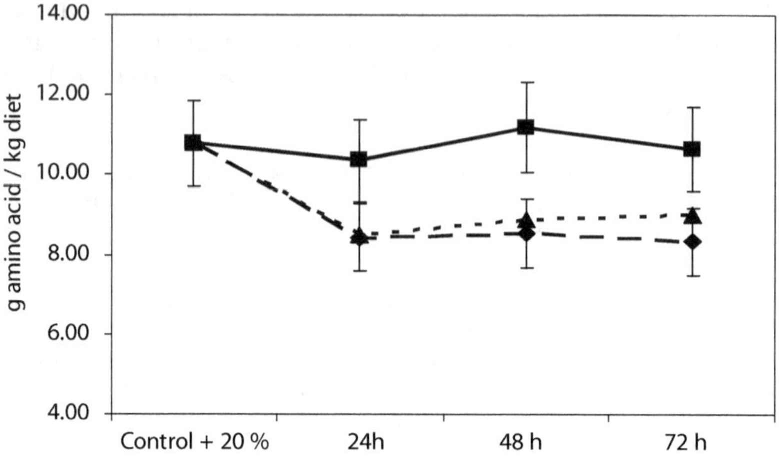

Figure 3. Contents of arginine in feed samples added with 20% synthetic amino acid and stored at 5 °C (–), 20 °C (---) and 30 °C (···). Error lines indicate 10% of individual values.

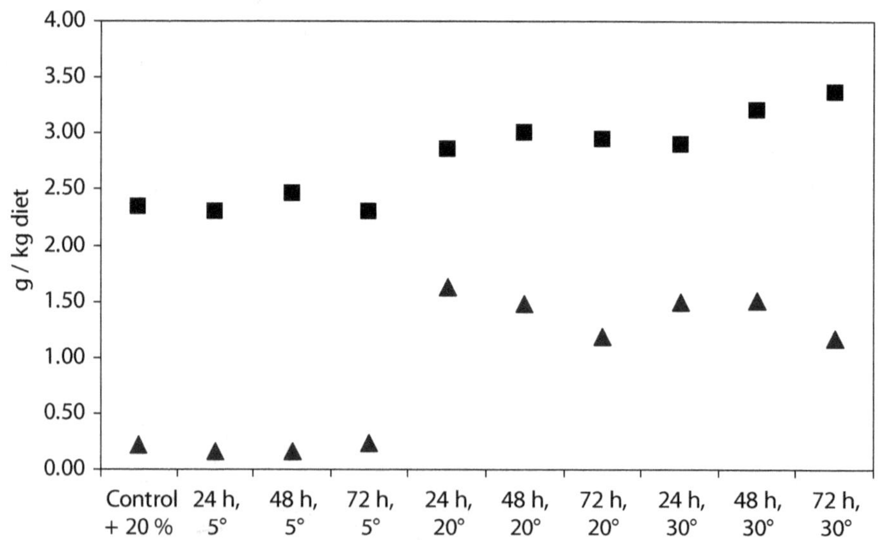

Figure 4. Contents of ornithine (▲) and ammonia (■) in mink feed samples stored at 5 °C, 20 °C or 30 °C.

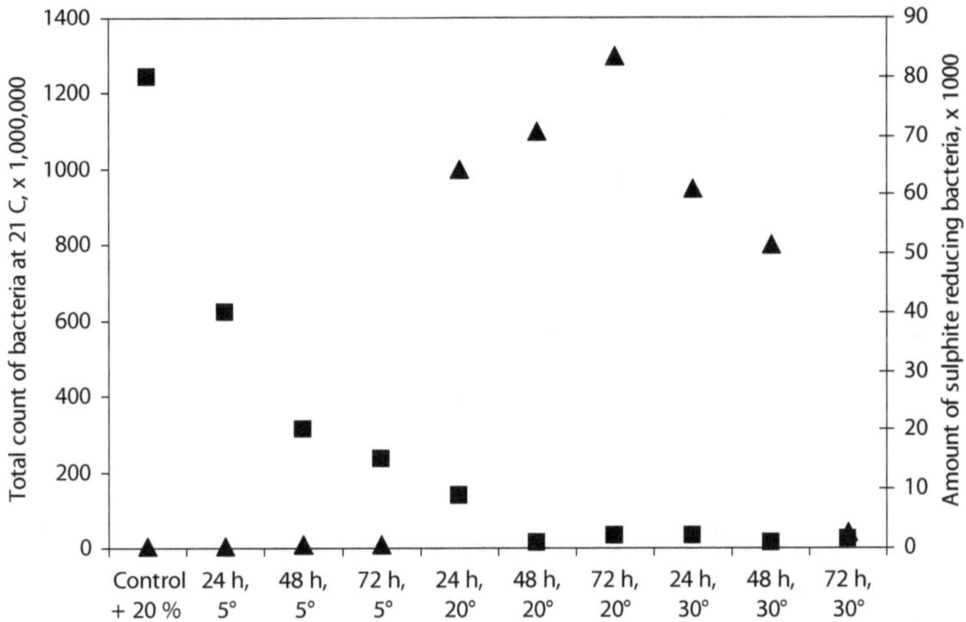

Figure 5. Contents of total bacteria at 21 °C (▲) and contents of sulphite reducing bacteria at 21 °C (■) in feed samples stored at 5, 20 or 30 °C.

References

Lassén, M., 2007. Recommendations for feed composition 2008 and news on feed ingredients. (In Danish) Landscentret Pelsdyr. 23 pp.

Pedersen, A.Ø. and Jensen, B.B., 2005. Degradation of synthetic amino acids during fermentation of liquid feed. (In Danish) Erfaring nr. 0501. Landsudvalget for svin og Videncenter for svineproduktion.

Rasmussen, D.K., 2011. High retrieval of free amino acids in mineral supplements. (In Danish) Meddelelse nr. 905. Videncenter for svineproduktion.

Appendix 1. Contents of amino acids in feed samples

Contents of amino acids in feed samples stored at three temperatures and for three days. Feed are added with 20% of each amino acid. Results are indexed to the control group.

	Control + 20%	24 h, 5 °C	48 h, 5 °C	72 h, 5 °C	24 h, 20 °C	48 h, 20 °C	72 h, 20 °C	24h, 30 °C	48 h, 30 °C	72 h, 30 °C
Cystine	100	102	101	101	101	99	101	104	111	113
Methionine	100	104	103	103	99	101	106	100	113	116
Aspartic acid	100	99	104	103	99	97	96	99	103	108
Threonine	100	100	107	104	100	100	100	102	106	109
Serine	100	98	108	101	94	89	87	91	92	96
Glutamic acid	100	101	107	104	102	100	95	102	103	100
Glycine	100	89	99	88	98	98	99	96	111	107
Alanine	100	96	104	98	102	105	109	105	119	124
Valine	100	100	107	104	102	102	102	103	109	116
Isoleucine	100	101	108	106	101	102	102	103	108	115
Leucine	100	102	107	107	101	102	101	103	106	112
Tyrosine	100	101	103	115	98	89	82	87	83	93
Phenylalanine	100	101	106	108	102	101	101	102	103	114
Histidine	100	102	107	108	102	101	96	101	103	104
Ornithine	100	74	72	104	725	662	526	665	671	526
Lysine	100	101	105	106	102	101	97	103	104	105
Ammonia	100	98	105	98	122	128	126	124	137	143
Arginine	100	96	104	99	78	79	77	79	82	83
Hydroxyproline	100	70	80	67	81	78	83	86	103	81
Proline	100	94	101	95	97	100	100	99	111	112
Tryptophan	100	109	111	101	92	83	82	89	85	85

Mortality in pair- and group-housed mink after weaning

J. De Rond [1] *and F.C. Kleyn Van Willigen*[2]
[1]*Dutch Fur breeders' Federation, Research Farm Edelveen, Postbox 488, 6600 AL Wijchen, the Netherlands,* [2]*Animal Health Service, Small Farm Animals, Postbox 9, 7400 AA Deventer, the Netherlands; j.derond@edelveen.com*

Abstract

From 2009 until 2011 the mortality from weaning until pelting of pair- and group housed mink (4-5 per group) was recorded at the Research Farm Edelveen. In 2009 the mortality in pairs and groups was 1.3% respectively 1.1%, in 2010 these figures were 1.4% and 1.8% and in 2011 1.7% and 1.9%. The three differences were not significant. It is concluded that group housing and pair housing do not differ concerning mortality. These result differ from Danish results which indicated higher mortality among Danish group housed mink than among pair housed mink.

Keywords: mortality, mink, group housing, standard cage, climbing cage

Introduction

On Danish farms the annual mortality rate of growing mink is 0.12% per week, resulting in appr. 2.5% during the growing period after weaning, but in group housing the mortality was considerably higher than in pair housing (Steen Møller, 2011). In the Netherlands figures in KWIN mention 2.6% and the (farmers) estimated mortality is 2-2.5% (±0.1% per week after weaning). Although group housing became common during the past 20 years, no indications were found that pair- and group housed mink differ in this respect were found (De Jonge, unpublished data from the Spelderholt). Since mortality during the growing period was not a problem, and because many possible causes of death complicates the analysis, systematic studies were not carried out. Due to the lack of published Dutch data, we present here the mortality rates at Edelveen among pair- and group housed mink, from 2009 onwards, in a new flock introduced in 2009.

Materials and methods

Animals and health status

The stock was the offspring of 1,500 females (glow/brown) which arrived at the farm in march 2009 as young females. There AD status was Iodine-free. In October 2010 the iodine test was switched to the Elisa test and at the end of 2010 the breeding animals had a low Elisa status.

Feeding

Feed was supplied by the feed kitchen Flevo. The mink were fed with the automatic feeding system Farm Pilot, such that the feeding places were 2 hours per day empty.

Housing

Pairs were housed in standard cages (w×l×h: 30×85×45 cm). Groups were housed in climbing cages, consisting of a standard cage connected with a smaller cage (30×57×40 cm) on the top. Both standard and climbing cages had one nest box (20×20×15 cm). The two cages will be mentioned STC and CLC.

Group composition

In STC these pairs were housed: mother-sun, empty female-young male, brother-sister. Further STC were filled with three mink: mother-2 suns or 3 young females. In CLC these groups were housed: 1 male and 3 females, 2 males and 2 females, 3 males and 1 female, 4 females and 5 females. Over 90% of the groups came from one and the same litter.

Records

The mortality is registered from July 15th until November 7th. Causes of death were not always assessed and not given in the figures.

Results

Table 1 shows the mortality rates from July 15th until November 7th, per year and per group type and in the bottom line the average per year of the whole stock. Statistics (F tests) did not reveal any significant difference between mink in pairs and mink in groups. Further, Table 1 and the statistics do not reveal any preference for group composition as far as mortality is concerned.

Table 1. Mortality in mink during the growing period per cage occupation at Edelveen 2009-2011.

Cage occupation	Between 15th of July and 7th of November					
	Cage	n/year	2009	2010	2011	2009-2011
Pair: 1 ♂ / 1 (old) ♀	STC	2,350	1.3%	1.4%	1.7%	1.5%
2 ♂/ 1 (old/empty) ♀	STC	1,465	1.4%	1.5%	2.4%	1.5%
young 3 ♀	STC	510	1.9%	1.3%		1.5%
Group young 1 ♂ / 3 ♀	CLC	840	0.7%	2.6%	1.5%	1.7%
Group young 2 ♂ / 2 ♀	CLC	3,660	1.0%	1.6%	2.2%	1.7%
Group young 3 ♂ / 1 ♀	CLC	510	1.2%	2.8%	1.1%	1.4%
Group young 4 ♀	CLC	920	1.2%	1.9%	2.0%	1.8%
Group young 5 ♀	CLC	500	1.4%	1.8%		1.7%
Pair	STC	2,350	1.3%	1.4%	1.7%	1.5%
Small group	STC	1,975	1.5%	1.5%	2.4%	1.5%
Group	CLC	6,430	1.1%	1.8%	1.9%	1.7%
Total		10,755	1.3%	1.6%	1.9%	1.6%

Table 2. Mortality of mink during the growing period (from 15 July – 7 Nov 2011).

	July	August	September	October	November	Total
Mortality	0.2%	0.3%	0.5%	0.8%	0.1%	1.9%

Table 2 shows the average mortality per month. STC and CLC are lumped. Data analysis made clear that rates of males, and young and adult females did not differ from each other. Although the causes of death were not recorded systematically, some remarks can be made. In July and August most casualties of dead were due to health deficiencies such as kidney/blather stones. In September and October some casualties were also related with health deficiencies (excessive or declining feed intake, kidney/blather stones, fatty infiltrated liver). Besides some casualties were related with aggressive behaviour. Further, the fur change may be very impactful.

Discussion

During this study, Edelveen did not suffer on severe problems due to weather, food quality or diseases and the mortality rates were normal in the opinion of Dutch mink farmers. An obvious result was that mortality in pairs did not differ from that in groups. So, this study did not reveal an argument in the discussion on group housing.

Still, intuitively one expects that group housing stimulates aggressive behaviour in groups, especially so in autumn when young in the wild leave their families and disperse. If this happened, it did not result in higher mortality rates at Edelveen. On the other hand pelt characteristics of group housed mink at Edelveen do not show a higher number of low graded pelts and even a better quality grading as mink from pair housing (De Rond, 2008). We do not reject the possibility that other stocks differ in this respect. If so, then the problem can be managed by refusing aggressive behaviour for breeding.

References

ASG (Animal Science Group Wageningen University), 2010. KWIN-Animal Husbandry 2010/2011, technical results mink, pp. 369.

Møller, S.H., 2011. Mortality, wounds and injuries in mink kits and juveniles. NJF no 450, Knistva, Sweden, Finland.

De Rond, J. and De Jonge, G., 2008. Group housed mink in flat and climbing cages. IX IFASA congress, Halifax Canada

Life cycle assessment of mink and fox pelts produced in Finland

F. Silvenius, N. Koskinen, S. Kurppa, T. Rekilä, J. Sepponen and H. Hyvärinen
MTT Agrifood Research, Finland; frans.silvenius@mtt.fi

Abstract

The functional unit of the study was a single modified and sewn mink or fox skin. The life cycle standards ISO 14040 and ISO 14044 were followed. The inventory based on questionnaires from the whole fur production chain. The greenhouse gas emissions resulting from fecal matter processing have been calculated following the IPCC and the ammonia model from the Finnish Environment Institute. The fecal matter loads (N_2O and NH_3 and nutrients) from fur farms have been allocated to fur production, and after being spread in fields, loads have been regarded as crop load. The fecal matter nutrients were regarded as substitute for mineral fertilizers. The methane emissions produced by the digestion of animals have been calculated on the basis of an expert knowledge from MTT. Economic allocation was made for calculating environmental impacts of by-products from slaughtering. The omission of nutrients from the Baltic Sea as a result of the fishing of Baltic herring for feeding fur animals has regarded. The carbon footprint of the production chain of a single piece of mink (28 kg CO_2-eqv/pelt) or fox pelt (83 kg CO_2-eqv/pelt) is at the same level as the carbon footprint resulting from one to three days' average consumption of a consumer. The single largest factor that increases the carbon footprint is the nitrous oxide in fecal matter. The fur production chain has a decreasing impact on the amount of eutrophying emissions because it exploits the fish from maintenance fishing. It is an important issue in Finnish circumstances, because the Baltic Sea is very sensitive to eutrophication. The acidifying emissions of the fur production chain result especially from the ammonia evaporating from the fecal matter of the fur animals.

Keywords: ammonia, feaces, environment

Introduction

Life cycle assessment of mink and fox pelts produced in Finland was funded by Finnish Fur Breeders Association and Finnish Fur Sales and was made by MTT Agrifood Research Finland in 2010-2011. The studied impact categories were greenhouse gas emissions, acidification and eutrophication and ecotoxicity was studied qualitatively. The functional unit in our study was a single piece of modified and sewn mink or fox skin. The life cycle standards ISO 14040 and ISO 14044 were followed.

The aim of the study was to find out the main ecological impacts of Finnish fur farming. One key option was to include the consequences which fur farming has in utilising side-products of meat and seafood production. As the Baltic Sea is very sensitive to eutrophication and one of the main components of the fur animal feed is Baltic Herring captured from the Baltic Sea, one special consequence was to point out the significant amounts of captured nutrients in the feed fish fishing. One aim was also to find the most important possible improvements, which reduce the environmental impacts caused by fur farming.

Materials and methods

The system boundaries of the investigation were the whole product chain of mink and fox pelts, and dressing was also included. The parts of the included production stages where agriculture inputs, agriculture functions and emissions of the soil, fishing activities, feed production, farm activities, pelting, dressing, fur auction and sewing. Production and maintaining functions of the infrastructure and machines and use and manufacturing of pesticides was left outside of the study (Figure 1).

The greenhouse gas emissions resulting from fecal matter processing have been calculated following the IPCC and the ammonia model from the Finnish Environment Institute. The fecal matter loads (N_2O and NH_3 and nutrients) from fur farms have been allocated to fur production, and after being spread in fields, loads have been regarded as crop load of the cultivated plant. The fecal matter nutrients were regarded as substitute for mineral fertilizers. The methane emissions produced by the digestion of animals have been calculated on the basis of an expert knowledge (A. Ali, personal communication). Economic allocation was made for dividing environmental impacts between by-products and main products from slaughtering. The omission of nutrients from the Baltic Sea as a result of the fishing of Baltic herring for feeding fur animals has regarded. The characterisation factors were based on Solomon *et al.* (2007) and Seppälä *et al.* (2004, 2006).

The inventory based on questionnaires from fur farms, feed kitchens and skinning, pelt dressing and dressmaker facilities and the answers were caught from 43 fur farms and seven feed kitchens

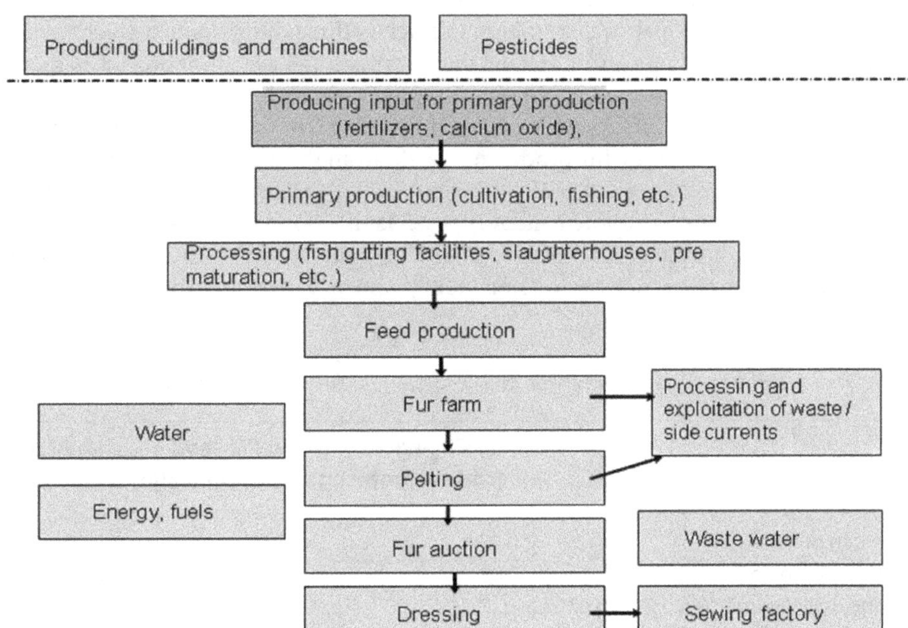

Figure 1. The system boundaries of the study.

of Finland. The investigation was based on production figures in 2009 and average feed recipes of Finland were used. Feed consumptions were modeled by Koskinen *et al.* (2010, 2011a,b,c) and Koskinen and Tauson (2010) and the feed consumption was almost the same as the answers from the questionnaires. The main components of the feed were by-products of slaughtering of pig, cattle and hen (48%), Baltic Herring and side-products of fish slaughtering (20%), barley (14%), protein supplement (6%) and water (10%). The conducting of data collection was completed by visiting feed kitchens, fur animal farms and dressing. The environmental impacts of pelting and dressing were based on primary data and impacts of sewing from a questionnaire to Greek sewing factory. The data of fuel consumption of Baltic Herring fishing was obtained from Finnish Game and Fisheries Research Institute, the data of rainbow trout farming from ongoing LCA-investigation of MTT Agrifood Research Finland, Finnish Game and Fisheries Research Institute and Finnish Environment Institute, agricultural functions on Saarinen *et al.* (2011) and modelling of MTT and data of Norwegian fishing and fish processing on Winther *et al.* (2009). Some comparisons were also made between fur coats and artificial coats, which contained cotton, polyester and acrylic coats and the sources were Ecoinvent database, Kalliala and Nousiainen 1999), Steinberg *et al.* (2009) and van Dijk (2002)..In the comparisons it was assumed that fur products are ten times longer time in use than artificial products. The end of life and washing of the clothes were not taken into account.

Results and discussion

The carbon footprint of the production chain of a single piece of mink was 28 kg CO_2-eq/pelt and of a fox pelt it was 83 kg CO_2-eq/pelt. The carbon footprint of the mink pelt was on the same level as the greenhouse gas emissions of one average consumption of a consumer and the carbon footprint of fox pelt was at the same level as three days' average consumption of a consumer. (Table 1) The single largest factor that increases the carbon footprint is the nitrous oxide in fecal matter (Figures 2 and 3). The fur production chain has a decreasing impact on the amount of eutrophying emissions because it exploits the fish from maintenance fishing. The part of the acidifying emissions of the fur production chain from the ammonia evaporating from the fecal matter of the fur animals was 87% for fox skin and 91% for mink skin. The environmental impacts in local scale were not studied and it is possible that fur farms have local environmental impacts caused by acidification. The local eutrophying impacts to the water systems have been decreasing because of the strict permission policy and it is possible to recover 99% of the

Table 1. The main results in relation to average daily emissions of Finland and driving activities.

	kg CO_2-eq (carbon footprint)	kg PO_4-eq (eutrophication)	kg AE-eq (acidification)
Product chain of mink pelt	28	-0,17	0,23
Product chain of fox pelt	83	-0,22	0,55
The average daily emissions of Finnish citizens	27	0,01	0,05
100 km driving by a medium size private car	18		0,009

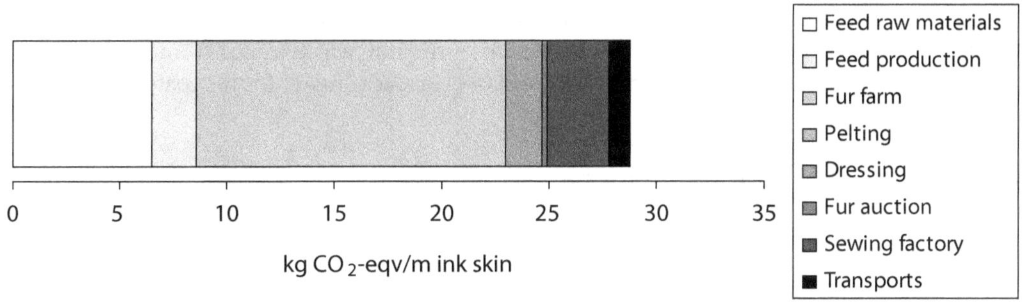

Figure 2. Greenhouse gas emission of product chain of mink pelt.

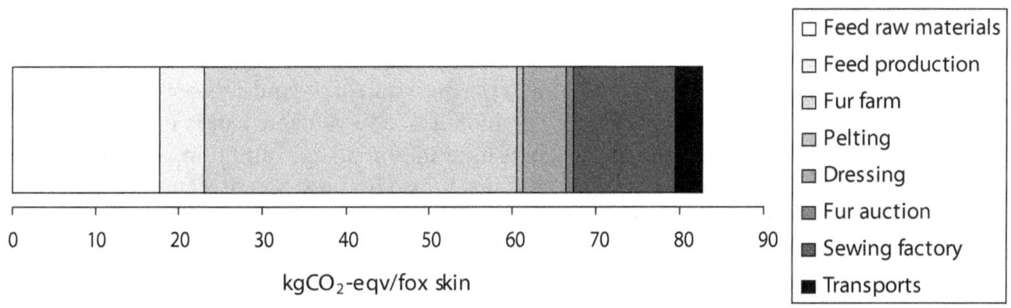

Figure 3. Greenhouse gas emissions of fox pelt product system.

waterborne emissions of nitrogen and almost 100% of the emissions of phosphorus by using different technologies and litters (Rekilä *et al.*, 2010). When one mink pelt is produced, the positive eutrophication impact is five times and for fox pelts four times as much as the impact of the eutrophying emissions of the fur farms

The challenging point of the investigation was to calculate nitrous oxide and ammonia emissions of fecal matter. Figures from greenhouse gas inventory of Finland and the ammonia model of the Finnish Environment Institute (Grönroos *et al.*, 2009) were used, but it has been claimed that ammonia evaporation from the manure can be even higher and new investigations are ongoing about the issue. The omission of the nutrients of the Baltic Herring fishing is 3.6% of the total phosphorus leach and 1.4% of the total nitrogen leach of Finland. The actual amounts are 155 tons phosphorus per year and 1080 tons nitrogen per year and they are much more than for example the nutrient loads of Viikinmäki central Wastewater Treatment plant of Helsinki. Ammonia emissions of Finnish fur farming are 7% of the total ammonia emissions of Finnish agriculture.

The results of the comparisons to artificial clothes were different for different impact classes. The carbon footprint of the studied product were a little bit higher for fur products according to the results, but because of the uncertainty of the significance of different using time of the clothes and because differences in thermal insulation capacity was not taken into account, there was a big uncertainty to define differences the carbon footprints of the studied products. On the other hand, the artificial products don't have the benefit of the nutrient removal from the Baltic Sea, which the fur products have. Acidification impacts were clearly higher for fur coats than artificial coats.

Conclusions and recommendations

The greenhouse emissions of one pelt equals to one to three average daily consumption of Finnish citizens. Main greenhouse gas source is nitrous oxide emissions of fecal matter, and the most important improvement is to develop methods to reduce nitrous oxide emissions of fecal matter. The best improvement method is to reduce acidifying emissions also related to fecal matter.

On the contrary, the significance of acidification is not so remarkable nowadays, but in may cause local environmental impacts.

The main environmental benefit of fur farming is the omission of the nutrients of Baltic Sea from feed fish fishing. Even more benefit will be obtained if not so valuable fish is used as feed raw material, because only a minor part of the fish is used in human consumption. A useful benefit of the fur production sector is also the possibilities to use by-products of slaughtering as feed material of the fur animals. It is also possible to use by-products as raw materials of energy and fertilizers, but because presents prices of fur products are high, it is economically worthwhile to use the by-products as animal feed.

References

Van Dijk, M., 2002 Milieugerichte levenscyclusanalyse van de productie van nertsenbont en imitatiebont. (Environmental life cycle analysis of the production of mink fur and artificial fur) *Rijkuniversiteit Groningen*. ISBN 90-367-1622-5.

Grönroos, J., Mattila, P., Regina, K., Nousiainen, J., Perälä, P., Saarinen, K. and and Mikkola-Pusa, J.., 2009. Development of the ammonia emission inventory in Finland. Revised model for agriculture. Helsinki Finnish Environment Institute.

IPCC, 1997. Houghton, J.T., Meira Filho, L.G., Lim, B., Treanton, K., Mamaty, I., Bonduki, Y., Griggs, D.J. and Callander, B.A. (eds.). Revised 1996 IPCC Guidelines for National Greenhouse Gas Inventories. Vol 1-3. London: IPCC, OECD and IEA.

IPCC, 2000. Penman, J., Kruger, D., Galbally, I., Hiraishi, T., Nyenzi, B., Emmanuel, S., Buendia, Martinsen, T., Meijer, J., Miwa, K. and Tanabe, K. (eds.). Good Practice Guidance Management in National Greenhouse Gas Inventories. Hayama: IPCC, IGES and IEA.

Kalliala, E.M. and Nousiainen, P., 1999 Life cycle assessment. Environmental Profile of cotton and poly-ester cotton fabrics. AUTEX Research J. 1: 8-20.

Koskinen, N., Mikkola, T., Lassén, T.M., Sepponen, J. and Rekilä, T., 2011a. Hapotetun rehun pitkäaikaisvaikutukset minkillä. Rehunvalmistajien luentopäivät, Kalajoki 27.-28.1.2011. STKL ry.

Koskinen, N., Sepponen, J. and Rekilä, T., 2010a. Raw fibre concentrate in the nutrition of mink and blue fox. Scientifur 33, 4: 80.

Koskinen, N. and Tauson, A.-H., 2010b. Energy metabolism of growing blue foxes, Alopex lagopus. In: ISEP 2010 – 3rd EAAP International Symposium on Energy and Protein. EAAP publication 127, Wageningen Academic Publishers, the Netherlands, pp. 515-516.

Koskinen, N., Tupeli, R., Sepponen, J., Rekilä, T. and Tauson, A-H. 2011b. Siniketun energia-aineenvaihdunta, jatkotutkimus. Rehunvalmistajien luentopäivät, Kalajoki 27.-28.1.2011. STKL ry.

Koskinen, N., Tupeli, R., Sepponen, J., Rekilä, T. and Mikkola, T., 2011c. Sinikettujen rehun rasva- ja energiapitoisuuden vaikutus taloudellisiin tuotanto- ominaisuuksiin ja terveyteen. Rehunvalmistajien luentopäivät, Kalajoki 27.-28.1.2011, STKL ry.

Rekilä, R., Vertanen, P. and Rekilä T., 2010. Turkisalan ympäristökäsikirja. https://portal.mtt.fi/portal/page/portal/mtt/mtt/julkaisut/turkistilanymparistokasikirja/715223EF033C99C8E040A8C0033C464E.

Saarinen, M., Kurppa, S., Nissinen, A. and Mäkelä, J., 2011. Aterioiden ja asumisen valinnat kulutuksen ympäristövaikutusten ytimessä. Suomen ympäristö 14 / 2011. Helsinki 2011. Ympäristöministeriö (in Finnish).

Seppälä, J., Posch, M., Johansson, M. and Hettelingh, J-P., 2006. Country-dependent characterisation factors for acidification and terrestrial eutrophication based on accumulated exceedance as an impact category indicator. International Journal of Life Cycle Assessment 11(6): 403-416.

Seppälä, J., Knuuttila, S. and Silvo, K., 2004. Eutrophication of aquatic ecosystems. A new method for calculating the potential contributions of nitrogen and phosphorus. International Journal of Life Cycle Assessment 9(2): 90-100.

Solomon, S., Qin, D., Manning, M., Alley, R.B., Berntsen, T., Bindoff, N.L., Chen, Z., Chidthaisong, A., Gregory, J.M., Hegerl, G.C., Heimann, M., Hewitson, B., Hoskins, B.J., Joos, F., Jouzel, J., Kattsov, V., Lohmann, U., Matsuno, T., Molina, M., Nicholls, N., Overpeck, J., Raga, G., Ramaswamy, V., Ren, J., Rusticucci, M., Somerville, R., Stocker, T.F., Whetton, P., Wood, R.A. and Wratt, D., 2007. Technical Summary. In: Climate Change 2007: The Physical Science Basis. Contribution of Working Group I to the Fourth Assessment Report of the Intergovernmental Panel on Climate Change. Cambridge University Press, Cambridge. pp. 19-91.

Steinberg, J.K., Damien, F., Jolliet, O. and Erkman, S., 2009 A spatially explicit life cycle inventory of global textile chain. International Journal of Life Cycle Assessment 14: 443-455.

Winther, U., Ziegler, F., Skontorp Hognes, S., Emanuelsson, A, Sund, V. and Ellingsen, H., 2009. Carbon footprint and energy of Norwegian seafood products. SINTEF Fisheries and aquaculture. December 2009.

Comparative digestibility of nutrients and energy in ferrets (*Mustela putorius furo*), mink (*Neovison vison*) and cats (*Felis catus*)

L. Sundling[1], *Ø. Ahlstrøm*[2] *and A-H. Tauson*[1,2]
[1]*Department of Basic Animal and Veterinary Sciences, Faculty of Health and Medical Sciences, University of Copenhagen, Grønnegårdsvej 3, 1870 Frederiksberg C, Denmark;* [2]*Department of Animal and Aquacultural Sciences, Norwegian University of Life Sciences, Box 5003, 1432 Ås, Norway; aht@life.ku.dk*

Abstract

Digestibility was evaluated in three strict carnivore species, the ferret (*Mustela putorius furo*), the mink (*Neovison vison*) and the cat (*Felis catus*) in a Latin square design experiment using a homemade ferret diet, a commercial wet mink diet and a grocery store canned cat diet. Six animals per species were used, but only two cats concluded the experiment. Ferrets and mink were housed in metabolism cages whereas cats were kept in their home environment. The experiment was performed with 3-days adaptation periods followed by 4-days periods of recording of feed intake and quantitative collection of faeces. Generally, the cats digested the major nutrients better than did ferrets and mink. Ferrets and mink were mostly close in digestibility, but the mink digested fat and energy in the cat diet and protein and energy in the mink diet more poorly than the ferret. The ferret diet had the highest average digestibility whereas the cat diet had the poorest.

Keywords: strict carnivore, digestibility coefficient, diet

Introduction

Ferrets (*Mustela putorius furo*) are strict carnivores, and during the last 20-30 years they have become increasingly popular as household pets. Domestication of ferrets is believed to have started more than 2,000 years ago and the ferret has a history of being a working pet with many tasks. Even though ferrets have been domesticated for very long time the knowledge of their digestive capacity and nutritional requirements is still inadequate. With increasing numbers of pet ferrets there is an urgent need to provide diets that sustain the animals' nutritional requirements throughout the life cycle. The number of commercial ferret diets available on the market is small, and these diets are often based on nutritional requirements of other strict carnivores such as cats and mink. The mink (*Neovison vison,* previously *Mustela vison)* is another strict carnivore, belonging to the same subfamily as the ferret. The mink is the closest relative to the ferret with which thorough dietary studies have been conducted, and it is therefore tempting to investigate if the mink can serve as a model for the ferret. The mink has a very short digestive tract, about 4 times the body length (Szymeczko and Skrede, 1990) to be compared to about 5 times body length in the ferret (Evans and Quoc An, 1998) and 4-5 times body length in cats (Dyce *et al.*, 2002). The small intestine makes up the major part of the digestive tract and the large intestine, consisting of colon, rectum and anus, is only about 10 cm in the ferret and mink. Furthermore, both species lack ceacum, appendix, taenia coli, haustra coli and no external anatomic division between ileum and colon can be seen (Kainer, 1954; Poddar and Murgatroyd, 1976). The feed passage rate in mink and ferrets is high, and has been measured to an average of about 3 hours in both species (Bleavins and Aulerich, 1981) and to be about 3.5 hours in the mink (Szymeczko and Skrede, 1990). The cat, however, has a small ceacum and feed transit time is considerably longer than in mink and ferret, and amounting to be between 26.5 and 35.7 hours depending on age (Peachey *et al.*, 1999), but transit times of up to 48 hours have also been reported (Fucci *et al.*, 1995).

In practise, diets for cats are often recommended for pet ferrets (Fox and Mc Lain, 1998), but it may be questioned if they are the optimal choice. Because of the profound differences in feed transit time between cats on one hand and ferrets and mink on the other it may be assumed that the cat has somewhat better ability to digest nutrients than ferrets and mink.

Recommendations on energy and nutrient supply based on these species' requirements for different life processes are available for the mink and cat, but for the ferret no such knowledge based recommendations exist. As a first step to evaluate if knowledge on the nutritional requirements of mink and cats can be applied on ferrets, possible differences in basic digestive functions need to be investigated and comparative apparent digestibility data collected for diets composed for the respective species. If such data suggest that the mink and/or cat digest nutrients and energy similar to ferrets, the bulk of data on nutrient digestibility of various feedstuffs available for these species could be used when formulating ferret diets. The main objective of this experiment was to compare apparent nutrient and energy digestibility in the ferret with those of the mink and with indicative values for the cat, testing the hypothesis that the mink would provide a suitable model for nutrient digestibility in the ferret.

Materials and methods

Animals

The experiment was designed as a Latin square with six 2-year old male mink (2.31±0.67 kg body mass), and six 2-year old male ferrets (1.73±0.25 kg body mass) and six cats. However, only two cats (one 1-year old male and one 2-year old female, both neutered; 4.8±0.4 kg body mass) accepted the diets and completed the experiment. All animals were in good health and not under medical treatment. No clinical disease or mortality occurred during the experiment. The mink and ferrets were kept individually in metabolic cages. The animals adapted well to the cages, they did not show any stress behaviour and they accepted food from the first day. The temperature in the laboratory during the trial was between 6 and 10 °C. The cats were privately owned, both were domestic shorthaired cats, and they remained in their respective homes during the trial. The room temperatures were not measured but estimated to be about 20 °C.

Diets

Three diets were used, one standard mink feed, produced by Sjællands Pelsdyrfoder Amba (Højby, Denmark). A ferret feed composed with the mink feed as a basis (60% mink feed was added 22% whole chicken, 10% liver, 5% fish meal and 3% maize oil), and a commercial wet cat feed was bought from a grocery store (Whiskas® Original Beef, Mars Inc., Denmark). The chemical composition of the three diets appears from Table 1. According to the label the

Table 1. Analysed chemical composition of the diets.

Diet fraction	Diet		
	Mink feed	Ferret feed	Cat feed
Dry matter (DM), %	32.4	36.7	17.8
In % of DM			
Organic matter	86.5	89.9	85.1
Crude protein	44.8	50.8	56.3
Crude carbohydrate	21.5	10.3	8.7
Crude fibre	3.27	2.00	1.57
Crude fat	20.3	28.7	20.1
Ash	13.5	10.1	14.9
Gross energy, MJ/kg DM	21.94	24.73	21.95

ingredient composition of the cat food was: meat and meat by-products, grain/barley, protein extracts of vegetable origin, minerals.

All the feed was prepared on a single occasion prior to the experimental start and then stored deep-frozen in daily potions. The different diets were mixed in a paddle mixer (Bear Varimixer 40 L, Wodschow & Co, Denmark) to a homogenous paste. The frozen diets were taken out of the freezer the day before feeding and thawed overnight in the laboratory (6 to 10 °C). The cats did not accept the wet food, and therefore the food was freeze-dried (Heto FD 8 freeze dryer) to approximately 94% dry matter before fed to the cats. The daily feed supply was calculated to give about 800 kJ metabolisable energy (ME) to the ferrets and 1,200 kJ ME to the mink. However, the ration was increased by about 30% during the trial to get as close as possible to *ad libitum* feeding. The cats were fed based on the owners' evaluation of the respective cat's appetite.

Experimental design

During the trial period the ferrets and mink were housed in metabolism cages to ensure total collection of faeces. The cat faeces were collected from their litter boxes, and any litter material clinging to the faeces was carefully removed. The mink and ferrets had a three days adaption period and the cats had a two days adaption period to each new diet, and the change between diets was abrupt. After the adaption period a collection period of four days started. The collection period was directly followed by a new adaption period until all three diets had been used. The collection period for the cats was three or four days due to feed shortage. The mink and ferrets had *ad libitum* access to drinking water through individual bottles connected to drinking nipples. They were fed once daily in bowls. The cats had access to water at all times in bowls and were fed multiple times during the day.

Data collection, chemical analyses and calculations

The animals were weighed at the start and end of each collection period. The feed was weighed and faeces and feed refusals were collected and weighed each morning between 09:00 and 12:00 h during the collection periods. The collected faeces were stored deep-frozen. After the collection periods the faeces was thawed and mixed and samples were taken before the faeces were freeze-dried for further analyses (Heto FD 8 freeze dryer). Chemical analyses were carried out, according to standard methods, on samples of the three diets and on faeces from all animals from all three collection periods. The feed refusals were not analyzed because of minor quantities. Wet samples were analyzed for dry matter (DM) and nitrogen (N). Gross energy (GE), crude ash, crude fat, crude fibre (CF) and DM analyses were carried out on freeze-dried samples. All analyses were performed in duplicate except fat analyses of the diets that were carried out in triplicate. The crude carbohydrate (CHO) content was calculated as the difference between dry matter and sum of CP, crude fat and ash. Organic matter (OM) was calculated as the difference between dry matter and ash. The apparent digestibility (AD) of the nutrients was calculated for as AD % = ((ingested nutrient – nutrient in faeces)/(ingested nutrient)) × 100. Digestible energy (DE) was calculated as GE – energy in faeces (FE).

Statistical analyses

Analyses of variance were carried out according to the GLM procedure of SAS® (Statistical Analysis Systems Institute, 1985). The fixed effect of diet and species and their interaction effect were analyzed according to the following model:

$$Y_{ijkl} = \mu + a_i + b_j + ab_{ij} + c_k(a_i) + \varepsilon_{ijkl}$$

Where Y_{ijkl} is the *ijkl*th observation, μ is the general mean, a_i is the fixed effect of species, b_j is the fixed effect of diet, ab_{ij} is the interaction between species and diet, $c_k(a_i)$ is random animal within species and ε_{ijkl} is random error. Results are presented as least-squares means (LS-means) achieved according to the described model and the root mean square error (RMSE) is given as a measure of variance. Differences between LS-means were tested using the PDIFF option and were considered significant if $P<0.05$ and as a tendency if $0.05<P<0.10$.

Results

The apparent digestibility (AD) of the nutrients for the three animal species is reported in Table 2. One cat on the ferret diet was excluded from the calculations because the values were unacceptable, and consequently the table only includes one cat on the ferret diet. The effects of diet and species on AD were significant for all nutrients. There was also a significant interaction between species and diet, the exception being CHO. The individual animals also had a significant effect on the digestibility of crude ash, OM, and CP.

Effect of species

There were some general differences between species: The cats had the highest average digestibility of all measured nutrients and energy, and the difference in AD between the cats and the other two species was significant, the exception being crude fat. The ferrets and mink were generally closer in their digestibility values, than they were to the cats, but except for CHO, AD in ferrets was significantly higher than in mink. The AD of organic matter (OM) was 83.0, 74.2 and 71.8% in cats, ferret and mink, differences between all species being significant (P<0.01). The AD of CP was very high in cats (87.2%) but significantly lower (P<0.001) in ferrets (77.0%) and mink (75.8%), differences between ferrets and mink also being significant (P<0.05). Crude fat was the most digestible of the major nutrients, AD values being 93.2, 92.3 and 84.5% for cats, ferret and mink, and only the AD of ferrets and mink differed significantly (P<0.01). Crude carbohydrate was generally poorly digested, the values being 49.0, 34.1, and 33.1% in cats, ferret and mink, cats being superior (P<0.001) to the other species. Energy digestibility, finally, was rather high (84.1% in cats, 78.0% in ferrets and 74.7% in mink) and differences between individual species were significant (P<0.01) (Table 2).

Effect of diet

There were also some general differences between diets: The mean AD was highest on the ferret diet for OM (82.2%), CP (85.5%), crude fat (97.3%) and energy (86.5%), differences in digestibility between the ferret diet and the two other diets being highly significant (P<0.001) for OM, CP and energy, whereas the crude fat digestibility was similar (P>0.05) in the ferret and mink diets. The OM (72.4%), CP (74.4%), crude fat (78.7%) and energy (72.9%) of the cat diet were most poorly digested, and the digestibilities of CP, crude fat and energy were significantly lower than those of the ferret and mink (CP: 80.0%, crude fat: 93.9%; energy: 77.4%) diets (P<0.01). The digestibility of CHO was only significantly higher on the cat diet (44.2%) than on the ferret (31.7%) diet (P=0.001) (Table 2).

Interaction between species and diet

Interaction between species and diet occurred when the digestibility did not follow the general pattern. The cat was superior to the other species in digesting OM of all diets (P<0.05). The cat diet was furthermore, more poorly digested by the mink than the ferret (P<0.05) whereas mink and ferrets digested OM in the mink and ferret diets similarly. Interaction between species and diet was further observed for the digestibility of protein. Among species it was the mink that digested the protein of the mink diet most poorly, whereas ferrets and mink digested the CP of the cat and ferret diets similarly. The digestibility of fat was very similar both between the diets and species, but on the cat diet the mink had a significantly lower (P<0.05) fat digestibility than the other two species. Energy in the cat diet, finally, was most efficiently digested by cats which differed significantly to ferrets (P<0.05) which in turn were more efficient than mink (P<0.05). The energy of the ferret and mink diets was equally digestible for all species (Table 2).

Table 2. Apparent digestibility (%) for the three diets and three species (LSmeans). Root mean square error (RMSE) and effect of diet, species, species×diet (S×D) and animal within species (animal (S)).

Species and nutrient	**Diets**			**RMSE**	***P*-value: effect of**			
	cat feed	**ferret feed**	**mink feed**		**diet (D)**	**species (S)**	**S×D**	**animal (S)**
Organic matter				2.20	<0.001	<0.001	<0.01	<0.05
Cat	82.2^{Aa}	88.7^{Bb}	78.2^{Ba}					
Ferret	70.3^{Ba}	80.7^{Ab}	71.6^{Aa}					
Mink	64.7^{Ca}	80.1^{Ab}	70.5^{Ac}					
Crude protein				1.45	<0.001	<0.001	<0.001	<0.01
Cat	84.5^{Ba}	91.6^{Bb}	85.5^{Aa}					
Ferret	69.1^{Aa}	82.8^{Ab}	79.1^{Bc}					
Mink	69.8^{Aa}	82.0^{Ab}	75.5^{Cc}					
Crudefat				8.06	<0.001	0.02	<0.01	0.33
Cat	88.4^{A}	98.0	93.2					
Ferret	85.9^{Aa}	96.8^{b}	94.1^{ab}					
Mink	61.9^{Bb}	96.9^{a}	94.5^{a}					
Crude carbohydrate				6.39	<0.01	<0.001	0.61	0.62
Cat	52.7^{A}	45.7^{B}	48.7^{B}					
Ferret	41.9^{ABa}	25.6^{Ab}	34.9^{Aa}					
Mink	38.0^{Ba}	23.7^{Ab}	37.5^{Aa}					
Energy				3.25	<0.001	<0.001	<0.01	0.09
Cat	81.7^{Aab}	90.6^{a}	80.0^{b}					
Ferret	72.9^{Ba}	84.5^{b}	76.7^{a}					
Mink	63.9^{Ca}	84.6^{b}	75.6^{c}					

^{a,b,c}: Values in a row with no common superscript differ significantly (*P*<0.05) between diets.

^{A,B,C}: Values within a trait and column with no common superscript in capitals differ significantly (*P*<0.05) between species.

Discussion

This experiment was performed so that feeding and housing were similar to the animals' daily life conditions, and handling of the mink and ferrets was restricted to weighing twice a week. It can therefore be anticipated that the experiment meant no or minimum discomfort to the animals. Concerning the ferrets and the mink it was a strictly controlled Latin square design study with homogenous groups of animals which would suggest that these results are highly reliable. The cat data are, however, not of the same quality because only two cats completed the trial, and one cat on one diet was excluded from calculations because of unacceptable results from this period. Furthermore, the cats were not kept in metabolism cages but in their home environment.

Faeces were collected from the litter boxes and therefore some contamination with litter material occurred, but the amounts were so small that they were considered negligible.

The values for protein and fat digestibility were generally in good agreement with results from literature: the mean digestibility for cats fed eight different cat and dog foods were for crude protein 82%, fat 76%, nitrogen-free extract 67% and energy 79% (Kendal *et al.*, 1982a). Similar mean apparent digestibility values were found in a study on commercial cat foods: crude protein 78%, acid ether extract 77% and nitrogen-free extract 69% (Kendal *et al.*, 1982b). In mink, Ahlstrøm and Skrede (1998) found mean apparent digestibility of protein to be 75.3%, fat 93.0% and carbohydrates 75.4%. However, the mink in the present study had a very low digestibility of protein (69.8%) and fat (61.9%) when fed the cat diet, which may be caused by the laxative effect the cat food had on the mink. The carbohydrate digestibility found in our study was, however, lower than literature values. However, the measured CHO digestibility of the mink diet (34.9% to 48.7%) was in acceptable agreement with digestibility values calculated from table values (Ahlstrøm *et al.*, 2004) and ingredient composition (38.4%). The cats' CHO digestibility of the ferret diet (45.7%) was also close to the calculated digestibility (43.0%) whereas the digestibility values for the mink and ferrets were unexpectedly low with 23.7% and 25.6%, respectively. However, when the dietary CHO content is as low as in the present experiment, estimates of CHO digestibility based on difference calculations may be inaccurate, because all analytical errors will add up on the CHO fraction (Neil, 1987).

There were some general differences between *species* in the total tract apparent digestibility, and these indicate that the cats had a greater ability to digest nutrients than the mink and ferrets. This may be explained by a lower feed passage rate in cats (Fucci *et al.*, 1995; Peachey *et al.*, 1999) compared with mink and ferrets (Bleavins and Aulerich, 1981; Szymeczko and Skrede, 1990) thus allowing more time for enzymatic digestion of nutrients. The cats had the highest mean digestibility of OM and major nutrients (CP, CF and CHO) and energy. These findings are not completely consistent with the study by Fekete *et al.* (2005) who showed no significant difference in average apparent digestibility of nitrogen-free extract in ferrets (82.3%) and cats (83.5%) and that ferrets had a significantly higher fat digestibility (95.8%) than the cats (89.4%) whereas, similar to in our study, the cats had a significantly higher protein digestibility (75.9%) than ferrets (69.6%). The discrepancy between the two studies may be explained by differences in ingredient and chemical composition of the diets used and analytical methods. The diets used by Fekete *et al.* (2005) generally were lower in crude protein and considerably higher in carbohydrate than our diets, and two of the diets were low in fat. Fat was furthermore analysed as ether extract which will overestimate fat digestibility because faecal fatty acids saponified with minerals are lost in the analysis. In the present experiment there was, for example, no significant difference in carbohydrate digestibility between the cats and ferrets when on the cat diet, which was very low in carbohydrate. On the two other diets with higher carbohydrate content though, the cats had significantly higher digestibility of carbohydrates than the ferrets. This demonstrates how nutrient composition of diets and choice of analytical methods can affect results, and this must be considered when comparing results from different studies.

Even though the ferrets and mink generally were closer in their digestibility values, than they were to the cats, the ferrets had a significantly higher mean digestibility than the mink, the exception being CHO. When comparing the different diets it was clear that mink digested the cat diet significantly more poorly than ferrets, the exception being protein. The cat diet had the lowest digestibility of most nutrients for both the ferrets and mink and therefore this finding suggests that ferrets are better at digesting nutrients with low digestibility than the mink, while they are equally good at digesting nutrients with a high digestibility. This needs, however, to be verified in further investigations. The reason for the ferrets' slightly higher digestive capacity is difficult to explain from gastrointestinal tract anatomy and physiology which is similar in the two species. Collectively, the species comparisons suggest that the mink can serve as a model for the ferret in digestibility studies, although mink digestibility values may under-estimate ferret digestibility slightly. Despite the weak cat material our results indicate that the cat not is a suitable model for ferret digestibility, a conclusion that concurs with Fekete *et al.* (2005).

Conclusion

The present study clearly demonstrates that even though the cat, ferret and mink all are strict carnivores they have differences in capacity for nutrient digestibility. Although the cat data were less precise than those of the ferret and mink, the results indicate a greater ability to digest nutrients in cats than mink and ferrets. This superior capacity is most likely a result of the longer feed transit time through the digestive tract. The ferret and the mink were mostly close in digestibility values, but when fed the diet with the poorest digestibility, the ferret exhibited higher digestibility values, indicating that the ferret may have better capacity to digest diets of low quality. These results suggest that the mink may be used as a suitable model for digestibility in ferrets, whereas the cat is not a suitable model animal for such purposes. More comparative data are, however needed, ranging from diets and/or feedstuffs with very high to very low digestibility.

References

Ahlstrøm, Ø. and Skrede, A., 1998. Comparative nutrient digestibility in dogs, blue foxes, mink and rats. Journal of Nutrition 128: 2676S-2677S.

Ahlstrøm, Ø., Aldén, E., Børsting, C.F., Dahlman, T., Elnif, J., Hansen, N.E., Mäkelä, J. Pölönen, I. and Skrede, A. 2004. Handbok for fôrmidler til pelsdyr. [Handbook on feedstuffs for fur bearing animals]. NJF-utredning/rapport no. 502, Ås, Norway.

Bleavins, M.R. and Aulerich, R.J., 1981. Feed consumption and food passage time in mink (*Mustela vison*) and European ferrets (*Mustela putorius furo*). Laboratory Animal Science 31: 268-269.

Dyce, K.M. Sack, W.O. and Wensing, C.J.G., 2002. Textbook of Veterinary Anatomy. 3rd edition. Saunders Elsevier, USA.

Evans, E.H. and Quoc An, N., 1998. Anatomy of the Ferret. In: Fox, J.G. (ed.) Biology and diseases of the ferret. 2nd edition. William and Wilkins, USA, pp. 19-70.

Fekete, S.G., Fodor, K., Proháczik, A. and Andrásofszky, E., 2005. Comparison of the feed preference and digestion of three different commercial diets for cats and ferrets. Journal of Animal Physiology and Animal Nutrition 89: 199-202.

Fox, J.G. and Mc Lain, D.E., 1998. Nutrition. In: Fox, J.G. (ed.) Biology and diseases of the ferret. 2nd edition. William and Wilkins, USA, pp. 149-172.

Fucci, V. Pechman, R.D. Hedlund, C.S. and Venugopalan, C.S., 1995. Large bowel transit time using radiopaque markers in normal cats. Journal of the American Animal Hospital Association 31: 473-477.

Kainer, R.A., 1954. The gross anatomy of the digestive system of the mink. American Journal of Veterinary Research 15: 83-97.

Kendall, P.T., Holme, D.W. and Smith, P.M., 1982a. Comparative evaluation of net digestive and absorptive efficiency in dogs and cats fed a variety of contrasting diet types. Journal of Small Animal Practice 23: 577-587.

Kendall, P.T., Holme, D.W. and Smith, P.M., 1982b. Factors affecting digestibility and *in vivo* energy content of cat foods. Journal of Small Animal Practice 23: 538-554.

Neil, M., 1987. Är kolhydraternas smältbarhet beroende av inblandningsnivån? Nordiske Jordbruksforskeres forening Seminar nr. 128. NJF's pelsdyrseminar, pp. 1-9.

Peachey, S.E., Dawson, J.M. and Harper, E.J., 2000. Gastrointestinal transit times in young and old cats. Comparative Biochemistry and Physiology, Part A 126: 85-90.

Poddar, S. and Murgatroyd, L., 1976. Morphological and histological study of the gastro-intestinal tract of the ferret. Acta anatomica 96: 321-334.

Statistical Analysis Systems Institute, 1985. SAS user's guide: statistics, 5th edition. SAS Institute Inc, Cary, NC, USA.

Szymeczko, R. and Skrede, A., 1990. Protein digestion in mink. Acta Agriculturae Scandinavica 40: 189-200.

Evaluation of the choline status in mink fed different levels and sources of choline

M.S. Hedemann[1], *B.M. Damgaard*[1], *T.N. Clausen*[2] *and P.F. Larsen*[2]
[1]*Department of Animal Science, Aarhus University, 8830 Tjele, Denmark;* [2]*Kopenhagen Forskning, Agro Food Park 15, 8200 Aarhus N, Denmark; mette.hedemann@agrsci.dk*

Abstract

Choline is an essential nutrient but the daily need for choline in mink has never been determined. Two experiments were performed to evaluate the choline status in mink kits and full-grown mink fed different levels of choline. In the first experiment mink kits were fed a synthetic diet with choline added either as choline chloride or as soy lecithin to provide four different levels of choline for 18 days. In the second experiment full-grown mink were fed a synthetic diet with varying choline content for three weeks. Blood samples were collected at the end of the experimental periods. All plasma samples were extracted to analyze the content of free choline and betaine by liquid chromatography-mass spectrometry. Plasma choline was only increased in mink kits fed the highest level of choline whereas plasma choline did not change in full-grown mink irrespectively of choline level in the feed. Plasma betaine increased when the level of choline in the diet increased. The changes in plasma betaine were most prominent in mink kits. The present study shows that plasma betaine may be a more reliable marker of choline status than plasma choline, especially in mink kits.

Keywords: betaine, lecithin, phosphatidylcholine, LC-MS

Introduction

Choline is an essential nutrient and it serves as a precursor for several important metabolites such as phosphatidylcholine, the methyl donor betaine, and the neurotransmitter acetylcholine. Consequently, choline deficiency is a serious condition and may lead to fatty liver and steatosis, cellular membrane fragility, DNA damage and apoptosis, altered gene expression, and cognitive impairments (Shin *et al.*, 2010).

The choline balance is determined by choline supply and choline utilization. Choline can be taken up from the diet or synthesized de novo (the phosphatidylethanolamine N-methyltransferase pathway) and the depletion pathways are choline oxidation to betaine and biliary phosphatidylcholine excretion (Li and Vance, 2008). De novo synthesis cannot meet the requirement and therefore uptake of choline from the feed is necessary. The daily need for choline in mink has never been determined. Consequently, choline chloride is added to the feed without knowing the needs for this. Mink feed contains appreciable amounts of choline as phosphatidylcholines (lecithin) but the bioavailability of choline in this form is unknown. The aim of the present study was to evaluate the choline status of mink kits fed different levels of choline either as choline chloride or as soy lecithin as well as the choline status of full grown mink kits fed different levels of choline chloride. The choline status was evaluated as the concentration of choline and betaine in plasma.

Materials and methods

Experiment 1. The effect of choline level and source on choline status of mink kits

A total of 75 mink kits were included in the experiment. The kits were allocated to eight experimental diets (9-10 kits per experimental diet). The diets were synthetic diets with 400, 1000, 2,500 or 4,000 mg choline per kg diet from either choline chloride or soy lecithin (Table 1). The amino acid composition of the diets was adjusted according to the optimal protein requirement for mink established by Sandbøl *et al.* (2004). The kits were included in the experiment when they were seven weeks old and they received the diets for 18 days. The mink kits were weighed at the beginning and the end of the experiment and feed intake was registered daily. On day 18, a blood sample was taken by venipuncture of vena cephalica antibrachii.

Table 1. Composition of the experimental diets for Experiment 1.

	Choline chloride				**Soy lecithin**			
Estimated choline level (mg/kg feed)	**400**	**1000**	**2,500**	**4,000**	**400**	**1000**	**2,500**	**4,000**
Ingredients (% of wet feed)								
Starch	7.67	7.66	7.64	7.61	7.58	7.44	7.12	6.82
Cellulose	6.10	6.09	6.07	6.05	6.03	5.92	5.66	5.42
Soy oil	5.16	5.15	5.13	5.11	4.76	4.18	2.81	1.56
Lard	5.16	5.15	5.13	5.11	4.76	4.18	2.81	1.56
Amino acids	16.87	16.85	16.80	16.74	16.68	16.37	15.65	14.99
Vitamin/mineral mixture[1]	0.38	0.38	0.38	0.38	0.38	0.37	0.35	0.34
Soy sauce	0.80	0.80	0.80	0.79	0.79	0.78	0.74	0.71
$NaHCO_3$	0.80	0.80	0.80	0.79	0.79	0.78	0.74	0.71
Soy lecithin					1.92	4.72	11.27	17.27
Choline chloride	0.09	0.22	0.56	0.89				
Water	56.98	56.90	56.71	56.52	56.31	55.28	52.85	50.62
Energy (kcal/100 g feed)	173.9	183.1	187.6	189.8	190.4	196.7	210.3	230.3
Distribution of energy (protein:fat:carbohydrate)	36:48:16	32:52:16	35:51:14	34:52:14	34:52:15	32:53:14	30:56:14	26:61:13

[1] Supplement 4 from Agro Korn, Videbæk, Denmark without choline.

Experiment 2. The effect of choline level on choline status of full-grown mink kits

Full-grown mink were allocated to four dietary treatment groups (12 mink per treatment). The diets were synthetic mixture with choline added at 0.004%, 0.04%, 0.1% or 0.5% of the feed (Table 2). The amino acid composition of the diets was adjusted according to the optimal protein requirement for mink established by Sandbøl *et al.* (2004). The mink were fed the experimental diets (400 kcal per day) for three weeks (from October 24 to November 12, 2011). The mink were weighed on day 1 and day 21. Feed intake was registered on daily basis. On day 22 the mink were fed 500 kcal feed, residues were removed on day 23 and the mink were fasted until day 24. On day 24 the mink were euthanized and a blood sample was taken by heart puncture.

Sample preparation

The blood samples were collected in Na-heparin tubes, stored on ice and centrifuged within 1.5 h at 3,000 rpm for 10 min at 4 °C. Plasma was separated from the blood cells and stored at -80 °C until analysis.

Choline and betaine in plasma was determined using a modification of the method described by Bruce *et al.* (2010). Plasma was thawed and aliquots (50 μl) were taken and placed into 2 ml centrifuge tubes. To the aliquots was added 150 μl of internal standard solution (10 μM of

Table 2. Composition of the experimental diets for Experiment 2.

	Choline chloride (mg/kg feed)			
	40	**400**	**1000**	**5,000**
Ingredients (% of wet feed)				
Starch	9.91	9.91	9.90	9.85
Cellulose	5.91	5.91	5.90	5.87
Soy oil	6.17	6.17	6.16	6.13
Lard	6.17	6.17	6.16	6.13
Amino acids	8.53	8.53	8.52	8.48
Vitamin/mineral mix[1]	0.37	0.37	0.37	0.37
Choline chloride	0.005	0.05	0.13	0.67
Soy sauce	0.80	0.80	0.80	0.80
$NaHCO_3$	0.80	0.80	0.80	0.80
Water	61.33	61.30	61.25	60.92
Energy (kcal/100 g feed)	179.1	178.4	179.3	181.1
Distribution of energy (protein:fat:carbohydrate)	18:61:21	19:60:21	19:60:21	19:60:21

[1] Supplement 4 from Agro Korn, Videbæk, Denmark without choline.

D_9-betaine and D_9-choline in methanol). After vortexed, the samples were placed at 4 °C for 20 min, centrifuged at 13,200×*g* for 10 min at 4 °C. The supernatant was pipetted into 200 µl vials for analysis.

Calibration curves

Calibration curves were prepared from choline and betaine standards with internal standards (7.5 µM of D_9-betaine and D_9-choline). The concentrations ranged from 0.39-100 µM and 3.9-1000 µM for choline and betaine, respectively. The calibration curves were linear with the mean r^2 value of 0.999 for nine choline and betaine standards.

Mass spectrometry

Mass spectrometry was performed on a MicroTOF-Q quadropole time-of-flight mass spectrometer (Bruker Daltonics, Bremen, Germany) equipped with an electrospray source operating in positive and negative mode. MS data were acquired in positive and negative electrospray (ESI) mode in a scan range from 50 to 1000 m/z at a sampling rate of 2 Hz. ESI parameters were as follows: capillary: 4.5 kV; nebulizer pressure: 1.8 bar; dry gas flow: 8 l/min; dry gas temperature: 200 °C. Lithium formate solution (5 mM) was used for automatic calibration and injected at the beginning of each chromatographic run. Blank runs and additional quality control (QC) samples were interspersed between the samples under investigation.

Results

Experiment 1

During the experimental period, two mink kits died in the group receiving 2,500 mg choline chloride/kg feed, and one mink kit died in each of the groups 400 mg soy lecithin/kg feed, and 400 and 1000 mg choline chloride/kg diet.

The body weight gain of the mink kits during the 18 days of experiment was significantly affected by the choline source ($P<0.0001$) and the choline level ($P<0.007$) (Table 3). The average body weight gain of the mink kits receiving choline chloride was 28.5 g/day and the average body weight of the mink kits fed the diets containing soy lecithin was 41.2 g/day. The mink kits fed 400 mg choline/kg feed gained 32.8 g body weight per day whereas the mink kits fed 1000, 2,500 or 5,000 mg choline/kg feed had a weight gain of 35.0-35.9 g/day. An interaction ($P=0.03$) between choline level and choline source was observed for the daily feed intake. Mink kits fed diets containing choline chloride had lower daily feed intake than the mink kits fed diets containing soy lecithin and within the groups fed diets containing soy lecithin the mink kits fed 400 mg choline/kg feed had a lower feed intake than the mink kits fed 1000 or 2,500 mg choline/kg feed (Table 3).

The plasma choline concentration was affected by the level of choline in the diets (Table 3). The plasma choline concentrations were 16.5, 16.4, 18.1, and 22.4 µM for mink kits fed diets containing 400, 1000, 2,500, and 4,000 mg choline/kg feed, respectively. Only the plasma choline concentration after feeding 4,000 mg choline/kg feed differed from the others. The plasma

Table 3. Start body weight (g), final body weight (g), body weight gain (g/day), feed consumption (g/day) and plasma concentration (μM) of choline and betaine on day 18 of mink kits in experiment 1.

	Choline chloride				Soy lecithin				*P*-value[1]		
Estimated choline level (mg/kg feed)	**400**	**1000**	**2,500**	**4,000**	**400**	**1000**	**2,500**	**4,000**	**Source**	**Level**	**S×L**
Start weight (g)[2]	936	945	924	934	941	863	945	898	0.33	0.70	0.41
	34	34	36	32	34	31	31	31			
Final weight (g)	1,387	1,395	1,402	1,396	1,663	1,739	1,792	1,726	0.0001	0.35	0.56
	40	40	43	37	40	36	36	36			
Weight gain (g/day)	27.5	28.4	29.2	28.9	38.1	41.6	42.7	42.3	0.0001	0.008	0.37
	1.0	1.0	1.0	0.9	1.0	0.9	0.9	0.9			
Feed consumption (g/day)[4]	157^{d}	151^{d}	157^{d}	153^{d}	217^{c}	255^{a}	250^{ab}	236^{bc}	0.0001	0.08	0.03
	7.4	7.4	7.9	6.9	7.4	6.6	6.6	6.6			
Plasma concentration:											
Choline (μM)	16.0	15.4	18.2	24.4	17.2	17.3	18.0	20.5	0.79	0.0001	0.18
	1.7	1.3	1.6	1.4	1.5	1.3	1.3	1.3			
Betaine (μM)[3]	70	99	357	652	84	129	331	628	0.30	0.0001	0.34
	54-92	81-122	278-458	523-812	67-106	105-159	269-407	510-774			

[1] Source = choline chloride or soy lecithin; Level = estimated choline level in the diet; S×L = the interaction between source and level.

[2] Values are least squares means with SEM.

[3] Values are back-transformed least square means (lower and upper limits of the 95% confidence interval).

[4] Groups with different superscripts are significantly different.

betaine concentration was affected by the level of choline in the diets as well. The plasma betaine concentrations were 77, 113, 344, and 640 μM for mink kits fed diets containing 400, 1000, 2,500, and 4,000 mg choline/kg feed, respectively. The concentrations were all significantly different.

Experiment 2

Due to low feed intake, two mink were excluded from the group receiving the diet containing 400 mg choline chloride/kg feed and one mink was excluded from the group receiving the diet containing 1000 mg choline chloride/kg feed.

The body weight gain and the feed consumption during the experimental period did not differ between the groups (Table 4). No difference in the plasma choline concentration was observed between the experimental groups. In contrast, the plasma betaine concentration was significantly higher in the group receiving the diet containing 5,000 mg choline chloride/kg feed compared to the other groups.

Table 4. Start body weight (g), body weight gain (g/day) and feed consumption (g/day) during the experimental period, and plasma concentrations (μM) of choline and betaine at the end of experiment 2.

	Choline chloride (mg/kg feed)				*P*-value
	40	400	1000	5,000	
Start weight (g)[1]	3,082	3,047	3,374	3,235	0.15
	105	115	110	105	
Final weight (g)	2,718	2,716	3,004	2,955	0.22
	119	130	124	119	
Weight gain (g/day)	-17.3	-15.8	-17.6	-18.4	0.86
	2.1	2.3	2.2	2.2	
Feed consumption (g/day)	252	249	261	289	0.39
	18	20	19	18	
Plasma concentration					
Choline (μM)[2]	14.9	15.5	11.4	16.0	0.43
	10.9-20.5	11.0-21.9	8.2-15.8	11.7-21.9	
Betaine (μM)[3]	85[b]	106[b]	102[b]	832[a]	0.0001
	69-106	84-134	82-128	672-1,030	

[1] Values are least squares means with SEM.

[2] Values are back-transformed least square means (lower and upper limits of the 95% confidence interval).

[3] Groups with different superscripts are significantly different.

Discussion

The daily need for choline in mink has never been determined, but based on experience from other animal species, especially dogs, it is recommended to use 20-40 mg choline chloride/kg body weight. Mink feed contains appreciable amounts of natural choline (phosphocholine, phosphatidylcholine, glycerophosphocholine, and sphingomyelin) and rich sources of natural choline are: liver, fish meal, soy, egg, blood meal, fish and meat. However, as the exact content of natural choline in the feed is usually not known choline chloride is added prophylactically to avoid choline deficiency.

In experiment 1, the body weight gain of the mink kits and the feed consumption were highly affected by the choline source. Mink kits fed diets containing soy lecithin had the highest feed intake. Mink fed synthetic diets usually have a low feed intake. It has previously been shown that addition of chicken breast to a synthetic diet increased the feed intake (Schulin-Zeuthen *et al.*, 2009) and the higher feed intake in the groups fed diets containing soy lecithin may be because of a better palatability of these diets. In experiment 2, all mink lost weight during the experimental period as the feed intake was below the requirement for maintenance of the animals. The feed intake was, however, based on experiences with synthetic diets at an acceptable level (Schulin-Zeuthen *et al.*, 2009).

The present study shows that measurement of plasma choline in order to evaluate choline status may be problematic. In young mink kits (experiment 1) plasma choline was low even at a concentration of 2,500 mg choline/kg feed, and in full-grown mink kits plasma choline did not differ between animals fed 40 and 5,000 mg choline chloride/kg feed. Human studies have shown that plasma choline never drops below 50% of normal, probably because of phosphatidylcholine and acetylcholine catabolism which prevents concentrations of choline from falling further (Sha *et al.*, 2010). In the present study even high levels of choline in the diet did not result in increasing plasma concentrations of choline. The major fate of choline after absorption is conversion to phosphatidylcholine (Li and Vance, 2008) which may be considered as the long-term storage form of choline. Furthermore, excretion of biliary phosphatidylcholine is one of the major depletion pathways (Li and Vance, 2008). Another major depletion pathway is the oxidation of choline to betaine which takes place in the liver and kidneys (Zeisel, 1990). In the present study, the plasma concentration of betaine increased when the amount of choline in the diet increased. This suggests that choline may readily be oxidized to betaine in mink. Young mink kits (experiment 1) seem to have a higher capacity for this oxidation as the betaine concentration is increasing with increasing choline content in the feed, whereas in full-grown mink the plasma betaine concentration only increased when the choline level in the feed reached 5,000 mg/kg feed (experiment 2). Alternatively, the young mink kits have a more efficient absorption of choline and hence have more substrate for the oxidation process.

The present study shows that measurement of plasma choline does not provide a reliable indication of the choline status. Plasma betaine may be a better indication for choline status especially in young mink kits.

References

Bruce, S.J., Guy, P.A., Rezzi, S. and Ross, A.B., 2010. Quantitative measurement of betaine and free choline in plasma, cereals and cereal products by isotope dilution LC-MS/MS. Journal of Agricultural and Food Chemistry 58: 2055-2061.

Li, Z. and Vance, D.E., 2008. Thematic review series: glycerolipids. phosphatidylcholine and choline homeostasis. Journal of Lipid Research 49: 1187-1194.

Sandbøl, P., Clausen, T.N. and Hejlesen, C., 2004. Idela protein for mink (*Mustela vison*) in the growing and furring periods. Annual Report 2004, 111-119. Danish Fur Breeders Research Center, Holstebro, Denmark.

Schulin-Zeuthen, M., Hvam K. and Sandbøl, P., 2008. Palatability of a synthetic dit in mink. Annual Report 2008, 109-112. Danish Fur Breeders Research Center, Holstebro, Denmark.

Sha, W., da Costa, K.A., Fischer, L.M., Milburn, M.V., Lawton, K.A., Berger, A., Jia, W. and Zeisel, S.H., 2010. Metabolomic profiling can predict which humans will develop liver dysfunction when deprived of dietary choline. FASEB Journal 24: 2962-2975.

Shin, W., Yan, J., Abratte, C.M., Vermeylen, F. and Caudill, M.A., 2010. Choline intake exceeding current dietary recommendations preserves markers of cellular methylation in a genetic subgroup of folate-compromised men. The Journal of Nutrition 140: 975-980.

Zeisel, S.H., 1990. Choline deficiency. Journal of Nutrional Biochemistry 1: 332-349.

Part 2. Health and disease

Evaluating DNA extraction kits for PCR detection of Aleutian mink disease virus

S.K. Cross, A.H. Farid and I. Hussain
Department of Plant and Animal Sciences, Nova Scotia Agricultural College, Truro, Nova Scotia, B2N 5E3, Canada; crosss@nsac.ca

Abstract

The success of viral DNA amplification by PCR depends on the amount and quality of DNA, and is influenced by the extraction method, particularly when viral copy number is low or samples contain inhibitory substances. We compared four commercial nucleic acid extraction kits, Dynabeads viral extraction kit (DB) from Invitrogen, QIAamp DNA mini kit (QI) from Qiagen, MiniPrep kit (AX) from Axygen and the viral extraction kit (ZR) from Zymo Research for quality and yield of Aleutian mink disease virus (AMDV) DNA. The quality of extracted DNA was assessed by the success of PCR amplification and yield was measured by quantitative real-time PCR (qPCR). DNA from spleen and fecal samples of infected mink was extracted in triplicate and 45 PCR reactions were performed for each kit. Compared with DB, DNA extracted by QI, AX and ZR was amplified by PCR in 95.0%, 53.3% and 62.2% of reactions, respectively, suggesting that DB and QI were comparable and outperformed AX and ZR. DB and QI kits were further compared by extracting DNA from identical samples in three trials. In trial 1, infected spleen, fecal matters, saliva swabs and plasma were used and viral recovery was measured by qPCR (64 reactions). In trial 2, spleen homogenate containing 10.5×10^4 AMDV copies per µL was diluted 14 times in the range of 1×10^0 to 1×10^{-4} and tested by PCR (435 reactions). In trial 3, DNA extracted from nine of the above dilution series was tested by qPCR (146 reactions). The mean viral copy number ranged between 31.2 and 10.5×10^4 per µl of sample. DB and QI kits performed equally for viral recovery in all trials. The findings indicated that the extraction method influenced PCR amplification success, and DB and QI were comparable when input and output volumes were similar.

Keywords: Aleutian mink disease virus, DNA extraction, samples

Introduction

The ability to detect Aleutian mink disease virus (AMDV) DNA in animals or the environment is extremely important for AMDV management. Similar to many viruses, only a few copies of AMDV are needed to cause infection, therefore the sensitivity of detection is extremely important. In recent years, because of its speed and reliability, PCR has become a common method for viral detection (Abbaszadegan *et al.*, 1997; Burgener *et al.*, 2003), including AMDV (Mañas *et al.*, 2001; Farid *et al.*, 2010). Although highly regarded, the sensitivity of PCR can be variable and is dependent on the quantity and quality of sample DNA. Commercially available DNA extraction kits facilitate the procedure, improve reproducibility, and increase extraction speed. Because each kit is unique and its ingredients are proprietary, they differ in nucleic acid recovery, removal of PCR inhibitors, and thus downstream PCR amplification success (Cler *et al.*, 2006; De

Vries *et al.*, 2009). Presently there are no published reports identifying superior DNA extraction kits for AMDV detection. The goal of this study was to evaluate four commercial nucleic acid extraction kits: Dynabeads viral extraction kit (DB) from Invitrogen, QIAamp DNA mini kit (QI) from Qiagen, MiniPrep kit (AX) from Axygen, and the viral extraction kit (ZR) from Zymo Research, for DNA yield and quality in the detection of AMDV by PCR. The yield was measured by quantitative real-time PCR (RT-qPCR) and the quality of extracted DNA was assessed by the success of PCR amplification.

Materials and methods

Samples of spleen homogenate (SH), fecal matter (FM), plasma (P) and saliva (S) were obtained from AMDV-positive mink. DNA extractions were performed according to the manufactures' instructions in triplicate. Standard PCR was performed using three volumes of DNA, and RT-qPCR amplifications were performed in triplicate. In trial one, DNA was extracted from three SH and two FM samples by all four kits (DB, QI, AX, and ZR), and amplified by standard PCR (45 reactions/kit). The DB and QI kits, which outperformed AX and ZR, were compared in the next three trials. In trial two, the DNA from seven samples (three SH, two FM, one P and one S) was extracted by the DB and QI kits, and were quantified by RT-qPCR. In trial three, the detection limit of extracted DNA was tested using 14 dilutions of a spleen homogenate (approximately 3×10^1 to 10.5×10^4 viral copies per/2 µl). All extracted DNA was amplified by standard PCR. In trial four, the quantity of extracted DNA from nine of the dilutions from trial three was determined by RT-qPCR.

Results

The results from trial one samples extracted by the DB, QI, AX and ZR kits were amplified in 100%, 95.6%, 53.3% and 62.2% of PCR reactions, and the Chi-square test showed that differences were significant (P=0.02). The DB and QI kits were comparable (P=0.831) for PCR success, and consequently the AX and ZR kits were removed from further testing. When elution volumes were considered, there was no significant difference between the DB and QI kits for the viral DNA yield as measured by RT-qPCR in trial two (P=0.97). In trial three and four, using diluted spleen homogenate samples, no significant difference was observed between the two extraction kits. Probit analysis of standard PCR showed that dilutions equal to 4,678 and 3,941 viral copies/ 2 µl of extracted DNA would produce positive amplification for DB and QI, respectively, with 95% probability. The DB and QI were comparable for DNA yield from diluted spleen homogenates measured by RT-qPCR in trial four. The combined results of the two kits revealed that as few as 31±19 viral copies in 2µL of extracted DNA can be detected by RT-qPCR (Figure 1).

Conclusion

In conclusion, the results of this experiment indicate that there are differences among commercially available DNA extraction kits. Dynabeads and QIAamp outperformed the MiniPrep and the Zymo research kits in PCR success. The Dynabeads and QIAamp kits were comparable for yield and quality of extracted DNA for viral detection by PCR.

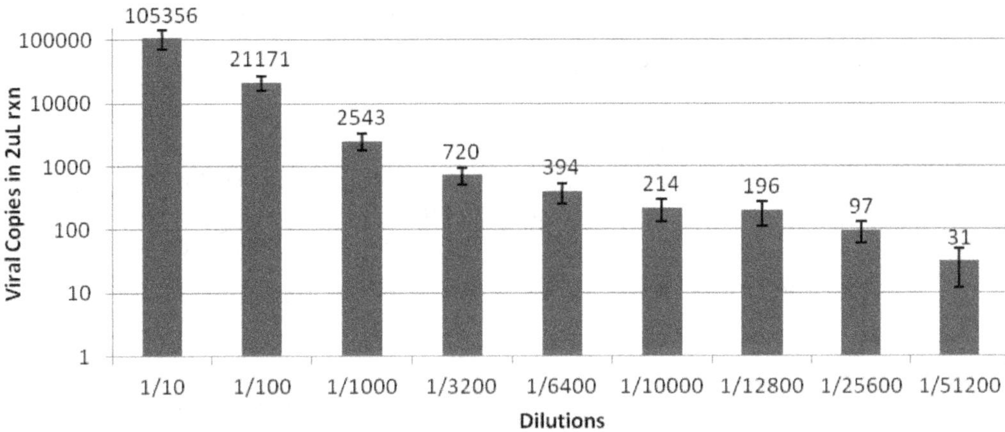

Figure 1. Mean viral copies in 2 μl of extracted DNA by the DB and QI methods in diluted spleen homogenates. The number above bar represents the mean and error bars represent standard deviation.

References

Abbaszadegan, M., Huber, M.S., Gerba, C.P., and Pepper, I.L., 1993. Detection of enteroviruses in groundwater with the polymerase chain reaction. Applied and Environmental Microbiology 59: 1318-1324.

Burgener, M., Candrian, U., and Gilgen, M., 2003. Comparative evaluation of four large-volume RNA extraction kits in the isolation of viral RNA from water samples. Journal of Virological Methods. 108: 165-170.

Cler, L., Bu, D., Lewis, C., and Euhus, D., 2006. A comparison of five methods for extracting DNA from paucicellular clinical samples. Molecular and Cellular Probes 20: 191-196.

deVries, J.J.C., Claas, E.C.J., Kroes, A.C.M., and Vossen, A.C.T.M., 2009. Evaluation of DNA extraction methods for dried blood spots in the diagnosis of congenital cytomegalovirus infection. Journal of Clinical Virology 46: S37-S42.

Farid, A.H., Rupasinghe, P., Mitchell, J.L., and Rouvinen-Watt, K., 2010. A survey of Aleutian mink disease virus infection of feral American mink in Nova Scotia. Canadian Veterinary Journal 51: 75-77.

Mañas, S., Ceña, J.C., Ruiz-Olmo, J., Palazón, S., Domingo, M., Wolfinbarger, J.B., and Bloom, M.E., 2001. Aleutian mink disease parvovirus in wild riparian carnivores in Spain. Journal of Wildlife Diseases 37: 138-144.

In vivo MALDI-TOF markers for early detection of Aleutian disease (AD) among the AD virus infected mink

A. Cepica[1], A. Zargar[1,2], B. Mohamed[1] and I. Iwamoto[1]
[1]Department of Pathology and Microbiology, Atlantic Veterinary College, University of Prince Edward Island, 550 University Ave., Charlottetown, P.E.I., Canada, C1A 4P3, [2]Department of Aquatic Animal Health, Faculty of Veterinary Medicine University of Tehran Qareeb Str., Azadi Ave., Tehran, Iran; acepica@upei.ca

Abstract

A significant proportion of infected mink of non-Aleutian genotype naturally resists progression to immune complex disease. The nature of the AD resistance is presumably genetic. While hyper-gammaglubulinemia can be insensitively detected by an iodine test, more sensitive diagnosis of hyper-γ globulinemia and of other serum protein parameters associated with an immune-complex disease is currently cost prohibitive, time consuming, and laborious. Thus, practical, cost effective, sensitive, and high throughput tests for *in vivo* markers of disease vs. health in AD virus infected animals are currently not available. Such markers could have potential practical use in selection of AD virus infected animals for disease resistance. In this study, initially, parameters for serum protein MALDI-TOF serum analysis were established, and subsequently serum proteins of infected mink (CIEP positive) and mink from ADV uninfected farms were profiled by MALDI-TOF spectrometry. We determined the albumin-γ globulin (A/γ G) ratios and albumin-C reactive protein (A/C) ratios associated with health and disease, and furthermore we identified a 76Kda protein, associated exclusively with disease, i.e. with pathological $A/\ \gamma\ G$ ratio. We concluded that MALDI-TOF profiles of CIPE positive animals provide practical and cost effective tool for selection of breeders for AD resistant phenotype, which could in turn facilitate 'on farm resistance breeding programs'.

Keywords: selection, breeding, resistance, albumin, γ-globulin.

Introduction

Severe economic impact of Aleutian disease virus (ADV), a member of amdovirus genus *parvoviridae* (Cho and Ingram, 1974; ICTV, 2011), on mink farming, stems from infection of all age groups. The of losses can be categorized into three groups. The losses due to reproductive problems of dams infected before pregnancy, resulting in reduced number of born kittens, and infections in mid-pregnancy cause fetal death, resorption, or abortion. Secondly, infections of newborn kittens without maternal antibody are followed by acute infection with high rate of mortality (Alexandersen, 1986). Finally, kittens born with maternal antibody and older mink may develop chronic progressive Aleutian disease (AD) (Bloom *et al.*, 1994). In these animals, the disease is clinically characterized by progressive wasting and death, and pathologically by splenomegaly, lymphadenopathy, polyclonal hyper-γ globulinaemia, (Lodmell *et al.*, 1973; Porter *et al.*, 1980), and acute interstitial nephritis.

The currently practiced control methods, i.e. detection of infection followed by removal or depopulation rarely accomplish meaningful and lasting eradication (T. McLellean, J. Mullen, M. Bosco, personal communication). The main problem with 'detection/removal' approach, is that the premise that must now be reconsidered in spite of 35 years of acceptance, as the virus is *not* primarily harbored by the infected animals, rather it persists in the environment after contamination with highly resistant parvovirus containing feces, saliva, and urine (The Merck Veterinary Manual, 2011; Gorham *et al.*, 1965). Under husbandary conditions where excreta fall on dirt floors, the environment becomes highly contaminated, well before the antibody develops in the initially infected animals. The virus can seep into the deeper layers of ground and into aquifers, and it can then be unearthed at later times. For economic reasons, depopulated and disinfected facilities are not generally allowed to sufficiently rest to ensure virus inactivation, where disinfectants could not reach. Restocking animals may be erroneously identified as virus free, e.g. while being in the incubation period, still without specific antibody, or even worse, if CIEP negatively animals from an infected farm are sold as virus free. Such practice has been admitted by some farmers. Contaminated feed, people, and fomites, as well as wildlife, can all be further sources of virus reintroduction. In conclusion, while raising mink without ADV would be ideal, achievement and maintenance of virus free status has had it serious challenges. In practice, the failures of detection/removal are often being blamed on unspecified and unproven breaches of bio-security. We submit that the rationale for detection/removal that was first formulated when the extreme resistance of the ADV and its shedding patterns were not known. The premise of this control method needs to be redressed now that we know that the virus is not harbored by the infected animals, rather the large amount of the virus are shed by infected animals in feces, saliva, urine, as well as during whelping, into the environment. Even in Denmark, where the control efforts have been most vigorous, facilitated by compulsory reportability since 1999, as well as by strict regulations regarding containment of the outbreaks, persistent clusters of farms infected with AD remain in Vendsyssel peninsula, and the 2002-2003 outbreak occurred within the previously eradicated area (Themudo *et al.*, 2011). Taken together, this illustrates that detectioneradication efforts are extremely difficult, they frequently fail, and they are very costly. Therefore, alternative methods of AD control should be pursued. In the absence of plausibility of vaccine development, raising 'herd resistance' through breeding for disease resistance seems a reasonable proposition, particularly when this approach was successfully applied to development of resistant lines of chickens to Marek's diasease and avian leukosis (Bacon *et al.*, 2000). Any future method of *in vivo* identification of AD resistant phenotype needs to be practical in terms of allowing the high number of samples to be processed in a timely and inexpensive manner.

It is the chronic subclinical form of AD that is a problem with respect to *in vivo* diagnosis, for the purposes of selection of 'resistant breeders', as mere detection of the virus infection or of specific antiviral antibody does not indicate the disease or freedom from it (Hadlow *et al.*, 1983). In every population, some of the infected mink develop chronic non-progressive AD during their lifetime, and thus exhibit disease resistant phenotype. Morbidity and mortality from AD depends generally to a lesser degree on the virus strain, and to the larger extent on the mink strain (Hadlow *et al.*, 1983). While ADV strains of very high pathogenicity, e.g. Utah-1, were reported (Porter *at al.*, 1969), the vast majority of the field ADV strains have been of relatively lower pathogenicity in commonly farmed non-Aleutian genotypes of mink (Henson *et al.*, 1976). A complicating factor in elucidating of the relative roles of a ADV and a host in ADV pathogenicity, may be the reported coexistence of more than one type of ADV DNA in *in vivo*-derived isolates that were not subjected to cell culture propagation (Alexandersen *et al.*, 1994).

In the absence of any satisfactory control measures that achieve lasting freedom from AD associated losses, should mink be tested for disease resistance, it will be initially based on the phenotype of the absence of the disease in the face of the infection. The proportions of animals, in any given population, that develop chronic disease over time, is known to depend to the large extent on the mink genotype. Aleutian genotype mink, e.g. saphire, is the most disease susceptible, in which all ADV strains caused over 90% mortality within 7 months after infection (Hadlow *et al.*, 1983). This is the reason why these mink, affected with Chediak-Highashi syndrome (Kanjanapongkul, 2006), autosomal recessive disorder that affects multiple systems of the body, are rarely farmed. Because of this genetic defect, selection for disease resistance in Aleutian genotype mink can unlikely be done, or if it can be done, it would be inefficient. The proportion of the disease resistant animals varies among non-Aleutian mink for all virus strains, and there is currently no *in vivo* marker to assess resistance vs. susceptibility. Chronically affected mink with AD succumb due to immune complex (IC) disease or due to secondary infections ADV induced immunosuppression (Ryan, 1979; Lodmell *et al.*, 1973), often without any premonitory signs. This makes identification of infected, but disease resistant animals through clinical examination virtually impossible.

As a general rule, pathogenicity of any viral infection is a result of interplay of both virus and host factors. In vast majority of field ADV infections, it is the host factors that seem to play more important role in determining the outcome of most of the ADV infection (Hadlow *et al.*, 1983). Most field AD virus strains are of relatively low pathogenicity, with sufficient number of animals resisting development of the progressive form ADV infection, in order to be exploited for resistance breeding. However, currently there is no practical, sensitive and sufficiently disease specific method of *in vivo* detection of such resistant individuals.

Currently, conventional electrophoresis methods can be used to separate serum proteins by size and charge into five groups; albumin, alpha-1, and 2 globulins, beta globulins, and γ globulins. However, this is laborious, lengthy, expensive, and not suitable for high throughput testing, as required for the AD resistance testing programs. In recent years, Matrix-assisted laser desorption/ionization-time-of-flight (MALDI-TOF) has been implemented in routine laboratories and utilized as a new approach for the identification of bacteria and yeast (Wieser *et al.*, 2012). MALDI is a two step process. First, desorption is triggered by a UV laser beam. Matrix material heavily absorbs UV laser light, leading to the removal of upper layer of the matrix material. A hot plume produced contains neutral and ionized matrix molecules, protonated and deprotonated matrix molecules, matrix clusters and nanodroplets. The second step is ionization protonation (deprotonation) of analyte molecules. In time-of-flight (TOF) mass spectrometry ions are accelerated by an electric field of known strength. The velocity of the ion depends on the mass-to-charge ratio. The time that it takes for the particle to reach a detector is measured. This time will depend on the mass-to-charge ratio of the particle (heavier particles reach lower speeds).

The method has become increasingly subject to intense research interest for cancer diagnosis by analysis of sera for cancer related peptides (Liao *et al.*,2012). These applications focus on small mol. mass proteins, and because of this the samples require laborious preparation to remove predominant protein, e.g. albumin, presence of which could mask the targets. We set to test MALDI-TOF for profiling large molecular weight constituents of entire mink serum, with hope that we would be somehow able to monitor γ globulin concentration. We were not aware of any reports that would apply MALDI-TOF for analysis of high molecular mass proteins of entire sera, as would be required for diagnosis of hyper-γ globulinemia accompanied by hypo-γ globulinemia, which are regular features of progressive ADV infection associated with disease.

Here, we report on establishment of parameters for MALDI-TOF profiling of untreated serum samples for albumin-globulin ratios ($A/\gamma G$), for detection of the main clinical pathology feature of AD, i.e. hypergammaglubulinemia. In the context of this work, behavior of other serum protein in relation to albumin, i.e. was also examined. We established cutoff point for normal and pathological A/D ratio values, as established by MALDI-TOF, by the analysis of sera from a farm with 100% prevalence of the infection as tested by CIEP, and controls from ADV free facility.

However, once the selection in a population is established, correlating genetic markers can be sought. Adoption of the program of breeding for disease resistance will at this stage have to be accompanied by discontinuation of 'detection/removal' control, since the power of selection at any given round will be directly related to the proportion of ADV infected animals. Obviously, this will be in the first production year associated with increased losses. With respect to other selection tools that can accompany the application of the phenotypic marker of resistance, females with reproductive underperformance should be culled, as this represents a reliable clinical marker of disease susceptibility. The remainder of females should be tested for laboratory markers of subclinical progressive chronic Aleutian disease, as should be all the males, because their genetic participation in female performance can't be determined because of the practice of multiple mating (Thom *et al.*, 2004).

Material and methods

MALDI-TOF. Serum aliquots were thawed at room temperature. For each sample, 1 µl serum was transferred to a 200 µl PCR tube. Then, proteomic grade water (Biorad, USA) was added to obtain serum dilutions of 1% (v/v). Each sample dilution was vigorously vortexed. Next, 1 µl from each serum dilution was taken and added to PCR tubes containing 1 µl of saturated sinapinic acid matrix solution (sinapinic acid in HPLC grade water: acetonitril: trifluroacetic acid 500:500:1, all from Sigma Aldrich, Germany). Again each sample was vigorously vortexed. 1 µl of sample matrix mixture was spotted onto MSP 96 target ground steel (Bruker Daltonics, Germany). Matrix assisted laser desorption ionization time of flight manufactured by Bruker Daltonics (MALDI-TOF) spectra were acquired using a Flexcontrol 3.3 software (Bruker Daltonics, Germany). Each spectrum was the sum of 700 laser shots of 100% intensity. High protein mass range was selected. Shots' sites were controlled manually. Instrument calibration was done by mixture of protein standards from the manufacturer (Bruker Daltonics). MALDI-TOF spectra were analysed on Flexanalysis 3.0 software (Bruker Daltonic, Germany). Albumin peaks were identified according to molecular weight. Gamma globulin peak was identified by molecular weight and confirmed by MALDI-TOF testing of Mink IgG isolated using HiTrap Protein G HP Column (GE Healthcare Life Sciences). Albumin: γ globulin peak heights ratios were calculated.

Serum protein electrophoresis

Small number of MALDI-TOF *A/ γ G* profiled sera were compared to the standard electrophoretic analysis. These analyses were done by the laboratory of Diagnostic Services of the Atlantic Veterinary College, Charlottetown, P.E.I., Canada, using Titan Gel Serum Protein Electrophoresis System (Helena Laboratories, Beaumont, Texas, USA) according to the manufacturer's instructions.

Counterimmuneelectrophoresis (CIEP)

CIEP was performed using a commercial antigen (Antigen Laboratory of the Research Foundation of the Danish Fur Breeders' Association, Glostrup, Denmark) by following the manufacturer's instructions, in the Weymouth Aleutian Disease Laboratory, Weymouth, Nova Scotia, Canada.

Serum samples

CIEP positive and negative sera were obtained from one year old Demi Wild mink by cardiac puncture, following placement for five minutes in 100% carbon dioxide. All the CIEP antibody positive animals were randomly selected from a farm with known high prevalence of ADV antibody, and indeed 100% of serum samples tested CIEP positive. The CIEP negative sera were obtained from animals in ADV free facility, and their negative status was confirmed. Only sera without visible hemolysis were subjected to MALDI-TOF analysis.

Statistical analysis

To assess the reproducibility, each MALDI-TOF ratio value represents three completely independent tests, performed at separate times. Means, maximum, and minimum values, together with the average of SD, and a range of SDs were calculated. Degrees of correlation coefficient between *A/ γ G* and other ratios were determined by calculating r values of correlation coefficient.

Results

Under the optimized test conditions, the MALDI-TOF procedure yielded good resolution of the proteins of interest in the area of molecular weight between 40 kDA and 180 kDA, with good visualization of the two important proteins of our interest, namely γG and albumin (Figure 1).

From individual MALDI-TOF profiles, ratios of albumin to four proteins were calculated; a) albumin-γ globulin (*A/ γ G*), b), albumin-putative C reactive protein (A/CRP?), c) albumin-α1 globulin (A/α1), and d) albumin-total globulin (A/tG). Even though *A/ γ G* ratios were in excellent correlation with the other ratios (A/CRP?, A/α1, and A/tG; r=0.914, r=0896, and r=0.874, respectively.), they were considered to be of most interest, primarily because of the biological relevance of γ globulin in the IC disease. In addition *A/ γ G* ratios were selected as the discriminating tool, because they yielded wides range of values thus facilitating more accurate quantification. The profiles of normal mink with normal serum profile could be also differentiated from CIEP positive mink serum with hyper-γ *globulin*emia accompanied by hypoalbuminemia,

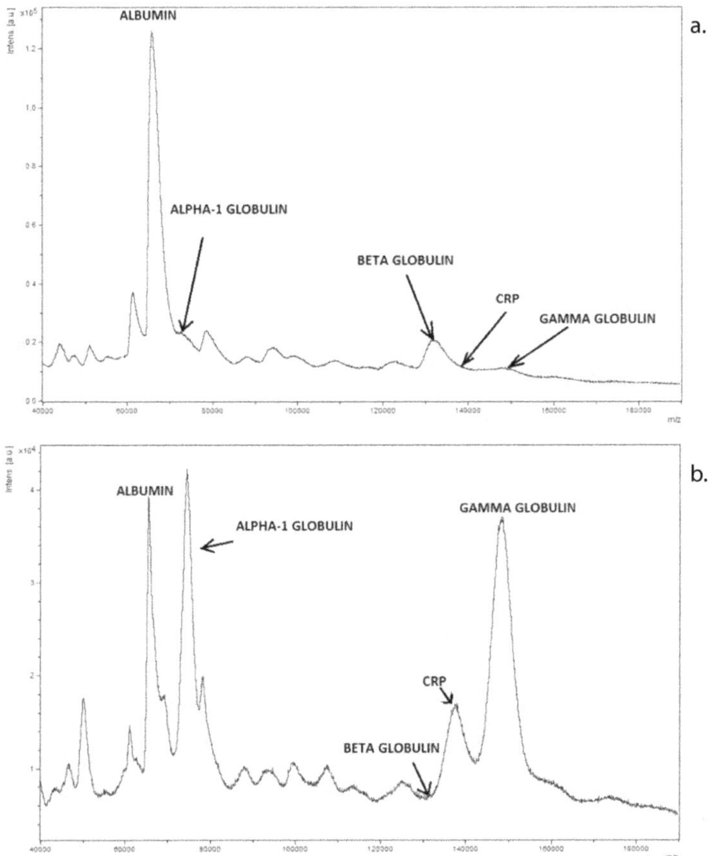

Figure 1. Representative MALDI-TOF profiles in CIEP positive healthy and sick mink. (a) CIEP positive serum (healthy animal) -normal serum protein profile. (b) CIEP positive serum (sick animal)-hypergammaglobulinemia, and extra peaks, of the mol. weight suggestive of C reactive protein (CRP?) and another in alpha1 globulin region.

based on the $A/\gamma G$ ratio (Figures 2 and 3). In addition, the profiles revealed two extra protein peaks in sera of diseased animals, which were not present in 'healthy' animals ($A/\gamma G>8$), (alpha 1 globulin ~75kDa) and (putative CRP~138 kDa) (Figure 1). The appearance of these two peaks, but particularly of the 75 kDa peak, correlated excellently with hyper-γ globulinemia, as they were never present in sera with $A/\gamma G$ higher than eight. The nature of the proteins behind the peeks remains unidentified.

The mean values for $A/\gamma G$ values presented, are based on triplicate independent processing and testing of all CIEP negative and positive samples. Based on the triplicate testing, among the *CIEP negative animals* (n=22), the mean values of ratios of albumin to the other target proteins, were as follows; albumin-γ globulin ($A/\gamma G$) 11.0 SD 0.91 (Figure 2), albumin-putative C reactive protein (A/CRP?) 10.4 SD 0.731), albumin-α1 globulin (A/α1) 5.42 SD 0.17), albumin-total globulin (A/

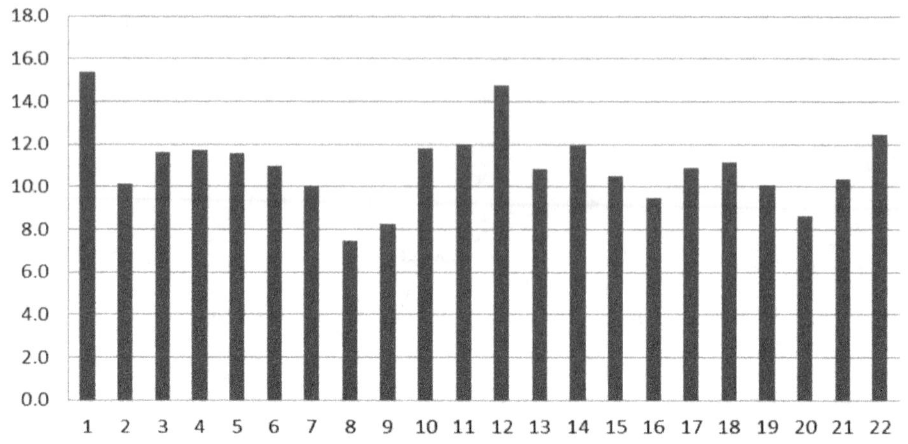

Figure 2. Albumin to gammaglobulin ratios (A/γ G) of CEIP negative mink. x-axis – animal number; y-axis – A/γ G; interrupted line indicates cutoff A/γ G between healthy and diseased serum.

tG) 0.716 SD 0.032). The maximum and minimal values for individual ratios among the CIEP negative animals were as follows; *A/ γ G* 15.4/7.5, A/CRP? 1.9/0.06, A/α1 6.9/4.6, A/tG 0.91/0.60.

Among the *CIEP positive animals* (n=111) (Figure 3), mean *A/ γ G* was 9.71 SD 0.92. The spread of *A/ γ G* values among CIEP positive mink was considerably larger compared to CIEP negative samples, with maximum and minimum values of 21.8 and 1.1, respectively. In addition, 39

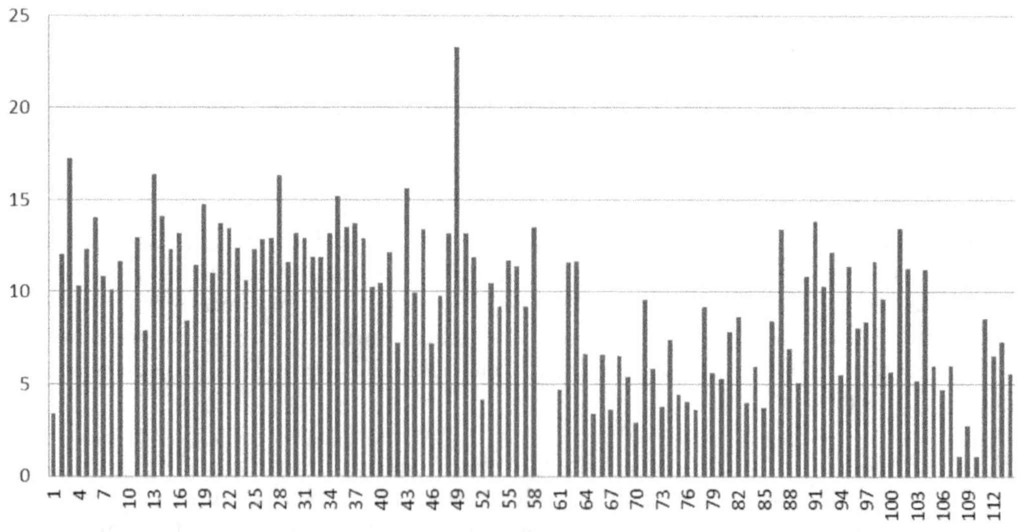

Figure 3. Albumin to gammaglobulin ratios (A/γ G) of CEIP positive mink. x-axis – animal number; y-axis – A/γ G; interrupted line indicates cutoff A/γ G between healthy and diseased serum.

animals out of the 111 had *A/ γ G* lower than the cut of point of eight, indicating serum pathology, while only one of the 22 of CIEP negative animals had a *A/ γ G* value slightly below eight. The mean SD among CIEP positive animals was 0.93, and the spread of SD from the triplicate testing among the CIEP positive animals ranged from 3.5 for *A/ γ G*=21.8 to 0.008 for *A/ γ G*=1.07.

Comprehensive analysis of the MALDI-TOF profiles with respect to the above mentioned ratios, and with respect to appearance of the peaks not seen in the uninfected mink (75kDa) and (138 kDa), as well as comparison of the MALDI-TOF profiles to electrophoresis profiles (Appendix 1), led us to define the cutoff point of five for *A/ γ G* ratio, for the sample to be considered positive for serum protein pathology associated with 'progressive AD disease. Animals with the ratios between five and eight were designated as ambiguous, and ratios above eight was designated as 'healthy'. Distribution of animals into the three categories according to *A/ γ G* ratios among CIEP positive and negative animals is presented in Table 1. There was strong positive correlation between *A/ γ G* ratio and all the other ratios, A/CRP?, A/α1, and A/tG, with r=0.914, r=0896, and r=0.874, respectively. Because of the known clinical significance of hypergammaglobulinemia in the pathogenesis of progressive AD, we chose to focus on *A/ γ G* ratios as an indicator of serum protein pathology. The lowest recorded among the CIEP negative mink was 7.5. Because of this, and also, because of the association of the *A/ γ G* ratios lower than eight with appearance of abnormal MALDI-TOF serum peaks, the cut off for potentially diseased MALDI-TOF profiles was set at *A/ γ G* ratio of lower than eight. Table 1 shows distribution of animals with respect to these categories among CIEP negative and positive groups. Among the animals with *A/ γ G* ratios lower than eight, there was significantly more males than females (*P*=0.01). Figures 2 and 3 display the *A/ γ G* ratios for ADV uninfected and ADV infected (CEIP positive mink).

Table 1. Distribution of A/ γ G *ratios among CIEP positive and negative animals.*

CIEP positive	**Number of animals**	**%**
<5 (diseased)	16	14.41
5-8 (ambiguous)	23	20.72
>8 (healthy)	72	64.86
Total	111	100
CIEP negative		
<5 (diseased)	0	0
5-8 (ambiguous)	1	4.5
>8 (healthy)	19	95.5
Total	22	100

Discussion

Originally, quantitative protein analysis by MALDI-TOF was viewed as implausible because of the variability of incorporation of the analyte into the co-crystallized matrix-analyte complex.

This variability can be at least partly circumvented, however, by pre-mixing the sample and matrix solutions and by facilitating faster crystallization times, thus generating smaller crystals and a more homogeneous incorporation of the analytes into the crystalline lattice. Fortunately, the relative signal intensities within the mass spectrum of a complex sample are constant. Furthermore, if the interest is the ratio of two proteins within complex sample like serum, the changes in relative signal intensities, in our case albumin vs. γ globulin, presumed CRP, alpha-1 globulin, and total globulin, both proteins in the ratios are influenced by the constant background, and thus the ratio is not affected. Such ratios can be taken as quantitatively reproducible (Duncan *et al.*, 2008). This feature makes MALDI-TOF as an excellent tool for diagnosis of hyper γ-globulinemia, by assessing it in a ratio with most abundant serum protein, albumin. In addition, hypergammaglobulinemia induces hypo-albuminemia through homeostatic regulatory mechanisms of albumin synthesis (Noda *et al.*, 2011; Keshgegian, 1984; Kyrtsonis *et al.*, 2001). Thus, assessing hyper-γ globulinemia in a ratio with albumin provides additional advantage in sensitivity. It needs to be stated that shifted *A/ γ G* ratio is a powerful diagnostic tool that goes well beyond our objective of detecting subclinical chronic progressive AD. *A/ γ G* an not only for progressive ADV infections, but several other chronic viral (Hunzinger *et al.*, 2003; De Milito *et al.*, 2003; Jones *et al.*, 1988), bacterial (Ko *et al.*, 2011), and protozoal infections (Buhles, 1974), as well as autoimmune disorders (Hunziker *et al.*, 2003). Incidentally, several chronic viral infections, including ADV, do induce autoimmune disorder in addition to the specific antiviral response (Hahn and Hahn, 1983).

A question could be asked with respect to how much advantage in detection of hypergammaglobulinemia MALDI TOF offers compare to iodine agglutination test (IAT). IAT was reported to become positive when gammaglobulin is at 2 grams / 10ml of serum and A/G is 1 (Gorham *et al.*, 1965). MALDI TOF on the other hand is capable of quantitative detection of A/G throughout its range, and we report here that the pathological A/G begins already at 8.

Not only assessment of ratio of two serum proteins like γ globulin and albumin by MALDI-TOF, in one serum sample, can be made reproducibly, but the the sensitivity of detection is increased by the use of the ratio of two proteins that move in opposite direction (hyper-γ globulinemia is automatically accompanied by hypo-albuminemia). When the concentration of γ globulins exceeds normal level, whether of monoclonal (myeloma), oligoclonal (virus), or polyclonal nature (autoimmune disorders, or virus combined with autoimmune disorders), then albumin concentration is decreased by inhibition of synthesis. These homeostatic mechanisms developed to control hyperviscosity of blood due to hyper-γ globulinemia or generally due to hyperproteinemia (Noda *et al.*, 2011) by reducing albumin synthesis by the liver (Keshgegian, 1984; Kyrtsonis *et al.*, 2001). Because of this linkage, sensitivity of determination of the *A/ γ G* ratio in immune-complex has the promise of being higher than determination of hyper-γ globulinemia alone by quantitative methods. It is true that in our application, theoretically, since hyper-γ globulinemia can arise due to other causes, e.g. chronic infections and autoimmune disease, other conditions could interfere with the determination of AD resistance of infected animals. In reality, the probability of 'ADV infection nonspecific decreased *A/ γ G*, compared to the frequency of this condition induced by ADV, is extremely low (unpublished data).

In this paper, we investigate *in vivo* method of diagnosis of progressive form of AD, as opposed to merely diagnosis of ADV infection. Recently, scientific interest was turned to selection of progressive AD resistant phenotype of mink, in order to increase the 'herd resistance' to AD, with the prospect of maintaining productivity without ADV elusive eradication. No currently available laboratory method, has the potential to facilitate such selection. Neither CIEP, a method of detection of specific antibody, nor PCR or real time PCR, methods of detection of the ADV can be used, as they do not differentiate animals whose innate and adaptive immune systems prevent the virus infection from inducing disease, from animals whose immune system allows them to stay healthy. Certainly, whether an animal became progressively infected, or not, they produce antibody. With respect to the virus (PCR, or real time PCR), we tested thousands of animals on several infected farms, and had to abandon the use of PCR, because the intermittent presence of the virus in blood and fecal swabs (unpublished data). Attention of some in the field of AD control, has also turned to quantitative assessment of specific anti-ADV antibody, often by quantitative ELISA. While this seems on the surface reasonable, it is based on the incorrect assumption that the levels of specific ADV antibody and the levels of total hyper-γ globulinemia are in correlation, as it could be expected only if the specific anti-ADV antibody were participating in the disease. In reality, the polyclonal ADV induced hyper-γ globulinemia of autoimune specificities is significant (Hahn and Hahn, 1983). To our knowledge, no evidence has been reported that the levels of specific ADV antibody during AD indeed correlate with the levels of total serum γ globulin, which would justify the premise of quantitative ELISA use for identification of animals with progressive AD. On the other hand, MALDI-TOF determined *A/ γ G* ratios, quantitatively establishes serum protein pathology due to progressive ADV infection, regardless whether the participating antibodies are of viral, or polyclonal autoimmune nature.

Because one of the features of the adult chronic form of AD is decreased reproduction, there is no doubt that this can be one of the selection criteria for disease resistance in females. Only progeny of females with acceptable litter size should be kept as breeders. All other females, and males intended for breeding, raised in the environment in the presence of ADV without 'detection/removal', could be sorted according to *A/ γ G* ratios into two groups; animals with significant serum protein pathology, presumably progressing to disease, and animals with normal *A/ γ G* ratios, presumably resistant to the disease. The power of such selection would be proportionate to the prevalence of ADV infected animals. The prevalence of ADV infections on farms rarely reached those levels while 'detection/removal' was practiced. Usually, such farms were depopulated. Should the few generations needed for the increase of the resistance be not economically possible, the selection might be feasible in quarantined facilities on a small scale, and once smaller numbers of the resistant animals would be produced, they could be expanded.

Genetic resistance to virally induced disease is very difficult to study, as it is multigenic trait governed most profoundly by adaptive immune functions, i.e. by humoral and cellular immune responses (Zekarias *et al.*, 2002). The immune functions are further affected by the host hormonal influences (Wu *et al.*, 1994). Further resistance mechanisms to viral infections also involve a variety of non-immune factors, e.g, presence and density of virus receptors. and vigor of interferon and other cytokine responses, some of which affect adaptive immune functions (Robinson *et al.*, 2011). However, since such resistance undoubtedly exists in at least part of every population-ADV combination, we can exploit its existence without understanding it.

Acknowledgements

Dr. A. Zargar, collaborated on this project while seconded to the Atlantic Veterinary College for part of his postdoctoral study from the Department of Plant and Animal Sciences, Nova Scotia Agricultural College, 58 River Rd., Truro, Nova Scotia, Canada, B2N5E3.

References

Alexandersen S., 1986. Acute interstitial pneumonia in mink kits: experimental reproduction of the disease. Veterinary Pathology 23: 579-88.

Bacon L.D., Hunt H.D., and Cheng H.H. 2000. A review of the development of chicken lines to resolve genes determining resistance to diseases. Poultry Science 79: 1082-93.

Bloom M.E., Kanno H., Mori S. and Wolfinbarger J.B., 1994. Aleutian mink disease: puzzles and paradigms. Infectious Agents and Disease 3: 279-301.

Buhles W.C. Jr, Huxsoll D.L. and Ristic M., 1974. Tropical canine pancytopenia: Clinical, hematologic, and serologic response of dogs to Ehrlichia canis infection, tetracycline therapy, and challenge inoculation. Journal of Infectious Diseases 130: 357-67.

Cho, H. J., and D. G. Ingram., 1974. The antigen and virus of Aleutian disease in mink. Journal of Immunological Methods 4: 217-228.

De Milito A., Nilsson A, Titanji K., Thorstensson R., Reizenstein E., Narita M., Grutzmeier S., Sönnerborg A. and Chiodi F., 2003. Mechanisms of hyper-γ globulinemia and impaired antigen-specific humoral immunity in HIV-1 infection. Blood 103: 2180-6.

Duncan, M.W., Roder, H. and Hunsucker, S.W., 2008. Quantitative matrix-assisted laser desorption/ionization mass spectrometry. Brief Funct Genomic Proteomic 7: 355-70.

Eterpi M., McDonnell G. and Thomas V., 2009. Disinfection efficacy against parvoviruses compared with reference viruses. Journal of Hospital Infection 73: 64-70.

Gorham J.R., Leader R.W., Padgett G.A., Burger D. and Henson J.B,. 1965. Some Observations of the Natural Occurrence of Aleutian Disease. In: Slow, latent, and temperate virus infections. Gajdousek D., Carleton G., Gibbs C.J.Jr., and Alpers M., NINDB Monograph No. 2. U.S.P.H.S. Publ. No. 1378, pp. 279-285.

Henson, J. B., Gorham J. R., McGuire T. C., and Crawford T. B., 1976. Pathology and pathogenesis of Aleutian disease, In R. H. Kimberlin (ed.), Slow virus diseases of animals and man. Elsevier/North-Holland Publishing Co., Amsterdam. pp. 175-205.

Hadlow W.J., Race R.E. and Kennedy R.C., 1983. Comparative pathogenicity of four strains of Aleutian disease virus for pastel and sapphire mink. Infection and Immunity 41: 1016-23.

Hahn E.C. and Hahn P.S., 1983. Autoimmunity in Aleutian disease: contribution of antiviral and anti-DNA antibody to hypergammaglobulinemia. Infection and Immunity 41: 494-500.

Hunziker L, Recher M, Macpherson AJ, Ciurea A, Freigang S, Hengartner H and Zinkernagel RM., 2003. Hyper-γ globulinemia and autoantibody induction mechanisms in viral infections. Nature Immunology 4: 343-9.

International Committee on Taxonomy of Viruses (ICTV), 2011. Virus Taxonomy. Available at: http://www.ictvonline.org/virusTaxonomy.asp?version=2011.

Jones J.F., Shurin S., Abramowsky C., Tubbs R.R., Sciotto C.G., Wahl R., Sands J., Gottman D., Katz B.Z. and Sklar J., 1988. T-cell lymphomas containing Epstein-Barr viral DNA in patients with chronic Epstein-Barr virus infections. New England Journal of Medicine 318: 733-41.

Kanjanapongkul S., 2006. Chediak-Higashi syndrome: report of a case with uncommon presentation and review literature. Journal of the Medical Association of Thailand 89: 541-4.

Keshgegian A.A., 1984., Hypoalbuminemia associated with diffuse hyper-γ globulinemia in chronic diseases: lack of diagnostic specificity. American Journal of Clinical Pathology. 81: 477-81.

Ko H.J., Yang H., Yang J.Y., Seo S.U., Chang S.Y., Seong J.K. and Kweon M.N., 2011. Expansion of Tfh-like cells during chronic Salmonella exposure mediates the generation of autoimmune hyper-γ globulinemia in MyD88-deficient mice. European Journal of Immunology. Nov 22.

Kyrtsonis M.C., Vassilakopoulos T.P., Angelopoulou M.K., Siakantaris P., Kontopidou F.N., Dimopoulou M.N., Boussiotis V., Gribabis A., Konstantopoulos K., Vaiopoulos G.A., Fessas P., Kittas C. and Pangalis G.A., 2001. Waldenström's macroglobulinemia: clinical course and prognostic factors in 60 patients. Experience from a single hematology unit. Annals of Hematology 80: 722-7.

Liao J., Zhang R., Qian H., Cao L., Zhang Y., Xu W., Li J., Wu M. and Yin Z., 2012. Serum profiling based on fucosylated glycoproteins for differentiating between chronic hepatitis B and hepatocellular carcinoma. Biochemical and Biophysical Research Communications Mar 7.

Lodmell, D.L., Bergman, R.K. and Hadlow, W.J., 1973. Antibody-forming cells and serum hemolysis responses of pastel and sapphire mink inoculated with Aleutian disease virus. Infection and immunity. Infection and Immunity 8: 769-774.

Noda S, Tsunemi Y, Ichimura Y, Miyamoto A, Tamaki Z, Takekoshi T, Asano Y, Hoashi T, Sugaya M, Inoue T and Sato S., 2011. Hyperviscosity syndrome associated with systemic plasmacytosis. Clinical and Experimental Dermatology 36: 273-6.

Porter, D. D., Larsen A. E., and Porter H. G., 1969. The pathogenesis of Aleutian disease of mink. I. *in vivo* replication and the host antibody response to viral antigen. J. Exp. Med. 130: 575-593.

Porter, D. D., A. E. Larsen, and H. G. Porter., 1980. Aleutian disease of mink. Advances in Immunology 29: 261-286.

Robinson B.A., O'Connor M.A., Li H., Engelmann F., Poland B., Grant R., DeFilippis V., Estep R.D., Axthelm M.K., Messaoudi I. and Wong S.W., 2012. Viral interferon regulatory factors are critical for delay of the host immune response against rhesus macaque rhadinovirus infection. Journal of Virology 86: 2769-79.

Ryan M.J., O'Connor D.J. and Nielsen S.W., 1979. Staphylococcus aureus mastitis in nursing mink affected with aleutian disease. Journal of Wildlife Diseases 15: 533-5.

The Merck Veterinary Manual, 2011. Aleutian disease (plasmacytosis). In: Viral Diseases. Merck Sharp and Dohme Corp., a subsidiary of Merck and Co., Inc.Whitehouse Station, NJ USA.

Themudo, G.E., Østergaard, J. and Ersbøll, A.K., 2011. Persistent spatial clusters of plasmacytosis among Danish mink farms. Preventive Veterinary Medicine 102: 75-82.

Thom, M.D., Macdonald, D.W., Mason, G.J., Pedersen, V and Johnson, P.J., 2004. Female American mink, *Mustela vison*, mate multiply in a free-choice environment. Animal Behaviour 67: 975-984.

Wieser A., Schneider L., Jung J. and Schubert S., 2012. MALDI-TOF MS in microbiological diagnostics-identification of microorganisms and beyond (mini review). Applied Microbiology and Biotechnology 93(3): 965-74.

Wu J.Y., Robinson D., Kung H.J. and Hatzoglou M., 1994. Hormonal regulation of the gene for the type C ecotropic retrovirus receptor in rat liver cells. Journal of Virology 68(3): 1615-23.

Zekarias, B., Ter Huurne, A.A., Landman, W.J., Rebel, J.M., Pol, J.M. and Gruys, E., 2002. Immunological basis of differences in disease resistance in the chicken. Veterinary Research 33(2): 109-25.

Preliminary results of field and experimental investigations of pre-weaning diarrhea in mink

A.S. Hammer[1], K. Ullman[2], T.H. Jensen[5], K.O. Hedlund[3], G. Czifra[2], H.H. Dietz[4] and C. Baule[2]
[1]Copenhagen University, Department of Disease Biology, Faculty of Health and Medical Sciences, Ridebanevej 3, 1870 Frederiksberg C, Denmark, [2]National Veterinary Institute, Department of Virology, Travvägen 20, 75189 Uppsala, Sweden, [3]Swedish Institute for Infectious Disease Control, Enheten för Speciell Diagnostik, Nobels väg 18, 17182 Solna, Sweden, [4]University of Copenhagen, Department of Large Animal Sciences, Hojbakkegaard Allé 5, 2630 Taastrup, Denmark; [5]Aalborg Zoo, Mølleparkvej 63, 9000 Aalborg, Denmark; hammer@life.ku.dk

Abstract

Diarrhea in the pre-weaning period (also known as 'greasy kits', ' sticky kits' or pre-weaning diarrhea syndrome) in mink is known in all mink-producing countries in the northern hemisphere, and has been observed on commercial mink farms in Denmark for more than 50 years. The definition of 'wet kits' is when mink kits develop a greasy, sticky exudate on the skin surface especially in the neck, and tail, present a red and swollen perianal region, severe diarrhea, and kits often exhibit distressed behaviour. During the period 2002-2009 several field and experimental investigations were performed with the aim to investigate the etiology of the pre-weaning diarrhea (greasy kits syndrome) in mink kits, and with focus on investigation of astrovirus and calici virus as important triggering factors. These investigations were performed in collaboration between the Swedish and Danish National Veterinary Institutes and the Danish Fur Breeders Association. The investigations included longitudinal studies systematically following the course of disease on farms during outbreaks as well as the development of an experimental model, including gross pathological and histopathological studies of organ samples from mink kits. Here were summarize results of several studies confirming the association between astroviral and caliciviral infections and pre-weaning diarrhea in mink kits and discuss the impact of the results.

Keywords: diarrhea, pre-weaning, greasy kits syndrome, astrovirus

Introduction

Pre-weaning diarrhea is a recurrent problem in mink farms in Europe. A case-controlled study revealed astrovirus as a significant risk factor for presentation of the disease (Englund *et al.*, 2002). Astrovirus has been detected in mink with diarrhea (Mittelholzer *et al.*, 2003b; Ullman *et al.*, 2004). Caliciviruses were also found in diseased as well as in healthy mink. The disease is highly infectious and spreads fast in an affected farm. Infected animals show fever, profuse diarrhea that can last up to 10 days, wet skin due to excessive excretion of the apocrine glands, and dehydration in severe cases (Clausen and Dietz, 2004). The disease can be further complicated by secondary bacterial infections that require antibiotic treatment. Virus is shed in faeces and can be detected by the polymerase chain reaction (PCR) for up to 7 weeks after infection. Gross pathological findings are not exclusive for astrovirus infections, but virological and histopathological findings are consistent with those of astroviral enteric infections in juvenile animals of other species, where the virus infection is associated with osmotic diarrhea without signs of inflammation and cell death (Hammer *et al.*, 2006).

Mink astrovirus is a small single-stranded positive sense RNA virus of 6.6 Kb in size, comprising three open reading frames (ORFs) flanked by 5' and 3' untranslated regions (Mittelholzer *et al.*, 2003a; Blomstöm *et al.*, 2010). While ORF1a and ORF1b encode non-structural proteins, ORF2 codes for the capsid protein that contains the antigenic determinants of the virus and elicits antibodies upon infection (Geigenmuller *et al.*, 2002; Mendez *et al.*, 2003). Different genetic variants of mink astrovirus have been identified based on the sequences of ORF2 (unpublished data).

Besides the gastrointestinal track of mink with the pre-weaning diarrhea syndrome, mink astrovirus has been found in the central nervous system of mink affected by the shaking mink syndrome (Blomström *et al.*, 2010). A recent study has shown genetic similarity between astroviruses found in faecal samples from children and species of mink astrovirus (Kapoor *et al.*, 2009). Also, an astrovirus identified in a case of encephalitis in a boy was found associated with mink astrovirus (Quan *et al.*, 2010). Altogether, these findings indicate that mink astrovirus causes different disease syndromes and shows a potential for interspecies transmission.

Mink caliciviruses were first isolated from normal mink with a history of hemorrhagic pneumonia, but no disease was attributed to their presence (Evermann *et al.*, 1983; Long *et al.*, 1980). Caliciviruses were detected in fecal samples collected from healthy and diarrheic young mink by electron microscopy as well as by PCR (Jorgensen *et al.*, 1996; Guo *et al.*, 2001). Caliciviruses are also small viruses with a plus-sense RNA genome of 7.3-8.3 kb in length. The genome has three ORFs: ORF1 encoding the non-structural proteins, and ORF2/ORF3 encoding the structural proteins.

The pre-weaning diarrhea syndrome has been reported in all mink producing countries. The magnitude of the losses is not only limited to high mortality rates. It is far more extensive considering that an entire year's turnover can end up costing 25-50 DKK less for all skins at the auction. This is often the case farmers envisage as the disease causes poorer skin quality generally for the whole livestock. With an average of 12.000 skins per year per farm, the losses can reach more than 3-600,000 DKK, as estimated in Denmark (Source: Market survey 'astrovirus and calicivirus capsid proteins as vaccine candidates for pre-weaning diarrhea syndrome in minks', Salvator, April of 2011). The losses due to this disease in the mink industry are estimated to 750 million DKK in Scandinavia.

Due to the need to find a solution for this problem, ongoing research in the collaborative project between SVA, DTU and KF has focused on studying the pathogens found associated to the pre-weaning syndrome in mink, on development of diagnostic assays to detect these pathogens and on development of vaccines for prevention of the disease. In this work we describe studies performed to understand the clinical picture of astrovirus and calicivirus infections, and in the first steps towards the development of a vaccine for astrovirus.

Materials and methods

Longitudinal field studies to define the pathogens relevant in pre-weaning diarrhea syndrome

We systematically followed the course of disease on three Danish farms during outbreaks of 'greasy kits' syndrome during the pre-weaning period 2006 and recorded all clinical signs of the disease. Faecal samples were collected from all 105 litters investigated and the samples were analysed with a mink astrovirus specific PCR and electron microscopy (EM).

Preparation of the astrovirus capsid protein expressed in bacteria for immunization

The difficulty to grow astroviruses from different species has limited the possibility to make conventional vaccines based on live attenuated virus. Therefore molecular biology based methods were the alternative used to develop vaccine candidates based on recombinant proteins. The production of the mink astrovirus capsid protein in bacteria has been described (Baule *et al.*, 2004). For immunization, the protein was purified under denaturing conditions using the ProBond Purification system (Invitrogen), according to the manufacturer's instructions. Subsequently, the protein was dialysed in four steps against 100x volume 10mM Tris buffer with raising pH (pH 5.0, 6.0, 7.0. 8.0) to refold the protein. The final protein concentration was 1mg/ml as measured in NanoDrop.

Experimental design

Pregnant female mink were purchased from a farm that has never experienced the mink diarrhea. Faecal samples were checked for astrovirus with negative results both with real-time RT-PCR (PCR) and electron microscopy (EM). The mink were divided in three groups, as follows:

- Group A: vaccinated females, challenged kits. Six female mink were immunized twice subcutaneously, three weeks apart, with 100 μg of purified recombinant astrovirus capsid protein four and one weeks before the birth using Freund's complete and incomplete adjuvant. The litters were challenged orally at the age of 10 days with a sample containing astrovirus, as determined by EM and PCR.
- Group B: mock-vaccinated females, challenged kits. Five female mink were injected twice subcutaneously, three weeks apart, with PBS containing Freund's complete and incomplete adjuvant. The litters from the females were challenged orally at the age of 10 days with a sample containing astrovirus, as determined by EM and PCR.
- Group C: control group. In this group, the litters originating from four female mink were given orally a sample that did not contain astrovirus nor calicivirus, as determined by EM and PCR.

Data recording

Clinical symptoms of the disease were recorded daily by examining every mink kit in all litters. Faecal samples and blood samples from the mothers. Faecal samples were collected from mink kits before and every third days in the period after challenge. The faecal samples were analysed with PCR and selected blood samples were analysed in an in-house ELISA (Ullman, personal communication).

Results

Longitudinal field studies to define the pathogens relevant in pre-weaning diarrhea syndrome

Mink astrovirus infection was frequently detected in mink kits on Danish farms experiencing outbreaks of pre-weaning diarrhea. The highest prevalence was found among kits aged 20 to 25 days. In many litters virus shedding continued for at least 20 days. Typical symptoms of the disease in two out of the three Danish farms appeared 7-10 days after the detection of the virus, while in one farm diarrhea and other signs of the disease were present several days ahead of detection of mink astrovirus in the faecal samples. EM analysis of the samples collected during this mink astrovirus free period showed calicivirus in the faeces of few diseased kits. This finding confirmed another early assumption that mink kits may become infected with calicivirus during the pre-weaning period and a combination of pathogens is an aggravating factor of pre-weaning diarrhea, that can further be complicated by bacterial infection.

Immunization of mink with a recombinant capsid protein of mink astrovirus expressed in bacteria

Four days after challenge, mink kits in litters from *mock-vaccinated* mothers (group B) showed severe symptoms characterised by diarrhea, greasy skin, black nails, and redness in the perianal region. The kits recovered only five days later.

Three out of six litters in the *vaccinated* mothers (group A) did not show any symptoms of disease during the same period, while kits in two litters showed milder form of the disease for three and five days respectively. Kits in one litter had severe diarrhea similar to those in the mock-vaccinated litters. Mink astrovirus was detectable both in the diseased and the apparently healthy litters during the whole disease period, by means of PCR.

Two of the four litters in the *control* (group C) were healthy for six days after challenge with the virus-free material. But the typical signs appeared 2-3 days after the disease had been observed in the two other groups suggesting a spread of the virus from the infected groups that were in the same room. Milder symptoms of transitory diarrhea were observed in the other two litters starting six days after the 'challenge', lasting for 2 and 4 days respectively.

Mink astrovirus was detectable with PCR a few days before the symptoms were observed. Positive and negative PCR results were confirmed by EM analysis. No other virus has been detected in the samples in this trial. Presence of specific antibodies to the mink astrovirus capsid protein was demonstrated in the sera of vaccinated mothers already two weeks after the first immunization, with an average OD of 2.4, as measured by ELISA. In the mock-immunized group B, the OD readings were bellow 0.4 after both injections, same as in the control group.

Discussion

Mink kits are affected by diarrhea during the pre-weaning period. Previous studies have shown the association of mink astrovirus infection with risk of development of pre-weaning diarrhea syndrome (Englund *et al.*, 2002). In addition, they pointed also for the involvement of calicivirus in the problem. In order to characterize the role of these pathogens in manifestation of pre-weaning diarrhea, longitudinal field studies were carried out in three Danish farms. The field studies confirmed that mink astrovirus infection is commonly found on Danish mink farms experiencing the 'greasy kits' syndrome. The highest prevalence of mink astrovirus was observed in 20 to 25 days old mink kits. On two out of three farms mink astrovirus was detected two to three days before typical signs of the pre-weaning diarrhea were observed in the affected litters, confirming a role of astrovirus infection in presentation of symptoms of pre-weaning diarrhea syndrome. In many litters shedding of mink astrovirus continued for at least 20 days.

On the third farm calicivirus was detected with EM in diseased but mink astrovirus-free kits before the mink astrovirus infection started. The finding suggests that calicivirus must also be considered while investigating possible causes of the pre-weaning diarrhea. It was also concluded that diarrhea or other clinical signs have no differential diagnostic value, therefore laboratory investigations are necessary to identify the causal agent. The findings have led to consider both astrovirus and calicivirus in the development of preventive measures for the disease complex.

The vaccine development initiated with development of a vaccine for astrovirus, based on the capsid protein that carries the antigenic determinants o the virus and induces neutralizing antibodies (Geigenmuller *et al.*, 2002; Mendez *et al.*, 2003). In the kits originating from control and mock-immunized females, typical signs of pre-weaning diarrhea syndrome were recorded in mink kits four days after challenge with mink astrovirus-containing material. The course of the disease was similar to natural cases observed on mink farms. The mink astrovirus spread very quickly within a litter and was shown to be transmissible from cage to cage. This was shown by detection of mink astrovirus also in samples from kits from the control group females. After two to four days, clinical signs followed the detection of virus also in this group. The fact that it was possible to reproduce the disease by administration of astrovirus containing sample, confirmed the ability of astrovirus to induce the pre-weaning syndrome, and established a basis for a challenge model. In the kits originating from the immunized females, half of the litters did not show signs of disease, while two litters manifested mild disease symptoms and the last litter showed severe disease as litters from the mock immunized or the control group. These findings show that a degree of protection is obtained form the capsid protein of astrovirus expressed in bacteria. However, it shows also that other factors are relevant for increased protection. It is plausible that passive transfer was not effective in all immunized females, reducing the protection of their off-spring, or that high antibody levels are not the only indicators of protection ability.

The results of the immunization experiment indicated that disease symptoms may be controlled through targeted vaccination strategies. The capsid protein was shown to stimulate a high level of antibodies in immunized pregnant female. Passive transfer of antibodies to the kits caused reduction of clinical symptoms and of virus load following challenge of kits. However, the difficulty to purify the protein expressed in bacteria in one hand limited the possibility of its reliable production in large scale. On the other hand, the fact that some challenged kits from vaccinated females showed clinical signs indicated partial protection, and that factors other than antibody level are relevant for good protection. Such partial protection has also been reported in chickens vaccinated with a recombinant capsid protein of chicken astrovirus produced in baculovirus system (Sellers *et al.*, 2010).

As proteins that preserve antigenic properties are a prerequisite for development of effective subunit vaccines, improvements have been made in the vaccine to target the immunogenic part of the capsid protein. To this end vaccine candidates for astrovirus have been developed in a system that facilitates scaling up and are currently under testing and validation.

Though traditionally associated with gastroenteric infection, recently, mink astrovirus has been detected in mink suffering from a central nervous system disease categorized as shaking mink syndrome (Blomström *et al.*, 2010). Therefore astrovirus associated with both gastroenteric and central nervous system disease should be taken into account in the design of preventive measures. Ongoing research is focusing on experimental testing of the vaccine for astrovirus under controlled conditions in lab Animal Houses at SVA (mice) and DTU (mink). The next step will be to characterise the effects of the vaccine candidate of astrovirus on farm conditions, and its ability to protect against the enteric and the encephalitis disease caused by mink astrovirus.

References

Baule C, Ullman C, Hellman U, Dietz H-H, Englund L and Czifra G. 2004. Diarre i dieperioden hos minkhvalpe II. Kloning og ekspression af Mink Astrovirus (MiAstV) Kapsidprotein. Faglig Årsberetning 2004.

Blomstrom A-L, Widen F, Hammer AS, Belak S and Berg M. 2010. Detection of a novel astrovirus in brain tissue of mink suffering from shaking mink syndrome by use of viral metagenomics. J Clin Microbiol 48: 4392-6.

Clausen TN and Dietz, HH. 2004. Wet kits in mink, a review. In: VIII International scientific congress in fur animal production, De Ruwenberg, Hertogenbosch, Netherlands, 15-18 September 2004., pp. 87-90.

Englund L, Chriél M, Dietz HH and Hedlund K-O. 2002. Astrovirus epidemiologically linked to pre-weaning diarrhoea in mink. Vet Microbiol 85: 1-11.

Evermann JF, Smith AW, Skiling DE and McKeirnan AJ. 1983. Ultrastructure of newly recognized caliciviruses of the dog and mink. Arch Virol 76: 257-261

Geigenmuller U, Chew T, Ginzton N and Matsui S. M. 2002. Processing of nonstructural protein 1a of human astrovirus. J Virol 76: 2003-8.

Guo M, Evermann JF and Saif LJ. 2001. Detection and molecular characterization of cultivable caliciviruses from clinically normal mink and enteric caliciviruses associated with diarrhoea. Arch Vir 146: 479-493.

Hammer AS, Ullman KG, and Czifra, G., 2006. Diarre i dieperioden hos minkhvalpe IV. Astrovirus infektion hos minkhvalpe, virologiske og patologiske fund i fedtede hvalpe indsendt til diagnostisk undersøgelse. Faglig Årsberetning 2006: 165-170.

Jorgensen M, Scheutz F, Strandbygaard B. 1996. *Escherichia coli* and virus isolated from 'sticky kits'. Acta Vet Scand 37: 163-169.

Long GG, Evermann JF and Gorham JR. 1980. Naturally occurring picornavirus infection of domestic mink. Can J Comp Med 44: 412-417.

Kapoor A, Li L, Victoria J, Oderinde B, Mason C, Pandey P, Zaidi SZ and Delwart E. 2009. Multiple novel astrovirus species in human stool. J Gen Virol 90: 2965-2972.

Mendez E, Salas-Ocampo MP, Munguia ME and Arias CF. 2003. Protein products of the open reading frames encoding nonstructural proteins of human astrovirus serotype 8. J Virol 77: 11378-11384.

Mittelholzer C, Englund L, Hedlund KO, Dietz H-H and Svensson L. 2003a. Molecular characterization of a novel astrovirus associated with disease in mink. J Gen Vir 84: 3087-3094.

Mittelholzer C, Englund L, Hedlund KO, Dietz H-H and Svensson L. 2003b. Detection and sequence analysis of Danish and Swedish strains of mink astrovirus. J Clin Micro 41: 5192-5194.

Quan PL, Wagner TA, Briese T, Torgerson TR, M. Hornig, A. Tashmukhamedova, C. Firth, G. Palacios, A. Baisre-De-Leon, C. D. Paddock, S. K. Hutchison, M. Egholm, S. R. Zaki, J. E. Goldman, H. D. Ochs, and W. I. Lipkin. 2010. Astrovirus encephalitis in boy with X-linked agammaglobulinemia. Emerg Infect Dis 16: 918-925.

Sellers H, Linneman E, Icard AH and Mundt E. 2010. A purified recombinant baculovirus expressed capsid protein of a new astrovirus provides partial protection to runting-stunting syndrome in chickens. Vaccine 28: 1253-1263.

Ullman K, Hedlund KO, Hammer AS, Isaksson M, Dietz HH, Englund L and Czifra G. 2004. Diarre i dieperioden hos minkhvalpe I. Påvisning af Mink Astrovirus (MiAstV) i klininske prover med Real-time RT-PCR. Faglig Årsberetning 2004.

Outbreaks of influenza A virus in farmed mink (*Neovison vison*) in Denmark: molecular characterization of the viruses

L.E. Larsen[1], S.Ø. Breum[1], R. Trebbien[1], K. Bradstad[2], L.P. Nielsen[2], M. Chriél[1], T.H. Jensen[1], C.K. Hjulsager[1], K. Handberg[1], P.H. Jørgensen[1], J. Le Févre Harslund[1], L. Rangstrup-Christensen[1], B. Pedersen[3] and A.S. Hammer[1]

[1]National Veterinary Institute, Technical University of Denmark; [2]Statens Serum Institute, Copenhagen, Denmark; [3]Holstebro Veterinary Clinic, Holstebro, Denmark; march@vet.dtu.dk

Abstract

Influenza in mink (*Neovison vison*) is assumed to be rare, but several outbreaks have been described during recent years in Europe and the North America. In 2009, influenza A of the subtype H3N2 was detected in several Danish mink farms with respiratory symptoms. Full-genome sequencing showed that the virus was a human/swine reassortant, with the H and N gene most related to human H3N2 viruses circulating in 2005. The remaining 6 genes were most closely related to H1N2 influenza viruses circulating in Danish swine. This virus had not previously been described in swine, mink or humans. PCRs assays specifically targeting the new reassortant were developed and used to screen influenza positive samples from humans and swine in Denmark with negative results. Thus, there was no evidence that this virus had spread to humans or was circulating in Danish pigs. In 2010 and 2011, influenza virus was again diagnosed in diseased mink in a few farms. The genetic typing showed that the virus was similar to the pandemic H1N1 virus circulating in humans and swine. The H3N2 virus was not detected in 2010 and 2011. Taken together, these findings indicate that mink is highly susceptible for influenza A virus of human and swine origin and may therefore act as a potential host/reservoir for influenza A viruses.

Keywords: influenza virus, genetic typing, mink

Introduction

Natural influenza virus infection associated with clinical disease has only sporadically been described (Klingeberg *et al.*, 1985; Berg *et al.*, 1990; Gagnon *et al.*, 2009; Yoon *et al.*, 2012). More recently, infection of mink with the pandemic H1N1 virus (pdm09H1N1) has been described in the Netherlands and in Norway (Akerstedt *et al.*, 2009). Furthermore, new reassortments between the pdm09H1N1 subtype and circulating swine viruses have also been described (Tremblay *et al.*, 2011). Taken together, these reports showed that mink are susceptible to influenza A viruses of mammalian and avian origin. There is a lack of passive and active surveillance of influenza A virus infections in mink and therefore the prevalence of natural occurring influenza infections in mink may be underestimated.

In 2009, influenza A virus was found in clinically disease mink in Denmark. The present paper describes the genetic composition of the identified virus and the results of follow-up studies performed in mink, swine and humans.

Materials and Methods

Animals, clinical finding and necropsy findings

An unusually large number of submissions with acute hemorrhagic pneumonia were received for necropsy during autumn 2009. A total of 54 farms submitted new cases. The affected farms were restricted to one area of Jutland and to farms with one common feed supplier. The epidemiology and clinical manifestations of the outbreaks in 2009 are described in details elsewhere by Chriel *et al.* (2012).

Post mortem examination was conducted and samples of lungs for RNA were collected and stored at -80 °C until test.

RNA extraction and RT-PCR

Total RNA was extracted from the lung tissue or nasal swab samples by the RNeasy Mini kit (Qiagen, Hilden, Germany) according to the manufacturer's protocol. The recovered RNA was stored at -80 °C until analysis. The initial RT-PCR for influenza A M-gene detection was performed using the FB-AI-M52C and -M253 primers for M-gene detection as previously described (Fouchier *et al.*, 2000).

Sequencing and phylogenic analysis

RT-PCR positive fragments were sequenced using the PCR primers at an external site (DNA Technology a/s, Aarhus, Denmark). Sequence assembly, multiple alignment and phylogenetic analysis were performed with CLC DMA Workbench software (CLC BIO, Aarhus, Denmark).

Sequences analysis was performed by DNA Technology a/s, Aarhus, Denmark. Sequence assembly, multiple alignment and phylogenetic analysis were performed with CLC DMA Workbench (CLC BIO, Aarhus, Denmark) (Slomka *et al.*, 2010).

The HA, NA and M genes were sequenced on clinical and cultivated viruses. Subsequently full length sequencing of all 8 influenza A genes were performed on one positive sample as previously described (Bragstad *et al.*, 2008). Sequences from each segment were aligned with the most closely related sequences in GenBank.

Results

RT-PCR test

A total of 60 farms (149 mink) were tested with a RT-PCR for influenza A virus and 25 farms (63 mink) tested positive.

Sequencing

Full-genome sequencing was performed by the Sanger methods on one of the RT-PCR positive samples as described (6). The isolate was named A/Mink/Denmark/52-1141/2009 and the sequences deposited in the GenBank database.

Phylogentic analysis of the sequences revealed that N and H genes were closely related to the H and N genes of human H3N2 influenza viruses circulating in Denmark in 2005/6 whereas the remaining 6 genes were most closely related to genes of H1N1/H1N2 viruses circulating in Danish swine.

Discussion

This is the first report of natural occurring influenza A virus in Danish farmed mink. Investigations indicate that the virus found in the mink was a novel strain carrying internal genes with a composition indicating that they have been derived from pigs and that HA and NA genes in the influenza virus were of human origin. There is therefore a potential risk that the virus could be transmitted to humans, but have been no reports of transmission of the influenza A virus strain from mink to humans and the virus have not been found in humans, swine or mink since the outbreak in 2009.

The similarity and consistency of the time, location, clinical presentation, microbiology, epidemiology and pathology suggest that the infection on different farms share the same source. Further studies and collection of information to identify link between the infected farms are needed because all affected farms had many factors in common. Thus, It remains unknown whether the outbreaks was due to continuous external infection (e.g. through food) or if the virus moved through the farm and infected susceptible animals due to direct contact rate.

Conclusion

The present report of hemorrhagic pneumonia in Danish farmed mink, caused by a novel reassorted influenza A virus, clearly supports the need for surveillance and further investigation of natural occurring influenza outbreaks in mink.

Acknowledgements

We thank Susanne Jespersen, Hanne Røigaard-Petersen, Susanne Stubbe from the National Veterinary Institute, Aarhus and Anne Bertelsen from Copenhagen Fur for excellent technical assistance.

References

Åkerstedt J, Valheim M, Germundsson A, Moldal T, Lie KI, Falk M and Hungnes O., 2012. Pneumonia caused by influenza A H1N1 2009 virus in farmed American mink (*Neovison vison*). Vet Rec. 170: 362.

Berg M, Englund L, Abusugra IA, Klingeborn B and Linne T., 1990. Close relationship between mink influenza (H10N4) and concomitantly circulating avian influenza viruses. Arch Virol 113: 61-71.

Bragstad K, Nielsen LP and Fomsgaard A., 2008. The evolution of human influenza A viruses from 1999 to 2006: a complete genome study. Virol J 5: 40.

Chriél M, Jensen TH, Hjulsager C, Larsen LE, Jørgensen PH, Harslund JF, Rangstrup-Christensen L, Petersen B and Hammer AS, 2012. Consequences of outbreaks of Influenza A Virus in Farmed Mink (*Neovison vison*) in Denmark in 2009 and 2010. Proceedings IFASA, Copenhagen 2012.

Fouchier RA, Bestebroer TM, Herfst S, Van Der KL, Rimmelzwaan GF and Osterhaus AD, 2000. Detection of influenza A viruses from different species by PCR amplification of conserved sequences in the matrix gene. J Clin Microbiol 38: 4096-101.

Gagnon CA, Spearman G, Hamel A, Godson DL, Fortin A, Fontaine G and Tremblay D., 2009. Characterization of a Canadian mink H3N2 influenza A virus isolate genetically related to triple reassortant swine influenza virus. J Clin Microbiol 47: 796-799.

Klingeborn B, Englund L, Rott R, Juntti N and Rockborn G., 1985. An avian influenza A virus killing a mammalian species--the mink. Brief report. Arch Virol 86: 347-51.

Tremblay D, Allard V, Doyon JF, Bellehumeur C, Spearman JG, Harel J and Gagnon CA, 2011. Emergence of a new swine H3N2 and pandemic (H1N1) 2009 influenza A virus reassortant in two Canadian animal populations, mink and swine. J Clin Microbiol 49: 4386-90.

Yoon KJ, Schwartz K, Sun D, Zhang J and Hildebrandt H., 2012. Naturally occurring Influenza A virus subtype H1N2 infection in a Midwest United States mink (*Mustela vison*) ranch. J Vet Diagn Invest 24: 388-91.

Lessons learned from typing *Pseudomonas aeruginosa* isolates from hemorrhagic pneumonia in mink using pulsed field gel electrophoresis

C.M. Salomonsen[1], N. Høiby[2], S. Molin[3] and A.S. Hammer[4]
[1]National Veterinary Institute, Technical University of Denmark, Hangoevej 2, 8200 Aarhus N, Denmark;[2]Department of Clinical Microbiology, Rigshospitalet, Blegdamsvej 9, 2100 Copenhagen Ø, Denmark; [3]Department of Systems Biology, Technical University of Denmark, Matematiktorvet, Building 301, 2800 Lyngby, Denmark; [4]Copenhagen University, Faculty of Life Sciences, Ridebanevej 3, Denmark; chms@vet.dtu.dk

Abstract

Hemorrhagic pneumonia is an acute and fatal disease in farm mink with outbreaks occurring almost exclusively in the autumn months in Denmark. The disease is caused by the bacterium *Pseudomonas aeruginosa* which is a well known pathogen in immune compromised humans. Serotyping has been the traditional method of typing *P. aeruginosa* isolates from mink, but in this experiment pulsed-field gel electrophoresis (PFGE) was utilized to acquire a more detailed typing of *P. aeruginosa* isolates from outbreaks of hemorrhagic pneumonia in Denmark from 2002 to 2011. The results indicate that outbreaks of hemorrhagic pneumonia are often caused by a local *P. aeruginosa* clone but that some clones or related clones now and then give rise to more than two outbreaks which can be geographically widespread. Our results show that *P. aeruginosa* can survive on the farm or in a contaminating source and remain virulent for mink for up to five years.

Keywords: *Pseudomonas aeruginosa*, pulsed field gel electrophoresis, mink, hemorrhagic pneumonia

Introduction

Pseudomonas aeruginosa has been known as a cause of hemorrhagic pneumonia in mink since 1953 (Knox, 1953). Outbreaks of disease occur, almost exclusively, from September to early December in Denmark. Hemorrhagic pneumonia is an acute, contagious and fatal disease in mink farming. It is characterized by sudden deaths among the mink which are often found dead with blood around nostrils and mouth. The disease can cause an epidemic on the farm level with mortalities ranging from 1% to 50 (Honda *et al.*, 1977; Knox, 1953). No known underlying cause has been identified preceding *P. aeruginosa* pneumonia in mink, which make mink the only animal known to develop spontaneous and contagious lung infections with *P. aeruginosa*.

P. aeruginosa is widespread in both environment and disease habitats and earlier work has demonstrated that in humans no specific clones are associated with certain habitats and that most *P. aeruginosa* possesses the genes required for infection (Alonso *et al.*, 1999; Morales *et al.*, 2004; Römling *et al.*, 1994).

One published study (Hammer *et al.*, 2003) shows that outbreaks of hemorrhagic pneumonia in Denmark in 1998-2001 are caused either by successful clones, only recovered in one year and causing almost 20% of the outbreaks in that year, or unique *P. aeruginosa* clones causing single outbreaks or small clusters of local outbreaks. This article continues this work by typing *P. aeruginosa* from 2002 to 2011 by pulsed field gel electrophoresis (PFGE). PFGE is considered the 'gold standard' for discriminatory typing of *P. aeruginosa* (Grundmann *et al.*, 1995; Johnson *et al.*, 2007; Renders *et al.*, 1996).

Materials and methods

Bacterial isolates

189 isolates representing 116 outbreaks of hemorrhagic pneumonia on 105 mink farms were typed by PFGE. Only *P. aeruginosa* isolated from lungs of mink dying of hemorrhagic pneumonia were included in the study. The mink were submitted to our laboratory for diagnostic investigations from Danish mink farms during the period 2002-2011 and amount to 85% of the recorded outbreaks of hemorrhagic pneumonia in this period. Two outbreaks from 2001 were included in the study to obtain information on recurrence of *P. aeruginosa*, since these farms experienced outbreaks in the following years. 68% of the outbreaks were caused by serotype 6 (Difco^{TM} polyclonal serotyping) while serotype 5 were recovered in 7% and serotype 7/8 in 18% of the cases. Serotyping of the last 7% was either not done or not possible.

The isolates were freeze-dried at the time of isolation and stored at room temperature before they were revived by incubation in liquid broth for 24 hours at 37 °C and plated on blood agar plates.

Pulsed field gel electrophoresis

The PFGE procedure has been described elsewhere (Nauerby *et al.*, 2000) and was followed with some modifications. The bacteria were washed and molded into 1% agarose plugs (Chromosomal Grade Agarose, Bio-Rad, Hercules, CA, USA) before being digested with 0.1 mg/ml proteinase K for two hours in a shaking water bath (56 °C, 200 rpm). Thin slices (approximately one mm thick) were cut from the plugs and digested with SpeI (BioLab, Ipswich, MA, USA) for four hours at 37 °C. The restriction fragments were separated on a 1% agarose gel (Pulsed-Field Certified Agarose, Bio-Rad) in a 0.5×TBE buffer (0.045 M Tris, 0.045 M boric acid, 0.05 M EDTA) using a CHEF-DR® III system (Bio-Rad). The ramping times were 7-10 s for 10 hours and 20-40 s for ten hours at 6 V/cm at 12 °C. Lambda Ladder PFG markers (BioLab) were run with the samples on the gels. The gels were stained with ethidium bromide and photographed under UV-light by GelDoc-It Imaging System (AH Diagnostics, Aarhus Denmark) with the software VisionWorks LS (UVP, Upland, CA, USA).

The resulting band profiles were analyzed using Bionumerics (Applied Maths, ver. 4.50) with Dice band based comparison and a position tolerance set to 1.7% as suggested by Carriço *et al.* (2005).

Results

We recovered 88 distinct PFGE patterns from the 116 outbreaks. Isolates form two or more mink were typed in 50 (43%) of the outbreaks. In 43 (86%) of the outbreaks with two or more isolates typed, the PFGE patterns were indistinguishable (Table 1).

Nine PFGE-types (S10, S15, S45, S27, S8, S37, S40, S100 and S101) grouped in a cluster of PFGE profiles with similarity values of 85-97% called 'Cluster 1' (Figure 1 and 2). The members of this cluster caused 12 outbreaks; they were geographically widespread and occurred from 2002-2011.

Table 1. PFGE types recognized in more than one outbreak.

PFGE type	No. outbreaks	Year	App. distance between farms
S2	2	2002	1 km
S3	2	2002 and 2003	0 km
S5	2	2002	1 km
S12	2	2002 and 2005	0 km
S19	2	2003 and 2004	0 km
S28	2	2004	8 km
S30	2	2004	1 km
S31	2	2004	180 km
S39	2	2005 and 2010	0 km
51	2	2006	7 km
S52	2	2006	1 km
S53	2	2006	18 km
S62	2	2006	1 km
S83	2	2009	10 km
S87	2	2009	1 km
S100	2	2011	40 km
S31	3	2004	From 170 to 350 km
S103	3	2011	From 1 to 3 km
S25	4	2004	From 1 to 7 km, 1 farm 300 km
S66	4	2006	From 25 to 170 km
S80	8	2009 and 2011	From 0 to 25 km

Discussion

Cluster 1, which consists of nine PFGE-types, has a widespread geographical distribution with PFGE-types occurring in six years. It is possible that these nine types have developed independently from a common ancestor or that one or several of the types occurring in 2002 has developed into the types encountered in 2003, 2004, 2005, 2006 and 2011. The widespread

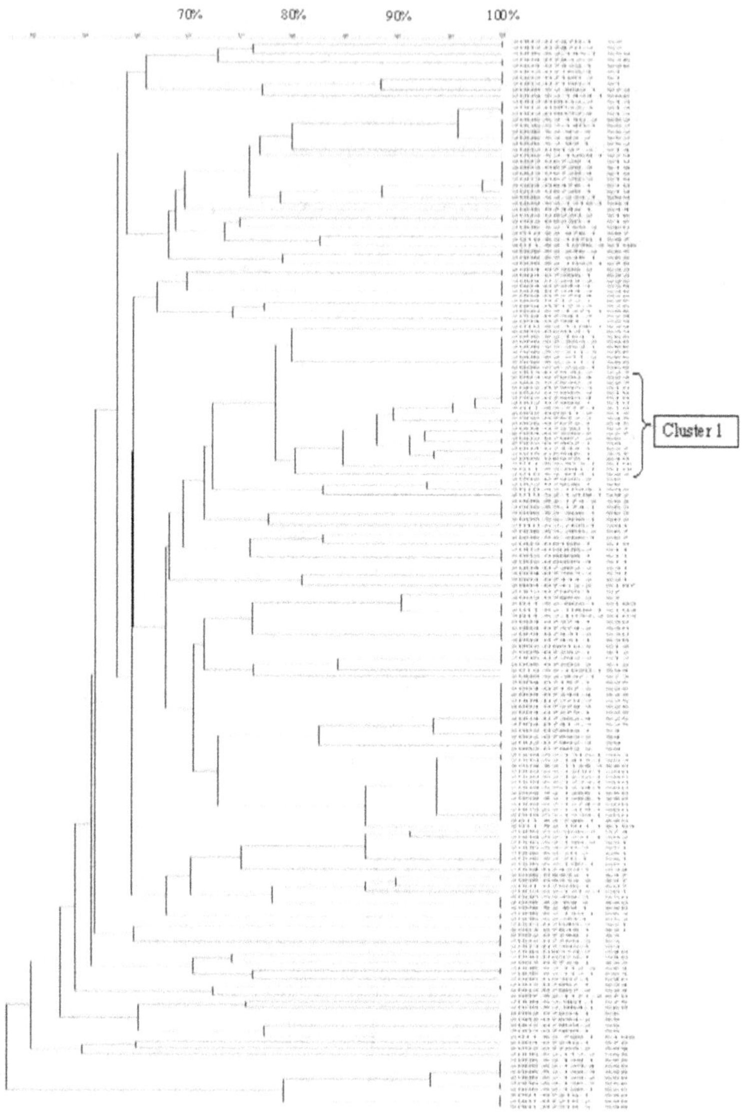

Figure 1. Relationship among P. aeruginosa *isolated from hemorrhagic pneumonia in mink.*

geographical distribution makes it difficult to make any assumptions regarding this, particularly since little is known of the farms' relationships with each other. The oldest isolate could be ancestor of the more recent isolate or they could share a common ancestor. To elucidate this, further analysis of the bacterial genomes are required.

Ten farms experienced repeated outbreaks in the study period. On five of these farms indistinguishable genotypes were found on the farm level with one to five years between the

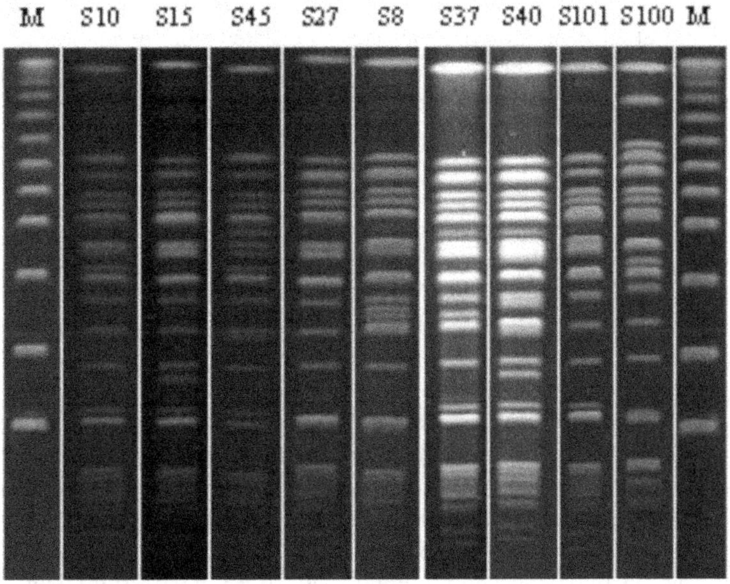

Figure 2. PFGE patterns of 'Cluster 1'.

outbreaks. Little similarity in PFGE-patterns was found between the isolates found on the other five farms where the time between the two outbreaks ranged from one to ten years.

Only 10% of the farms included in this study witnessed more than one outbreak in the study period. Since some of the farms probably vaccinate against the disease after having experienced an outbreak, only little can be said of the ability of *P. aeruginosa* to re-infect farms in succeeding years and how well it thrives in the environment on the mink farm. The data shows that the same PFGE-type has caused outbreaks both one and five years apart on the same mink farm which supports previous work in this field (Hammer *et al.*, 2003). It is clear that *P. aeruginosa* is able to survive in the environment on the mink farm or in a contaminating source without losing virulence but due to the use of vaccination it is difficult to assess to which degree this happens. Both feed, water troughs, equipment, personnel, air and feed manufacturers has been suggested as contamination sources (Gierløff, 1980; Hammer *et al.*, 2003; Knox, 1953; Shimzu *et al.*, 1974). Investigations including genetic typing of *P. aeruginosa* in both animals and suspected sources have never been tried and hence no confirmed contaminating source has been identified. The PFGE-patterns recognized from two farms were very often located close together geographically emphasizing that some kind of local transmission is plausible.

The isolates in cluster 1 accounts for twelve of the recorded outbreaks (10%) from 2002 to 2011. Furthermore twenty-two outbreaks (19%) were caused by five unrelated genotypes causing more than two outbreaks. This means that 70% of the recorded outbreaks of hemorrhagic pneumonia in mink are caused by more or less unique genotypes probably transmitted to the mink from the farm environment i.e. cages, personnel, feed, water, soil and handling equipment. It also means that 30% of the outbreaks can be attributed to infection with genotypes that for some reason are more prevalent in the population of *P. aeruginosa* isolated from hemorrhagic pneumonia in mink. Whether this is due to enhanced virulence or a higher prevalence in the environment has at present not been investigated but having identified these more prevalent genotypes, we now have an opportunity to further examine their genomes and their virulence. This will hopefully lead to a better understanding of the pathogenesis and disease transmission of this disease.

The relatively low degree of relationship among the population of *P. aeruginosa* causing hemorrhagic pneumonia in mink supports previous work in this field (Hammer *et al.*, 2003). The results are to some degree comparable to what has been found in human *P. aeruginosa* infections, where multiple unique and a number of highly clonal isolates are identified as the cause of various infections (Curran *et al.*, 2004; Maatallah *et al.*, 2011; Tramper-Stranders *et al.*, 2008).

PFGE is considered a very discriminative typing method and may in some settings be too discriminating making it hard to identify related clones (Johnson *et al.*, 2007; Kiewitz and Tümmler, 2000; Maatallah *et al.*, 2011; Morales *et al.*, 2004). Since horizontal transfer and genetic recombination of the accessory genome are believed to be the major factors of evolution in *P. aeruginosa* (Kiewitz and Tümmler, 2000; Klockgether *et al.*, 2008; Larbig *et al.*, 2002) few genetic events can cause significant changes in the PFGE-pattern if the bacterium gains or losses large pieces of genetic material in the accessory genome, while the core genome remains unchanged and shared among related species.

Further identification of clonal complexes in mink isolates might be possible using slightly less discriminative methods for typing, e.g. the *P. aeruginosa* Genotyping Kit AT-Biochip (Clondiag Chip Technologies, Germany) which is based on single nucleotide polymorphisms (SNP) or multi-locus sequence typing (MLST) (Curran *et al.*, 2004; Johnson *et al.*, 2007).

It is concluded that the majority of outbreaks of hemorrhagic pneumonia in mink are caused by unique *P. aeruginosa* clones. Some outbreaks are caused by more prevalent clones or related clones. These clones might have a higher virulence for mink or may be over-represented in the environment. The same *P. aeruginosa* clone has been found in two outbreaks on a farm five years apart, demonstrating that *P. aeruginosa* can persist in the environment and remain virulent for mink.

Acknowledgements

The technical assistance and guidance of Lis Nielsen, Herdis Berg Johansen, Jane Andersen and Ulla R. Johansen is gratefully acknowledged. This work was partly supported by a grant from Nordvacc and The Danish Fur Breeders Association.

References

Alonso, A., F. Rojo, and J.L. Martinez, 1999. Environmental and clinical isolates of *Pseudomonas aeruginosa* show pathogenic and biodegradative properties irrespective of their origin. Environmental Microbiology 1: 421-430.

Carrico, J. A., F. R. Pinto, C. Simas, S. Nunes, N. G. Sousa, N. Frazao, H. de Lencastre, and J. S. Almeida, 2005. Assessment of band-based similarity coefficients for automatic type and subtype classification of microbial isolates analyzed by pulsed-field gel electrophoresis. Journal of Clinical Microbiology 43: 5483-5490.

Curran, B., D. Jonas, H. Grundmann, T. Pitt, and C.G. Dowson, 2004. Development of a multilocus sequence typing scheme for the opportunistic pathogen *Pseudomonas aeruginosa*. Journal of Clinical Microbiology 42: 5644-5649.

Gierløff, B., 1980. *Pseudomonas aeruginosa*. IV. IV. Pyocine typing of strains isolated from the blue fox (*Alopex lagopus*), mink (*Mustela vison*), and dog (*Canis familiaris*) and from their environment. Nordisk Veterinaermedicin 32: 147-160.

Grundmann, H., C. Schneider, D. Hartung, F. D. Daschner, and T. L. Pitt, 1995. Discriminatory Power of Three DNA-based Typing Techniques for *Pseudomonas aeruginosa*. Journal of Clinical Microbiology 33: 528-534.

Hammer, A.S., K. Pedersen, T.H. Andersen, J.C. Jorgensen, and H.H. Dietz, 2003. Comparison of *Pseudomonas aeruginosa* isolates from mink by serotyping and pulsed-field gel electrophoresis. Veterinary microbiology 94: 237-243.

Honda, E., J.Y. Homma, C. Abe, K. Tanamoto, H. Noda, and R. Yanagawa, 1977, Effects of the common protective antigen (OEP) and toxoids of protease and elastase from *Pseudomonas aeruginosa* on protection against hemorrhagic pneumonia in mink, Zentralblatt fur Bakteriologie, Parasitenkunde, Infektionskrankheiten und Hygiene.Erste Abteilung Originale.Reihe A: Medizinische Mikrobiologie und Parasitologie, 237: 297-309.

Johnson, J.K., A.M. Arduino, O.C. Stine, J.A. Johnson, and A.D. Harris, 2007. Multilocus Sequence Typing Compared to Pulsed-Field Gel Electrophoresis for Molecular Typing of *Pseudomonas aeruginosa*. Journal of Clinical Microbiology 45: 3707-3712.

Kiewitz, C. and B. Tümmler, 2000. Sequence diversity of *Pseudomonas aeruginosa*: impact on population structure and genome evolution. Journal of Bacteriology 182: 3125-3135.

Klockgether, J., D. Würdemann, L. Wiehlmann, T.T. Binnewies, D.W. Ussery, and B. Tümmler, 2008. Genome Diversity of *Pseudomonas aeruginosa*. In: Cornelis, P. (ed.) *Pseudomonas*: Genomics and Molecular Biology, Caister Academic Press, pp. 19-42.

Knox, B., 1953, *Pseudomonas aeruginosa* som årsag til enzootiske infektioner hos mink, Nordisk Veterinaermedicin 5: 731.

Larbig, K., C. Kiewitz, and B. Tümmler, 2002. Pathogenicity islands and PAI-like structures in *Pseudomonas* species. Current topics in Microbiology and Immunology 264: 201-211.

Maatallah, M., J. Cheriaa, A. Backhrouf, A. Iversen, H. Grundmann, T. Do, P. Lanotte, M. Mastouri, M. S. Elghmati and F. Rojo, 2011. Population structure of *Pseudomonas aeruginosa* from five mediterranean countries: evidence for frequent recombination and epidemic occurrence of CC235. PloS one 6: e25617.

Morales, G., L. Wiehlmann, P. Gudowius, C. van Delden, B. Tümmler, J. L. Martinez, and F. Rojo, 2004. Structure of *Pseudomonas aeruginosa* populations analyzed by single nucleotide polymorphism and pulsed-field gel electrophoresis genotyping, Journal of Bacteriology, 186: 4228-4237.

Nauerby, B., K. Pedersen, H. H. Dietz, and M. Madsen, 2000. Comparison of Danish Isolates of *Salmonella enterica* Serovar Enteritidis PT9a and PT11 from Hedgehogs (*Erinaceus europaeus*) and Humans by Plasmid Profiling and Pulsed-Field Gel Electrophoresis.Journal of Clinical Microbiology 38: 3631-3635.

Renders, N., U. Römling, H. Verbrugh, and A. van Belkum, 1996. Comparative Typing of *Pseudomonas aeruginosa* by Random Amplification of Polymorphic DNA or Pulsed-Field Gel Electrophoresis of DNA Macrorestriction Fragments. Journal of Clinical Microbiology 34: 3190-3195.

Römling, U., J. Wingender, H. Muller, and B. Tümmler, 1994. A major *Pseudomonas aeruginosa* clone common to patients and aquatic habitats. Applied and Environmental Microbiology 60: 1734-1738.

Shimzu, T., J. Y. Homma, T. Aoyama, T. Onodera, and H. Noda, 1974. Virulence of *Pseudomonas aeruginosa* and Spontaneous Spread of *Pseudomonas* Pneumonia in a Mink Ranch. Infection and Immunity 10: 16-20.

Tramper-Stranders, G.A., C.K. van der Ent, T.F. Wolfs, J.L. Kimpen, A. Fleer, U. Johansen, H.K. Johansen, and N. Hoiby, 2008. *Pseudomonas aeruginosa* diversity in distinct paediatric patient groups. Clinical Microbiology and Infection 14: 935-941.

Antimicrobials in mink – consumption and resistance patterns

M. Chriél[1], J.F. Harslund[1], T.H. Jensen[3], E. Jakobsen[2], A.S. Hammer[1] and A.M. Lassen-Nielsen[1]
[1]National Veterinary Institute, Technical University of Denmark, Bülowsvej 27, 1790 Copenhagen V, Denmark; [2]Danish Veterinary and Food Administration, Mørkhøj Bygade 19, 2860 Søborg, Denmark; [3]Kopenhagen Fur, Langagervej 74, 2600 Glostrup, Denmark; march@vet.dtu.dk

Abstract

Use of antimicrobials in the mink production is of vital importance for controlling diseases. However, the increasing risk of developing antimicrobial resistant bacteria as a consequence of the medical treatment requires special attention to ensure continuing availability of antimicrobials for which the bacteria are fully susceptible. Vetstat, a database for veterinary medicine registration, was implemented in Denmark in 2000. The aim of Vetstat was recording all antimicrobial prescriptions on farm level, including veterinarian involved, date of use, animal species, age group and diagnosis. During the period 2003-2010, Denmark experienced an almost linear increase (R^2=0.93) of the annual consumption of antimicrobials in the mink production of 135% as reported through Vetstat. This increase could not solely be explained by an increasing number of mink produced (9%). The seasonal consumption pattern of antimicrobials used in mink is generally characterized by very low levels used during winter, a slight increase in April, maximum levels used in May and intermediate levels used during late summer and autumn. However, during the last 4-6 years an increase in antimicrobial consumption used particularly in late summer and autumn seems accountable for the overall increase during 2003-2010. The proportion of resistant hemolytic and non-hemolytic *E. coli* and *Staphylococcus* spp. isolated from diagnostic samples to the National Veterinary Institute has increased in the same period. It may therefore be concluded that a change in the practice of antibiotics used in the mink farm is important to avoid excessive development of antimicrobial resistant bacteria.

Keywords: antimicrobials, *E. coli, Staphylococcus*, resisitant bacteria, mink

Introduction

All intensive animal production relies on the use of antimicrobials in order to cure infectious diseases and thereby ensuring animal welfare. The treatment had been initiated whether is of primary cause or secondary e.g. due to infections, feed quality or farm management An earlier study of gastro-intestinal disorders in mink the feed producer accounted for 80% of the variation of gastro-intestinal disorders (Rattenborg *et al.*, 1999). Use of antimicrobials does, however, select for resistance in the bacterial population and may result in reduced – or in worst case – no effect of the antimicrobial treatment. Many factors affect the use of medicines; the economic break-even between price for the medicine, the animal of concern and the route of administration will affect the decision for treatment or not (Chriel and Dietz, 2003). In the mink production the new generation of feed trucks with automated dosing equipment for medication has eased the use of e.g. antimicrobials.

The purpose of this study is to evaluate the resistance pattern in *E. coli* (hemolytic and non-hemolytic) and *Staphylococcus* isolated from diseased mink viewed in the perspective of the consumption pattern of antimicrobials in the mink production.

Materials and methods

Relevant organs from mink submitted for necropsy at the National Veterinary Institute at Technical University of Denmark (DTU-VET) were sampled for bacteriological examinations by cultivation Organ tissue materials were incubated blood agar overnight at 3 °C. Resistency patterns were determined on isolated colonies by semi-automatised Sensititre by dilution in micro titer plates.

This study shows the results from bacterial isolates of hemolytic *E. coli*, non-hemolytic *E. coli* and *Staphylococcus sp.* from mink necropsies in 2010. The samples originate from submissions through 72 veterinarians of 631 mink housed at 185 farms.

The consumption of antibiotics on each farm is recorded through Vetstat – a database containing all veterinary prescribed medicines used for production animals in Denmark. Data for the period 2003-2010 were extracted from all Danish mink farms.

Data on the number of breeding females in the period 2003-2010 were provided by Kopenhagen Fur (S. Gerslev, Kopenhagen Fur, personal communication, March, 2011).

Results

The proportion of resistant hemolytic *E. coli* isolates was highest for ampicillin, sulfamethoxazole, tetracycline, streptomycin and trimethoprim. Moderate (cephalothin, neomycin) to low (chloramphenicol, colistin, nalidixan) resistance was recorded for the remaining antimicrobials (Table 1).

The proportion of resistant non-hemolytic *E. coli* isolates was highest for ampicillin, sulfamethoxazole, tetracycline and streptomycin. Moderate (trimethoprim, cephalothin, chloramphenicol, nalidixan, and neomycin) to low (colistin, nalidixan, cefpodoxime, amoxicillin + clavulanic acid, ciprofloxacin, gentamicin, apramycin, ceftiofur, and florfenicol) resistance was recorded for the remaining antimicrobials (Table 2).

Generally, the proportion of resistant hemolytic *E. coli* isolates is higher than the non-hemolytic isolates. Only minor difference between the origins of the isolate was noted. *Staphylococcus* is often isolated from mink and especially from mink with cystitis, pre-weaning mink syndrome mastitis and sepsis. The highest proportions of resistant samples were noted for colistin, tetracycline, streptomycin, and penicillin. Moderate proportions of resistance were recorded for erythromycin, and ampicillin, whereas low was recorded for enrofloxacin, trimethoprim+sulfamethoxazole, and fucidin (Table 3).

Table 1. Antimicrobial resistance among hemolytic E. coli *isolated from Danish mink in 2010 (in percent resistant isolates).*

Antibiotic	Feces or intestine (n=39)	Lung (n=131)	Spleen or liver (n=57)	Urogenital tract or mammary glands (n=20)	Total (n=247)
Ampicillin	61.5	74.0	59.6	60.0	67.6
Tetracycline	51.3	52.7	36.8	30.0	47.0
Streptomycin	38.5	56.5	36.8	20.0	46.2
Sulfamethoxazole	35.9	57.3	40.4	5.0	45.7
Trimethoprim	20.5	40.5	17.5	20.0	30.4
Cephalothin	16.7 (n=12)	18.2 (n=22)	9.1 (n=22)	35.3 (n=17)	19.2 (n=73)
Neomycin	5.1	13.0	7.0	0	9.3
Chloramphenicol	5.1	6.9	5.3	0	5.7
Colistin	2.6	0.8	5.3	0	2.0
Nalidixan	0	2.3	0	0	1.2

During the period 2003-2010, Denmark experienced an almost linear increase (R^2=.93) of the annual consumption of antimicrobials in the mink production of 135% as reported through Vetstat. This increase could not solely be explained by an increasing number of mink produced because the breeding stock only increased 9% in the same period. After 2007 a reduction in the breeding stock was seen, but the consumption of antimicrobials continued to increase (Figure 1). However, the fact that the mink farms are increasing in size i.e. number of mink per farm as well as the size of the individual mink increase may account for a part of the increased consumption of antibiotic.

The resistance patterns found in the isolated bacteria are closely related to the relative distribution of the consumed groups of antimicrobials (Figure 2).

The use of antibiotics is closely related to the life cycle of the mink. The peak in consumption of antibiotic is observed in May in all years and practically no medicine is prescribed during the 4-month winter season in any of the years (Figure 3).

Discussion

The mink production is a very simple production system compared to other intensive animal productions due to the monocyclic nature of the animal and the strictly hierarchical structure with only few feed producers involved. An earlier study showed that 80% of the variation in the incidence of gastro-intestinal disorders was associated with the feed producer indicating that the quality of the feed is of vital importance for the health (Rattenborg *et al.*, 1999). However, farm management plays an important role for the use of antimicrobials, too.

Table 2. Antimicrobial resistance among non-hemolytic E. coli *isolated from Danish mink in 2010 (in percent resistant isolates).*

Antibiotic	Feces or intestine (n=100)	Lung (n=28)	Spleen or liver (n=68)	Urogenital tract or mammary glands (n=5)	Total (n=201)
Ampicillin	50.0	39.3	42.6	40.0	45.8
Sulfamethoxazole	38.0	39.3	42.6	20.0	39.3
Tetracycline	36.0	32.1	44.1	0	37.3
Streptomycin	35.0	35.7	36.8	20.0	35.3
Spectinomycin	25.0	10.7	26.5	0	22.9
Trimethoprim	16.0	21.4	20.6	0	17.9
Cephalothin	20 (n=20)	14.3 (n=14)	0	0	12.8 (n=47)
Chloramphenicol	15.0	7.1	8.8	0	11.4
Nalidixan	8.0	14.3	4.4	20.0	8.0
Neomycin	6.0	10.7	8.8	0	7.5
Colistin	6.0	3.6	2.9	0	4.5
Cefpodoxime	5 (n=20)	7.1 (n=14)	0	0	4.3 (n=47)
Cefotaxime	5 (n=80)	7.1 (n=14)	1.8 (n=57)	0	3.9 (n=154)
Amoxicillin + clavulanic acid	5.0	7.1	0	0	3.5
Ciprofloxacin	1.0	7.1	2.9	20.0	3.0
Gentamicin	4.0	3.6	0	0	2.5
Apramycin	2.0	3.6	1.5	0	2.0
Ceftiofur	2.0	3.6	1.5	0	2.0
Florfenicol	2.0	0	0	0	1.0

Despite substantial variations in the proportion of resistant isolates between samples for the various antimicrobials an overall increase were noted for all hemolytic and non-hemolytic *E. coli* isolates compared to the study carried out for the period 2000-2005 (Pedersen *et al.*, 2009). The proportion of resistant *Staphylococcus spp.* isolates has increased for all antimicrobials compared to the previous study carried out for the period 2000-2005 (Pedersen *et al.*, 2009).

Earlier studies have demonstrated that feces from healthy mink may contain hemolytic *E. coli* and *Staphylococci* without any clinical symptoms including diarrhoea (Vulfson *et al.*, 2001). All samples examined in this study were carried out on clinical cases submitted for necropsy from farms with clinical disease. Therefore, some of the mink may have been treated with antimicrobials before determination of the resistance patterns and thereby increasing the proportion of resistant isolates.

The consumption of prescribed antimicrobials used in mink increased with 135% during the period 2003 till 2010. The exact cause of the increase is not determined, but altered farm

Table 3. Antimicrobial resistance among hemolytic Staphylococcus *isolated from Danish mink in 2010 (in percent resistant isolates).*

Antibiotic	Liver (n=72)	Lung (n=31)	Urogenital tract or mammary glands (n=37)	Total (n=140)
Colistin	94.3 (n=35)	100 (n=23)	100 (n=17)	97.5 (n=75)
Tetracycline	45.8	54.8	45.9	46.6
Spectinomycin	26.4	22.6	37.8	29.7
Penicillin	34.7	22.6	21.6	29.1
Erythromycin	20.8	19.4	24.3	25.0
Clindamycin	20.8	19.4	21.6	24.3
Ampicillin	12.5	9.7	2.7	9.5
Enrofloxacin	1.4	0	2.7	1.4
Trimethoprim + Sulfamethoxazole	1.4	0	2.7	1.4
Fucidin	0	0	2.7	0.7

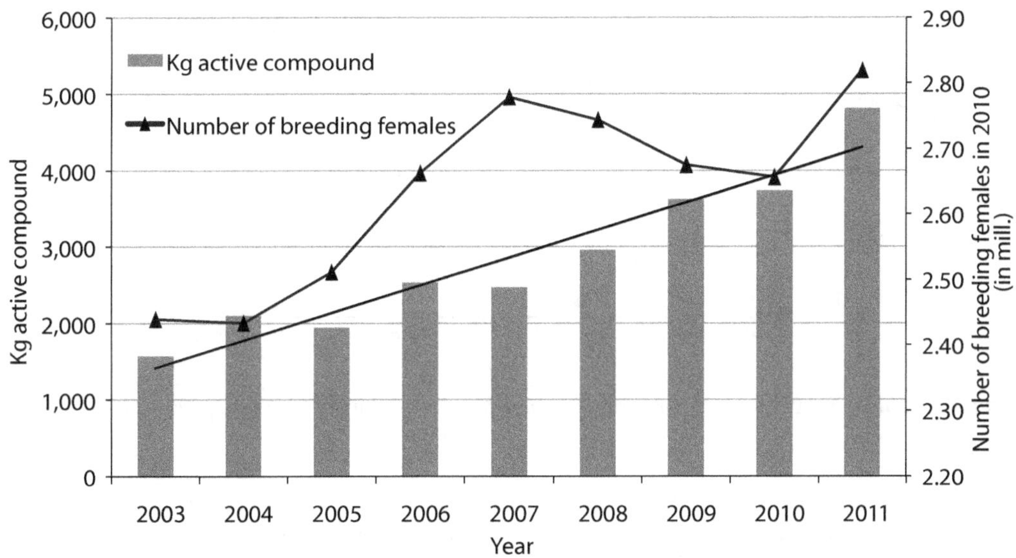

Figure 1. Consumed antimicrobials (kg active compound) recorded in Vetstat for Danish mink, 2003-2010.

management and feeding strategy, larger farm units and increase in farm size may influence the increase. There are no indications that higher use of antimicrobials will result in lower morbidity or mortality of infectious diseases in production animals (Chriel and Dietz, 2003).

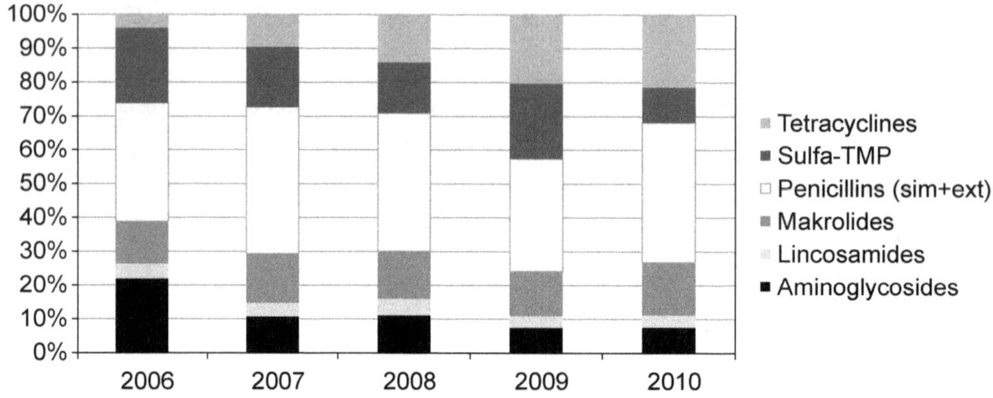

Figure 2. Relative distribution of consumed antimicrobials recorded in Vetstat for Danish mink, 2006-2010.

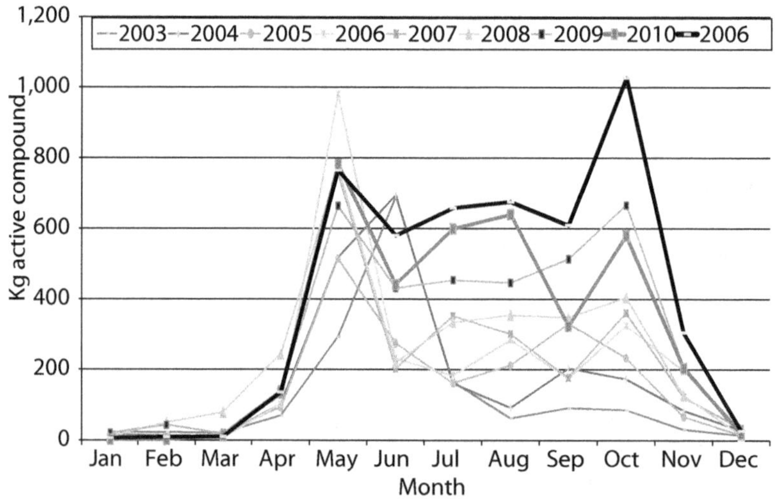

Figure 3. Monthly prescribed antimicrobials (kg active compound) for usage in mink, 2003-2010.

The increase can especially be noted in the autumn and in particular for year 2009 and 2010 where an epidemic of influenza gave a number of clinical outbreaks. In these farms was hemolytic *E. coli* associated pneumonia diagnosed and resulted in supportive treatment in a number of farms for a longer period.

The main goal for decreasing the usage of antimicrobials is to reduce the risk for developing resistant bacteria, so a continuing focus on prudent antibiotic usage is vital. However, good animal health is of major importance for animal welfare for which reason all animals should be treated adequately if diseased.

References

Chriel, M. and Dietz, H.H., 2003. Medication of production animals – cure of malfunctioning animals or production systems? Acta Vet Scand Suppl 98: 65-70.

Pedersen, K., Hammer, A.S., Sorensen, C.M. and Heuer, O.E., 2009. Usage of antimicrobials and occurrence of antimicrobial resistance among bacteria from mink. Vet Microbiol 133: 115-122

Rattenborg, E., Chriel, M. and Dietz, H.H., 1999. Influence of farm, feed-producer and season on incidence of gastrointestinal disorders in Danish farm mink. Prev Vet Med 38: 231-237.

Vulfson, L., Pedersen, K., Chriel, M., Frydendahl, K., Andersen, T.H. and Madsen, M., 2001. Serogroups and antimicrobial susceptibility among *Escherichia coli* isolated from farmed mink (*Mustela vison* Schreiber) in Denmark. Vet Microbiol 79: 143-153.

The influence of succinic acid in the formation of immunity against salmonellosis in foxes

O.Yu. Bespyatykh, A.E. Kokorina, T.V. Tebenkova, I.A. Domskiy and Z.N. Beltyukova
Russian Research Institute of Game Management and Fur Farming, 79, Engels st., Kirov, 610000, Russia, b__oleg@mail.ru

Abstract

The inclusion of succinic acid to the ration of young foxes leads to the highest level of immune response and optimization of metabolism after vaccination against salmonellosis. The application of the preparation 5 days before and 3 days after the immunization promotes hyperimmune reaction which is undesirable for animals. The application of 5 mg/kg LW of succinic acid 5 days before vaccination against salmonellosis appeared to be optimal for foxes.

Keywords: fox, immunobiochemical parameters of blood, vaccination, salmonellosis, succinic acid.

Introduction

Salmonellosis is one of the most serious infectious diseases of alimentary origin in foxes. Foxes are one of the fur animals most susceptible to salmonellosis. *Salmonella choleraesuis* and *Salmonella typhimurium* are two of the most common strains found in fur animals. In spite of proactive preventative measures, salmonellosis appears among farmed animals due to contamination from feeding and husbandry practices. Captivity and farming practices negatively affect the natural disease resistance and positive immune response of farmed animals resulting in reduced resistance to infectious diseases. The impact of negative factors caused by farming conditions results in a breach of plastic and energy exchange within the farmed organism. It is advisable to vaccinate farmed animals against diseases that have an influence on the metabolism of the organism. Succinic acid has been shown to positively affect an organism's immune response even in small doses. We examined the influence of succinic acid in the development of antibodies and immune response toward salmonellosis of foxes.

Materials and methods

The experiments were conducted on two month old red foxes from Ltd. 'Fur Farming 'Vyatka' (Kirov region) in 2008-2009. The fox kits were separated into groups of sixteen animals. The group 1 was the control group and did not receive succinic acid. In the group 2 we added a 5 mg/kg of LW dose of succinic acid to the feed of the animals five days before immunization. The group 3 was given the 5 mg/kg of LW dose of succinic acid five days before and three days after vaccination. All foxes were immunized with an inactivated vaccine against salmonellosis. The vaccine consisted of microbic suspension *Salmonella tyhimurium* and *Salmonella choleraesuis* (which was made by 'Armavirskaya biofactory' in August 2007, s/n 9, control 9).

We took a blood sample once a week from five animals during a month. The blood was taken from the plantar vein of the animals before their breakfast. In serum we sorted out: total protein, fruit sugar, aspartate aminotransferasw (AST), alanine aminotransferase (ALT), alkaline phosphatase (AP), lactate dehydrogenade (LDH) – on semiautomatic biochemical analyzer 'Biochem SA' (USA) with the help of the assay kit of the firm 'High Technology' (USA), protein fraction – by nephelometric method, total immunoglobulin – by the method of desalting by sodium sulfate, antiserum capacity to salmonella antigens – in the agglutination reaction, bactericidal activity of serum (BAS) and index of phagocytosis. The results were processed by statistical methods in 'Biostat' program.

Results and discussion

During the first week after vaccination, the immunities in the blood of the young red foxes increased: gamma globulin, total immunoglobulin, antiserum capacity to salmonella antigens and index of phagocytosis (Table 1, 2). The level of gamma globulin on the seventh day after immunization increased the least in the group 1-24.1%. Immunity increased the most in the group 3-72.4%. On the 21st day after vaccination, there was a decrease of the rates level by 24-42% from the maximal levels. However, it was still higher than initial findings of 17-26%.

The level of immunoglobulin, 7 days after vaccination, increased in the groups 1 and 2 by 41.8%, the largest increase was in group 3 – by 66.7%. On the fourteenth day after immunization, the substance of the total immunoglobulin decreased in the groups 2 and 3 to the initial rate. One week later the amount of total immunoglobulin almost reached the maximum. A week after vaccination, the antiserum capacity to *Salmonella* increased twice. The least increase of titres was the group 1, the largest increase was in the group 3. After the initial increases the rates began to recede.

BAS reached the highest in the group 1 on the 21st day, it was 23.2%. In groups 2 and 3 on the 28 day after inserting the antigen – accordingly by 78 and 3.4%, compared with initial data. The index of phagocytosis increased in a week after immunization: in the group 1 by 23.8%, in group 2 by 21.2%, in group 3 by 29%, compared with initial data. It then decreased slightly till the end of the observations in the groups 1 and 3 it increased, accordingly, by 18.3 and 2.2% and reached its maximum.

The amount of whole protein reached its maximum on the 14th day after vaccination in group 2 by 32.2%, on day 21 in the group 1 by 24% and in group 3 by 49.1%. To the end of the observation the level of whole protein decreased in the large degree in group 3 by 47.5%, in the least degree in the group 2 by 31.6% from the maximal level.

The first week after immunization there was a decrease of albumens from protein fractions and increase of the rest of the fractions. By the 21st day, on the contrary, albumens increased, the rest decreased. The large increase of albumens was in group 1 by 10.7%, the least in groups 2 and 3 by 6.7% compared with initial data. The substance of alpha globulin on the 7th day after vaccination did not change in the group 2 and increased in group 3 by 54.3% compared with initial data. The decrease of alpha globulin on the 21st day after immunization was the least in group 1 by 30.3%,

Table 1. Biochemical parameters of blood of young fox after vaccination against salmonellosis.[1]

Groups	Total protein, g/l	Protein fractions, %				Glucosa, mmol/l	AST, U/l	ALT, U/l	AP, U/l	LDH, U/l
		albumen	α-globulin	β-globulin	γ-globulin					
Before vaccination										
	61.9±4.5	65.3±2.4	16.2±1.0	12.6±1.3	5.8±0.4	9.90±1.04	63.7±1.7	77.8±7.8	136.0±13.0	473.7±37.2
Day 7 after vaccination										
1	66.5±3.2	56.2±1.6^{G}	18.6±1.4	17.2±1.5	7.2±1.4	6.17±0.22^{G}	61.6±5.1	60.7±4.3	166.2±22.1	361.0±37.3
2	64.1±4.1	51.2±2.0^{H}	16.6±1.2	21.6±1.8^{H}	8.5±1.2	7.09±0.47	64.8±2.1	64.9±2.7	190.2±19.8	411.0±38.2
3	58.7±5.6	48.7±2.0^{HD}	25.0±2.5^{GA}	16.7±2.0	10.0±1.3^{G}	7.33±0.42^{D}	61.0±2.2	66.5±2.8	157.3±4.0	375.6±23.4
Day 14 after vaccination										
1	67.6±7.6	63.1±4.8	16.2±1.5	14.6±1.7	6.1±0.6	5.72±0.31^{H}	66.0±5.3	78.3±11.0	156.8±27.1	266.5±34.0^{H}
2	81.9±8.4	62.5±2.5^{J}	15.7±2.0	15.5±1.4^{J}	6.2±0.7	8.89±1.02^{D}	67.6±1.8	65.8±8.5	230.4±10.0^{HD}	324.4±43.2^{G}
3	60.8±4.5	59.0±2.4^{J}	16.5±1.7^{J}	17.3±1.4	7.1±0.5	7.14±0.83	64.2±6.7	64.8±1.7	186.7±6.0^{GKA}	367.8±14.1^{GD}
Day 21 after vaccination										
1	79.4±7.0	72.3±1.8^{L}	13.1±1.1^{J}	8.7±1.3^{KM}	5.8±0.6	7.05±0.29^{GM}	68.7±5.0	78.4±11.3	173.0±18.9	423.3±6.4^{N}
2	76.8±6.0	69.7±2.8^{K}	11.0±0.6^{HKM}	13.8±1.8^{J}	6.0±0.7	7.24±0.80	66.1±4.7	73.1±3.5	127.1±8.4^{JQ}	388.0±24.1
3	92.3±4.0^{HKN}	69.7±1.0^{LN}	12.4±1.0^{GK}	11.4±0.1^{JN}	7.0±0.7	7.67±0.55	65.5±2.1	64.6±4.5	142.0±10.2^{N}	465.8±21.2^{JN}
Day 28 after vaccination										
1	53.4±4.8^{P}	72.6±2.0	13.2±0.9^{J}	7.1±1.0^{GKN}	7.1±1.4	8.70±0.50^{KNP}	66.5±5.3	69.3±5.5	123.2±15.0	467.2±18.3^{JN}
2	56.0±5.6^{MP}	64.6±3.6^{J}	13.3±1.3^{K}	14.5±0.7^{J}	7.2±1.1	8.03±0.91	61.5±5.9	71.7±6.4	152.9±3.0^{QP}	535.1±45.0^{MP}
3	48.5±2.0^{GMS}	64.5±1.0^{LPD}	14.4±2.0	14.3±1.0^{PE}	6.8±0.7	10.30±1.01^{JMP}	70.7±6.0	73.2±6.6	146.3±10.0^{M}	372.8±29.1^{PDA}

[1] The differences are reliable between groups at one period after immunization: A with group 2, accordingly $P<0.05$; D with group 1, accordingly $P<0.05$; the differences are reliable in every group at different periods after immunization: G, H with the initial rates, accordingly $P<0.05$, $P<0.01$; J, K, L with the 7th day after immunization, accordingly $P<0.05$, $P<0.01$, $P<0.001$; M, N, Q with the 14th day after immunization, accordingly $P<0.05$, $P<0.01$, $P<0.001$; P, S with the 21st day after immunization, accordingly $P<0.05$, $P<0.001$.

Table 2. The immune parameters of blood of young fox after vaccination against salmonellosis.[1]

Groups	Total immunoglobulin, g/l	Antiserum capacity to		BAS, %	Index of phagocytosis
		Sal.t.m.	Sal.h.s.		
Before vaccination					
	33.04±4.10	1:14.08	1:11.30	22.61±2.72	9.76±0.73
Day 7 after vaccination					
1	46.83±1.95 ^{G}	1:640.00	1:113.14	23.74±0.79	12.08±1.10
2	46.75±2.37 ^{G}	1:761.09	1:1522.19	21.90±3.66	11.83±0.73
3	55.00±2.89 ^{H}	1:1280.00	1:2152.69	17.62±1.34 ^{E}	12.60±1.33
Day 14 after vaccination					
1	15.67±0.67 ^{HL}	1:95.14	1:113.14	26.97±1.90	13.53±1.46
2	28.33±2.73 ^{KE}	1:160.00	1:95.14	21.43±1.80	11.60±0.92
3	35.30±3.71 ^{KE}	1:135.54	1:95.14	17.86±1.24 ^{E}	12.30±0.74
Day 21 after vaccination					
1	49.50±3.52 ^{G}	1:47.57	1:47.57	27.86±3.71	10.45±0.54
2	45.00±3.24 ^{GN}	1:40.00	1:67.27	22.50±2.18	11.73±1.00
3	49.00±1.46 ^{GM}	1:80.00	1:56.57	20.71±1.43	12.36±1.06
Day 28 after vaccination					
1	39.83±3.48 ^{Q}	1:40.00	1:33.64	25.00±1.80	14.30±1.15 ^{GP}
2	37.38±2.11 ^{JM}	1:33.64	1:56.57	40.24±3.85 ^{HAJNR}	10.17±0.82
3	37.00±2.20 ^{KR}	1:33.64	1:56.57	23.39±2.60 ^{D}	12.88±0.36 ^{H}

[1] The differences are reliable between groups at one period after immunization: A with group 2, accordingly $P<0.05$; D, E with group 1, accordingly $P<0.05$, $P<0.01$; the differences are reliable in every group at different periods after immunization: G, H with the initial rates, accordingly $P<0.05$, $P<0.01$; J, K, L with the 7th day after immunization, accordingly $P<0.05$, $P<0.01$, $P<0.001$; M, N, Q with the 14th day after immunization, accordingly $P<0.05$, $P<0.01$, $P<0.001$; P, R with the 21st day after immunization, accordingly $P<0.05$, $P<0.01$.

in the large degree in group 3 by 50% from the maximal level. The amount of beta globulin in a week after vaccination increased in the least degree in the group 3 by 31.5%, in the large degree in the group 2 by 70.9% from the initial data. On day 21 after vaccination the rate decreased in groups 2 and 3 by 32-37%, considerable decrease was on day 28 in group 1 by 58.7% from maximal meanings.

The amount of fruit sugar in the first week after immunization decreased in group 1 by 37.4%, in group 2 by 28.3%, in group 3 by 26.3% from the initial level. Then the substance of fruit sugar began to increase, but was lower than the initial level till the end of the examination. In the group 2 the maximal concentration of the rate was on day 14 after vaccination by 25.3%, on day 28 in groups 1 and 3 by 52.6% and 45% from the minimum accordingly.

During observation the activity of AST varied in different groups near the initial meaning. At the first week after immunization the substance of ALT decreased in group 1 by 21.9%, in group 2 by 16.6%, in group 3 by 14.4% from initial concentration and till the end of the test it was lower than the initial rate.

The substance of AP after vaccination increased and reached the maximal meaning on day 21 in group 1 by 27.2%, on day 14 in group 2 by 69.1%, in group 3 by 37.5% from the initial level. To the end of the observation the rates decreased almost till the initial meaning.

The LDH decreased on day 14 after immunization in the large level in group 1 by 43.7%, in the least level in group 3 by 22.4% from the initial meaning. Then the activity of LDH increased till the primary level on day 21 after vaccination in group 3, on day 28 in groups 1 and 2.

So, the immune system maximum reacts to the introduction of antigens almost in a week after vaccination because of the higher reactivity of the red fox organism. The inclusion of the succinic acid to the ration of animals leads to the formation of the higher level of immune response and optimization of metabolism in postvaccinal period. The using of this drug 5 days before and 3 days after immunization leads to hyperimmune reaction, and it's bad for animal's organism. It is optimal for young foxes to use the succinic acid only 5 days before vaccination.

Conclusion

The addition of succinic acid to the rations of young foxes in doses of 5 mg/kg of LW, administered five days before vaccination against salmonellosis, leads to the highest level of immune response and optimization of metabolism in the postvaccinal period.

Cage housed marmot's diseases

I.A. Plotnikov
Russian Research Institute of Game Management and Fur Farming, 79 Engels Street, Kirov, 610000, Russia; bio.vniioz@mail.ru

Abstract

Marmot is a new fur animal breeding species. An sickness rate analysis was carried out among marmots under in-cage management and the following infectious diseases, pasteurellosis (hemorrhagic septicemia) and staphylococcosis, were found. Endoparasites leave the host due to the lowering of the marmot's body temperature down to 2 °C in the beginning of the hibernation phase. Among the non-contagious diseases we most often encountered improper front teeth attrition. We reviewed issues of feeding and keeping hygiene. We have tested and recommended some prevention measures against infectious, invasive and non-contagious diseases.

Keywords: fur farming, prevention, veterinary hygiene, keeping, feeding

Introduction

Marmot is a new in-cage fur animal breeding species. These phytophagous little animals hibernate for 5-8 months a year having no food. Along with the development of new animals for fur farming we are faced with problems concerning different biological aspects in relation to these animals. Habitat and nutrition changes, stress factors and other reasons result in changes in animals' organisms and promote illnesses, unusual for these species or illnesses which in free animal progress slowly (Plotnikov and Zabolotskikh, 2000). When we started marmot farming we observed various changes that were dangerous to the animals' health.

Material and methods

The objects of our research were the Bobak Marmot (*Marmota bobak bobak* Muller, 1776) and the Black-capped Marmot (*Marmota camtschatica camtschatica* Pallas, 1811). Our objective was to carry out an analysis of marmots' sickness rate under in-cage management and offer diseases prevention and treatment measures. The research was carried out in State Scientific Institution the Russian Research Institute of Game Management and Fur Farming of Russian Agricultural Academy on the basis of two fur farms: 'Vyatka', Kirov Region, and 'Pushkinsky', Moscow Region, Russia.

Results and discussion

It is known that marmots are amenable to various infectious diseases, like pasteurellosis (hemorrhagic septicemia), pseudotuberculosis, salmonellosis, erysipeloid, listeriosis, tularemia (rabbit disease), distemper, leptospiroses, rickettsiosis, toxoplasmosis, and Powassan encephalitis (Bibikov, 1989).

Among the infectious diseases occurring to marmots under in-cage management we recorded pasteurellosis and staphylococcosis. Pasteurellosis infection was transferred to marmots from birds, which dwelled in large amounts nearby the neighboring farm. The main transmission route for the infectious disease was grass contaminated with birdlime and mowed down close by the marmot fur farm. Infected marmots became sluggish, with low activity and refused eating. Declination of body temperature was also present in some marmots. Body temperatures of healthy animals ranged within 35.5-37.6 °C, while the body temperature of active, sick animals fell down to 22 °C at ambient temperature of 17 °C and to 16 °C at ambient temperature 12 °C respectively. Food poisoning and inflammatory diseases were also accompanied with body temperature declination. Hibernation, independent of season was the organism's defense reaction on the disease. There were some lethal outcomes among marmots sick with pasteurellosis. Autopsy showed changes typical for pasteurellosis, which are similar to those noted in other species. Sick animals suffered from hematolysis. The same situation took place at other diseases connected with body temperature declination. The provisional diagnosis was confirmed with microbiological tests. Serum and antibiotic therapy in conventional doses was efficient. The disease was localized after the source of infection had been eliminated. Thus, grass growing close by a breeding farm cannot be used for feeding animals.

Staphylococcosis outbreak among marmots arose during winter, when the animals were hibernating. On the 1st October, animals went into hibernation without any symptoms of the disease. Since that time they did not eat, nor were in contact with specimens from other cages. Around mid-December an hypodermic abscess, 1 to 5 cm in diameter, caused by a *Staphylococcus* pathogenic strain, was detected among 50% of total population. Most of abscesses were located on rumps and necks, and fewer – on backs and sides of sick animals. Certain species had severe local lesion on paws. Sick animals were coming out of hibernation. Their body temperature rose from 4 °C up to 25-38 °C. Marmots were losing their weight quickly and were uneasy. Several cannibalism cases were noted during this disease. Therapy included bathing of open abscesses with hydrogen peroxide solution and poudrage of wound with tricilline. Positive effect was obtained with intramuscular injection of bicilline I in a dose up to 200,000 units of activity per head and trivitamin up to 2 ml per head. Main preventive measures are proper feeding before hibernation and, when symptoms of disease appear, disinfection in houses and runs before hibernation and exclusion of litter reuse.

We found the following endoparasites in Bobak marmots: cestodes, nematodes and flukes. Black-capped marmots were infested with 3 types of helminthes: nematodes *Ascaris tarbagan, Citellina schulzi* and filariae, suborder *Onchocercaria* gen. sp. (Mashkin, 1997). In early hibernation some species of endoparasites leave the host, because of their bodies' temperature declination down to 2 °C. By autopsy of marmots, perished in hibernation, we discovered no helminthes. External parasites were found on marmots in small numbers. They are fleas, gamasid mites (*Gamasidae*) and ixodic ticks (*Ixodidae*). Marmots' disinfection in an isolation period allows excluding introduction of external parasites to the farm.

Among non-contagious diseases most often we encountered improper front teeth attrition. An intensive growth of separate cutters (up to 1 cm per month) results from marmots' traumas and feeding process breaches. Outgrown cutters traumatize the soft tissue and prevent marmots from eating. Such front teeth must be shortened to normal size.

It is important to take into account the hygienic issues of feeding marmots. Wet-mash feeding has many advantages. Firstly, the diet can be easily balanced in feed-stuff composition. Feed-stuff amount can be reduced or some ingredients can be added. Secondly, it is not of a great difficulty to add the wet-mash feeding vitamins and minerals, and even therapeutic agents when necessary. A disadvantage is that such feeding has short storage period. An intensive bacterial growth takes place because of moisture. At +18-26 °C plate count of the mash rises in hundred times (from 100,000-300,000 to 13-46×10^6 bacteria per 1 gram). Feed-stuff pellets keep better. Its ten days storage in a hopper feeder does not influence its quality indexes.

A number of diseases is connected with unbalanced feeding. Inadequate amount of green crops in a feed-stuff without any additional vitamin complex results in vitamin deficiency. As a result partial fall-off, thinning out or absence of fur on a rump occurs. Marmots cannot be given only one plant culture without combining it with other feed-stuff. When carrots or rape herbage exceeded 60% in nutritional value of all feed-stuff, cases of digestion disorder were detected.

Management technology influences marmots' health. Long term management in cages with floor made of metal net traumatizes marmots' paws. Skin on paws flakes off and chaps that leads to open wounds formation. Bobak marmots get thinner fur on rumps and tails. Black-capped marmots have a stronger hair-coat, which wears out less. Marmots often injure noses and mouth cavities against the edges of a net that stick out.

Conclusion

Habitat and nutrition changes, stress factors and other reasons result in changes in young marmots' organisms under their in-cage management and forward pathogenesis. Presently the marmots' management technology does not meet adaptive abilities of animals to the full extent and it must be improved in accordance with biological features of these unique animals.

References

Bibikov, D.I., 1989. Marmots. Moscow, Agropromizdat, 255 pp. (in Russian).

Mashkin, V.I., 1997. The European *Marmota bobac*: ecology, conservation and use. Kirov, 166 pp. (in Russian).

Plotnikov, I.A. and Zabolotskikh, Yu.S., 2000. Peculiarities of keeping, feeding and breeding of steppe marmots (*Marmota bobak* Mull.) and black-capped marmots (*M. camtschatica* Pall.). Scientifur, 24, 4., pp.136-138.

Influenza on Dutch mink farms in 2011

F.C. Kleyn van Willigen and R. Dijkman
Animal Health Service, P.O. Box 9, 7400 AA Deventer, the Netherlands;
f.kleyn.v.willigen@gddeventer.com; r.dijkman2@gddeventer.com

Abstract

In the middle of May 2011 three farms in the Netherlands claimed increase of mortality of the young pups round the age of 15 days old. The three farms were scattered over The Netherlands. Employees of the diseased farms even as the farmer and his family were suffering from influenza like symptoms. The pups at the 3 farms feel bad and show pumping breath like the acute form of plasmacytosis. The mortality of the pups at the 3 farms run from 11% to 30%. The lungs of the pups at the obduction were very red and were sinking in water. Histological research prove that the pups had an acute interstitial pneumonitis, perhaps caused by a viral infection. Bacteria were not found. Immunohistochemical stains for influenza type A were positive in the lungs of these pups. The PCR influenza A H1N1 was positive in all three cases. In the ELISA test for influenza antibodies from blood samples of ad random chosen adult females at the diseased farms run from 21.8% to 97.3% positives. The ELISA test demonstrated that 72% of 25 selected farms scored positive in this test. The positive females run from 2.7% to 97.3%. Sixty percent of these farms had no clinical symptoms with the pups. Retesting in August of 25 females who lost one or more pups at the diseased farms, demonstrated that mothers were 64% to 96% positive reactors in the ELISA test. Positive reacting pups in the ELISA test were running from 6% to 12%. Till this moment it is not yet clear where the infection at the stroked farm came from and why the farms with minks with positive Elisa for influenza had no clinical problems.

Keywords: mortality, plasmacytosis, pneumonitis, serology, zoonosis

Introduction

Influenza A viruses are highly contagious pathogens that have been isolated from a wide variety of animals, including humans, birds, pigs, horses, minks, seals, whales, cats, dogs and most recently from skunks (Yassine *et al.*, 2009; Britton *et al.*, 2010). The segmented nature of the influenza A virus genome (8 individual segments of single stranded RNA) enables genetic reassortment which can lead to the generation of novel strains of the virus (Yassine *et al.*, 2009).

Different strains can induce respiratory problems in minks (*Neovison vison*). In Japanese experiments (Matsuura *et al.*, 1979; Okazaki *et al.*, 1983a,b) it was possible to infect minks with influenza A strains H3N8, H3N2, H11N4, H7N7, H8N4 and H5N3 (1,2,7). During an outbreak in 1990 of respiratory disease in mink in the south east of Sweden (Berg *et al.*, 1990; Englund, 2000) was influenza A H10N4 and H10N7strain isolated.

During the outbreak in 2009 at 11 mink farms in North-West Jutland of Denmark with respiratory problems was the morbidity 0.2-1.3% and the mortality 0.1-2.35%. Registered respiratory problems 2 days before dying were sneezing and coughing. The dead animals were in good condition. On necropsy the lungs showed acute pneumonitis sometimes related with *E. coli* bacteria. At 9 farms the PCR influenza A H3N2 was positive (Criél *et al.*, 2010). A recent publication on a 2010 H3N2 influenza outbreak on a 15,000 head mink farm in the Midwest United States describes a severe respiratory distress and nose and/or mouth bleeding, with losses averaging 10 animals per day. Both adults and kids were affected (Yoon *et al.*, 2012).

Outbreak influenza in the Netherlands

In the middle of May of 2011 when the mink pups were between 10-15 days old suddenly the mortality raised at 3 farms in the Netherlands. Two farms were located in the south and 1 farm in the west of the Netherlands. Two farms got food from one feed producer and 1 farm from another feed producer. The employees of the 3 farms claimed respiratory problems. One employee had severe respiratory problems and spend 2 days in hospital.

Before the pups died they showed a severe pumping, costo-abdominal, breath. Two farms had a history of plasmacytosis (Aleutian disease). One farm was free of plasmacytosis and got new animals in 2011. The farmers claimed an amount of empty females between 20%-30%. The females showed no clinical respiratory symptoms. The problems at the 3 farms lasted round about 21 days with an total loss of the kids respectively 11%, 23% and 30%. All the kids with clinical signs died. The first impression was acute plasmacytosis because the first pathological histological findings showed acute pneumonitis, caused by a virus infection. The immunohistochemical staining for distemper was negative.

The farmer of one of the plasmacytose positive farms told that the problems were started in one corner of the farm and spread from that corner over the farm. In that corner were many wild birds. After these information was further research started with influenza as one of the possible causes of the problems at these farms. In the Netherlands influenza was not diagnosed in minks with respiratory problems before 2011.

Pathological examination

Submitted mink kids were examined grossly. On macroscopy lungs were heavy, oedemateus and red. Other organs were not affected. Diverse tissues were stored in 10% buffered formalin and in -70 °C storage. Formalin-fixed tissues were embedded in paraffin and stained with heamatoxylin and eosin stain (HE-stain) for histopathological examination. Histopathology showed severe interstitial pneumonia with mild necrotizing bronchiolitis. An immunohistochemical staining for pan-influenza A virus was performed (M-x-INF A; clone 1331).These showed positivity in alveolar type 1 cells, alveolar macrophages, and scarcely in bronchiolar epithelial cells. No bacteria were cultured from these lungs.

Material serological screening

One heparin micro tubes blood was sampled per animal by cutting of a toe nail for the serological screening June 2011. After centrifugation was the serum re-set on Elisa test paper. At the 3 hidden farms from 25 females who lost the most of their pups and her remaining pups, 2 cc blood was sampled out the vena jugularis in August 2011. More blood was necessary if it should be wishfully, if serum samples were positive in the Elisa test, to type the influenza antibodies in the HAR test.

Tests

The Elisa test gave only information about the H antigen of Type A influenza. It could be every influenza type who belonged to A type. Differensation of Influenza was not possible with this test. With PCR it was possible to find the influenza virus type. For typing of influenza antibodies in serum is the HAR (Haeam Agglutination Reaction test) used. The Counter Immuno Electro Phoresis (CIEP test) test is used to detect antibodies against plasmacytosis.

Screening results

Twenty five farms were in June 2011 examined with the Elisa test on antibodies against influenza. Per farm were 100 samples collected. In the Elisa were in 72% (18 farms) positive on antibodies against influenza. 28% (7 farms) of the examined farms had not antibodies against influenza. It was not possible for all the farms to find back the empty females and the females who lost all their kids and the females who lost a more than 50% of their litter (Table 1). There was a great difference between the percentages of positive females between the farms. Farm W, X and Y were the farms with outbreak of influenza.

In August 2011 was blood screened in the Elisa test from the thee affected farms. Twenty five females per farm, who lost a great part of their pups, and her remaining pups (Table 2). In October 2011, 99 selected females at the 3 hidden farms were tested in the Elisa screening.

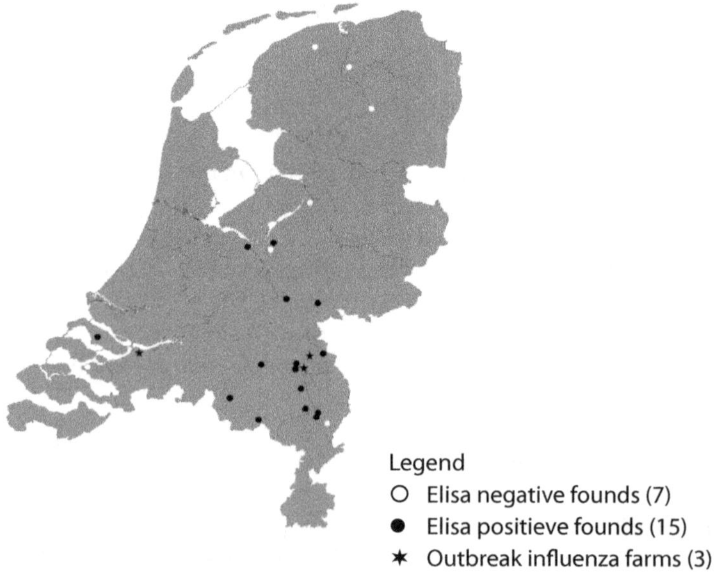

Figure 1. Localisation screened farms June 2011 in the Netherlands.

Table 1. Percentage positive Elisa blood samples at 22 selected farms and 3 affected farms.

Farm	Total	Empty females and zero litters	Females with pups
A	21.2	24.0	12.5
B	29.0		
C	19.0	14.7	32.0
D	8.3		
E	5.0		5.0
F	11.8		
G	0.0		
H	0.0	0.0	0.0
I	0.0		
J	2.6		
K	23.0	18.7	36.0
L	0.0	0.0	0.0
M	0.0	0.0	0.0
N	35.6	35.5	36.0
O	35.8	6.3	16.7
P	11.1		
Q	0.0	0.0	0.0
R	11.0		
S	10.1		
T	0.0	0.0	0.0
U	8.0		8.0
V	2.7	1.3	5.4
W	21.8		
X	45.0	45.0	
Y	97.3		

Table 2. Percentage positive females and kids in the Elisa test at the affected farms August 2011.

Farm	Females (n=25) + pups (n=100)	Females (n=25)	Pups (n=100)
W	19	64	6
X	22	48	6
Y	30	96	12

Table 3 shows a review about the results after 3 times screening research in June, August and October 2011 at the affected farms.

Table 3. Percentage positive females in the Elisa test at the affected farms.

Farm	June: females (n=100)	August: females (n=25)	October: selected (n=99)
W	22	64	54
X	45	48	9
Y	97	96	30

Conclusions and discussion

The first farm visit and the first microscopically results gave the impression of acute viral inflammation perhaps according to plasmacytosis. Plasmacytosis could not be likely because two farms were positive on plasmacytosis and one farm not. Females with sick kids at the plasmacytosis free farm scored negative the CIEP test for antibodies against plasmacytosis. There must be another acute viral infection. To exclude distemper, in relation to the acute pneumonitis, was the immunohistochemical staining for distemper done. This staining was negative.

After the information from the farmer of one of the affected farms, that he had problems with many birds in the corner where the respiratory problems started, new fresh lungs from all the tree hidden farms were positive tested on influenza virus type A in the Immunohistochemical staining. In the PCR were the lungs of the kids from the three affected farms positive of H1N1, the humane variant agree to Mexican flu.

Influenza was an zoonosis and employees of three farms had clinical respiratory problems. It was not excluded that mild Influenza virus types in animals can transform in more aggressive types for human people (Gonzales *et al.*, 2012).

Reason to start a blood screening programs at 22 a-selected mink farms in the Netherlands. The results were important to react adequate at questions and rules formulated by the Dutch Government for the mink production in the Netherlands. In 2012 the Elisa blood screening program will be continued and blood of positive reacting minks tested on influenza virus type in the HAR test.

To control the possible clinical influenza symptoms of human people at and round all the Dutch minks farms were the farmer, his family and his employees advised to vaccinate themselves yearly against influenza.

References

Berg M., Englund.L., Abusgra IA, Klingeboom B and Linné T, 1990. Close relationship between mink influenza (H10N4) and concomitantly circulating avian influenza viruses Arch Virology 113: 61-71

Britton A.P., Sojonky K.R., Scouras A.P. and Bikulka J.J., 2010. Pandemic (H1N1) 2009 in Skunks, Canada. Emerging Infectious Diseases 16: 1043-1045.

Chriél M, Jensen T.H, Hjulsager C. K, Larsen L.E, Harslund J, Rangstrup-Christensen L, Pedersen B and Hammer A.S., 2010. Outbreak of influenza A virus in farmed mink (*Neovison vison*) in Denmark. OFFLU Annual Annual Technical Meeting. Rome, Italy, 2010.

Englund L., 2000. Studies on influenza viruses H10N4 and H10N7 of avian origin in mink. Veterinary Microbiology 74: 101-107.

Gagnon C.A., Spearman G., Hamel A., Godson D. L., Fortin A., Fontaine, G. and Tremblay D., 2009. Characterization of a Canadian Mink H3N2 Influenza A virus isolate genetically related to triple reassortant swine influenza virus. Journal of Clinical Microbiology 47: 796-799.

Gonzales J.L, Elbers A.R.W, Bouma A., Koch G., de Wit J.J. and Stegeman J.A., 2012. Transmission characteristics of low pathogenic avian influenza virus of H7N7 and H5N7 subtypes in layer chickens. Veterinary Microbiology 155: 207-213, 388-391.

Yoon, K.Y., Schwartz K., Sun D., Zhang J. and Hildebrandt H., 2012. Naturally occurring Influenza A virus subtype H1N2 infection in a Midwest United States mink (*Mustela vison*) ranch. J Vet Diagn Investigation 24: 388-391.

Matsuoka Y., Lamirande E.W. and Subbarao K., 2009. The ferret model for influenza. In: Curr Protec Microbiol. 2009, Chapter 15: Unit 15G.2.

Matsuura Y., Yanagawa R. and Noda H., 1979. Experimental infection of mink with influenza A virus. Archives of Virology 62: 71-76.

Okazaki K., Yanagawa R. and Kida H., 1983a. Contact infection of mink with 5 subtypes of avian influenza virus. Archives of Virology 77: 265-269

Okazaki K., Yanagawa R. and Kida H., 1983b. Human influenza virus infection in mink: Serological evidence of infection in summer and autumn. Veterinary Microbiology 8: 251-257.

Yassine H.M., Khatri M., Zhang Y..J., Lee C.W., Byrum B.A., O'Quin J., Smith K.A. and Saif Y.M., 2009. Characterization of triple reassortant H1N1 influenza viruses from swine in Ohio. Veterinary Microbiology 139: 132-139.

Consequences of outbreaks of influenza A virus in farmed mink (*Neovison vison*) in Denmark in 2009 and 2010

M. Chriél[1], T.H. Jensen[1], C. Hjulsager[1], L.E. Larsen[1], P.H. Jørgensen[1,2], J. Le Févre Harslund[1,2], L. Rangstrup-Christensen[1], B. Petersen[3] and A.S. Hammer[1]
[1]National Veterinary Institute, Technical University of Denmark; [2]Danish Veterinary and Food Administration; [3]Holstebro Veterinary Clinic, Holstebro, Denmark; march@vet.dtu.dk

Abstract

Influenza in mink (*Neovison vison*) is assumed to be rare, but outbreaks have previously been reported in farmed mink. The first report was from Swedish mink farms in 1984 which was caused by influenza A virus H10N4 of avian origin. In 2009 and 2010 outbreaks of respiratory disease were seen in several Danish mink farms. In all of the farms, the clinical symptoms were upper respiratory tract symptoms with sneezing and coughing as the most dominant symptoms. Peracute deaths were seen in mink without any clinical symptoms. Influenza H3N2 was found detected by PCR in the lungs from diseased mink. The mean mortality rate was 1.20% (95% confidence intervals: 0.58-1.82) during the outbreak period. Young mink and especially males were reported to be more likely to die. The outbreak in the farms varied from two to ten weeks. During the outbreak period most farms treated all mink with antimicrobials and four of these farms used feed medication in three weeks. The farmers, however, noted that the medication had little or no effect. The most plausible way of transmission of the influenza is from the raw untreated pig waste containing lungs used in the production of mink feed. Because the first clinical symptoms were observed few weeks after the raw pig waste was added to the wet mink feed.

Keywords: mortality, influenza A virus, medication, mink

Introduction

Influenza in mink (*Neovison vison*) is assumed to be rare, but outbreaks have previously been reported in farmed mink. The first report of respiratory disease caused influenza infection was Swedish farmed mink in 1984, this outbreak was caused by influenza A virus H10N4 A/Mink/Sweden/390084 of avian origin (Englund *et al.*, 1986). In 2007, an outbreak of influenza A virus H3N2 (A/Mink/Nova Scotia/1055488/07) was reported from Canadian mink farms. This virus was genetically related to a triple reassortant swine influenza occurring in pigs and the route of infection was speculated to be through feeding of fresh, untreated slaughter offal from pigs (Gagnon *et al.*, 2009).

In other mammals (e.g. humans, pigs and horses), the infection is acute self-limiting infection characterized by the rapid onset of disease, a relatively brief period of symptoms, and resolution within days, and accompanied by early production of infectious virions and elimination of infection by the host immune system (Reperant *et al.*, 2009). Mink have different population dynamics than pigs and humans, as all offspring are born within 14 days in the spring and no new individuals are included in the population beside this time. Therefore, an acute outbreak in a mink population will be characterized as an outbreak in the most susceptible individuals relatively fast and then the infection will die out.

In 2009 and 2010, Danish mink with respiratory symptoms were submitted for necropsy to the National Veterinary Institute. Post-mortem examination of the diseased mink revealed hemorrhagic pneumonia associated with hemolytic *Escherichia coli*.

In 2009, influenza A (H3N2) was diagnosed in mink from 18 Danish farms. It is the first time influenza virus has been diagnosed in Danish mink. Investigations showed that the virus probably originated from circulating swine influenza. The H and N genes were identical to H and N genes in the human seasonal influenza circulating in 2005 which indicate a potential risk of virus transmission to humans. Furthermore, the changes in the H gene also meant that the seasonal influenza vaccination campaign in humans in 2009 not necessarily would protect against the mink associated virus infection (Larsen *et al.*, 2012).

In late August and early September 2010 influenza A (vH1N1) was diagnosed in four mink farms. This is the first time H1N1v has been detected in mink.

Veterinarians submitting mink for necropsy indicated that the problem was more extensive that indicated by the number of submissions. However, the outbreak was limited to a minor region in the western part of Jutland, and both in 2009 and 2010 the infected mink farms all received feed from one feeding kitchen using fresh, untreated slaughterhouse waste from pigs containing inter alia lungs and trachea. So the most likely explanation for the outbreaks in mink is that they were infected via the feed, although this could not be documented as no feed were found to be contaminated with influenza.

Materials and methods

In spring-2010 all 18 farmers in which influenza was diagnosed in 2009 were contacted for an interview. The questionnaire asked for information onset of clinical symptoms, observed symptoms, mortality during the outbreak, duration of the outbreak, morbidity, fatality distribution between sex, age, and genotypes, and efficacy of used medicines.

The total number of mink at risk was calculated for each farm. The total number of mink at risk was the sum of the number of breeding females and the average number of mink kits per female reported by Kopenhagen Fur in 2009 times the number of females. The mortality was calculated using the reported number of dead mink for each farm during the period out of the total number of mink at risk.

Results

In total, 11 farmers completed the questionnaire. All interviewed farmers reported sneezing, coughing and blood from the nose. The symptoms were most prominent when the mink were disturbed e.g. feeding or inspection of the nest boxes. Two farmers noted that if one mink found dead in the morning the other mink in the same cage was found dead the next day. Ten outbreak farms reported the number of dead mink due to influenza (Figure 1). The mean mortality rate was 1.20% of the live mink (95% confidence interval: 0.58-1.82) during the outbreak period. Young mink and especially males were reported to be more likely to die. The outbreak varied from two to ten weeks.

During the outbreak period most farms gave antimicrobial treatment to avoid secondary bacterial infections and four farms used medicine for three weeks. The farmers, however, noted that the medication had little or no effect on the clinical outbreak.

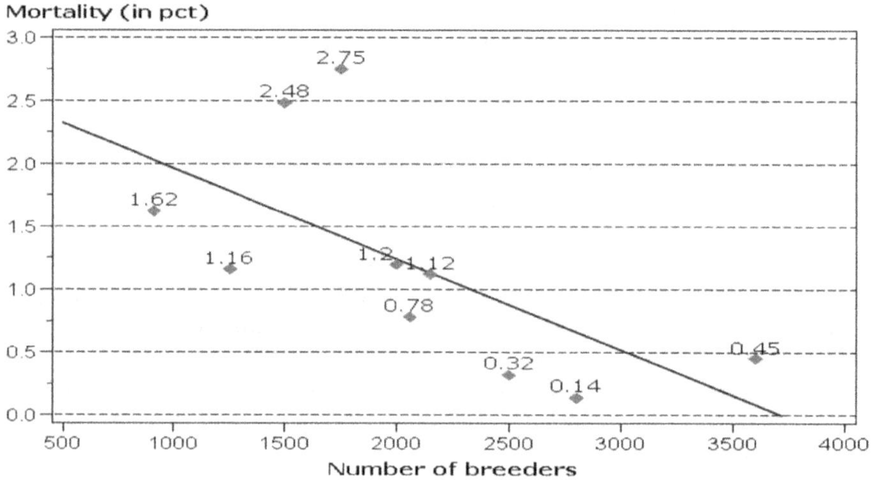

Figure 1. The reported mortality rate (in pct) due to Influenza A in 2009 in ten Danish farms (Pearson correlation r=-0.65).

Discussion and conclusion

The outbreaks were all notices by the farmers in the same period and very similar symptoms were observed on the farms. The duration of the outbreak in the farms was approximately 3 weeks with high morbidity but low mortality. It is unknown whether the extended course was due to continuous external infection (e.g. through food) or if the virus slowly moved through the farm and infected susceptible animals due to low direct contact rate. The low mortality was negatively associated with the farm size. The reason for this is not known but might be influenced by reporting bias, housing systems, or other management practices.

The outbreaks of mink influenza in Danish farm in 2009 and 2010 were all receiving feed from one feeding kitchen using fresh, untreated slaughterhouse waste from pigs containing inter alia lungs and trachea, so the most likely explanation for the outbreaks in mink is that they have been infected via the feed, although it was not confirmed in tests carried out in feed samples. Subsequently, all slaughterhouse offal shall be heat treated before used in mink feed.

No indication that H3N2 has spread to humans or circulating in pigs. It cannot be ruled out that influenza in mink is more common than previously thought, but the National Veterinary Institute does not have an opportunity to assess the spread of influenza virus in Danish mink because no ongoing monitoring is carried out. Influenza in mink is not notifiable either the EU or Denmark.

References

Englund L, Klingeborn B and Mejerland T., 1986. Avian influenza A virus causing an outbreak of contagious interstitial pneumonia in mink. Acta Vet Scand 27: 497-504.

Gagnon CA, Spearman G, Hamel A, Godson DL, Fortin A, Fontaine G and Tremblay D., 2009. Characterization of a Canadian mink H3N2 influenza A virus isolate genetically related to triple reassortant swine influenza virus. J Clin Microbiol 47: 796-799.

Larsen LE, Breum SØ, Trebbien R, Bradstad K, Nielsen LP, Chrié, M, Jensen TH, Hjulsager C, Handberg K, Jørgensen PH, Harslund JF, Rangstrup-Christensen L and Hammer AS, 2012. Outbreak of Influenza A Virus in Farmed Mink (*Neovison vison*) in Denmark: Molecular characterization of the virus and follow up studies. Proceeding IFASA, 2012.

Reperant LA, Rimmelzwaan GF and Kuiken T., 2009. Avian influenza viruses in mammals. Rev Sci Tech 28: 137-59.

Application of Ultrasonography for *in vivo* diagnosis of fatty liver disease in mink (*Neovison vison*)

A.S. Hammer[1] and T. Clausen[2]
[1]Copenhagen University, Faculty of Life Sciences, Ridebanevej 3, 1790 Frederiksberg, Denmark; [2]Kopenhagen Forskning, Agro Food Park 15, 8200 Aarhus N, Denmark; hammer@life.ku.dk

Abstract

High abundance of fat stored within the liver, referred to as fatty liver disease (FLD) is frequently associated with mortalities in mink during the growth period. Studies have indicated that increased tendency for development of FLD may be one of the consequences of reduced protein contents in mink food. For studies on FLD in mink, quantitative analysis of hepatic fat in live mink using ultrasonography would be an appealing non-invasive study modality. The objective of this study was to develop and evaluate a technique for quantifying hepatic fat content in mink by ultrasonography, applying techniques previously validated for use in humans and cats. This study shows that infiltration of the mink liver with fat may be evaluated non-invasively using ultrasonography. This diagnostic imaging technique may prove useful for research in diffuse parenchymal organ disease in mink, though local hair removal prior to application of ultrasonography of organs is a considerable disadvantage in some settings.

Keywords: diseases, fatty liver disease, mink, *Neovison vison*, ultrasonography

Introduction

Hepatic lipidosis or fatty liver disease (FLD) is an important disease in mink characterized by excess fat accumulation in the liver (Clausen, 2006, 2010). Also studies have indicated that one of the consequences of lowered protein contents in mink feed is increased tendency to development of FLD (Clausen and Sandbøl, 2005; Damgaard *et al.*, 1994, 1998a,b). Therefore there is a need for tools to investigate FLD in mink.

In other species radiological methods such as ultrasound (Yeager and Mohammed, 1992; Nicoll *et al.*, 1998; de Lima *et al.*, 2008; Dias *et al.*, 2008; Feeney *et al.*, 2008) have been used to detect and to follow-up experimental diffuse liver disease models. Qualitative ultrasonography (US) has become the most common evaluation method of liver diseases in clinical practice both in animals and in humans, remains to be validated as a technique for the non-invasive diagnosis of such diseases in mink. US examination has been extensively used FLD can be diagnosed by the observation of an increased hepatic echogenicity, characterized by hyperechoic liver parenchymal texture compared to the renal cortex, and decreased visualization of deeper structures, such as the diaphragm and small peripheral vessels (Partington and biller, 1995; Mwanza *et al.*, 1997). Recent studies of the use of US in the follow-up of murine diffuse liver disease models has shown, that this approach is feasible and efficient in this species (Lessa *et al.*, 2010).

The purpose of our study was to examine the potential of US in the assessment of FLD in mink and comparison to histological results by following up a mink model of hepatic disease induced by protein deficiency. The potential implication of this study is to provide a non-invasive method that allows follow-up studies of FLD of individual mink for investigations of the disease entity and for studies of the possible association between protein or amino acid balance in the feed and development of FLD in mink. This poster presents preliminary results of these investigations.

Materials and methods

Animals

US were perfomed on a total of 22 male mink from the Research farm in Holstebro (Kopenhagen Research). The animals were 7 months of age. After examination all animals were submitted to necropsy and liver biopsies were sampled from the liver for histopathological examination. Hair was removed from an area in the midline of the abdomen (just caudal to the sternum).

Histopathology

The liver was removed from the abdominal cavity for macroscopic post mortem analysis. For microscopic examination, liver tissue was fixed overnight incubation in 10% buffered formalin solution and embedded in paraffin. Liver samples were sectioned (5 μm) and stained with hematoxilin and eosin (H&E) according to standard protocols.

Ultrasonographic imaging

Ultrasonography was performed using mobile ultrasonographic equipment (MicroMaxx from SonoSite). In one ultrasound image, liver tissue were visualized. Imaging was performed by an experienced radiologist. One standard image, with image depth 4 cm was used for analysis.

Results

Healthy mink showed homogeneous liver parenchyma, with medium level echogenicity and regular liver surface. FLD in mink was characterized by homogeneous and diffusely increased echogenicity ('bright liver'), reduced visualization of small peripheral vessels, and by no changes in liver surface.

Discussion and conclusions

In this work we demonstrated that it was possible to diagnose FLD in mink using ultrasonography. Ultrasound imaging has many advantages to histopathological diagnosis, namely its low cost, extensive availability and the fact that it can be easily done. Moreover, it is an non-invasive technique and it is the most commonly used method for the investigation of liver diseases in humans and small animals (Yeager and Mohammed, 1992; Bedogni *et al.*, 2009; Partington and Biller, 1995; Lamb, 1990; Schalm, 1997). Potential down sides of the method is, that sedation or immobilization in a appropriate cage which allows access to the abdominal region of the mink are required. Also abdominal shaving is required and this is a problem in settings were the fur quality is an important factor, either of economical reasons or if fur quality is included as a parameter in the research project.

Although ultrasound is an essentially subjective diagnostic method, its validity as an imaging technique for liver evaluation has been extensively described in the literature for humans (Lamb, 1990; Schalm, 1997), dogs (Partington and Biller, 1995; Mwanza *et al.*, 1997; Biller *et al.*, 1992) and cats (Yeager and Mohammed 1992; Partington and Biller, 1995) and also murine models (Lessa *et al.*, 2010). In our study, we focused our attention on mink FLD.

Healthy mink showed homogeneous liver parenchyma, with medium level echogenicity and regular liver surface, as previously described in humans (Dick, 1998), dogs (Partington and Biller, 1995; Biller *et al.*, 1992, England, 1996) and cats (Nicoll *et al.*, 1998; Partington and Biller, 1995). FLD in mink was characterized by homogeneous and diffusely increased echogenicity ('bright liver'), equal or greater to the right renal cortex, reduced visualization of the diaphragm and of small peripheral vessels, and by no changes in liver surface, as described in dogs (Partington and Biller, 1995; Biller *et al.*, 1992), and cats (Yeager and Mohammed, 1992; Partington and Biller, 1995).

These preliminary results of an *in vivo* study on hepatic fat ultrasonography in mink showed that routine ultrasound images may be used to assess hepatic fat content in mink, though local hair removal prior to application of ultrasonography of organs is a considerable disadvantage in some settings. The method needs to be improved and validated before being used as clinical research modality. Also reliability should be improved by more protocolized procedures and if possible application of quantitative image analysis.

References

Bedogni, G., Bellentani, S., Miglioli, L., Masutti, F., Passalacqua, M., Castiglione, A.and Tiribelli C. 2006. The Fatty Liver Index: a simple and accurate predictor of hepatic steatosis in the general population. BMC Gastroenterol. 6: 33.

Biller, D.S., Kantrowitz, B. and Miyabayashi, T., 1992. Ultrasonography of diffuse liver disease. A review. J Vet Intern Med. 6: 71-76.

Clausen, T.N. and Sandbøl, P., 2005. Fasting of mink kits feed different feed rations and its effect on liver fat content, plasma metabolites and enzymes. NJF seminar nr 337, pp. 5, Uppsala, Oct. 5-7.

Clausen, T, 2006. Hvad dør mink af genem et produktionsår (In Danish). In: Møller, S.H. (ed.), Store mink – store udfordringer. Produktion af højtydende mink uden uønskede følgevirkninger, Intern Rapport, Husdyrbrug nr. 2. Danmarks Jordbrugsforskning, Tjele, Denmark, pp. 68-71.

Clausen, T.N., 2010. Kit death from birth to August the first (In Danish) Dødsårsager fra fødsel til 1. august. In: Sandbøl, P. (ed.), Annual Report 2009, Danish Fur Breeders Research Center.

Damgaard, B.M., Clausen, T.N. and Henriksen, P., 1994. Effect of protein and fat content in feed on plasma alanine aminotransferase and hepatic liver infiltration in mink. J. Vet. Med. A. 41: 620-629.

Damgaard, B.M., Clausen, T.N. and Dietz, H.H., 1998a. Effect of dietary protein level on growth performance, mortality rate and clinical blood parameters in mink (*Mustela vison*). Acta Agric. Scand., Sect. A, Animal Sci. 48: 38-48.

Damgaard, B.M., Clausen, T.N. and Børsting, C.F., 1998b. Effect of dietary supplement of essential amino acids on mortality rate, liver traits and blood parameters in mink (*Mustela vison*) fed low-protein diets. Acta Agric. Scand., Sect. A, Animal Sci. 48: 175-183.

De Lima, V.M., Oliveira, C.P., Alves, V.A., Chammas, M.C., Oliveira, E.P., Stefano, J.T., de Mello, E.S., Cerri, G.G., Carrilho, F.J. and Caldwell, S.H., 2008. A rodent model of NASH with cirrhosis, oval cell proliferation and hepatocellular carcinoma. J Hepatol. 49: 1055-1061.

Dick, R., 1998. The liver and spleen In: Textbook of radiology imaging. Chapter 6. Sutton D. (ed.). China: Churchill Livingstone, pp. 981-1028.

Dias, J.V., Paredes, B.D., Mesquita, L.F., Carvalho, A.B., Kozlowski, E.O., Lessa, A.S., Takiya, C.M., Resende, C.M., Coelho, H.S., Campos-de-Carvalho, A.C., Rezende, G.F. and Goldenberg, R.C., 2008. An ultrasound and histomorphological analysis of experimental liver cirrhosis in rats. *Braz J Med Biol Res.* 41: 992-999.

England, G.C.W.,1996. Renal and hepatic ultrasonography in the neonatal dog. Vet Radiol Ultrasound. 37: 374-382.

Feeney, DA, Anderson KL, Ziegler LE, Jessen CR, Daubs BM and Hardy RM., 2008. Statistical relevance of ultrasonographic criteria in the assessment of diffuse liver disease in dogs and cats. Am J Vet Res. 69: 212-221.

Lamb, C.R., 1990. Abdominal ultrasonography in small animals: examination of the liver, spleen and pancreas. J Small Anim Pract. 31: 5-8.

Lessa, A.S., Paredes, B.D., Dias, J.V., Carvalho, A.B., Quintanilha, L.F., Takiya, C.M., Tura, B.R., Rezende, G.F., Campos de Carvalho, A.C., Resende, C.M. and Goldenberg, R.C., 2010. Ultrasound imaging in an experimental model of fatty liver disease and cirrhosis in rats. BMC Vet Res. 29: 6.

Mwanza, T., Miyamoto, T., Okumura, M., Kadosawa, T. and Fujinaga, T., 1997. Ultrasonography, biochemical and hematological profiles in liver disease caused by intravenous administration of dimethylnitrosamine in dogs. Jpn J Vet Res. 45: 153-161.

Nicoll, R.G., Jackson, M.W., Knipp, B.S., Zagzebski, J.A., Steinberg, H. and O'Brien, R.T., 1998. Quantitative ultrasonography of the liver in cats during obesity induction and dietary restriction. Res Vet Sci. 64(1): 1-6.

Partington, B.P. and Biller, D.S., 1995. Hepatic imaging with radiology and ultrasound. Vet Clin North Am Small Anim Pract. 25: 305-335.

Schalm, SW., 1997. The diagnosis of cirrhosis: clinical relevance and methodology. J Hepatol. 27: 1118-1119.

Yeager, A.E. and Mohammed, H., 1992. Accuracy of ultrasonography in the detection of severe hepatic lipidosis in cats. Am J Vet Res. 53: 597-599.

The ixodic tick borreliosis monitoring

D. Zhuravlev, M. Perevozchikova, I. Domsky and Y. Berezina
Prof. B.M. Zhitkov Russian Research Institute of Game Management and Fur Farming, Russia; juravlev_d@bk.ru

Abstract

This survey summarizes results of ixodic tick Borreliosis investigations made by foreign researchers in collaboration with scientists of our country, and the research workers' practices of the Prof. B.M. Zhitkov Russian Research Institute of Game Management and Fur Farming.

Keywords: diagnostics, wild animals, infection rate, ticks, *Borellia*

Introduction

At present time the wide spread of Borreliosis hot spots with high sickness rate in people and domestic animals signifies the urgency of this issue for many countries in America, Europe, Asia and Australia (Ananjeva, 2006). The number of ticks, the infection rate, and consequently the number of people and animals bitten, especially in suburban areas steadily grows in Russia in spite of acaricide treatment. In this situation the necessity of an annual monitoring, of which the results have significance for both application studies and fundamental science, is beyond doubt.

Work materials and methods

Microscopic, cultural, biomolecular and serological research methods were used in the work. The tick collection was carried out according to conventional methods in public places used for recreation: gardens, parks, and suburban areas from late April until early June. *Borrelia* in ticks was detected by direct microscopy, culture on media and PCR methods.

Serum samples for serological tests were taken from following species: 95 mountain hares (*Lepus timidus* L.), 24 foxes (*Vulpes vulpes* L.), 37 raccoon dogs (*Nyctereutes procyonoides* Gray), 67 elk (*Alces alces* L.), 11 wild boars (*Sus scrofa* L.), 2 wolves (*Canis lupus* L.), 24 dogs (*Canis familiaris* L.), 34 grouse (*Lyrurus tetrix* L., *Tetrao urogallus* L., *Tetrastes bonasia* L.) and 24 woodcocks (*Scolopax rusticola* L.). In total 318 head were used.

Bearing in mind the principle of genetic affinity we used immunofluorescence serums for productive and domestic animals to detect antibodies in wild animals. For bacteriological research we used internals (heart, kidney, spleen and liver) from the following species: mountain hare (5), wild boar (3), Eurasian badger (*Meles meles* L.) (7), woodcock (2). In total 17 head were used.

Results

The tick studies (n=707) showed that the *Borrelia* infection rate among ticks was 40.4% in 2007, 31.1% in 2008 and 41.9% in 2009. The highest infection rate with ixodic tick Borreliosis was detected in mountain hare: 52.7%, 20.0% and 47.0% in 2006, 2007 and 2008, respectively, and in fox 38.4%, 37.5% in 2006 and 2007, respectively. This rate was somewhat lower in raccoon dog – 12.5% in 2006 and elk – 10.0%, 7.4% and 27.2%, repectively in 2006, 2008 and 2009. The lowest infection rate was detected in grouse – 5.8% in 2006. In sera obtained from woodcock blood no *Borrelia* antibodies were detected.

Among 54 bacterial BSK-H broth platings of internals from free-living animals *Borrelia* was found in the heart of the Eurasian badger, and the liver and spleen of wild boar, that is in 5.5% of the samples.

Discussion

In last three years the *Borrelia* infection rate in ticks in the Kirov region area averages 37.8%. According to Obert *et al.* (2001) an infection rate of over 25% is regarded as high. The results showed that ixodic ticks parasitize the mountain hare, fox, raccoon dog, grouse, wood grouse, badger, and wild boar, implicating that these species serve as reservoirs for *Borrelia* infection in nature.

The considerable infection rate in mountain hares can be explained by the higher probably of its contact with ticks through bites due to factors, such as size conforming to the ticks concentration field layer size, a thick multilayered coat, high activity and mobility in spring (the rutting season) and summer time.

The analysis allows us to assume that a wide range of reservoir hosts for the agent, which includes besides mouse-like rodents (ixodic ticks in preimaginal phase parasitize them) also many species of game on which adult (imago) ixodic ticks live, agues for a widening of species composition to be included in survaillance for *Borrelia* antibodies in wild animals.

The comparative analysis of ixodic ticks and wild animals infection conducted by our group showed that the *Borrelia* tick infection rate is high in areas where infected animals occur. We believe that the abundance index of mountain hare can be used as a parameter to predict the ixodic tick Borreliosis epizootic situation in subsequent years, as even minor changes in reservoir hosts abundance can influence the number and infection rate of carriers as it increases possibility of *Borrelia* horizontal transmission. While reduction in numbers of reservoir hosts will result in carrier's population decline and consequently in decline of the agent's number. Similar data on the parasitic system dynamics in ixodic tick Borreliosis hot spots are given by Kovalevsky (2002).

References

Ananjeva, L.P., 2006. Lyme Disease. The Physician's Guide 1: 29-33.

Kovalevsky, Y.V., 2002. Long-term dynamics of epizootic activity of ixodic tick Borreliosis hot spots in western Ural. Tick Borreliosis: records from research and practice conference. Izhevsk, 2002, pp. 149-151.

Obert, A.S., Drozdov, V.N. and Rudakova, S.A., 2001. Ixodic tick Borreliosis: nozogeographical and medico-ecological aspects. Novosibirsk: Nauka, 110 pages.

Field evaluation of CIEP and PCR detection/removal control methods of Aleutian mink disease (AD) in Canada

A. Cepica[1] *and T. Iwamoto*
Department of Pathology and Microbiology, Atlantic Veterinary College, University of Prince Edward Island, 550 University Ave., Charlottetown, Prince Edward Island C1A 4P3, Canada; acepica@upei.ca

Abstract

Detection/removal control method of Aleutian disease by CIEP detection of ADV-antibody and farm or barn depopulation/repopulation, have been the recommended approaches of AD control since mid 1970s. The detection/removal was, at least under common N. American husbandry conditions, unsuccessful in controlling AD in a sustainable manner. Recently, attention was turned towards virus detection by PCR, and its use for AD eradication by removal of positive individual animals. In view of the common failures of CIEP to facilitate AD eradication, we were skeptical about premature acceptance of PCR-detection/removal, as the recommended control method. The frequent failures of CIEP test/removal were often blamed on breaches of bio-security. However, we believed that this method has been based on fundamentally wrong premise that the virus is primarily harbored by the infected animals. In reality, this sturdy parvovirus is harbored primarily in the contaminated environment through feces, saliva, urine, whelping, as well as through blood during bleeding for testing. While the use of both CIEP and PCR for monitoring of farms free of the virus remains certainly a valid approach, the data obtained in this study indicate that detection/removal by neither of the methods could facilitate real and lasting freedom from the virus, under the conditions of the study.

Keywords: Aleutian disease, control method, premise

Introduction

Control of AD has been a priority to mink industries all over the world since the appearance and spread of the disease in late 1040s and early 1950s (Gorham *et al.*, 1965). Hypergammaglobulinemia has been identified early as an important feature of the progressive form of the disease, and iodine agglutination test (IAT) has been the first 'laboratory test' to advance the diagnosis, albeit nonspecifically (Henson *et al.*, 1962). However, IAT identification of mink with AD related hypergammaglobulinemia, followed by culling, has led to very unsatisfactory degree of mitigation of AD related losses, since it only detects extreme cases of hypergammaglobulinemia, where gammaglobulin is at 0.2 g/1 ml, and albumin/gammaglobulin ratio is below 1 (Gorham *et al.*, 1965).

CIEP detection/removal AD control era started with the report of Cho and Greenfield (1978), on eradication of Aleutian disease of mink by eliminating positive counterimmunoelectrophoresis test reactors. In this report three farms were subjected to CIEP testing/cull. Tested took place over 4 years in November (pelting) and February (before breeding). Unqualified success was reported on two farms, and on one there was one positive animal reported in each of the final two years. Undoubtedly, farmers and veterinarians alike noticed over the last more than 30 years of following this practice, that their experience with the program was generally disappointing. This recommended control method for AD has not generally been followed by lasting eradication, often in spite of the most rigorous biosecurity methods to prevent reintroduction (T. McLellean, J. Mullen, M. Bosco, personal communication). The inadequacy of this method is also indirectly acknowledged by recent interest in employment of the polymerase chain reaction (PCR) for the same purpose (Stahl *et al.*, 2010). This time, once again, a simplistic assumption was made. This assumption ignored the available knowledge regarding the long time presence of the virus in the hosts, the nature of virus shedding (Bloom *et al.*, 1994), and the resistant nature of the virus (Eterpi *et al.*, 2009; Sauerbrei and Wutzler, 2009). It was assumed that mere switch to PCR virus detection must accomplish what CIEP clearly did not accomplish through antibody detection. No wonder that this 'new hope' generated an excitement. In N. America, the main proponent was being invited to speak on his 'experience' with PCR to mink producer organization in the USA and Canada. The assumptions, once again, ignored the modes and frequency of virus shedding, as well as the resistant nature of the virus shed into the environment, encouraging premature and unfounded hopes. Such was the genuine conviction of M. Stahl (Pensylvania, USA), that he filed for patenting of his ideas regarding PCR and the AD control (Stahl *et al.*, 2010). Once again, on its surface, just like with CIEP, this approach of 'weeding out' the virus from mink operation seemed valid and supported by the voluminous body of literature attesting to specificity and sensitivity of PCR for virus detection. In order to save the Canadian mink industry expenditures and frustrations associated with premature implementation of a control method with unproven efficacy, we put in place a pilot study in which several aspects of ADV detection by PCR were tested. The primer selection for PCR and the procedure was developed and tested for specificity and sensitivity, the suitability of samples for the highest detection rate was investigated, several farms were tested by PCR for ADV incidence following the CIEP round of detection/cull, and finally five CIEP positive animals were followed for three months with respect to consistency of the PCR detection.

Materials and methods

Animals

For comparison of PCR detection effectiveness in blood, organ tissues (spleen and kidney), saliva, and rectal swabs, samples were collected on the same day from fifty 8 months old CIEP positive mink, which had been confirmed positive for over two months. All other mink were between six to twelve months of age when tested by PCR. These animals were previously CIEP tested negative during the round of CIEP testing when positive animals had been removed. The time elapsed between CIEP test/removal and rectal swab PCR testing is included with the results. Mink farms from Prince Edward Island and Nova Scotia, Canada, were included, that observed reasonable biosecurity precautions, including the wildlife proof fencing, and did not introduce animals from outside for a minimum of one year before the last CIEP testing that preceded the PCR testing.

Samples and sample processing

Blood was obtained by standard procedure of toenail cutting, followed by collection into glass capillary tubes. After centrifugation, the samples were shipped in cooled containers. Serum was collected on reception next day, and frozen at -75 °C until testing. We chose for monitoring of ADV presence after 'CIEP detection/cull' animals housed in common building and common area. The tests were perfomormed individually, or in some cases they were first pooled by groups of tens, with the intention that only the positive groups would be retested. This was based on the experimental determination that dilutions of samples 1:10 had no effect on the vast majority of positive samples.

Rectal and oral swabs saturated with collected and saliva, respectively, were made with sterile 'Cotton tipped applicator' (Puritan Medical Products Company, LLS, Guilford, Maine, U.S.) from farmed mink 7 to 12 months of age. Rectal swabs that had solid fecal material were discarded and taken again, as the DNA extracted from the fecal material often contained inhibitory substances for PCR. The swabs were individually placed in sterile plastic bags, shipped to the laboratory in a cooled container, and kept at 4 °C.

Counterimmuneelectrophoresis (CIEP). CIEP was performed using a commercial antigen (Antigen Laboratory of the Research Foundation of the Danish Fur Breeders' Association, Glostrup, Denmark) according to the manufacturer's instructions, in the Weymouth Aleutian Disease Laboratory, Weymouth, Nova Scotia, Canada.

DNA extractions

The samples were processed within 72 hours after collection. For testing of individual samples of rectal or saliva swabs, the swabs were immersed in 0.65 ml/swab of PBS and incubated at 4 °C for 24 hours. Then, each tube was wortexed and centrifuged at 12,000 rpm in a VWR microcentrifuge tube (VWR, IL USA) for 1 minute. Then 200 μl of the supernatant from processed rectal and oral swabs were recovered for DNA extraction from each of the individual swab sample. For group testing (groups of 10 samples), swabs were at first individually processed, and the recovered supernatants of ten animals (20 μl each) were pooled into 1.5 ml VWR microcentrifuge tube (VWR, IL USA).

Spleen, kidney, and blood, used for comparison of PCR ADV detection effectiveness for PCR were collected from 50 eight months old CIEP positive animals.

DNA extraction was performed according to the manufacturer's instructions by QIAGEN DNeasy Blood and Tissue kit for DNA extraction (Qiagen Inc., Gaithersburg, MD, USA). Final elution was done in 100 μl Elution Buffer.

Blood for PCR testing was collected into heparinized capillaries, and whole blood was subjected to DNA extraction. Spleen and kidney tissues were processed according to the manufacturer's instructions.

Polymerase chain reaction (PCR)

All nucleotide sequences of ADV that have been deposited to the Genebank were retrieved and aligned using CLC sequence viewer (http://www.clcbio.com) in order to design PCR primers in the conserved region of VP2 gene. PCR amplification was performed in a 10 μl reaction volume containing 1× PCR Buffer, 0.2 mM of each dNTP, 0.5 μM of each ADV F3444 (5'-GCAGAGATGAACAGARTTAGAC-3') and ADV R3831 (5'-CCTCATCTTCAAAGTGTGTG-3'), 0.25U of HotStarTaq Plus DNA polymerase (QIAGEN) and 2.5 μl of extracted DNA. The reactions were subjected to an initial incubation at 95 °C for 5 min followed by 40 cycles of 95 °C for 30 sec, 52 °C for 30 sec and 72 °C for 30 sec, and a final elongation at 72 °C for 10 min. Amplification products (10 μl) were separated using electrophoresis on 1.5% agarose gels and visualized by staining with ethidium bromide and photographed under UV illumination.

Results

Comparison of PCR detection of ADV genome was performed on blood, saliva, rectal swabs, and pooled suspensions of spleen and kidney of 50 CIEP positive animals and 20 animals from ADV free facility (Table 1). The percentages of positive samples among blood, saliva, rectal swabs, and organ suspensions was 48%, 86%, 88%, and 86%, respectively. No positive samples were found among 20 animals from CIEP negative farm (see Figure 1 for an example of the test results).

Table 2 shows PCR results from rectal swab monitoring of animals from 5 farms, which the animals had been CIEP tested, and the positive animals were removed immediately after the testing. The number of PCR positive animals arising in farms where culling of CIEP positive animals took place range from 3.5% to 65%. Two rows that are highlighted represent group of

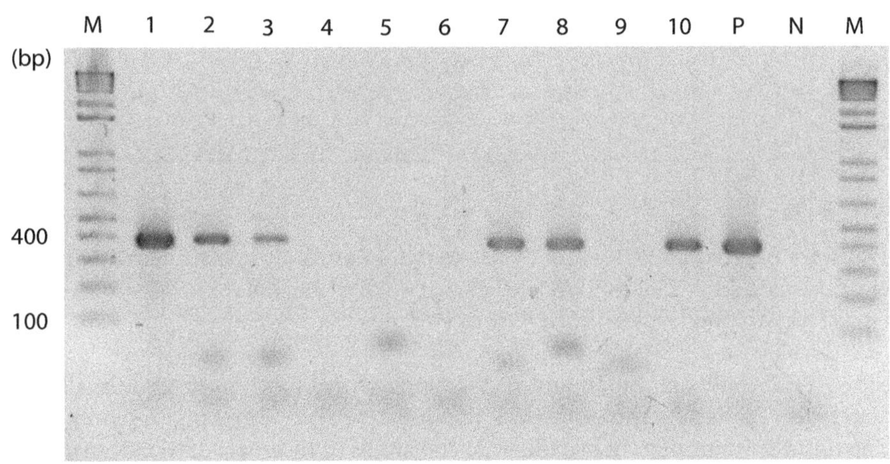

Figure 1. Example of results from group testing of pooled rectal swabs. Each group consist of 10 animals. Lane 1-3, 7, 8 and 10 were positive; P = ADV CDNA clone (positive control), N = DNA elution buffer used for extraction (negative control), M = 100 bp ladder.

Table 1. Simultaneous ADV PCR detection in blood, organ tissues (spleen and kidney), saliva, and rectal swabs in CIEP positive and negative animals.

Mink #	Blood	Saliva	Rectal swab	Spleen/ kidney	Mink #	Blood	Saliva	Rectal swab	Spleen/ kidney
1	-	+	+	+	26	+	+	+	+
2	-	-	-	+	27	+	+	+	-
3	-	-	+	-	28	+	+	+	+
4	-	-	-	-	29	+	+	+	+
5	-	-	+	-	30	-	+	+	+
6	-	-	-	+	31	-	+	+	+
7	-	-	+	-	32	+	+	+	+
8	-	+	+	+	33	+	+	+	+
9	-	+	-	+	34	+	+	+	+
10	-	-	-	-	35	+	+	+	+
11	+	+	+	+	36	+	+	+	+
12	-	+	+	+	37	+	+	+	+
13	+	+	+	+	38	-	+	+	+
14	-	+	+	+	39	-	+	+	+
15	-	+	+	+	40	-	+	+	+
16	-	+	+	+	41	+	+	+	+
17	-	+	+	+	42	+	+	+	+
18	-	+	+	+	43	-	+	+	+
19	-	+	+	+	44	+	-	+	+
20	-	+	+	+	45	-	-	+	-
21	-	+	+	-	46	-	+	+	+
22	-	+	+	+	47	+	+	+	+
23	+	+	+	+	48	+	+	+	+
24	+	+	+	+	49	-	+	+	+
25	+	+	+	+	50	-	+	+	+
51-70	-	-	-	-					

animals that were retested by PCR within close to two months of time. The first PCR test took place in early January, and the second in late February. Following the removal of 26% of animals on the first test, the second test identified 65% of PCR positive animals.

Table 3 shows PCR results on blood and rectal swabs of five CIEP positive animals, in three monthly intervals. The results shows the lack of consistent shedding, as both blood and rectal swab testing indicated intermittent shedding. Rectal swabs of three out of the five animals were positive at all three testing times. Two animals were rectal swab negative on one testing. As for blood, two animals were positive at all three testing times, two were positive at two times, and one animal was positive at one time.

Table 2. PCR monitoring of ADV incidence after CIEP detection/removal.

Farm	PCR test	Previous CIEP test and cull	PCR sample	Individual vs. group sample	No. of samples tested	No. of PCR pos. mink	% of PCR posit. mink
A	Dec	Nov	RS	I	50	23	46.00
A	*Jan*	*Nov*	*RS*	*I*	*187*	*48*	*25.67*
A	*Feb*	*Jan*	*RS*	*I*	*137*	*89*	*64.96*
B	Feb	Nov	RS	G	100	61	61.00
C	Feb	Nov	RS	G	75	36	48.00
C	Feb	Nov	RS	I	362	163	45.03
D	Mar	Nov	RS	G	51	11	21.57
D	Mar	Nov	RS	G	68	21	30.88
D	Mar	Nov	RS	I	210	41	19.52
E	Apr	Feb	RS	G	10	0	0.00
E	May	Feb	RS	G	105	35	33.33
E	Jun	Feb	RS	I	135	10	7.41
A	Jun	Feb	RS	G	26	11	42.31
A	Jun	Feb	RS	I	60	18	30.00
A	Jun	Feb	RS	G	17	4	23.53
A	Jun	Feb	RS	I	40	2	5.00
E	Aug	May	RS	I	58	2	3.45

Note: RS – rectal swab; I – individual animal sample; G – Group sample (n=10); Shaded and italicized area highlight two subsequent tests from the same farm/barn where PCR detection/removal was performed in early January, and the PCR testing took place in February.

Table 3. Results of the blood and rectal testing of CIEP positive mink, in 3 monthly intervals.

Animal I.D. #	October		November		December	
	Blood	Rectal swabs	Blood	Rectal swabs	Blood	Rectal swabs
1	+	+	-	+	+	+
2	-	+	-	-	+	+
3	+	+	+	+	-	+
4	+	+	+	+	+	+
5	+	-	-	+	+	+

Discussion

Regarding suitability of the investigated sample types (Table 1), the relative low percentage of positives among the blood samples was surprising. It could be argued that the use of heparin as an anticoagulant could be blamed for the poor performance of the blood samples, as there have been some reports suggesting that under certain conditions heparin may be inhibitory to PCR (García *et al.*, 2002). We do not believe, however, that this was the reason for the relatively poor performance of blood samples, compared to saliva, rectal swabs and organ suspension, as the DNA extraction kit used, was explicitly marketed for work with heparinized whole blood. We favor an explanation that the results reflected the true availability of ADV DNA in samples. More frequent shedding through feces and saliva may be related to sequestered local virus replication, and if true, this would constitute evolutionary advantage for the virus by increasing the chances of animal to animal spread. This would particularly be advantage to the virus in the natural habitat, since mink are in wild territorial animals. Apart from the unexpected higher detectability rate from rectal swabs (88%), compared to blood (48%), there were also other considerations that made us decide against the use of blood or saliva in farm monitoring. Perhaps the most important consideration was the potential of blood sampling to spread ADV and possibly other blood born pathogens. Droplets of blood splattering by the animals reacting to discomfort contaminate handlers' protective gloves and clothing which can not be disinfected or changed between individual animals. Furthermore, the stress associated with toenail clipping may be exacerbating stress induced conditions. Rectal swabs were collected with minimal stress to animals, and cross-contamination of rectal swabs can be better prevented, than cross-contamination of blood samples. While any contamination of toenail clippers would be very inefficient in antibody cross-contamination, it would very efficient in PCR contamination because of the extreme sensitivity of PCR virus detection.

Counter-immunoelectrophoresis (CIEP) gained wide use since the publication by Cho and Ingram (1972). The specificity of this test with respect to causative agent, was a major advance compared to the past iodine agglutination test (IAT), which detected a major consequence of progressive AD, hypergammaglobulinemia, albeit non-specifically. Serological detection of the virus brought great clarity into the process of determination whether ADV gained entry on a given ranch. Unfortunately, a great leap of thinking was made regarding the CIEP's potential to facilitate eradication of the virus, mainly because of the poor understanding of the virus and of the pathogenesis of the disease at that time. After the report of eradication with the use of CIEP 'detection/removal' (Cho and Greenfield, 1978), this method became promoted as a dogma in AD control. It can't be determined retroactively whether in those first reports true and lasting eradication did indeed occur, but we can say that over the period of 30 plus years of implementation of CEIP detection/removal the results were generally disappointing (T. McLellean, J. Mullen, and M. Bosco, personal communications). In our opinion, any appearance of eradication may have been fictitious, in that while temporarily there were no CIEP positive animals present, the virus remained to be present in some form on these 'eradicated' farms, only to resurface later when it gained access to animals. This could have been in the form of animals in the incubation period, false negatively testing animals, or this very resistant virus could have been present in the ground and later uncovered, or on the equipment, etc.

Throughout the period since introduction of CIEP detection/removal, admission of failure was associated with suspicions of errors in execution of this method of control, with plethora of areas were a farmer could have been blamed for unverifiable mistakes. For example, testing was not performed often enough, animals in the early stages of infection may have been missed, biosecurity on a farm may have been breached, or the virus may have been re-introduced, etc. Unfortunately, these speculations added nothing to improving the situation. Subsequently, not all farmers had the honesty to admit failures, particularly because of the economic and other implications. The testing did generate some sense of 'doing at least something', emanating from the belief that elimination of CIEP positive animals eliminates at least the bulk of the virus load. This belief may have been reasonable at early times when the properties of ADV parvovirus and the pathogenesis of the infections were not well known. The extreme resistance (Sauerbrei and Wutzler, 2009), the shedding, and the pathogenesis of ADV have begun to be better understood in 1980s and 19890s, CIEP detection/removal method of AD control should have been reviewed, but because of unavailability of alternative control method, it remained unquestioned. Elimination of the CIEP positive animals on at least two rounds, at the time of pelting and before breeding, especially if only small percentage of the animals tested CIEP positive, was sincerely but incorrectly believed to mean a negative status of the facility. Any future incidence was explained by errors in execution, timing of testing, or breaches of biosecurity.

Such dogmatic interpretation of reality of CIEP facilitated AD control was clearly incorrect, and it should have been realized long time ago. Massive shedding of this very resistant parvovirus (Uttenthal *et al.*, 1990; Gregersen and Roth, 2012) by feces, urine, saliva, as well as blood due to collection of blood for CIEP testing, may occur throughout the lifetime of the infected animals, even well before CIEP positivity develops (Hadlow and Kennedy, 1983). Under the common farming conditions in N. America, where excreta fall underneath the cages on the ground, and partly eaten, saliva contaminated food, become a lasting source of infection directly for other animals, and indirectly through personnel and fomites. Of course, in addition to these main modes of virus transfer within a facility containing infected animals, there are other highly probable means of spread of the virus, e.g. contacts during mating, contaminations through AD induced abortions and still births, handling of animals after weaning, movement of dust particles and/or aerosol droplets generated during power washing. Consequently, we believe, that 'CIEP detection/removal' method of ADV eradication has been based on faulty assumptions, and following the same assumptions, after switching the method of detection of infected animals from CIEP to PCR, would unlikely improve the outcomes of the 'detection/removal' control efforts.

In the past, the success rate of culling of CIEP positive animals, was monitored by CIEP again. Here, we monitored the rate of CIEP detectin/removal success by subsequent monitoring of the herds by PCR testing. Furthermore, in one facility, we followed the spread of the virus after culling of rectal swab PCR positive animals. Within less than two months, the incidence rate reached 65%, which can be partly explained by the failure to detect all infected animals, because of the intermittent shedding (Table 2). It could be argued that perhaps combination of CIEP and PCR testing/cull, could achieve better results. We do not believe that is likely, because of the previously stressed severe environmental contamination by this extremely resistant virus. We believe, that while depopulation and repopulation, following cleaning, disinfection, and facility resting can be successful, but the threat of introduction will remain. At this time, we believe that 'on farm breeding for AD resistance' will be a reliable future method of control of AD.

While CIEP testing is highly satisfactory for monitoring of the CIEP status on ranches, it is our view that the recommendation of 'CIEP facilitated detection/removal eradication' should be discontinued as ineffective, and providing false security. It could be argued that 'CIEP detection/removal' is useful even if it does not accomplish eradication, by lowering the incidence of ADV infection on a farm. While this may be the case in a short term, we have documented rather rapid spread of ADV infections on several farms that practiced this method of presumed eradication.

Arguing against attainability of ADV eradiation from farm environment, we have shown that all on all ranches that we looked at, after a round of CIEP detection/removal, the incidence of ADV infection as monitored by PCR rapidly rose. This was rather unexpected, even though PCR can can identify both animals with advanced infection and animals in the early stages, as well as and animals in the incubation period. The high rates of PCR positivity can perhaps be also partly explained by difference of CIEP and PCR regarding their ability to react with the respective targets when they are a part of an immune complex. Immune complex is an important part of the pathogenesis of AD (Porter *at al.*, 1973). When immune complex is formed, only the excess of antibody in the blood is available to the CIEP test, and the complexed antibody is unavailable. Should all of the specific ADV antibody be complexed to the virus, the CIEP test would be negative. This is not the case for the PCR target, the viral DNA. DNA extraction procedure recovers viral DNA with equal effectiveness, whether it is complexed to antibody or not.

Setting aside the indisputably certain role of highly resistant AD parvovirus in the environment in difficulties with CIEP detection/removal, we examined in this work whether PCR on fecal swabs, blood, or saliva, had better potential as a tool for eradiation by 'detection/removal', than CIEP. While it might appear from the results presented (Table 2) that the rate of detection was very high attesting to a diagnostic power of the method, we do not know how this would compare with CIPE, if applied on samples taken at the same time. However, clearly, it is not reasonable to expect that PCR will identify every animal harboring replicating the virus, since the virus viral replication is restricted at the level of the individual cell and can be detected in only a small population of macrophages and follicular dendritic cells (Bloom *et al.*, 1984).

However, when applied under average farm conditions with reasonable biosecurity practices, in our hands neither CIEP nor PCR 'detection/removal' accomplished significant and lasting drop in incidence of the infection. Failures of CIEP are clear from the overall look at the PCR results performed after the CIEP detections/culls only few months prior to PCR testing (Table 2). In spite of the relatively short periods between CIEP and PCR, the percentages of PCR positive animals are quite high (Table 2). We discontinued the attempts to use PCR 'detection/removal' once we obtained the evidence of intermittent shedding in considerable number of animals (Table 3). Thus, in addition to heavy environmental contamination, another significant obstacle to eradication has been only intermittent presence of the virus in the samples (blood and rectal swabs). Thus in addition to the high cost associated with PCR testing, every round of testing leaves behind enough infected animals to maintain significant environmental contamination on the farms.

In conclusion, while either CIEP or PCR can be used to monitor appearance of ADV on farms previously free of the virus, application of either of the test to 'detection/removal eradication' program, at least under the conditions of this study, remained elusive. During the last 35 years, CIEP detection/removal was widely recommended in veterinary and farmers' literature, as effective. Frequent field failures have been blamed on presumed, but often unconfirmed and unprovable breaches of biosecurity. This situation led to difficulties with acquiring ADV free breeding stock, as Canada and USA, buyers must rely on unverifiable sellers' information regarding the ADV herd status. Often the economic incentive, combined with the lack of understanding lead to provision of supposedly ADV free breeding stock that in reality carries the virus into depopulated farms.

References

Bloom M.E., Kanno H., Mori S. and Wolfinbarger J.B., 1994. Aleutian mink disease: puzzles and paradigms. Infectious Agents and Disease 3: 279-301.

Cho H.J. and Ingram D.G., 1972. Antigen and antibody in Aleutian disease in mink. I. Precipitation reaction by agar-gel electrophoresis. Journal of Immunology 108: 555-7.

Cho H.J, and Greenfield J., 1978. Eradication of Aleutian disease of mink by eliminating positive counterimmunoelectrophoresis test reactors. Journal of Clinical Microbiology 7: 18-22.

García M.E., Blanco J.L., Caballero J. and Gargallo-Viola D.J., 2002. Anticoagulants interfere with PCR used to diagnose invasive aspergillosis. Clinical Microbiology 40: 1567-8.

Gregersen J.P. and Roth B., 2012. Inactivation of stable viruses in cell culture facilities by peracetic acid fogging. Biologicals. Mar 15.

Gorham J.R., Leader R.W., Padgett G.A., Burger D. and Henson J.B,. 1965. Some Observations of the Natural Occurrence of Aleutian Disease. In: Slow, latent, and temperate virus infections. Gajdousek D., Carleton G., Gibbs C.J.Jr., and Alpers M. (eds.) NINDB Monograph No. 2. U.S.P.H.S. Publ. No. 1378, pp. 279-285.

Hadlow W.J., Race R.E. and Kennedy R.C., 1983. Comparative pathogenicity of four strains of Aleutian disease virus for pastel and sapphire mink. Infection and Immunity 41: 1016-23.

Henson J.B., Gorham J.R., and Leader R.W., 1962. A field test for Aleutian disease. Preliminary Report. National Fur News 34: 8.

Porter D.D., Larsen A.E. and Porter H.G., 1973. The pathogenesis of Aleutian disease of mink. 3. Immune complex arteritis. American Journal of Pathology 71: 331-44.

Sauerbrei A and Wutzler P. 2009. Testing thermal resistance of viruses. Archives of Virology 154: 115-9.

Stahl M., Stahl J., and Stahl R., 2010 Patent application. Available at: http://www.faqs.org/patents/app/20110086339.

Uttenthal A, Larsen S, Lund E, Bloom ME, Storgård T and Alexandersen S., 1990. Analysis of experimental mink enteritis virus infection in mink: in situ hybridization, serology, and histopathology. Journal of Virology 64: 2768-79.

Part 3. Breeding, genetics and reproduction

Sequencing of the mink genome: plans and perspectives

R. Anistoroaei[1]*, K. Christensen*[1] *and B. Benkel*[2]
[1]*University of Copenhagen, Faculty of Health Sciences, Dept. of Animal and Veterinary Basic Sciences, Division of Animal Genetics and Bioinformatics, Groennegaardsvej 3, 1870 Frederiksberg C,Denmark;* [2]*Nova Scotia Agricultural College, Department of Plant and Animal Sciences, P.O. Box 550,Truro, B2N 5E3 Nova Scotia, Canada; ran@life.ku.dk*

Abstract

The assembled sequence of the whole genome is a must nowadays in any industrially important species. The aim of this research program is to provide the industry and the community with a robust mink genome sequence assembly. It shall be complemented with the identification of widely distributed polymorphic single nucleotide polymorphisms (SNPs) organized into an array designed for undertaking linkage and association studies necessary for mapping, identifying and characterizing genes underlying any specific traits of interest in American mink. Genome assembly based on the sequences of individual clones from the mink Bacterial Artificial Chromosome (BAC) library would result in a draft genome of high quality and accuracy. We explore an alternative approach, called BAC-NGS (from next generation sequencing) that utilizes the BAC library as a solid backbone, complemented with shotgun sequences of the whole genome, for filling gaps for the final assembly. We tested different settings of the new approach in several pilot projects in order to assign the most efficacious approach to sequence assembly. Approximately 15% of the mink genome has so far been generated and assembled by means of this method. This sequencing project represents a long-term investment in the future mink fur industry and farming management, enhancing the development of applications in academic and industrial research.

Keywords: American mink, genome sequencing project, SNPs, BAC-NGS

Introduction

Genomics is one of the corner stones of biology, opening new frontiers in product development and the improvement of animal production while sequence information is becoming essential for understanding the biology of any organism. Whole genome sequence assemblies (WGSA) are essential for understanding the genetics and biology of any organism, since they enable investigations of not only the gene sequences themselves, but also the role of surrounding noncoding DNA and the elucidation of complex/multigenic traits. They also serve as a tool in discovering genomic variation within the species (polymorphisms). Besides the high-quality human genome, assemblies are available for several complex eukaryotic species (e.g. dog-BUILD 3.1, cattle-BUILD 6.1, pig – Sscrofa 9.2; 10.2), but they are commonly known by their users as being incomplete and inaccurate, and in some crucial regions difficult to (re-)sequence. Despite advances in technologies that have accelerated the (re)sequencing complete genomes (Hillier *et al.*, 2008; Wheeler *et al.*, 2008), the need for high-quality assemblies remains (Lewin *et al.*, 2009). Current methods employed for *de novo* genome assemblies are: (1) Classical BAC to BAC method involving Sanger shotgun sequencing of individual BAC clones. This method is time consuming and very expensive in relation to new sequencing standards. (2) Next Generation Sequencing methods (NGS) (e.g. 454 Roche, Illumina) generate large amounts of sequence data from a genome, avoiding the step of molecular cloning in a vector (e.g. BAC library production). However, genome assemblies generated from NGS have so far fallen short of those obtained with the classical approach. NGS methods are regarded as very effective when re-sequencing a genome for which there is an available draft assembly. Whereas data generated with NGS alone can be readily used for a wide range of research investigations, it has proven difficult to use them to generate high-quality *de novo* genome assemblies of vertebrate genomes, which are large and repeat-rich. Although algorithms have been developed for improving the assembling of *de novo* sequences derived from NGS (Gnerre *et al.*, 2010), without a solid, physically anchored scaffold, the accuracy of the assembly (particularly at segmental duplications or long transposable elements) is compromised. To our knowledge, BAC-based physical maps combined with extensive NGS of BACs have not been exploited for WGSA of any mammalian species.

Additionally, development of specific SNPs is very important for fine mapping of traits and for comparing regions of the genome between cohorts in genome wide association studies and investigations determining genetic diversity and phylogenetic relations. They are highly abundant and randomly distributed throughout the genome and because they are bi-allelic, they are easily assayed. SNP maps are used for identification of genes (aka QTLs) underlying complex traits – areas in which linkage maps have generally proven to be ineffective. SNP arrays and, in some cases, SNP chips (e.g. human – Affymetrix; mouse – Illumina; pig – Illumina) are available for most domesticated species (e.g. dog – Karlsson *et al.*, 2007; horse – Chowdhary and Raudsepp, 2008; cattle – Matukumalli *et al.*, 2009) but not for mink.

Fur animal breeding raises ethical and economic issues, both of which must be addressed at the same time. The most efficient and sustainable way to deal with these issues is by close analysis of information from genomics, proteomics, transcriptomics, non-coding DNA, etc. This information can be obtained by generating a complete, robust, accessible and manageable whole genome sequence assembly (WGSA) supplemented with a fine map of single nucleotide polymorphisms (SNPs) for the American mink (*Neovison vison*). These are the most useful tools researchers from all the biological fields investigating a species can have at their disposal.

This project has been designed to test the novel BAC-NGS hybrid method in order to facilitate the building of a cheap, sufficiently annotated and reliable WGSA for the American mink which ultimately shall be complemented with the identification of widely distributed SNP markers organized into an array (a high resolution genetic map).

Materials and methods

DNA

DNA extraction, purification, libraries preparation and sequencing of the BACs (from the CHORI-231) were performed in different settings both in our laboratories and by outsourcing parts of the pilot-projects. Details of these settings are proprietary. Paired-end libraries were prepared for the Illumina HiSeq2000 platform and reads of 100 n were analysed.

Biocomputing

Gradual *de novo* assembly for sequences of each BAC batch was performed using Velvet (Zerbino and Birney, 2008) and SOAPdenovo (Li *et al.*, 2010) software platforms. They were optimized with respect to processing speed and other settings throughout the development (such as the possibility of connecting contig nodes by the pair ends). GapFiller and PHRAP were employed for the assembly and gap filling at this level. Sets of obtained contigs were organized on a gapped scaffold along the genome. Given that the size of a BAC is 170 kb (Anistoroaei *et al.*, 2011), the number of BACs being sequenced from a specific pool was estimated based on their relative genomic position against the dog genome. Additionally, the existing ZooFISH-based mink/dog/human data (Graphodatsky *et al.*, 2000) and the existing genetic and physical maps (Anistoroaei *et al.*, 2009) facilitated the positioning of the individually assembled contig sequences along the mink chromosomes by linearity (Anistoroaei *et al.*, 2011), indicating breakpoints between mink and human resp. mink and dog.

Results and discussion

The novel strategy *BAC-NGS*, tested in several different pilot projects has enabled the assembly of approximately 15% of the mink genome at various locations as derived from the BAC clones sequencing and assembling. The difference in this BAC-NGS approach compared to NGS *de novo* whole genome sequencing strategies is that the sequencing and *de novo* assembling is performed in parts. By using the CHORI-231 BAC library (Anistoroaei *et al.*, 2011), each part will include a selected number of clones simultaneously NGS analysed at individual iterations, thereby facilitating the working with the data. Another substantial advantage is that the 'per batch' assembling is based on one strand of DNA only (BAC inserts are mono-stranded) thus avoiding the ambiguities given by the heterozygous strands of DNA in a stringent assembling process using genomic DNA as the starting point. Compared to the classic BAC-to-BAC methodology, the method tested here is much cheaper and faster, and the end product will be at least of equal quality. The derived 'per batch' contigs sequences were gradually assembled and integrated. Because the human genome is fully sequenced and comparatively well annotated, we focused on comparison between human and mink. Comparative studies were also performed using the dog genome sequences since is the species phylogenetically closest to mink, that has an assembled genome of an agreeable quality (BUILD 3.1).

The optimal hybrid approach to be utilized for further sequencing of the American mink genome is not yet readily available for publication but the schematic representation of the approach to be embraced is schematized in Figure 1.

Relevance and perspectives of the project

Method and reference

A cheap, resource efficient and at the same time high-quality method for mink WGSA can serve as a model for assemblies and employment in other species. In addition, existing BAC libraries available for several other species could be utilized for WGSA by the BAC-NGS method.

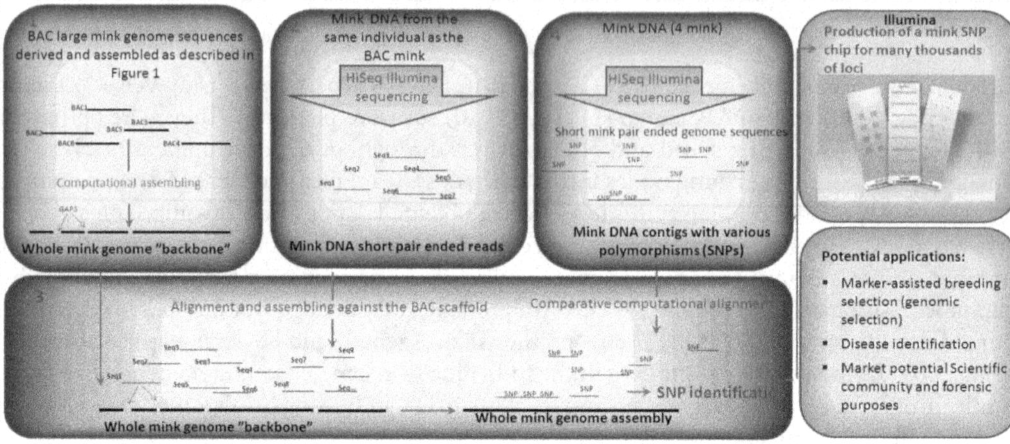

Figure 1. Representation of essential steps (dark frames) taken for producing a WGSA and an array of SNPs for the mink genome. Light frames represent the long term goal of such a project.

Fine association and mapping

An array of SNPs supported by the WGSA is of considerable importance for linkage and genome-wide association studies necessary for identifying genes responsible for mono/poly genic traits and/or diseases and for defining the molecular variation between them.

Characterization of genes

Characterization of genes of interest, their allelic variants and functional investigation will be approached by combined methods utilizing the newly generated genome sequences, which will provide access to the genes for the investigated traits.

Exploring and exploiting the uniqueness of the species

Being able to identify specific and unique genes of mink as well as their complexity would be of interest for biological and medical investigations (e.g. metabolism, immunity), as well as for mink breeding. The mink nomenclature for fur colors differs from that of other animals (Nes *et al.*, 1988), thus understanding its background, would represent an important contribution to explaining the genetics of color.

Serving the industry and animal well-being

WGSA is critically important for positional cloning and identification of the QTLs that were recently identified in the mink genome (Thirstrup *et al.*, 2012). Although positional cloning by interspecific extrapolation is in some cases possible without WGSA, the assembly will greatly accelerate the pace of important research discoveries. It will facilitate implementing sustainable breeding programs and genomic selection. In the farming conditions, traits such as aggressivity have a direct link to the animal welfare and general stress. They are genetically determined and seem to show variation (Cagan – personal communication). Thus, profiling the specific genes/traits can enable selection of suitable animals that are more adapted. It can also aid ecological studies in protecting wild life. In regard to the investigation of genes of interest, combined strategies can be used: (1) exploitation of the previously established maps substantially supported by the newly generated SNP markers and array to perform fine mapping, association studies and genome scan in informative families to identify regions and/or genes involved in the development of phenotypic variation; (2) exploitation of functional and comparative knowledge about genes and (3) further investigations of the individual genes and their functional background utilizing the WGSA like in all the other eukaryotic large genome projects.

Additional resources

A solid WGSA (with functional annotation) of the mink will enable development of resources for genomic and proteomic research. For example, research on the biological course of Aleutian Disease Virus (ADV), the most vicious disease in mink, and the development of host immunity to infections will be aided greatly by the mink genome sequence information to be generated in this project. It will enable the development of molecular tools (e.g. expression arrays, proteins, antibodies) that are virtually non-existent at present, in an expeditious manner.

Surrogate systems for human experimentation

WGSA can provide investigational potential for disease models such as deafness (*unpublished data*) or Chédiak-Higashi syndrome (Anistoroaei *et al. in press*), which are prevalent on some commercial mink farms.

Forensics

Given the increasing number of robbery cases of furs from the farms, a SNP array (eventually supported on a SNP chip) would serve as an easy to use tool in forensic investigations establishing ownership and origin (Gill, 2001).

Expanding our understanding of evolutionary processes in general and of mink evolution in particular

The mink WGSA will enable the examination of differences in the structures of genes, diversity of gene families, and the accumulation and distribution of repetitive elements among genomes at several levels of evolutionary divergence, and to relate these differences to characteristics of gene expression, function, and epigenetic regulation.

To conclude, the long term prospect of a WGSA project is to support sustainable mink breeding programs leading to improved animal welfare and economy in mink farms globally. It will not only maintain and strengthen the mink research it also benefits genetics in general, with implications in all natural science research areas.

References

Anistoroaei, R., Ansari, S., Farid, A., Benkel, B., Karlskov-Mortensen, P. and Christensen, K., 2009. An extended anchored linkage map and virtual mapping for the American mink genome based on homology to human and dog. Genomics 94(3): 204-210.

Anistoroaei, R, ten Hallers, B., Nefedov, M., Christensen, K. and de Jong P., 2011. Construction of an American mink Bacterial Artificial Chromosome (BAC) library and sequencing candidate genes important for the fur industry BMC Genomics 12: 354.

Anistoroaei, R., Krogh, A.K. and Christensen, K., 2012. A frameshift mutation in the LYST gene is responsible for the Aleutian color and the associated Chédiak-Higashi syndrome in American mink, Animal Genetics, in press.

Chowdhary, B.P. and Raudsepp, T., 2008. The Horse Genome Derby: racing from map to whole genome sequence Chromosome Research 16: 109-127.

Gnerre, S., MacCallum, I., Przybylski, D., Ribeiro, F.J., Burton, J.N., Walker, B.J., Sharpe, T., Hall, G., Shea, T.P., Sykes, S., Berlin, A.M., Aird, D., Costello, M., Daza, R., Williams, L., Nicol, R., Gnirke, A., Nusbaum, C., Lander, E.S. and Jaffe, D.B., 2010. High-quality draft assemblies of mammalian genomes from massively parallel sequence data. PNAS 108: 1513-1518.

Gill, P., 2001. An assessment of the utility of single nucleotide polymorphisms (SNPs) for forensic purpose. Int J Legal Med 114: 204-210.

Graphodatsky, A., Yang, F., Serdukova, N., Perelman, P., Zhdanova, N. and Ferguson-Smith, M., 2000. Dog chromosome-specific paints reveal evolutionary inter- and intrachromosomal rearrangements in the American mink and human. Cytogenet Cell Genet 90: 275-278.

Hameister, H., Klett, C., Bruch, J., Dixkens, C., Vogel, W. and Christensen, K., 1997. Zoo-FISH analysis: the American mink (*Mustela vison*) closely resembles the cat karyotype. Chromos Res 5: 5-11.

Hillier, L.W., Marth, G.T., Quinlan, A.R., Dooling, D., Fewell, G., Barnett, D., Fox, P., Glasscock, J.I., Hickenbotham, M., Huang, W., Magrini, V.J., Richt, R.J., Sander, S.N., Stewart, D.A., Stromberg, M., Tsung, E.F., Wylie, T., Schedl, T., Wilson, R.K. and Mardis, E.R., 2008. Whole-genome sequencing and variant discovery in *C. elegans*. Nat Methods. 5: 183-188.

Karlsson, E.K., Baranowska, I., Wade, C.M., Salmon Hillbertz, N.H.C., Zody, M.C., Anderson, N., Biagi, T.M., Patterson, N., Pielberg, G.R., Kulbokas, E.J., Comstock, K.E., Keller, E.T., Mesirov, J.P., von Euler, H., Kampe, O., Hedhammar, A., Lander, E.S., Andersson, G., Andersson, L. and Lindblad-Toh, K., 2007. Efficient mapping of mendelian traits in dogs through genome-wide association. Nature Genetics 39: 1321-1328.

Lewin, H.A., Larkin, D.M., Pontius, J. and O'Brien, S.J., 2009. Every genome sequence needs a good map. Genome Res 19: 1925-1928.

Li, R, Zhu, H, Ruan, J, Qian, W, Fang, X, Shi, Z, Li, Y, Li, S, Shan, G, Kristiansen, K, Li, S, Yang, H, Wang, J and Wang, J., 2010. *De novo* assembly of human genomes with massively parallel short read sequencing. Genome Res. 20: 265-72.

Matukumalli, L.K., Lawley, C.T., Schnabel, R.D., Taylor, J.F., Allan, M.F., Heaton, M.P., O'Connell, J., Moore, S.S., Smith, T.P.L., Sonstegard, T.S. and Van Tassell, C.P., 2009. Development and characterization of a high density SNP genotyping assay for cattle. PLoS ONE 4(4): e5350.

Nes, N.N., Einarsson E.J. and Lohi O., 1988. Beautiful fur animals – and their colour genetics.Scientifur, Hilleroed, Denmark. 271 pp.

Wheeler, D.A., Srinivasan, M., Egholm, M., Shen, Y., Chen, L., McGuire, A., He, W., Chen, Y.-J., Makhijani, V., Roth, G.T., Gomes, X., Tartaro, K., Niazi, F., Turcotte, C.L., Irzyk, G.P., Lupski, J.R., Chinault, C., Song, X.-z., Liu, Y., Yuan, Y., Nazareth, L., Qin, X., Muzny, D.M., Margulies, M., Weinstock, G.M., Gibbs, R.A. and Rothberg, J.M., 2008. The complete genome of an individual by massively parallel DNA sequencing. Nature 452: 872-876.

Zerbino, D. and Birney, E. 2008. Velvet: Algorithms for de novo short read assembly using Bruijn graphs. Genome Res. 18(5): 821-829.

Identifying QTL for fur quality traits in mink (*Neovison vison*)

J.P. Thirstrup[1], R.M. Anistoroaei[2], B. Guldbrandtsen[1], R.S. Labouriau[1], K. Christensen[2], M. Fredholm[2] and V.H. Nielsen[1]
[1]Department of Molecular Biology and Genetics, Faculty of Science and Technology, Aarhus University, Denmark; [2]Department of Basic Animal and Veterinary Sciences, Faculty of Life Science, Copenhagen University, Denmark; janne.thirstrup@agrsci.dk

Abstract

Mapping of quantitative trait loci (QTL) affecting fur quality traits (guard hair length, guard hair thickness and density of wool) was performed in a 3-generation population (F_2-design). In the parental generation, Nordic wild mink were crossed reciprocally with American short nap mink. Twenty one wild mink were mated to 25 short nap mink. In the F_1-generation, 103 mink were mated. 934 mink were obtained in the F_2-generation. The mink were genotyped using 104 microsatellites covering all 14 autosomal chromosomes. The marker spacing was approximately 11 cM. Recordings of fur quality traits were made on all genotyped mink by Kopenhagen Fur. The QTL analyses were performed by least square regression implemented in the software GridQTL. Evidence was found for QTL for the fur quality traits on eight autosomal chromosomes (LOD score >3.0). QTL were detected for guard hair thickness on chromosomes 1, 2, 3, 4, 6, 11 and 13, for guard hair length on chromosomes 2, 3 and 6 and for wool density on chromosome 6, 7 and 13. Similar locations of QTL for guard hair length and guard hair thickness on chromosomes 2, 3 and 6 suggest that these traits are in part under the influence of the same genes.

Keywords: genetics, F_2-design, microsatellites, guard hair, wool.

Introduction

Fur quality is important in mink production. Fur quality is a composite trait, including e.g. skin length, color and purity as well as the structure of guard hair and density of wool. The identification of genes influencing fur quality traits and knowledge of the action of the genes in question provide a useful tool for improving fur quality. In mink, most of the work so far has been focused on identifying coat color genes (Anistoroaei and Christensen 2007, Anistoroaei *et al.*, 2008; Benkel *et al.*, 2009).

The mapping of quantitative trait loci (QTL) is the first step towards the identification of genes and causal mutations that affect traits of importance in agricultural production (Goddard and Hayes 2009). The purpose of QTL-analyses is to identify chromosomal areas, where loci affecting quantitative traits are harbored and provide a solid background for further analysis narrowing the chromosomal regions (Anderson 2001). It also provides the basis for understanding the genetic architecture underlying the characters (Anderson 2001) and for identifying chromosomal variability associated with a particular trait in a population in order to know if it would be possible to exploit this variability in breeding programs (Alfonso and Haley 1998).

A QTL project with the aim of identifying genomic regions harboring genes with effect on fur quality traits is being conducted. Here we present results of a search for QTL affecting guard hair length, guard hair thickness and wool density.

Materials and methods

Animal material

A reciprocal F_2-design between divergent lines was used. The lines were Nordic wild mink and American short nap mink. Mink from the Nordic line were large with coarse and long guard hair. They were brown with reddish wool of poor quality. The line has been selected for large size for several generations. American short nap animals were small with silky and short guard hair and dense wool of good quality. The mink was black, with short head and blunt snouts. The American line was established in 1998 from Patrick × Nova Scotia.

Forty-six mink from the two lines, five males from each line and 16 females from the Nordic line and 20 females from the American line, were selected for the parental generation. To establish the F_1-generation two 'blocks' were made. Short nap males were mated to Nordic wild females (block A) and Nordic wild males were mated to short nap females (block B). Five males were each mated to maximum five females. The F_2-generation was established by mating both within and between blocks. This mating was performed using 31 males and 108 females. To ensure sufficient power for statistical test (Weller *et al.*, 1990) the mink were mated in two succeeding years. It was attempted to minimize relationship between parents. Because of increased litter size in two year old females compared to one year old females, two year old females were selected in the parental generation. Mothers of the males were avoided and neither of the female mink were sibs.

Genotyping

In total 1,083 mink (parental, F_1 and F_2) were genotyped. DNA was extracted from spleens sampled at pelting from all breeding animals in the parental and the F_1-generation and from all animals in the F_2-generation.

Genomic DNA was extracted by the phenol-chloroform method. One hundred and four microsatellite markers positioned with a distance of approximately 11 cM were used for the study. Fluorescence labeling of the PCR amplicon was achieved by labeling the forward primer with FAM, HEX, or TAMRA. The 3-primer methodology (Schuelke *et al.*, 2000) has been employed as a cost-effective and efficient means of fluorescently labeling PCR fragments. Typing was produced on an ABI 3130xl sequencer (Applied Biosystems®). Scoring was conducted automatically using genotyper v. 2.5 (Perkin Elmer®).

Genetic linkage map

The published linkage map (Anistoroaei *et al.*, 2009) was improved with newly developed microsatellite markers using the pedigree information and genotypes from F_1- and F_2-generations. The software Crimap 2.4 (Green *et al.*, 1990) was applied for the analysis. A two-point analysis, with a cut-off threshold of log of odds (LOD) 3, was performed to identify linkage groups. A pair wise recombination fractions and LOD ratios were computed. The FLIPS and BUILD options were employed for assigning the final order of the marker. A total of 104 markers were mapped to 14 linkage groups corresponding to the mink autosomes. The average linkage map spans 1,457.7 centiMorgan (cM) with an average intermarker distance of 11 cM.

Phenotyping

Phenotypes were collected for all genotyped animals. The phenotypic assessments were made at Kopenhagen Fur. All mink from the three generations were phenotyped simultaneously. Twelve phenotypic traits were evaluated. Here, results for guard hair thickness, guard hair length and wool density are presented. Guard hair thickness was measured in four categories from finest (1) to thickest (4). The guard hair length is measured as the hair protruding from the wool. The guard hair length is measured in 7 categories from very short (1), which is also called velvet 3 and is shorter than the wool, to very long (7). The wool is evaluated in 6 categories from best (1), which has the highest wool density to the poorest (6), which has the lowest density of wool. Table 1 shows mean and standard errors for the different traits.

Data-analysis

The QTL analyses were performed by a least square regression method developed by Haley *et al.* (1994) for crosses between outbreed lines, and implemented in the software program GridQTL (Seaton, 2006). The option of fitting a single QTL was used. A significance level of 0.01 was set because of a risk for marginal significance by chance due to the large datasets (Weller *et al.*, 1990). Permutation tests with 1000 permutations were used for estimating the significance level (Churchill and Doerge, 1994). Log of odds threshold (LOD score) was set to 3.7 and suggestive QTL significant at 0.05 level was set to LOD score threshold of 3 based on the aforementioned significance level. A model including adjustments of the effects of sex and year was used. Confidence intervals were estimated using bootstrap analyses with 10,000 iterations (Visscher *et al.*, 1996).

A positive additive effect indicates that alleles from the parental Nordic wild mink increase the trait value on the categorical scale. A positive dominance effect of the same order of magnitude as the additive effect indicates that a heterozygote phenotypic is similar to the wild mink,

Table1. Mean and standard errors (SE) of guard hair thickness, guard hair length and wool density of Nordic wild mink, short nap mink, and F_1- and F_2-mink born in 2006 and 2007, respectively.

	Nordic wild mink	Short nap	F_1	F_2 (2006)	F_2 (2007)
Guard hair thickness	4	1	2.78 (0.06)	2.45 (0.04)	2.29 (0.04)
Guard hair length	6.3 (0.18)	1.43 (0.13)	4.44 (0.06)	4.03 (0.04)	3.85 (0.05)
Density of wool	4.24 (0.13)	2.43 (0.34)	3.05 (0.07)	3.12 (0.05)	3.25 (0.05)

indicating that the alleles originating from the wild mink is dominant to the alleles originating from short nap. A negative dominance effect but of the same order of magnitude as the additive effect indicates that a heterozygote phenotypic is similar to the short nap, indicating that the alleles originating from short nap is dominant to alleles originating from the wild mink. The phenotypic value of the heterozygote depends on the degree of dominance relative to the additive effect. Dominance effect around zero indicates that a heterozygote is phenotypic intermediate between to the parental lines. If the dominant effect is greater or less than the additive there is overdominance. This indicates that the heterozygote has better or worse trait values than either of the parental lines. Modes of gene action are assessed from additive and dominance effects.

Results

We found evidence for QTL affecting guard hair thickness on seven chromosomes, guard hair length on three chromosomes and wool density on three chromosomes. In all, QTL affecting the traits were found on eight out of the 14 mink autosomes.

QTL affecting guard hair thickness were found on chromosome 1, 2, 3, 4, 6, 11 and 13 (Table 2). QTL on chromosome 1, 4, 11 and 13 did not reach significance level of 0.01, corresponding to a LOD score >3.7, but are suggestive at significance level of 0.05, corresponding to a LOD score >3. Alleles originating from the wild mink increase guard hair thickness (shown by positive additive effect). On chromosomes 1, 2, 3, 4 and 11 there is no dominance effect and the alleles act additively. The additive alleles from the wild mink increase the trait by a factor 0.14 (0.08) – 0.30 (0.05) on the scale. On chromosome 6, the allele from the Nordic wild mink acts dominantly to the short nap allele (shown by a dominant effect in the order of magnitude as the additive effect). On chromosome 13, the results show overdominance (shown by a larger dominance effect relative to the additive effect). Thus the heterozygote has thicker guard hair.

Table 2. Chromosome number, chromosomal location with 95% confidence intervals (CI), Lod score, additive and dominance effect with standard errors (SE) for guard hair thickness.

Chromosome	Location cM (95% CI)	Lod-score[1]	Additive effect (SE)	Dominance effect (SE)
1	0 (0.0-100.0)	3.85**	0.22 (0.05)	0.0 (0.09)
2	58 (10.0-120.0)	6.48**	0.20 (0.05)	-0.06 (0.08)
3	9 (2.0-79.0)	7.29**	0.30 (0.05)	-0.03 (0.08)
4	49 (0.0-86.0)	3.12*	0.17 (0.05)	0.08 (0.08)
6	36 (19.0-53.0)	6.55**	0.27 (0.06)	0.21 (0.08)
11	78 (0.0-81.0)	3.48*	0.19 (0.05)	-0.08 (0.07)
13	27 (0.0-63.0)	3.68*	0.14 (0.08)	0.33 (0.09)

[1] * Significant at 0.05 level; ** Significant at 0.01 level.

QTL affecting guard hair length were detected on chromosome 2, 3 and 6 (Table 3). Alleles from the Nordic wild mink add with additive effect in the range of 0.22 (0.06) – 0.31 (0.05). On chromosomes 2 and 3 there is incomplete dominance where heterozygote are phenotypic more alike short nap. On chromosome 6, the alleles originating from the wild mink are dominant to the short nap alleles.

QTL affecting wool density were found on chromosomes 6, 7 and 13 (Table 4). QTL on chromosome 7 does not reach significance level of 0.01, but is suggestive at 0.05 level. Alleles from the wild mink contribute with additive effect in the range of 0.05 (0.06) – 0.28 (0.09). Alleles from the wild mink on chromosomes 7 and 13 are dominant to short nap alleles. On chromosome 6 there is overdominance, where the heterozygotes have much lower density of wool. There is practically no additive effect (0.05 (0.06)) of this allele on chromosome 6.

The QTL on chromosomes 2, 3 and 6 affecting guard hair length and thickness were detected at the same location or in very close proximity. There are more QTL affecting guard hair thickness compared to guard hair length. Guard hair thickness is also influenced by QTL on chromosome 1, 4, 11 and 13. An analysis of correlation between the traits (Spearmans correlation test) showed very high phenotypic correlation between guard hair length and thickness ($r=0.69$, $p=7.5\ e^{-31}$).

Table 3. Chromosome number, chromosomal location with 95% confidence intervals (CI), Lod score, additive and dominance effect with standard errors (SE) for guard hair length.

Chromosome	Location cM (95% CI)	Lod-score[1]	Additive effect (SE)	Dominance effect (SE)
2	58 (11.0-60.0)	9.26**	0.31 (0.05)	-0.11 (0.09)
3	9 (0.0-79.0)	4.23**	0.22 (0.06)	-0.14 (0.09)
6	36 (19.0-53.0)	7.38**	0.24 (0.06)	0.27 (0.08)

[1] ** Significant at 0.01 level.

Table 4. Chromosome number, chromosomal location with 95% confidence intervals (CI), Lod score, additive and dominance effect with standard errors (SE) for wool density.

Chromosome	Location cM (95% CI)	Lod-score[1]	Additive effect (SE)	Dominance effect (SE)
6	23 (11.0-40.0)	5.34**	0.05 (0.06)	0.31 (0.08)
7	46 (36.0-78.0)	3.18*	0.26 (0.09)	0.32 (0.13)
13	83 (0.0-83.0)	3.97**	0.28 (0.09)	0.27 (0.12)

[1] ** Significant at 0.01 level.

Discussion

We found QTL affecting guard hair length, guard hair thickness and wool density on 8 of the 14 mink autosomes. QTL affecting guard hair thickness were detected on chromosomes 1, 2, 3, 4, 6, 11 and 13. QTL affecting guard length were found on chromosomes 2, 3 and 6. QTL affecting wool density were found on chromosomes 6, 7 and 13. As expected, the alleles that increase the phenotypic values came from the Nordic wild mink. Alleles originating from the Nordic wild mink provide longer and thicker guard hair and lower density of wool. All of the QTL affecting guard hair length are located at the same positions or in very close proximity to QTL effecting thickness. These findings suggest that the traits are under influence of the same genes or that some of the genes affecting the traits might be linked. An analysis of correlation between the traits showed very high phenotypic correlation between guard hair length and thickness. This might have implications on selection programs, as it does not seem possible to select for long guard hair without also selecting for thick guard hair.

Cadieu *et al.* (2009) found that very few genes control guard hair length, coarseness and curly hair in dogs. We found that several QTL affect guard hair length and guard hair thickness. It is believed that in dogs short hair is dominant to long hair and that long hair is caused by mutations in the FGF5 gene (e.g. Housley and Venta 2006). The present study shows that the heterozygote are more alike the short nap mink, indicating that the alleles providing short hair are partly dominant to long hair. The FGF5 gene in dogs is located on chromosome 32, homolog to chromosome 11 in mink. We did not find a QTL at 5% significant level on chromosome 11 but a suggestive QTL (LOD = 2.58). In dogs RSPO2 on chromosome 13 is responsible for coarseness of the fur (Cadieu *et al.*, 2009). The position of this gene on chromosome 13 is homolog to the p arm of chromosome 4 in mink. We detected similarly a QTL affecting guard hair thickness on chromosome 4 (LOD = 3.48). In our analysis we found that that some of the QTL affecting guard hair length and guard hair thickness are located close to each other, or might even be the same genes. In the study of Cadieu *et al.* (2009) they found the genes affecting guard hair length and coarseness on different chromosomes. Thus our findings suggest a different genetic background for guard hair in mink and dogs.

References

Alfonso L. and Haley C.S., 1998. Power of different F_2 schemes for QTL detection in livestock. Animal Science 66: 1-8.

Anderson L., 2001. Genetic dissection of phenotypic diversity in farm animals. Nature Reviews Genetics 2: 130-138.

Anistoroaei R., Ansari S., Farid H., Benkel B., Karlskov-Mortensen P. and Christensen K., 2009. An extended anchored linkage map and virtual mapping for the American mink genome based on homology to human and dog. Genomics 94: 204-210.

Anistoroaei R. and Christensen K., 2007. Mapping the silver gene in mink and its association with the dilution gene in dog. Cytogenetic and Genome research 166: 316-618.

Anistoroaei R., Fredholm M., Kristensen K. and Leeb T., 2008. Albinism in the American mink (*Neovison vison*) is associated with a thyrosinase nonsense mutation. Animal Genetics 39: 645-648.

Benkel, B.F., Rouvinen-Watt, K., Farid, H. and Anistoroaei, R., 2009. Molecular characterization of the Himalayan mink. Mammalian Genome 20: 256-259.

Cadieu E., Neff M.W., Quignon P., Walsh K., Chase K., Parker. H.G., VonHoldt B.M., Rhue A., Boyko A., Byers A., Wong A., Mosher D.S., Elkahloun A.G., Spady T.C., André C., Lark K.G., Cargill M., Bustamante C.D., Wayne R.K. and Ostrander E.A., 2009. Coat Variation in the Domestic Dog Is Governed by Variants in Three Genes. Science 326: 150-153.

Churchill, G.A. and Dorge, R.W., 1994. Empirical threshold values for quantative trait mapping. Genetics 138: 963-971.

Goddard M.E. and Hayes B., 2009. Mapping genes for complex traits in domestic animals and their use in breeding programmes. Nature Reviews Genetics 10: 381-391.

Green P., Falls K. and Crooks S., 1990. Documentation for CRIMAP Version 2.4, Washington University School of Medicine, St Louis, MO, USA.

Haley C.S., Knott S.A. and Elsen J-M., 1994. Mapping Quantitative Trait Loci in Crosses Between Outbred Lines Using Least Squares. Genetics 136: 1195-1207.

Housley D.J.E. and Venta P.J. (2006). The long and the short of it: evidence that FGF5 is a major determinant of canine 'hair'-itability. Animal Genetics. 37: 309-315.

Schuelke M., 2000. An economic method for the fluorescent labelling of PCR fragments, Nature Biotechnology 18: 233-234.

Seaton G., Haley C.S., Knott S.A., Kearsey M. and Visscher P.M., 2002. QTL Express: mapping quantitative trait loci in simple and complex pedigrees. Bioinformatics 18: 339-340.

Weller J.I., Kasshi Y. and Soller M., 1990. Power of Daughter and granddaughter Designs for determining Linkage Between Marker Loci and Quantitative trait Loci in Dairy Cattle. Journal of Dairy Science 73: 2525-2537.

Visscher P.M., Thompsen R. and Haley C.S., 1996. Confidence intervals for QTL locations using bootstrapping. Genetics 143: 1013-1020.

Resolving the enigma of embryonic diapause, a forty-year scientific journey

B.D. Murphy
Centre de recherche en reproduction animale, Faculté de médecine vétérinaire, Université de Montréal, St-Hyacinthe, Québec J2S 7C6, Canada; bruce.d.murphy@umontreal.ca

Abstract

Embryonic diapause is an evolutionary strategy by which a reversible arrest in embryo development occurs. In its two forms, facultative and obligate, it assures that offspring are born when optimal maternal and environmental conditions are present to ensure maximal survival. We have explored obligate delayed implantation in the mink over four decades, first by evaluation of the environmental regulation, then by determination of the pituitary factors that maintain diapause and provoke implantation followed by evaluation of the ovarian contribution to the process. As the uterine environment is the proximal regulator of diapause, we employed a strategy of global gene analysis to discover differentially expressed pathways during embryo arrest and reactivation. These trials revealed that the synthesis of polyamines was increased in the uterus with reactivation of the embryo *in vivo*. Subsequent experiments demonstrated that the polyamine, putrescine, was capable of inducing escape of the embryo from obligate diapause, providing strong evidence that the paucity of polyamines induces developmental arrest and reactivation is coupled to renewed uterine and/embryonic synthesis of these polycations.

Keywords: embryonic diapause, regulation, uterine environment, polyamines, putrescine

Introduction

I first became enthused about obligate embryonic diapause as a graduate student interested in reproductive biology of wild species. This phenomenon, characterized by arrest in embryo development at the blastocyst stage, is found in several orders of mammals, but is particularly evident in the *Mustelidae* and *Phocidae* (Lopes *et al.*, 2004). The proposed evolutionary significance of embryonic diapause is to uncouple breeding and parturition, so that each occurs at the time of year most favorable for reproductive success. The trait appears to have arising independently in the marsupial and placental mammals, as many of the kangaroos display lactationally and seasonally-induced embryonic diapause (Renfree and Shaw, 2000). A recent report suggests that embryos of some traditionally non-diapausing species can undergo reversible developmental arrest (Ptak *et al.*, 2012.), suggesting that the potential for this condition is more widespread than previously believed.

The mechanism of environmental regulation of the duration of diapause

My initial plan was to study diapause in a species displaying a long period of developmental arrest, the long-tailed weasel (*Mustela frenata*), a common carnivore in western Canada. After a frustrating winter of live trapping in which I captured only five animals, three of which were males, it was clear that the ranch mink (*Neovison vison*) was a more tractable model. The first experiments focused on exploration of the role of environmental light and its mechanisms by which it acted. The results demonstrated that short days (<12 h of light) maintained the embryos in diapause, and their emergence was associated with photoperiod of the vernal equinox (Murphy and James, 1974). Intact photoreceptors were required, and, denervation of the pineal gland provided the first clue of the essential role of melatonin in maintaining diapause. This finding was confirmed by other investigators (Bonnefond *et al.*, 1990), and by our much later experiments with melatonin implants (Murphy *et al.*, 1990).

The essential role of the pituitary

The next question that we addressed was the how the photoperiodic signal was transmitted, i.e. what neural and pituitary hormones were involved in the termination of embryonic diapause in the mink. The first trials clearly demonstrated that the hypophysis was essential for the activation of the corpus luteum and the continuation of gestation (Murphy and Moger, 1977). The strategy of hypophysectomy was then used to address the problem in the domestic ferret (*Mustela putorius furo*), a closely related species that does not display diapause (Murphy, 1979). Luteal function and embryo implantation was rescued in hypophysectomized ferrets by daily administration of only one pituitary hormone, prolactin. Pat Concannon first showed that prolactin regulated diapause advancing implantation in mink with prolactin injections and by treatment by treatment with the dopamine agonist, bromocryptine, to inhibit endogenous prolactin secretion (Papke *et al.*, 1980). Our subsequent investigations employing hypophysectomy demonstrated that in the mink (Murphy *et al.*, 1981), as in the ferret (Murphy 1979), prolactin alone could maintain the corpus luteum and support gestation. This finding was further confirmed by studies showing that dopamine antagonists induced precocious prolactin secretion, activated the corpus luteum and terminated diapause (Murphy, 1983). Later investigations (Murphy *et al.*, 1990), alluded to above, demonstrated that the role of melatonin. Thus, the short photoperiod that precedes the vernal equinox maintains diapause by inducing prolonged secretion of melatonin that, in turn, inhibits prolactin secretion from the pituitary.

The ovarian factors regulating diapause and its termination

In the mink model, ovulations and matings can occur as frequently as one week apart, the embryos then enter into diapause and all can contribute to the consequent litter (Lopes *et al.*, 2004). Following ovulation, corpora lutea are formed, and low levels of progesterone are secreted (Mead, 1993). During diapause, there is reduction in luteal cell volumes, and, in contrast to other species, the luteal cells are not terminally differentiated, i.e. they retain their mitotic potential (Douglas *et al.*, 1998). Prolactin is directly lutotropic in the mustelid CL, both *in vivo* (Murphy, 1979; Murphy *et al.*, 1981) and *in vitro* (McKibbin *et al.*, 1984). It reactivates the dormant corpus luteum and induces rapid increases in progesterone synthesis (Murphy *et al.*, 1993). Concurrent increase in the expression of prolactin receptors in luteal cells occurs, augmenting the abundance of the steroidogenic enzymes that drive progesterone formation. This differs from the role of prolactin in rodent corpora lutea where prolactin maintains luteinizing hormone receptors and inhibits the conversion of progesterone to less biologically active metabolites (Stocco *et al.*, 2007).

The direct luteal support of the steroidogenesis in mustelids appears to be a phenomenon unique to carnivores.

In rodents, a facultative diapause can be induced by ovariectomy and progesterone replacement, and emergence is induced by estrogen administration (Lopes *et al.*, 2004). This paradigm of treatment to terminate diapause has been tried and proven unsuccessful many times in mink. Our studies of luteal autotransplants demonstrated why this regimen of steroid treatment fails, there is a factor (or factors) emanating from the corpus luteum, other than steroid hormones that is required to induce implantation (Murphy *et al.*, 1983). The identity of this factor remains obscure.

The uterine control of diapause

Having traced the photoperiodic and hypophyseal, and to a certain extent, the ovarian effectors of diapause and reactivation, we turned our focus to understanding the interactions between the uterus and the embryo. In a classic study, M.C. Chang (Chang, 1968) reciprocally transferred of mink blastocysts in delay to the ferret uterus and ferret blastocysts, that do not display delay, to the mink uterus. The ferret embryos entered diapause and the mink embryos emerged from quiescence, demonstrating that diapause was not inherent to the embryo, rather, was induced by the uterus. Recently, there has been a remarkable extension of this finding, in which sheep embryos, transplanted to the mouse uterus in the diapause state, entered a state of reversible developmental quiescence (Ptak *et al.*, 2012).

To address the uterine control of diapause, we established an *in vitro* model by immortalizing mink uterine epithelial and stromal cells for co-culture with mink embryos in diapause (Moreau *et al.*, 1995). This paradigm resulted in extended survival of the embryos *in vitro*, and, in some cases, escape from their quiescent state. Using a candidate gene approach, we then explored the expression patterns of genes expected to be involved in diapause and implantation. Our studies showed concurrent elevations in leukemic inhibitory factor (LIF) (Song, *et al.*, 1998), cyclo-oxygenenase-2 (Song, *et al.*, 1998) and vascular endothelial growth factors and their receptors (Lopes *et al.*, 2003). In the meantime we characterized the emergence from diapause *in vivo* (Desmarais *et al.*, 2004), and attempted without success, to reactivate embryos *in vitro* with candidate factors (epidermal growth factor and prostaglandins, among others) (unpublished data).

The maturation of molecular technology provided a new experimental strategy, global gene analysis. The mink genome had not (and to date, has not) been sequenced, and no cognate microarray chips were available at that time. We therefore employed the subtractive suppressive hybridization (SSH) technique (Lefevre and Murphy 2009) to compare gene expression patterns between the embryo in diapause and the embryo following activation *in vivo*. Uterine gene expression was likewise compared (Lefevre *et al.*, 2011a). The consequent analysis revealed >300 genes in both the embryo and the uterus were differentially expressed by more than two-fold between diapause and activation. Our interest in the uterus was in secretory proteins, based on the concept that these may be responsible for reactivating the embryo (Lefevre *et al.*, 2011a). We therefore validated their differential expression in the uterus at the transcript level and by immunocytochemistry. Among these was growth and differentiation factor-3 (GDF-3) and secreted protein acidic and cysteine-rich (SPARC) (Lefevre *et al.*, 2011a). Incubation of embryos in diapause with these factors alone did not reinitiate development.

Polyamines: the key regulators of obligate embryonic diapause

Among the genes upregulated in activated uterus was a cluster that collectively coded for net increase in the quantity ubiquitous cationic compounds known as polyamines, spermine, spermadine and putrescine (Lefevre *et al.*, 2011b,c). These included the rate-limiting factor in conversion of ornithine to putrescine, ornithine decarboxylase (ODC), antienzyme inhibitor-1 (AZIN-1) that inhibits polyamine degradation, and spermidine-spermine N-acetyltransferse (SAT1), that, as its name suggests, is involved in the interconversion of polyamines (Lefevre *et al.*, 2011b,c). Further evidence of the importance of the polyamine pathway came from demonstration that ODC expression was upregulated i in the uterine endometrial stroma at the time the embryo was activated. Uterine extracts following activation were shown to have more than two-fold the concentration of putrescine relative to extracts from the delay uterus. Further, inhibition the action of ODC by treatment with a-difluoromethylornithine produced a reversible arrest in development of activated embryos *in vivo* and in mink trophoblast cells *in vitro* (Lefevre *et al.*, 2011b). The most definitive evidence, so far unpublished, came from experiments whereby treatment *in vitro* of embryos recovered in diapause with 0.5-1.0 μM putrescine induced resumption of development. The reactivation of the embryo phenocopied the events with reactivation *in vivo* (Desmarais *et al.*, 2004), including proliferation of both the trophoblast and inner cell mass cells, embryo expansion and, in some cases, embryo hatching (Moreau *et al.*, 1995). Together, these results provide strong evidence that the embryo is maintained in diapause by the paucity of polyamines of uterine provenance.

Conclusions and future directions

Over a long investigational journey, several facets of the regulation of obligate embryonic diapause in the mink have been revealed. The new findings identifying polyamines as the factor that is required for reactivation of the embryo in diapause provide in depth insight in to the mechanisms of uterine control. The evidence suggests that the maintenance of the diapause state is due to the absence of a factor necessary for continued development of the embryo. There is much left to learn, and investigation into how hypophyseal and ovarian factors regulate the expression of the polyamine gene cluster in the uterus is at the top of the list. There is ample evidence that the prolactin receptor is expressed in the endometrium and that this expression increases as the embryo in diapause is reactivated, but a direct role for prolactin in the uterus has not been demonstrated. The embryo itself expresses ODC at the time of activation, and this may be a mechanism to amplify the uterine polyamine signal. The molecular tools now exist to determine the elusive luteal factor required for termination of diapause. As these and many more questions remain unanswered, thus, the journey continues.

Acknowledgements

All of the original research on obligate diapause in my laboratory has been supported by the generous and continuous support of the Natural Sciences and Engineering Research Council of Canada. A number of talented students and postdoctoral fellows, dedicated technicians, undergraduate research students and colleagues participated in these investigations, and I gratefully acknowledge their invaluable contributions.

References

Bonnefond, C., Martinet, L. and Monnerie, R., 1990. Effects of Timed Melatonin Infusions and Lesions of the Suprachiasmatic Nuclei on Prolactin and Progesterone Secretions in Pregnant or Pseudopregnant Mink, *Mustela vison.* Journal of Neuroendocrinology 2: 583-91.

Chang, M.C, 1968. Reciprocal insemination and egg transfer between ferrets and mink. Journal of Experimental Zoology 168: 49-59.

Desmarais J.A., Bordignon., V., Lopes, F.L., Smith, L.C. and Murphy, B.D., 2004. The escape of the mink embryo from obligate diapause. Biology of Reproduction 70: 662-70.

Douglas DA, Song JH, Moreau GM and Murphy BD, 1998. Differentiation of the corpus luteum of the mink, *Mustela vison*: mitogenic and steroidogenic potential of luteal cells from embryonic diapause and postimplantation gestation. Biology of Reproduction 58: 1163-9.

Lefevre, P.L. and Murphy, B.D., 2009. Differential gene expression in the uterus and blastocyst during the reactivation of embryo development in a model of delayed implantation. Methods in Molecular Biology 550: 11-61.

Lefevre P.L., Palin, M.F., Beaudry, D., Dobias-Goff, M., Desmarais, J.A., Llerena Vargas, E. and Murphy, B.D. 2011a. Uterine signaling at the emergence of the embryo from obligate diapause. American Journal of Physiology-Endocrinology and Metabolism 300: E800-8.

Lefevre, P.L., Palin, M.F., Chen, G., Turecki, G. and Murphy, B.D., 2011b. Polyamines are implicated in the emergence of the embryo from obligate diapause. Endocrinology 152: 1626-1639.

Lefevre, P.L., Palin, M.F. and Murphy, B.D. 2011c. Polyamines on the reproductive landscape. Endocrine Reviews 32: 694-712.

Lopes, F.L., Desmarais, J., Gevry, N.Y., Ledoux, S. and Murphy, B.D., 2003. Expression of vascular endothelial growth factor isoforms and receptors Flt-1 and KDR during the peri-implantation period in the mink, *Mustela vison.* Biology of Reproduction 68: 1926-33.

Lopes, F.L., Desmarais, J.A. and Murphy, B.D., 2004. Embryonic diapause and its regulation. Reproduction 128: 669-78.

McKibbin, P.E., Rajkumar K. and Murphy, B.D., 1984. Role of lipoproteins and prolactin in luteal function in the ferret. Biology of Reproduction 30:1160-6.

Mead, R.A.,1993. Embryonic diapause in vertebrates. Journal of Experimental Zoology 266: 629-41.

Moreau, G.M., Arslan, A., Douglas, D.A., Song, J., Smith, L.C. and Murphy, B.D., 1995. Development of immortalized endometrial epithelial and stromal cell lines from the mink, *Mustela vison.* uterus and their effects on the survival *in vitro* of mink blastocysts in obligate diapause. Biology of Reproduction 53: 511-8.

Murphy, B.D., 1979. The role of prolactin in implantation and luteal maintenance in the ferret. Biology of Reproduction 21: 517-21.

Murphy, B.D., 1983. Precocious induction of luteal activation and termination of delayed implantation in mink with the dopamine antagonist pimozide. Biology of Reproduction 29: 658-62.

Murphy, B.D., Concannon, P.W., Travis, H.F. and Hansel, W., 1981. Prolactin: the hypophyseal factor that terminates embryonic diapause in mink. Biology of Reproduction 25: 487-91.

Murphy, B.D., DiGregorio, G.B., Douglas, D.A. and Gonzalez-Reyna, A., 1990. Interactions between melatonin and prolactin during gestation in mink *(Mustela vison)*. Journal of Reproduction and Fertility 89: 423-9.

Murphy, B.D. and James, D.A., 1974. The effects of light and sympathetic innervation to the head on nidation in mink. Journal of Experimental Zoology 187: 267-76.

Murphy, B.D., Mead, R.A. and McKibbin, P.E., 1983. Luteal contribution to the termination of preimplantation delay in mink. Biology of Reproduction 28: 497-503.

Murphy, B.D. and Moger, W.H., 1977. Progestins of mink gestation: the effects of hypophysectomy. Endocrine Research Communications 4: 45-60.

Murphy, B.D., Rajkumar, K., Gonzalez-Reyna, A and Silversides, D.W., 1993. Control of luteal function in the mink, (*Mustela vison*). Journal of Reproduction and Fertility Supplement 47: 181-8.

Papke, R.L., Concannon, P.W., Travis, H.F. and Hansel. W., 1980. Control of luteal function and implantation in the mink by prolactin. Journal of Animal Science 50: 1102-7.

Ptak, G.E., Tacconi, E., Czernik, M., Toschi, P., Modlinski, J.A and Loi, P., 2012. Embryonic diapause is conserved across mammals. Public Library of Science One 7: e33027.

Renfree, M.B. and Shaw, G., 2000. Diapause. Annual Review of Physiology 62: 353-75.

Song, J.H., Houde, A. and Murphy, B.D., 1998. Cloning of leukemia inhibitory factor, (LIF) and its expression in the uterus during embryonic diapause and implantation in the mink *(Mustela vison)*. Molecular Reproduction and Development 51: 13-21.

Song, J.H., Sirois, J., Houde, A. and Murphy, B.D., 1998. Cloning, developmental expression, and immunohistochemistry of cyclooxygenase 2 in the endometrium during embryo implantation and gestation in the mink *(Mustela vison)*. Endocrinology 139: 3629-36.

Stocco, C., Telleria, C. and Gibori, G., 2007. The molecular control of corpus luteum formation, function, and regression. Endocrine Reviews 28: 117-49.

Application of sequence-characterized amplified regions for detection of self-biting in the blue fox

E. Ren[1], Z. Liu[2], S. Wang[2], X. Xing[2] and F. Yang[2]
[1]Shijiazhuang Academy of Agricultural and Forestry Science, Shijiazhuang 050041, China; [2]Institute of Special Economic Animal and Plant Sciences, The Chinese Academy of Agricultural Sciences, Jilin 132109, China; yangfh@126.com

Abstract

The present research aimed to use molecular markers to study rapid detection of self-biting disease. We used RAPD marker to assess the level of genetic differentiation between self-biting and healthy blue fox group. A single RAPD marker S414 amplified a specific band of 768bp in self-biting of blue fox, which was designated as SRS414. The sequences of the bands exhibited 70% similarity to *Macaca mulatta* MHC class I region. The RAPD marker was converted into SCAR marker via cloning and sequencing of RAPD amplification and designing of 18-mer oligonuleotide primers. The RAPD and SCAR marker were validated in two groups respectively. χ^2 test showed significant difference ($P<0.05$) between the detection rate of the two groups. This indicated that SCAR-S414 can be used as a positive marker to detect self-biting in farm. Furthermore, the finding that self-biting linked with MHC genes has implications for the mechanism of this disease.

Keywords: blue fox, self-biting, random amplified polymorphic DNA

Introduction

Self-biting disease is a chronic disease that is quite common in carnivores with long tail, such as in fox, mink and racoon dog. They showed periodic excitation and bit themselves on certain part which was normally tail, hip, hindlimb and abdomen. It can degrade the fur quality and infuluence the fox breed industry. Self-biting disease is found in captive animals but rare in natural (Masion, 1991). The cage environment of farmed fur animals, sometimes considered as a poor environment. Although more attention is payed on environment, the role of the genetic origins cannot be overlooked either. A positive correlation between the occurrence of self-biting in parents and their offspring has been indicated in mink and other species (Hansen, 1993).

Genomics generally proceeds along a specific pathway of investigations in order to identify genes involved in specific traits (Andersson, 2001), and determining its mechanisms. Li *et al.* (2008) reported different fragments were amplified in the healthy and self-biting mink by the random amplified polymorphic DNA (RAPD) primer of S356, and fragment of 1000 bp only existed in self-biting mink groups. Liu *et al.* (2010) indicated that SCAR marker (SRA8-250) was useful for detecting self-biting disease in minks.

Bulked sergeant analysis (BSA) is a rapid method for identifying markers linked to any specific gene or genomic region. Now, BSA in combination with amplified fragment polymorphisms AFLP/

RAPD method has been used widely and proved to be a very powerful and efficient technique for identifying markers linked to specific gene (Agrama *et al.*, 2002). Especially, some AFLP and RAPD markers linked to the interested genes were converted into sequence-characterized amplified regions (SCAR) markers. SCARS are based on sequencing the polymorphic fragment derived from RAPD primers and designing longer primers of approximately 20 bases that will specifically bind to this fragment. SCAR markers are more advantageous than RAPD markers and are more specific. Their amplification is less sensitive to reaction conditions and more reproducible by increasing the specificity (Hernandez *et al.*, 1999).

In the present study, we used BSA in combination with RAPD to identify specific DNA sequences associated with self-biting disease and converted them into SCAR markers to distinguish self-biting foxes from healthy ones.

Materials and methods

Sample collection

The experiment was carried out at the Fur Animals Experiment Station on Institute of Special Economic Animals and Plants of the Chinese Academy of Agricultural Sciences. The healthy and self-biting blue fox were housed in individual standard rearing cages (100×70×70 cm). Total genomic DNA was extracted from blood sample using DNA Extraction Kit. Two sets of healthy and self-biting DNA bulks were composed for BSA, each bulk contained an equivalent amount of the total DNA for each group.

RAPD-PCR analysis of genomic DNA

100 random primers provided by Shanghai Sangon Biological Engineering Technology Company were used to run the RAPD-PCR analysis in two sets of DNA bulks. PCR amplification was preformed in the Eppendorf AG in a 15 μl total volume containing 1U TaKaRa Taq DNA polymerase (TaKaRa Biotechnology Co., Ltd., Dalian, China), 2.5μl of 10×PCR buffer (containing 500 mM KCl and 15 mM $MgCl_2$), 0.2 mM each dNTP, 6 μM each of primers, and 2 μl of DNA template. PCR cycling parameters were as follows: 30 cycles each consisted of 94 °C for 30 s, 60 °C for 1 min, and 72 °C for 1.5 min, with a initial hot start at 94 °C for 5 min and a final extension at 72 °C for 5 min. The PCR products were detected by 2% agarose gel electrophoresis. And more healthy and self-biting blue fox genomic DNA were used to test the RAPD marker, each group including 30 individuals.

Cloning and sequencing of the RAPD fragment

Once the RAPD fragment presented polymorphic between the two bulks, the band of amplified DNA corresponding to the RAPD primer, was excised from the agarose gel, purified using the Agarose Gel DNA extraction Kit and cloned into *E coli* JM109 strain by means of the PMD-18T vector. The sequences obtained were analyzed by BLAST program for the similarity search.

Design of primers and analysis of the SCAR marker

The SCAR primers (sense: 5'AACGGCCAGTGATTCGA3' and anti-sense: 5' CATTACCAGGCATCCATTCC 3') were designed based on sequence of RAPD fragment by Primer 5 software. PCR conditions for SCAR amplification were similar to those that were used in the RAPD analysis. Healthy and self-biting individuals were tested for the SCAR marker. χ^2 test for independence in 2×2 table were analyzed to determine if the relationship exists between RAPD and SCAR marker of two groups. SAS8.0 software was used to analyze the data.

Results

RAPD-PCR analysis

High-molecular weight genomic DNA was isolated from the fresh blood. To identify DNA markers linked to the self-biting blue fox, 100 RAPD primers were screened by using two bulks of DNA from two groups. The random primer S414 amplified an approximately 800 bp fragment (named as SRS414-800) from most self-biting group, but seldem in healthy group. Hence the fragment SRS414-800 was regarded the fragment specific to self-biting disease (Figure 1). The validity of primer pair S414 in detecting self-biting fox was tested on 30 individuals of the self-biting and healthy foxes. The χ^2 test indicated significant difference (P<0.01) between the two groups (Table 1).

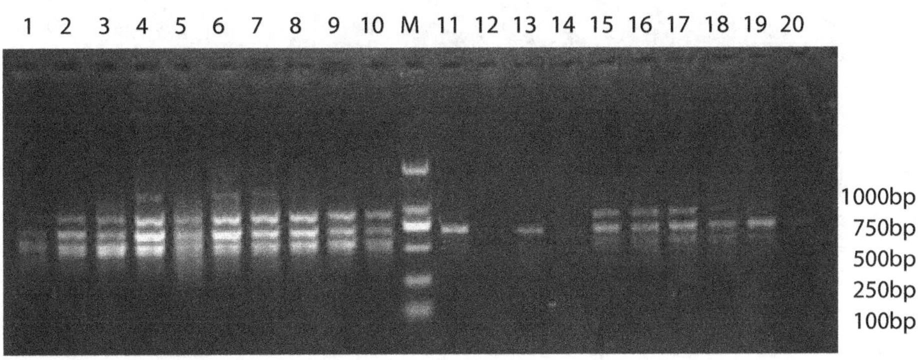

Figure 1. PCR amplified results by RAPD marker. M: Marker 1~10 self-biting fox; 11~20 healthy fox.

Table 1. Distribution of RAPD marker among healthy and self-biting fox.

		Genotype		**n**	**df**	**χ^2 test (P-value)**
		SRS414-800 present	**SRS414-800 absent**			
Phenotype	Healthy	13 (43%)	17(57%)	30	1	0.001
	Self-biting	27 (90%)	3(10%)	30	1	

Sequence analysis

The fragment SRS414-800 were excised from the agarose gel, recovered, purified, and sequenced. SRS414-800 was 781 bp in length and exhibited 70% similarity to *Macaca mulatta* MHC class I region, showing 137 bp homology. Query coverage and E values ranged from 33% to 38%, and from 1×10^{-31} to 2×10^{-28}, respectively.

Conversion of the RAPD marker into a SCAR marker

Based on the sequence of the specific RAPD fragment SRS414-800, a pair of primers named SCAR-S414 was designed to convert the RAPD marker into a single locus SCAR marker. As expected, the primer pair produced a 772 bp fragment from genomic DNA of most self-biting group, but seldom from the healthy (Figure 2). The validity of primer pair SCAR-S414 was tested on blue fox. The χ^2 test indicated significant difference ($0.01<P<0.05$) between the two groups (Table 2).

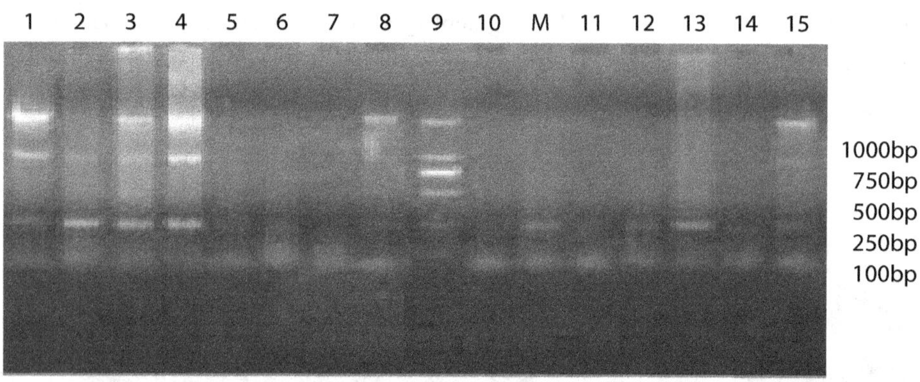

Figure 2. PCR amplified results by SCAR marker. M: Marker 1~8 self-biting fox; 9~15 healthy fox.

Table2. Distribution of SCAR marker among healthy and stereotyped fox.

		Genotype		**n**	**df**	**χ^2 test (*P*-value)**
		SCAR-S414 present	**SCAR-S414 absent**			
Phenotype	Healthy	3 (20%)	12 (80%)	15	1	0.0253
	Self-biting	9 (60%)	6 (40%)	15	1	

Discussion

Some animal welfare advocator formulated self-biting disease as stereotypy behavior that is an aimless and biological insignificance behavior, but there is an obvious difference between these two behaviors, self-biting disease can cause damage directly to the body, so I think it can not be concluded to a repeated insignificance behavior.

The etiology of self-biting in blue fox has different opinions, involving complex interactions among genetic, environmental, neurological, physiological, endocrinological and social factors. Zhao (1996) showed great relationship between heredity and self-biting in blue foxes. Therefore, the study on etiology of self-biting from gene aspect is feasible. Li *et al.* (2008) used the RAPD to amplify different fragments in healthy and self-biting mink, and showed that primer S356 can amplify different straps between healthy and sick mink.

PCR-based DNA markers are especially desirable because they are methodologically fast, easy, and reproducible. This strategy was employed in this study by combining the bulked sergeant analysis and RAPD method to rapidly identify molecular marker strictly linked to self-biting foxes. The RAPD method is easy to operate, but poor in repeatability. Therefore, self-biting specific marker was converted into stable SCAR markers. The distribution frequency of specific fragments SRS414-800 is 90% and 43% in self-biting foxes and healthy ones respectively, The $\chi 2$ test indicated significant difference ($0.01<P<0.05$), we make a conclusion that SRS414 can detect self-biting foxes and healthy ones. We design a pair of special primers to convert RAPD marker to SCAR marker successfully. The SCAR marker derived from a RAPD fragment had similarity of 70% to Macaca mulatta MHC class I region. MHC is a tightly linked cluster of genes encoding proteins that are essential for the establishment and regulation of immune response in most vertebrates. The canine MHC or dog leukocyte antigen (DLA) complex has been systematically studied. Eight functional genes and five pseudogenes have been identified to date for the class I and II regions. In particular, MHC class II molecules in the dog are constitutively expressed on the surface of dendritic cells (DCs), macrophages, B cells, T cells, thymocytes and vascular endothelium (Doveren *et al.*, 1985; Mueller *et al.*, 2004). DCs and MHC class II molecules play crucial roles in regulating immune responses in skin (Delves and Roitt, 2000). Enhanced MHC class II expression has been documented in dogs with pyoderma (Day, 1996), ectoparasitism (Stemmer *et al.*, 1996), immune-mediated dermatoses (Jackson *et al.*, 2004) and neoplasia (Affolter and Moore, 2002). On the other hand, genes in the class II region in humans have been associated with susceptibility and resistance to a variety of diseases ranging from autoimmune diseases such as diabetes mellitus and rheumatoid arthritis to allergic diseases such as atopic dermatitis. The linkage of this genomic region to the skin disease implies that MHC gene is potentially related to self-biting. This is supported by SCAR marker developed here.

All of our marginally positive findings are quite preliminary due to some limitations. The sample size, especially for SCAR test, is small for a genetic study. If we have adequacy sample for the test, p-values would have been lower than RAPD analysis. It should be noted that the SCAR marker was detected in about 20% of healthy blue fox. This might imply that there is a threshold in the blue fox central nervous system to behave self-biting even though they carry self-biting gene. Otherwise, the SCAR marker is dominant marker and not able to show the interaction between alleles. Since behavior is normally inherited in polygenic, additive manner, we are actually looking for many genes, and we wish to estimate the contribution of each of the genes to the phenotypic variation of self-biting disease. Weather the marker can be used as a genetic marker to cull self-biting individuals in early selection, we need to expand the sample numbers, minimize the individual difference to provide detecting means for non-self-biting foxes population to reduce the loss of fox breed industry.

Acknowledgements

The project was supported by Special Fund for Public Welfare Technology Research of Agricultural Industry (200903014).

References

Affolter V.K. and Moore P.F., 2002. Localised and disseminated histiocytic sarcoma of dendritic cell origin in dogs. Vet. Pathol. 39: 74-83.

Agrama, H.A., Houssin, S.F. and Tarek, M.A., 2002. Cloning of AFLP markers linked to resistance to Peronosclerospora sorghi in maize. Mol. Genet. Genomics 267: 814–819.

Andersson, L., 2001. Genetic dissection of phenotypic diversity in farm animals. Nat. Rev. Genet. 2: 130-138.

Day, M.J., 1996. Expression of major histocompatability complex class II molecules by dermal inflammatory cells, epidermal Langerhans cells and keratinocytes in canine dermatological disease. J. Comp. Pathol. 115: 317-326.

De Bellis, M.D., Keshavan, M.S., Clark, D.B., Casey, B.J., Giedd, J.N., Boring, A.M., Frustaci, K. and Ryan, N.D., 1999. Developmental traumatology Part II: Brain development. Biol. Psychiat. 45: 1271-1284.

Delves, P.J. and Roitt, I.M., 2000. Advances in immunology: the immune system – parts 1 and 2. N.Engl. J. Med. 343: 37-49, 108-117.

Doveren, R.F., Buurman, W.A., Schutte, B., Groenewegen, G. and Van der Linden, C.J., 1985. Class II antigens on canine T lymphocytes. Tissue. Antigens. 25: 255-265.

Hanse, C.P.B., 1993. Stereotypes in ranch mink: the effect of gene, litter size and neighbors. Behav. Process. 29: 165-178.

Hernandez, P., Martín, A. and Dorado, G., 1999. Development of SCARs by direct sequencing of RAPD products: A practical tool for the introgression and marker-assisted selection of wheat. Mol. Breeding. 5: 245-253.

Jackson, H.A., Olivry, T., Berget, F., Dunston, S.M., Bonnefont, C. and Chabanne, L., 2004. Immunopathology of vesicular cutaneous *lupus erythematosus* in the rough collie and Shetland sheepdog: a canine homologue of subacute cutaneous *lupus erythematosus* in humans. Vet. Dermatol. 15: 230-239.

Li, Y.M., Yao, J., Ma, L., Li, Z. and Bai, X., 2008. RAPD genetic analysis on etiological factor of Mink self-biting disease. Chin J Biotech, 24(4): 563-568.

Liu, Z.Y., Ning, F.-Y., Yang, H.-Y., Wei, L. and Bai, X.-J., 2010. Rapid detection of self-biting disease of mink by specific sequence-characterized amplified regions. J. Forest. Res. 22 (1): 123-126.

Mason, G.J., 1991. Stereotypies: a critical review. Anim. Behav. 41: 1015-1037.

Mueller, R.S., West, K. and Bettenay, S.V., 2004. Immunohistochemical evaluation of mononuclear infiltrates in canine lupoid onychodystrophy. Vet. Pathol. 41: 37-43.

Schore, A.N., 2002. Dysregulation of the right brain: A fundamental mechanism of traumatic attachment and the psychopatho genesis of posttraumatic stress disorder. Aust. NZ. J. Psychiat. 36: 9-30.

Stemmer, B.L., Arlian, L.G., Morgan, M.S., Rapp, C.M. and Moore, P.F., 1996. Characterization of antigen presenting cells and T-cells in progressing scabietic skin lesions. Vet. Parasitol. 67: 247-258.

Van Der Kolk, B.A., 1994. The behavioral and psychobiological effects of developmental trauma. In A. Stoudemire (ed.), Human behavior: An introduction for medical students New York: Lippincott, pp. 328-343.

Zhao, Y.K., 1996. The relationship of heredity and self-biting in blue foxes. J. Fur Animal Breeding. 3: 4-5.

Nucleotide sequence polymorphism of the growth hormone gene in American mink (*Neovison vison*)

J.J. Skorupski and M. Kmieć
West Pomeranian University of Technology in Szczecin, Department of Genetic and Animal Breeding, Doktora Judyma 6, 71-460 Szczecin, Poland; jakub@gajanet.pl

Abstract

The nucleotide sequence variation of the American mink (*Neovison vison*) growth hormone gene was analysed. For this purpose 1,113 bp region of the mink genomic DNA, encompassing four exons and three introns of somatotmropin gene, was amplified by nested-PCR and sequenced. The study included 116 individuals of wild mink (Nova Scotia, Canada), ranch mink (Scanblack, Wild, Sapphire, Pearl, Black Cross and Sapphire Cross; West Pomerania, Poland) and feral mink (Poland and Iceland). Sequencing allowed to determine the nucleotide composition of previously unrecognized non-coding regions, as well as exons and to identify 12 new SNP substitutions, one single nucleotide deletion and one ins/del polymorphism. The average incidence of SNPs in coding sequences was equal to 1/156.6 bp and in non-coding – 1/86.9 bp. Analysis of the observed number and location of substitutions revealed its non-random distribution and significant preponderance of transitions over transversions. The results also revealed a significant differences between originally wild mink and domesticated animals (both ranch and feral) in identified SNPs type and distribution. The results of the present study, supported by further analysis, can help in effective genetic monitoring of feral mink, as some differences were also found between ranch and feral mink.

Keywords: feral mink, genetic polymorphism, growth hormone gene, ranch mink, wild mink

Introduction

Growth hormone (GH, also called somatotrophin, STH) is a peptide hormone secreted by eosinophilic cells of anterior lobe of the pituitary gland (Machnik and Lechniak, 2000). GH is a very comprehensive and systemic hormone, affects target cells in direct and indirect manner (Kmieć *et al.*, 2007). Its most important function is to stimulate growth and development of cells, tissues and the whole body (Kmieć *et al.*, 2008). Functionally, growth hormone belongs to the somatotrophic axis, together with its receptor (GHR) and somatoliberin (GHRH), as well as prolactin (PRL), prolactin receptor (PRLR), insulin-like growth factor-I (IGF-I), transcription factor Pit-I and the transcription factor STAT5 (Zwierzchowski *et al.*, 2004).

Nucleotide sequence of mink growth hormone gene's cDNA was first reported by Shoji *et al.* in 1990. The open reading frame includes 648 bp and product of its translation – 216 amino acids with app. 24.5 dalton. Signal peptide consist of 26 amino acids and mature hormone – 190 (Shoji *et al.*, 1990).

The growth hormone gene shows high structural-functional similarity to genes encoding prolactin and placental lactogen (LH). All these genes belong to one gene family (Wallis, 1992; Ohta, 1993; Goffin *et al.*, 1995; Soares, 2004), evolved from one common gene, whose age is determined on the 400 million years (Miller and Eberhardt, 1983). It is worth noting that while the same nucleotide sequence of the GH gene has a relatively high degree of similarity to the PRL gene, their protein products show only 16% homology (Krzymowski and Przała, 2005).

The aim of this study was to investigate the nucleotide sequence of mink GH gene, as well as to analyse its polymorphism.

Material and methods

The study included 116 individuals – 22 of wild mink (Nova Scotia, Canada), 58 of ranch mink (six colour types – Scanblack, Wild, Sapphire, Pearl, Black Cross and Sapphire Cross from West Pomerania, Poland) and 36 of feral mink (from Poland and Iceland). Genomic DNA was isolated from muscle tissue, using the High Pure PCR Template Preparation Kit from Roche (Germany). Target fragment of mink growth hormone gene include 1,113 bp region of the mink genomic DNA, encompassing whole 3rd and 4th exon and whole 2nd, 3rd and 4th intron, as well as parts of 2nd and 5th exon.

In order to amplify the target fragment of the growth hormone gene, nested-PCR with outer primers, flanking 1,220 bp region, and inner primers, flanking 1,113 region was carried out. In both reactions DNA amplification was performed in a mixture with a volume of 15.0 μl, containing ready-to-use PCR mix (A&A Biotechnology), each outer primer, water and DNA template (genomic DNA for first reaction and first reaction mix, which was used as DNA template for the second reaction). The PCR reaction was performed in a T3-type oil-free thermocycler of the Biometra (Germany) company with time-temperature programme, especially optimized for these reactions. The specificity and efficiency of amplification was assessed by checking the PCR products by electrophoresis in a 1.5% agarose gel with ethidium bromide.

The nucleotide sequence of the amplicons, as well as its sequence variation was analysed by sequencing, performed using BigDye® Terminator v3.1 kit (Applied Biosystems, Life Technologies). Sequencing reaction products were separated on a capillary sequencer 3730xl DNA Analyzer and 3130xl Genetic Analyzer (Applied Biosystems, Life Technologies).

Results

Sequencing allowed determining nucleotide composition of previously unrecognized non-coding regions, as well as exons. These include three introns, at positions: intron 2nd, 171 bp; intron 3rd, 175 or 176 bp; and intron 4th, 272 or 290 bp. Also 14 new polymorphisms were identified: 12 SNP substitutions (one in 2nd intron: G→C, five in 3rd exon: G→A, T→C, three in 3rd intron: G→C, A→G, C→T, and three in 4th intron: A→G, C→G, T→C), one single nucleotide deletion (in 3rd intron: C/-) and one ins/del polymorphism (18 bp; in 4th intron). The average incidence of SNPs in coding sequences was equal to 1/156.6 bp and in non-coding – 1/86.9 bp. Analysis of the observed number and location of substitutions revealed its non-random distribution (more than 56% substitution is located in a central part of the amplicon) and significant preponderance of transitions over transversions, representing only app. 21% of all substitutions.

Comparing the genetic polymorphism of wild, ranch and feral mink the most important are differences were found in heterozygosity rate and presence of ins/del polymorphism in the 4th intron. Namely, average heterozygosity rate in wild mink was equal to 18.75%, in ranch mink – 15.87% and in feral mink – 23.38%. Ins/del polymorphism (deletion of 18 bp fragments) in the 4th intron was detected only in wild mink, and its prevalence was equal to 66.67%.

Discussion

The results revealed some significant differences between originally wild mink and domesticated animals (both ranch and feral) in identified SNPs type and distribution. Very interesting is investigation of the heterozygosity rates of mink from different environments. Firstly, smaller number of heterozygotes in farm animals, in comparison with wild ones, is an evidence of breeding work and selection pressure, under whose influence is the first of these groups. In turn, a significant increase in the heterozygosity rate in feral mink seems to indicate the influence of environmental factors in shaping genetic variability of animals under high natural selection pressure in new environments.

Further analysis are needed to determine the possibility of use the results of the present study in effective genetic monitoring of feral mink, as some differences were also found between ranch and feral mink. This could be used to develop an alternative to microsatellite genetic profiling and differentiation of wild, ranch and feral mink's populations.

Acknowledgements

This research project would not have been possible without the kind support of Prof. Hossain Farid from Nova Scotia Agricultural College, Canada and Mr Róbert Arnar Stefánsson and Ms Menja von Schmalensee from West Iceland Centre of Natural History, Iceland. The project was funded by the National Science Centre.

References

Goffin, V., Martial, J.A. and Summers, N.L., 1995. Use of a model to understand prolactin and growth hormone specificities. Protein Engineering 8: 1215-1231.

Kmieć, M., Polasik, D., Napierała, F. and Polasik, I., 2008. GH/HaeII and GH/MspI Restriction Polymorphism in a Herd of Polish Large White Sows. Journal of Animal and Veterinary Advances 7: 165-169.

Kmieć, M., Terman, A., Wierzbicki, H. and Zych, S., 2007. Association of GH Gene Polymorphism with Semen Parameters of Boars. Acta Veterinaria Brno 76: 41-46.

Krzymowski, T. and Przała, J. 2005. Fizjologia zwierząt. PWRiL. Warszawa.

Machnik, G. and Lechniak, D., 2000. Funkcje hormonu wzrostu (GH) w męskich procesach rozrodczych. Medycyna Weterynaryjna 56: 218-221.

Miller, W.L. and Eberhardt, N.L., 1983. Structure and evolution of the growth hormone gene family. Endocrine Reviews 4: 97-130.

Ohta, T., 1993. Pattern of nucleotide substitutions in growth hormone-prolactin gene family: a paradigm for evolution by gene duplication. Genetics 134: 1271-1276.

Shoji, K., Ohara, E., Watahiki, M. and Yoneda, Y., 1990. Cloning and nucleotide sequence of a cDNA encoding the mink growth hormone. Nucleic Acid Research 18: 6424.

Soares, M.J., 2004. The prolactin and growth hormone families: Pregnancy-specific hormones/cytokines at the maternal-fetal interface. Reproductive Biology and Endocrinology 51: 1-15.

Wallis, M., 1992. The expanding growth hormone/prolactin family. J. Molecural Endocrinology 9: 185-188.

Zwierzchowski, L., Dymnicki, E., Flisikowski K., Gajewska A., Grochowska R., Klauzińska M., Krzyżewski, J., Maj A., Oprządek J., Siadkowska E., Snochowski M., Strzałkowska N. and Szymanowska, M., 2004. Polimorfizm sekwencji kodujących i regulacyjnych genów: hormonu wzrostu, prolaktyny, ich receptorów i czynników transkrypcyjnych Pit-I i STAT 5 oraz ich wpływ na użytkowość mleczną i mięsną bydła. In: Postępy genetyki molekularnej bydła i trzody chlewnej, Świtoński M. (ed.), AR im. A. Cieszkowskiego w Poznaniu, Poznań, pp. 61-114.

Evidence for genetic variation in bite marks in group housed mink

S.W. Alemu[1], P. Bijma[2] and P. Berg[1]
[1]Institute of Molecular Biology and Genetics, Faculty of Science and Technology, Aarhus University. Postboks 50, 8830 Tjele, Denmark; [2]Animal Breeding and Genetics Group, Wageningen University, 6709 PG Wageningen, the Netherlands; setegnw.alemu@agrisci.dk

Abstract

There is evidence from several species (e.g. poultry) that selection can contribute to reduced aggression in groups, and selection thus could be one way of reducing aggression in group housed mink. On this background, a selection experiment was started, aiming at reducing the number of bite marks on the skin side. This study describes variation in bite marks in the three generations. The study includes evaluation of bite marks on a total of 1986 mink in group housing. We hypothesise that bite marks might be influenced by both direct genetic effects and social genetic effects. Direct genetic effects refers to the effect of the animals own genotype and social effects refer to the effects of the genotype of the other animal in the group. Both direct and social genetic effects contribute to the variation in bite-marks in mink. Social genetic effect and the covariance between direct and social genetic effect explains more than 5 times the variation in bite marks explained by direct genetic effect. This is true in bite marks for the neck, body, and tail region as well as for the average of these regions. Therefore selection which take in to account social genetic effect will be successful to decrease bite marks in mink.

Keywords: bite mark, social genetic, aggression behaviour, mink

Introduction

Group housing of mink has become common in mink production. This has several advantages from a production perspective, but it also has some drawbacks related to animal welfare. More bite marks on animals kept in group housing than in traditional pair wise housing (Hansen and Houbak, 2005; Hänninen *et al.*, 2008a,b; Jeppesen, 2009, Møller *et al.*, 2003).Bite marks are likely related to aggression and fearful behaviour and is thus an indicator of reduced animal welfare (EC, 2001, Hansen and Jeppesen, 2008, Møller *et al.*, 2003).

Most studies has focused on adapting the cage environment to reduce aggression in group housing, e.g. by allowing animals in group housing access to an extra nest box (Jeppesen, 2009). An alternative and complementary strategy is to select for animals with reduced aggression in group housing, animals that are better adapted to the more complex social environment in groups. Selection for group performance in egg laying hens resulted in large decreases in mortality, mainly from aggression, from 68% in generation 2 to 9% in generation 6 (Muir, 1996). The reason for this success is that aggression in chickens is affected by both direct and social genetic effects. In this paper we are going to test the hypothesis that aggressive behaviour in mink, as measured by the number of bite marks at pelting, is affected by both direct and social genetic effect.

In 2009 a selection experiment was started to select for reduced number of bite marks at pelting, at Research Centre Foulum in Denmark. The objective of this paper is to describe sources of variation in number of bite marks and to test if there is both direct and social genetic variation in number of bite marks measured at pelting.

Material

In this presentation we have analysed data from three generations of the selection experiment. A total of 1986 minks descending from136 sires and 349 dams were used in our analysis. They grouped in cages in either, 2, 3, or 4. All mink kits in group housing were pelted in November. At pelting, the number of bite marks were evaluated at the skin side of the pelt. The number of bite marks was subjectively measured on the scale described in Table 1.

A total of 1985 mink from group housing was scored according to the scale described in Table 1. The number of bite marks was scored in the Neck (from nose tip to shoulder/front leg), Body (from shoulder to 10 cm above the base of the tail) and Tail (from 10 cm above the base of the tail, incl. back legs). An average score was computed as the average of these three scores.

Theory and methods

In traditional quantitative genetics, the phenotypic value of individual i is the sum of a heritable component, A_i, known as breeding value, and a non-heritable residual E_i (Falconer and Mackay, 1996)

$$P_i = A_i + E_i \tag{1}$$

Table 1. Bite marks score used for subjectively measuring number of bite marks at pelting.

Bite mark score	Number of bite marks
0	0
1	1-5
2	6-10
3	11-15
4	16-20
5	21-25
6	26-30
7	31-35
8	36-45
9	More than 45

With social genetic effect, the phenotype of an individual stems from two unobserved effects: A direct effect originating from the individual itself, and the sum of social effects originating from its $n-1$ group mates (Griffing, 1967a).

$$P_i = A_{D,i} + E_{D,i} + \sum_{j=1}^{n-1} A_{S,j} + \sum_{j=1}^{n-1} E_{S,j} \quad (2)$$

where i denotes the focal individual, j a group mate, $A_{D,i}$ the direct genetic effect (DGE) of i, $E_{D,i}$ the corresponding non-heritable direct effect, $A_{S,j}$ the social genetic effect of group mate j, and $E_{S,j}$ the corresponding non-heritable social effect. Equation 2 contains two genetic effects, direct effects, A_D, and social effects, A_S. Hence, fitting Equation 2 involves the estimation of three genetic variance components; , and.

$\sigma^2_{A_D}$, $\sigma_{A_{DS}}$ and $\sigma^2_{A_S}$

Heritable variation

This section presents the heritable variation among interacting individuals in a group. Each individual interacts with n-1 group members, so the total heritable impact of an individual on the population mean:

$$TBV_i = A_{D_i} + (n-1)A_{S_i} \quad (3)$$

Where represents total heritable variation

The total breeding value is the heritable impact of an individual on trait values in the population, and is a generalization of the traditional breeding value to account for social genetic effect (Bijma 2011b). Therefore the total heritable variation available for response to selection will be:

$$\sigma^2_{TBV} = \sigma^2_{A_D} + 2(n-1)\sigma_{A_{DS}} + (n-1)\sigma^2_{A_S} \quad (4)$$

σ^2_{TBV} represents variance of total heritable variation.

Genetic analyses were based on the animals scored for bite marks in group housing. We hypothesise that bite marks might be influenced by both direct genetic effects and social genetic effects. Direct genetic effects refers to the effect of the animals own genotype and social effects refer to the effects of the genotype of the other animal in the group.

We fitted a univariate animal model:

$$y = Xb + Z_D a_D + Z_S a_S + Wg + e \quad (5)$$

Where Y, is a vector of bite mark scores on either Neck, Body, Tail or the average score, β is a vector of fixed effects of sex,year,shed, and their interaction is a vector of random direct breeding value with incidence matrix Z_D, a_S is vector of social breeding value with incidence matrix Z_S

linking phenotypic value of individual to the social breeding value of their social, Wg accounts for the non-heritable social effects, vector of residuals.

The covariance structure of the genetic terms in Equation 6 is

$$\begin{bmatrix} \mathbf{a}_D \\ \mathbf{a}_S \end{bmatrix} \sim MVN(\mathbf{0}, \mathbf{C} \otimes \mathbf{A}) \tag{6}$$

where $\mathbf{C} = \begin{bmatrix} \sigma^2_{A_D} & \sigma_{A_{D,S}} \\ \sigma_{A_{D,S}} & \sigma^2_{A_S} \end{bmatrix}$

This fully describe the genetic components of covariance between observations.

g is a vector of random group effects, with g~$N(0,I_g\sigma^2_g)$ and incidence matrix W linking records to groups, and e is a vector of residuals with e~$N(0,I_e\sigma^2_e)$, where I is an identity matrix.

Results and discussion

All of the fixed effects are significants, that is shed, year,sex are significant for all parts of the body.

From Table 2 a comparison of total breeding value variance with direct genetic variance shows the importance of social genetic effects for neck, body, tail and average bite mark traits. The direct heritability estimates are low, but when we take into account the social genetic effect the total heritability estimate is significantly higher. This is true for all parts of the body.

These results are in line with other experiments in pigs where the social effect contribute the vast majority of heritable variation for the trait growth rate and feed in take (Bergsma *et al.*, 2008), survival days in chickens where two third of the heritable variation is caused by social genetic effects (Bijma *et al.*, 2007).

Table 2. Direct genetic variance, total breeding value variance, phenotypic variance, direct heritability, and total heritability of bite mark scores in neck, body tail parts of the body, and their average.

Parts of the body	Direct genetic variance	Total breeding value variance	Phenotypic variance	Direct heritability	Total heritability
Neck	0.48	1.64	3.25	0.15	0.51
Body	0.38	2.82	5.19	0.074	0.54
Tail	0.27	2.14	5.36	0.05	0.40
Average	0.37	1.96	3.23	0.114	0.61

Conclusion

We conclude that both direct and social genetic effects contribute to variation in bite marks measured at pelting.

Social genetic effect explains and the covariance between direct and social genetic effect more than 5 times as much of the variation in bite marks as the direct genetic effect, irrespective of the part of the body analyzed. The total heritability is large for number of bite marks. Therefore selection which take into account social genetic effects can reduce number of bite marks and possibly socially aggressive behavior in mink. This indicates that aggressive behavior in group housed mink can be reduced by appropriate genetic selection.

References

Berg, P. and Møller, S.H., 2010. Mulighed for at selektere for reduceret aggression i gruppeindhusning. Faglig Årsberetning 2009. Pelsdyrerhvervets Forsøgs- og Forsknings Center, Holstebro, Danmark.

Bergsma, R., E. Kanis, E.F. Knol and P. Bijma, 2008. The contribution of social effects to heritable variation in finishing traits of domestic pigs (*Sus scrofa*). Genetics 178: 1559-1570.

Bijma, P., 2011. A General Definition of the Heritable Variation That Determines the Potential of a Population to Respond to Selection. Genetics 189: 1347-1359.

Bijma, P. and M.J. Wade, 2008. The joint effects of kin, multilevel selection and indirect genetic effects on response to genetic selection. Journal of Evolutionary Biology 21: 1175-1188

Bijma, P., W.M. Muir, E.D. Ellen, J.B. Wolf and J.A. Van Arendonk, 2007. Multilevel selection 2: Estimating the genetic parameters determining inheritance and response to selection. Genetics 175: 289-299.

European Commission, 2001. The Welfare of Animals Kept for Fur Production. Report of Scientific Committee on Animal Health and Animal Welfare, Adopted on 12-13 December 2001. DG Sanco, Directorate B – Scientific Health Opinions; Unit B3 – Management of scientific committees II. 211 pp.

Griffing, B., 1967. Selection in Reference to Biological Groups. I. Individual and Group Selection Applied to Populations of Unordered Groups. Australian Journal of Biological Sciences 20: 127.

Hansen, S.W. and Jeppesen, L.L., 2008. Bidmærker som velfærdsindikator hos mink. Faglig Årsberetning 2007, Pelsdyrerhvervets Forsøgs- og ForskningsCenter, pp. 13-23.

Hansen, S.W. and Houbak, B., 2005. To skridt frem og tre tilbage – gruppeind-husning af mink. Faglig Årsberetning 2004, Pelsdyr-erhvervets Forsøgs- og Forsknings-Center, pp. 39-47.

Hänninen, S., Mononen, J., Harjunpää, S., Pyykönen, T., Sepponen, J. and Ahola, L., 2008a. Effects of family housing on some behavioural and physiological parameters of juvenile farmed mink (*Mustela vison*). Applied Animal Behaviour Science 109: 384-395.

Hänninen, S., Ahola, L., Pyykönen, T., Korhonen, H.T. and Mononen, J. 2008b. Group housing in row cages: an alternative housing system for juvenile mink. Animal 2: 1809-1817.

Jeppesen, L.L., 2009. Ekstra redekasse i klatrebure og ekstra etage til standardbure. Effekt på adfærd, bidmærker og undersorter hos mink. Pelsdyrerhvervets Forsøgs- og ForskningsCenter. Faglig Årsberetning 2008, pp. 23-36.

Lindberg, H.M.K., Hansen, S.W., Alden, E. and Lidfors, L., 2007. Effect of climbing cages and group size on behaviour and production in juvenile mink. NJF-seminar 403, Kolding, Denmark. 6 pp.

Mononen, J., Kasanen, S., Harjunpää, S., Harri M., Pyykönen, T. and Ahola, L., 2000. A family housing experiment in mink. Scientifur, 24, pp. 114-117.

Muir W.M., 1996. Group Selection for Adaptation to Multiple-Hen Cages – Selection Program and Direct Responses. Poultry Science 75(4): 447-58.

Møller, S.H., Hansen, S.W. and Sørensen, J.T., 2003. Assessing animal welfare in a strictly synchronous production system: The mink case. Animal Welfare 12: 699-703.

Wade M.J., P. Bijma, E.D. Ellen and W. Muir, 2010. Group selection and social evolution in domesticated animals. Evolutionary Applications 3:453-465.

Wolf J.B., E.D. Brodie III, J.M. Cheverud, A.J. Moore and M.J. Wade, 1998. Evolutionary consequences of indirect genetic effects. Trends in Ecology and Evolution 13: 64-69.

The regulation role of behavior in coat color formation in mink (*Neovison vison*)

O.V. Trapezov
Institute of Cytology and Genetics, Siberian Branch of Russian Academy of Sciences, Novosibirsk, Russia, e-mail: trapezov@bionet.nsc.ru

Abstract

To date, 35 coat color mutations have been registered in minks. Market demands on fur, on novel coat color phases, have always been met by retention and breeding of color variations that arose by chance among captive bred individuals. Domestication has uncovered in domestic mink populations hidden genetic variability in coat color. This in turn gave clues for search and selection of new versions of coat color in mink. In other words, domestication suggests the technology appropriate for deliberately creating hither to rarely, if at all, observed novelties. To illustrate, at the Institute of Cytology and Genetics (Novosibirsk, Russia) a method has been developed for evaluation how the degree of domestic behavior may affect coat color pattern in mink. The *Karelian spotted* mink (S^K/+) was provided as an example.

Keywords: color mutation, genetic variability, behaviour

Introduction

The science of the visible trait coat color in different model organisms traces its beginnings to the nineties, and its history is replete with accumulating evidence and thought-provoking explanations (Sturm, 2006). The diversity of coat color patterns are due to mutations that change just one pigment melanin. However, melanogenesis proved to be a rather difficult issue because genes at over 60 loci affect pigmentation (Silvers, 1979). As regards the American mink, to date 35 mutations acting in a dominant, incompletely dominant, and recessive fashion have been identified in the course of its farm breeding (Evsikov, 1987; Nes *et al.*, 1988; Trapezov, 1997).

Analysis of the expression of these genes could be helpful in clarifying the genetic control of pigmentation of the hair coat. Hair coat pigmentation is under the control of dozens of gene. The aim of this work was to examine the regulatory effects of tame and aggressive behavior on the expression of the S^K gene that causes coat color pigmentation pattern in *Karelian spotted* mink.

Material and methods

Cross-breeding experiments were performed using male minks with the *Karelian spotting* phenotype (S^K/+) (Figure 1) and *dark brown standard* female minks (+/+) of tame and aggressive lines bred and maintained at the Experimental farm of this Institute (Figures 2, 3 and 4).

According to the classical description (Evsikov, 1987) minks with the *Karelian spotting* phenotype when heterozygous (S^K /+) show a distinctive pattern: white ventral areas contrasting with a

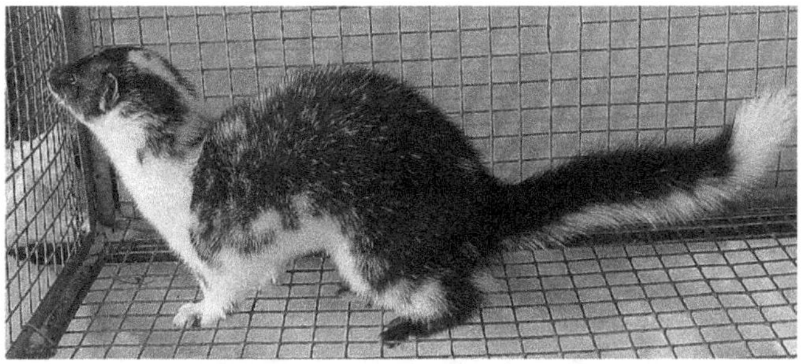

Figure 1. The Karelian spotting mink (S^K/+).

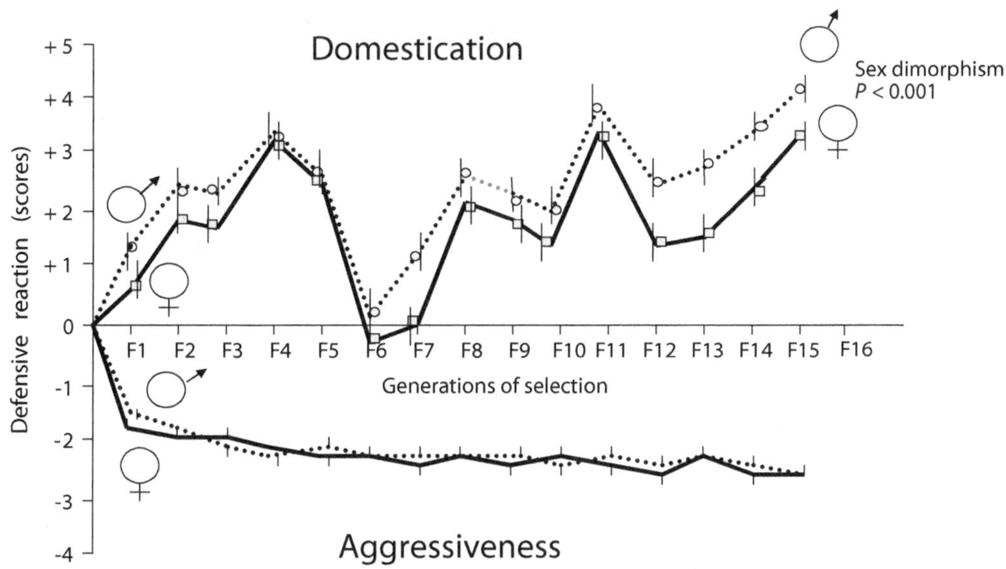

Figure 2. Reorganization of the defensive responses to human in minks in the course of the breeding program designed to the study the effects of selection on aggression and tameness.

Figure 3. Mink from tame line.

Figure 4. Mink from aggressive line.

completely pigmented back, the forehead is pigmented but with white spots (Figure 1). Dark brown color of the sides of the body, as a rule, gradually whitens. In many animals, areas of pigmented hairs are interspersed with gray hairs. The S^K mutation has been detected in a herd of minks at the 'Kutezhskaya' state fur farm in Russia (Evsikov, 1987). The mutation is incompletely dominant and it is a member of a series of multiple alleles at the Black cross locus: (*Shadow*) S^H > (*Karelian spotting*) S^K > (*Black cross*) S > (*Royal silver*) S^R > (*standard*) +. The observed ratio of phenotypes is 1 standard dark brown (+/+) : 2 spotting (S^K/+) : 1 white (S^K/S^K) among offspring of parents with the dilute phenotype (S^K/+ × S^K/+). All kits homozygous for this gene (S^K/S^K) die. Its effect becomes lethal in neonates aged 19-34 days and they succumb from liver and spleen malfunction (Evsikov, 1987).

Tame and aggressive lines of *standard* (+/+) minks differing by their defensive response towards human were developed by using a breeding program designed to study the effects of destabilizing selection (Trapezov, 1997) (Figure 2).

The presence of variability in the defense response to human in mink populations bred at commercial farms is the condition necessary for their successful taming. Based on the disclosed variability in this response, two lines by selection in the opposite directions of two populations of tame *standard* (+/+) minks and aggressive minks were developed (Figure 3 and 4).

Results

(S^K/+) males (Figure 1) of a farm population, which had not been subjected selection for behavior, were crossed to *standard dark brown* (+/+) females of lines selected for tame and aggressive behavior (Figure 3 and 4). The F_1 generation from this cross displayed 8 types of *Karelian spotting* patterns (S^K/+). Figure 6 shows how the types differ from each other in the degree of expression of specific pigmentation patterns. As seen in Figure 5, the percentage of pigmented areas reduced from almost complete, 90% (type I) to 5% (type VIII) with stray dark spots or specks remaining on an otherwise fully white background and with specific dark markings on the top of the head and around the nose.

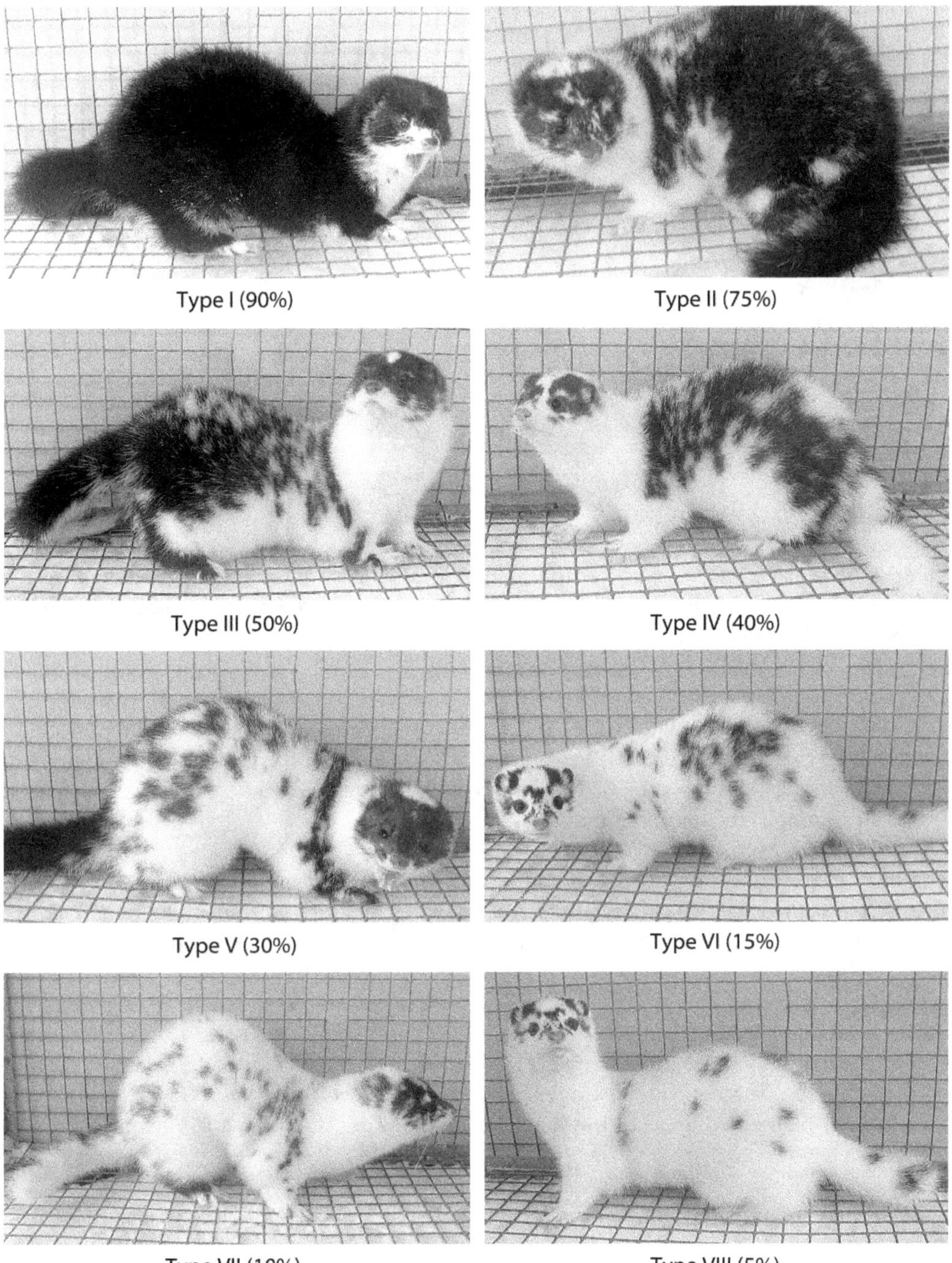

Figure 5. Eight types of pigmentation patterns in minks with the Karelian spotting phenotype ($S^K/+$). Percentage of pigmented areas of the body.

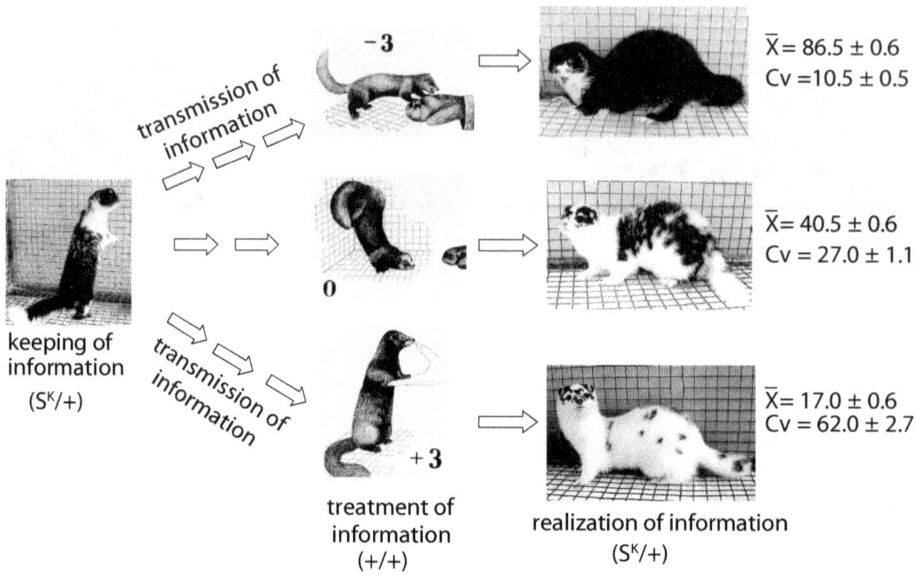

Figure 6. A model explaining the principle of operation: the same object may convey information of different kind. Depending on the system to which the signal is delivered, the object may convey meaning of one kind or of another completely different kind.

In the first set of crosses, males with the *Karelian spotting* mutation ($S^K/+$), not subjected to selection for behavior, were crossed to *standard* females (+/+) of the line selected for aggressive behavior (with the highest score -3). The data in Table 1 support Evsikov´s description (1987) of $S^K/+$ as an incompletely dominant mutation.

Offspring of 100 crosses between males with the *Karelian spotting* mutation and females assigned the lowest -3 score gave two classes of phenotype segregation at a 1:1 ratio, a half had wild type coats, the other half had mutant white spotting coats ($S^K/+$) ($\chi^2 = 0.9$, $P>0.05$).

As seen in Table 1, crosses between the *Karelian spotting* males (carriers of the $S^K/+$ gene), which had not been subjected to selection for behavior, and standard females of the aggressive line (low score -3 points) yielded 235 kits exhibiting low variability in coat color patterns ($\sigma^2 = 83.1$). Just two pigmentation patterns were distinguished, type I (181 kits) and type II (54 kits). The mean percentage of pigmented areas of the body for all offspring was $M = 86.5 \pm 0.6$ with variation coefficient $Cv = 10.5 \pm 0.5$.

In the second set of experiments, *standard* (+/+) females of the tame line whose response to human was scored zero were crossed to males with the *Karelian spotting* mutation not subjected to selection for behavior (Table 1). Here again, the cross yielded two classes of phenotypes at a 1:1 ratio, an equal number of mutation white spotting versus wild type among offspring ($\chi^2 = 1.14$, $P>0.05$).

Table 1. Segregation ratio of phenotypes in offspring from crosses between Karelian spotting males (S^K/+), not subjected to selection for behavior, to standard dark brown (+/+) females of lines selected for tame and aggressive behavior.

Crossing		**Offspring (♂♂ + ♀♀)**										
♂♂ (S^K/+)	**♀♀ (+/+)**	**No. crossing**	**No. kits**	***Standard dark brown***	**Types of pattern pigmentation in *Karelian spotting* minks (S^K/+)**							
Behavior	**Behavior**			**(+/+)**	**I**	**II**	**III**	**IV**	**V**	**VI**	**VII**	**VIII**
0±0.0 (farm bred population)	-3±0.0 (aggressive line)	100	491	256	181	54						
0±0.0 (farm bred population)	0±0.0 (tame line)	100	528	252		17	25	146	88			
0±0.0 (farm bred population)	+3±0.0 (tame line)	100	531	259			6	12	40	114	71	43

Crosses of *Karelian spotting* males (carriers of the S^K/+ gene), not subjected to selection for behavior, to *standard* females of the tame line, assigned zero score, produced 276 kits with white spotting phenotype that displayed the following 4 types of pigmentation patterns: type II, 17, type III, 25, type IV, 146, and type V, 88 kits (Table 1). Variability in coat color phenotype increased two-fold compared to that in the preceding cross (σ^2 = 116.8). The mean percentage of the pigmented area of the body was M = 40.0±0.6, with variation coefficient being Cv = 27.0±1.1. The modal class of minks was represented by pigmentation pattern of type IV.

Finally, males with the *Karelian spotting* phenotype (S^K/+), not subjected to selection for behavior, were crossed to *standard* females scored +3 for tame behavior. Two classes of phenotypes were distinguished among offspring at a 1:1 segregation ratio, the number of kits with *Karelian spotting* and wild-type patterns being equal (χ^2 = 0.31, P>0.05).

As Table 1 shows, in this cross, the number of pigmentation patterns rose to 6 among the 272 offspring with the white spotting phenotype: type III, 6, type IV, 12, type V, 40, type VI, 114, type VII, 71, and type VIII, 43 kits.

Variability in coat color phenotype increased sharply, σ^2 = 110, the variation coefficient Cv = 62.0±2.7. The mean percentage of the pigmented area of the body for all offspring was M=17.0±0.6.

Discussion

The experimental data obtained here with the model object (*Neovison vison*) demonstrated that the achieved diversity of pigmentation patterns was due to the regulatory effects of genes for behavior when the phenotypic expression of the *Karelian spotting* (S^K/+) (Figure 1) acquired an extremely wide range of variability compared to the classical Evsikov (1987) described. The salient finding was that the higher the mother's score for tame behavior, the more unusual and wide ranging was the pigmentation pattern in offspring compared to the classical description.

It was impressive that offspring from crosses between the $S^K/+$ mutant males and *standard* females with low -3 behavior scores displayed just two types of pigmentation patterns (I and II) while those to standard females with high +3 scores displayed as high as 6 types (III, IV, V, VI, VII, and VIII). Furthermore, in the genetic environment of the genes controlling aggressive behavior the modal class of phenotypes of offspring (genotype $S^K/+$) is represented by animals with pigmentation pattern of type I (77%), while in that of the genes for tame behavior it is represented by type VI (40%) (Figure 6).

Let us consider a scheme of a cybernetic system in which information, i.e. various signals, communications, messages, are stored, and transmitted. Genetic information is stored as $S^K/+$ symbol in this scheme. The information is processed in a black box that treats the past or present input information signal in the -3 to +3 range.

Implementation of information in this experiment (in its semantic sense) is represented as depigmentation effect when the value of the depigmented area of the body (its quantitative expression $\bar{X}$) changes from M = 86.5±0.6 (genetic background of aggressive individuals) to M = 17.0±0.6 (genetic background of tame individuals). Conceptually, in the scheme, the amount of information from the object ($S^K/+$) when interference increases (enhancement of aggressive behavior to the -3 score) tends to zero, whereas the amount of information from it when interference decreases (enhancement of tame behavior to the +3 score) tends to unity.

This conceptual scheme for the obtained experimental material may be represented by the informational-cybernetic principle: the same object may convey sometimes one information, sometimes another information. Depending on the system to which the signal is delivered, the object may convey one meaning or alternatively another absolutely different.

References

Evsikov, V.I., 1987. Genetical and evolutionary aspects of the problem of mammalian homeostasis (mink as a model). Russian Journal of Genetics 23(6): 988-1002.

Nes, N., Einarsson, E., Lohi, O., Jarosz, S. and Sheelje R., 1988. Beautiful Fur Animals and their colour genetics. Glostrup, Denmark. Published by Scientifur. 271 pp.

Silvers, W.K., 1979. The coat colors of mice: a model for mammalian gene action and interaction. New York: Springer-Verlag. 379 pp.

Sturm, R.A., 2006. A golden age of human pigmentation genetics. Trends in Genetics 20: 464-468.

Trapezov, O.V., 1997. Black Crystal: A novel coat color mutant in the American Mink. Journal of Heredity. 88: 164-166.

Genetic parameters of eye infection in Finnish blue fox (*Alopex lagopus)*

R. Kempe[1], T. Köykkä[1], N. Koskinen[2] and I. Strandén[1]
[1]MTT Agrifood Research Finland, Biotechnology and Food Research, Biometrical Genetics, FI-31600 Jokioinen, Finland. [2]MTT Agrifood Research Finland, Animal Production Research, Monogastric Research, Tervamäentie 179, 05840 Hyvinkää, Finland; riitta.kempe@mtt.fi

Abstract

The frequency of eye infections in the Finnish blue fox population has increased during the past decade. Eye infection may incur economic losses to producers due to reduced selection intensity. Ethical aspects also need to be considered, since eye infection reduces animal wellbeing. The purpose of this study was to estimate genetic parameters for eye infection (EYE) and to determine the genetic correlations between EYE, pelt character and animal size traits. The heritability estimate for EYE was moderate (0.22), suggesting that the blue fox is genetically susceptible to eye infection. EYE had a moderate antagonistic genetic correlation (-0.39) with pelt guard hair coverage (pGC) and a low negative genetic correlation with pelt length (pSI_{cm}) and grading size (gSI) (-0.23 and -0.27, respectively). Our results suggest that the antagonistic connections between EYE, pelt quality and size traits should be accounted in Finnish blue fox breeding value evaluation by using a multi-trait animal model.

Keywords: correlation, frequency, heritability, pelt traits, size traits

Introduction

The main goals in Finnish blue fox breeding are to increase litter and pelt size and to improve fur quality. Primary emphasis in selection is given to pelt size because of its major economic importance (Rekilä *et al.*, 2000; Peura *et al.*, 2004). The size of the pelt is known to have a high genetic correlation with animal grading size (0.74), and both are moderately heritable (0.16-0.34) (Peura *et al.*, 2005; Kempe *et al.*, 2008). Animal grading size, in turn, is highly correlated both genetically and phenotypically with feed intake. In practice, blue foxes are given as much food as they can eat during the growth period, in order to attain heavy animals with large pelt size (Rekilä *et al.*, 2000; Kempe *et al.*, 2008). Consequently, most production foxes are extremely fat at pelting time, which may increase the risk of obesity and related health problems (Kempe *et al.*, 2009, 2010).

Along with larger animal size, the frequency of eye infections has increased in the Finnish blue fox population over the past decade. Farmers and veterinarians have observed that eye infections are most frequent at pelting time in November-December and in January-February. The factors that have impaired eye health in the Finnish blue fox population are still unclear, but increased animal size and/or fatness as well as certain pelt quality traits like density are suspected to be predisposing factors.

Veterinary examinations indicate that the main abnormalities causing eye infections in the blue fox are structural defects of the eyelid. On the other hand, primary microbial infections caused by viruses, bacteria, fungi and protozoa are fairly rare. The most common eyelid problem in the blue fox appears to be entropion. This is an inversion of the eyelid, while ectropion is the reverse situation. Entropion, ectropion and keratoconjunctivitis sicca (dry eye), distichiasis, trichiasis and ectopic cilia, have all been found in the blue fox. These defects may cause irritation to the cornea or conjunctiva and expose the eye to secondary eye infection. Eye infection can be very painful for the animal and reduces its wellbeing, and so sick animals need to be culled immediately. This causes economic losses through loss of animals and reduced selection intensity for other selected traits. The genetics behind these eye defects remains unclear, but it can be presumed to be inheritable. Entropion has also been reported in other species such as dog, cat, horse and sheep. Yet, despite the relatively common occurrence of eyelid problems in dogs and cats, genetic evidence is sparse (OMIA 2012). In different sheep breeds, heritability estimates of entropion have varied from 0.08 to 0.21 (Sakul and Kellom, 1996).

The main objective of this study was to estimate genetic parameters for eye infection in the Finnish blue fox population, as well as to determine the potential for efficient selection against it. Culling against eye infection is currently carried out on very few farms, and it is based on the fox's phenotype without information about its relatives. However, the available data on eye infection obtained on the farms could be utilized in the BLUP (best linear unbiased estimation) index. Another objective was to study whether there are any antagonistic genetic and/or phenotypic correlations between eye infection, animal size and pelt character traits.

Material and methods

The experiment was carried out at the fur animal research station of MTT Agrifood Research Finland in Kannus. The details of its experimental design are in a previous article of Kempe *et al.* (2010). Live animal grading traits and eye health were evaluated on days November 7 to 16 in 2005 and on days October 23 to 26 in 2006. Eye health was evaluated subjectively on a scale from 1 to 5 (healthy) at the time of grading. Based on the severity of the infection it was classified into four categories: score 1 represented the worst case where both eyes were infected, swollen and excreting pus; score 2 meant that both eyes were infected, one swollen and excreting pus, score 3 that both eyes were infected; and score 4 that one eye was infected. The phenotypic distribution of the five categories of eye infection was very skewed, and so the trait was analyzed as a binary trait: 0=infected, 1=healthy. This binary eye health trait is denoted by EYE. In animal grading size (gSI), score 1 was the smallest and 5 the largest. Body weight was also recorded at the time of grading (BW_{Nov}).

Pelt character traits were evaluated both with an automatic grading machine (length and colour) and by a professional pelt grader at the auction house of Saga Furs. The pelt character traits classified according to the auction house grading system were recorded on a scale of 1 to 5 (top rating) to correspond to the system used in the Finnish blue fox breeding programme. Pelt size in cm (pSI_{cm}) was measured with the grading machine from the nose to the root of the tail. Pelt density (pDE) was graded into four categories according to underfur thickness or density. Pelt guard hair coverage (pGC) comprised several factors: length, evenness and coverage of guard hair. pGC was graded into five categories, where score 1 (samson) represented the worst quality, scores 2 to 3 were classified as woolly, score 4 was normal and the best score, 5, denoted optimal guard hair coverage. Moreover, pelts classified into the higher pGC categories had more guard hairs than the lower categories. Automated colour shading equipment was used to measure pelt colour, contrast and lightness in pixels. The studied traits were colour darkness (pDA_{pix}) and colour clarity (pCL_{pix}).

Statistical analysis

Preliminary analyses were performed using PROC GLM in the SAS system for Windows 9.1.3 through the SAS Enterprise Guide, Release 4.1 (SAS Institute Inc., Cary, NC, USA). Environmental effects and the interactions between fixed effects were studied using GLM by excluding random effects other than the residual. Only statistically significant ($P<0.05$) effects were included as fixed effects in the animal model used for genetic parameter estimation of EYE.

Variance components were estimated with the DMU program (Madsen and Jensen, 2000), which relies on the restricted maximum likelihood (REML) method. They were estimated using single- and multiple-trait animal models. The animal model was:

$$y = Xb + Wc + Za + e \qquad (1)$$

where y is a vector of observations, b is a vector of fixed effects, and c, a and e are vectors of random non-genetic litter effects, animal genetic effects and residual effects, respectively. Matrices X, W and Z are the corresponding incidence matrices. Fixed and random effects were the same in both single- and multiple-trait analysis. The fixed effects for each trait are presented in Table 1.

In the single-trait analyses, the litter (c), animal (a) and residual (e) effects were assumed to be independent, normally distributed random effects with zero mean and $\mathrm{var}(\boldsymbol{c})=\boldsymbol{I}\sigma^2_c$, $\mathrm{var}(\boldsymbol{a})=\boldsymbol{A}\sigma^2_a$ and $\mathrm{var}(\boldsymbol{e})=\boldsymbol{I}\sigma^2_e$, where $\boldsymbol{I}$ is the identity matrix, $\boldsymbol{A}$ is the additive genetic relationship matrix among animals, σ^2_c is litter environmental variance, σ^2_a is additive genetic variance and σ^2_e is residual variance. Heritability (h^2) and proportion of litter variation (c^2) for a trait were calculated as:

$$h^2 = \sigma^2_a/(\sigma^2_a + \sigma^2_c + \sigma^2_e) \text{ and } c^2 = \sigma^2_c/(\sigma^2_a + \sigma^2_c + \sigma^2_e).$$

Genetic correlations between EYE, size traits and pelt character traits were estimated with the multiple-trait animal model, three traits at a time. Consequently, several (co)variances were estimated for some traits. The estimated genetic correlations and their standard errors were obtained by averaging the estimates from multiple runs. Litter, animal and residual effects were assumed to be independent, normally distributed random effects with zero mean and $\mathrm{var}(\boldsymbol{c})=\boldsymbol{I}\otimes\boldsymbol{C}$, $\mathrm{var}(\boldsymbol{a})=\boldsymbol{A}\otimes\boldsymbol{G}$ and $\mathrm{var}(\boldsymbol{e})=\boldsymbol{I}\otimes\boldsymbol{R}$, where $\otimes$ is the Kronecker product, $\boldsymbol{I}$ is the identity matrix, $\boldsymbol{C}$ is the variance-covariance matrix for litter effects between traits, $\boldsymbol{G}$ is the genetic variance-covariance matrix between traits, and $\boldsymbol{R}$ is the variance-covariance matrix for residual effects between traits.

Table 1. Fixed effects in single- and multiple-trait analysis of the studied traits.

	EYE	pSIcm	pGC	pDE	pDA_{pix}	pCL_{pix}	gSI	BW_{Nov}
house		x	x					
year		x	x					
house-year	x				x	x	x	
sex[1]					x			x
pair[2]						x		x
time of birth[3]	x	x			x			x
age of dam[4]					x			
house×sex							x	
house×time of birth						x		
house×age of dam								
year×sex				x		x		
year×time of birth				x				
year×age of dam				x		x	x	x
sex×age of dam	x	x	x					
pair×time of birth								
pair×age of dam							x	
time of birth×age of dam			x	x			x	

[1] Sex (two classes: male or female).

[2] Pair (three classes: male-male, male-female or female-female pair).

[3] Time of birth (four classes: 104-129, 130-144, 145-160 or 161-180 days from beginning of year).

[4] Age of dam (three classes: 1, 2 or ≥3 years).

Results and discussion

The total number of observations, mean, standard deviation and coefficients of variation for the studied traits are in Table 2. Of the foxes in this two years' single farm study, 54.8% were healthy (n=2068), but the frequency of eye infection was as high as 45.2%. Among the sick animals, 28.2% had mild infection in one eye. The rest had infection in both eyes (15.2%) and the most severe cases had infected eyes that were swollen and excreting pus (1.8%).

Heritability estimates and litter variation

The heritability estimate for EYE was moderate (0.22), suggesting that susceptibility to eye infection has a genetic background in the blue fox (Table 2). Variation in EYE due to the common litter effect was quite low, 5% of the total variance. Because the heritability of EYE has so far not been reported in the literature for fur animals, no direct comparisons with previous results can be made. Genetic evidence of eye defects in dogs is also quite sparse (OMIA 2012). However, in sheep, entropion has been reported to have a genetic background, with heritability estimates varying from 0.08 to 0.21 (Sakul and Kellom, 1996).

Table 2. Total number of observations (n), mean, standard deviation (SD), coefficients of variation (CV), estimated phenotypic variances (σ^2_P), proportion of litter variation (c^2), heritability estimate (h^2) and standard errors (s.e.) of c^2 and h^2 for eye infection, pelt character traits and size traits in the Finnish blue fox.

Trait[1]	n	Mean	SD	CV	σ^2_P	c^2±s.e.	h^2±s.e.
EYE, (0/1)	2,066	0.45	0.50	-	0.23	0.05±0.02	0.22±0.06
pSI_{cm}, cm	1,762	122.46	7.42	6.06	43.71	0.05±0.03	0.51±0.12
pGC, (scores 1-5)	1,773	3.27	0.81	23.12	0.62	0.06±0.03	0.34±0.10
pDE, (scores 2-5)	1,774	3.51	0.58	16.49	0.31	0.00±0.02	0.37±0.10
pDA_{pix}, pixels	1,768	365.28	60.20	16.48	3,768.08	0.08±0.04	0.65±0.14
pCL_{pix}, pixels	1,765	8.69	0.71	8.16	0.30	0.13±0.03	0.19±0.08
gSI, (scores 1-5)	2,062	3.70	0.90	24.33	0.56	0.14±0.03	0.34±0.09
BW_{Nov}, kg	2,061	12.07	1.58	13.07	1.82	0.21±0.04	0.37±0.09

[1] EYE=eye infection, pSI_{cm}=pelt size, pGC=guard hair coverage, pDE=pelt density, pDA_{pix}=colour darkness, pCL_{pix}=colour clarity, gSI=grading size, BW_{Nov}=body weight in November.

The high heritability estimate of 0.51 obtained for pSI_{cm} (Table 2) is close to the heritability of fox body length (Kempe *et al.*, 2010). BW_{Nov} also had fairly high heritability, 0.37. The estimated heritability for the subjectively evaluated gSI was 0.34, which is lower than estimated for pSI_{cm} or BW_{Nov}. A common litter environment had a fairly large effect on BW_{Nov} and gSI, 21% and 14% of the total variance, respectively. However, the effect of a common litter environment was found to decrease as the fox grew larger, accounting for only 5% of the total variation in pSI_{cm}. The heritabilities for pSI_{cm} and gSI were considerably higher in our study than reported by Nikula *et al.* (2000) and Peura *et al.* (2005). One reason for this may be the model used in our study, into which the sex effect was included. Also, the more controlled environmental conditions of a single-farm study can lead to higher heritability estimates (Berg, 1993).

Among the pelt character traits, pDA_{pix} had the highest (0.65) and pCL_{pix} had the lowest (0.19) heritability. These heritability estimates are in line with those presented in Peura *et al.* (2005). Moderate heritability was obtained for both pDE and pGC. The heritability estimate for pDE (0.37) in our study was close to the range reported for mink (from 0.30 to 0.34) (Berg, 1993; Lagerkvist *et al.*, 1994). Variation due to the common litter effect was quite low (from zero to 8%) in the pelt character traits, except for pCL_{pix}. A common litter environment also had a fairly large effect (13%) on the cleanliness of the pelt.

Phenotypic and genetic correlations

Genetic correlations between EYE and the size traits pSI_{cm}, gSI and BW_{Nov} were quite low (from -0.09 to -0.27), with high standard errors (Table 3). These estimates are therefore only indicative. However, the results do give weak support for the conclusion that pSI_{cm} and gSI may have a low unfavourable genetic correlation with EYE. All corresponding phenotypic correlations between EYE and the size traits were also unfavourable, although low (from -0.12 to -0.14). Hence, large

Table 3. Estimated genetic correlations with standard errors (upper triangle) and phenotypic correlations (lower triangle) between eye infection, pelt character traits and size traits in the Finnish blue fox.

Trait[1]	EYE	pSI_{cm}	pGC	pDE	pDA_{pix}	pCL_{pix}	gSI	BW_{Nov}
EYE		-0.23±0.20	-0.39±0.22	-0.06±0.21	-0.09±0.21	0.06±0.25	-0.27±0.21	-0.09±0.23
pSI_{cm}	-0.12		0.18±0.21	0.11±0.19	-0.10±0.19	0.11±0.24	0.70±0.09	0.72±0.09
pGC	0.01	0.05		-0.26±0.20	-0.31±0.19	0.13±0.26	0.30±0.21	-0.02±0.22
pDE	-0.01	0.02	-0.07		0.08±0.19	0.15±0.24	0.25±0.19	0.07±0.19
pDA_{pix}	-0.07	0.01	-0.14	-0.03		0.04±0.25	-0.01±0.20	0.00±0.20
pCL_{pix}	-0.04	0.12	-0.01	-0.01	0.10		0.44±0.21	0.18±0.24
gSI	-0.12	0.46	0.01	0.09	0.08	0.10		0.80±0.08
BW_{Nov}	-0.14	0.59	-0.08	0.06	0.10	0.13	0.59	

[1] EYE=eye infection, pSI_{cm}=pelt size, pGC=guard hair coverage, pDE=pelt density, pDA_{pix}=colour darkness, pCL_{pix}=colour clarity, gSI=grading size, BW_{Nov}=body weight in November.

size may slightly impair eye health at pelting time in November-December. This may be because very fat foxes have subcutaneous fat deposits and possibly rolls of fat on the face, which, along with loose skin, can contribute to the development of entropion and predispose to eye infection (Kempe *et al.*, 2009; Linton and Collins, 1993).

The moderate antagonistic genetic correlation (-0.39) found between EYE and pGC (Table 3) indicates that the short, silky and even guard hair with good covering and deep underfur which is preferred in selection may be genetically connected to impaired eye health. The genetic correlation between EYE and pDE was surprisingly low and hardly differed from zero. However, both of these pelt character traits seem to have an unfavourable genetic connection to fox eye health. The phenotypic correlations of pelt character traits with EYE were close to zero in all cases. A negative genetic correlation was also observed between pGC and pDE, which is in line with the results of Blomstedt (1998) but inconsistent with those reported by Peura *et al.* (2005). Blomstedt (1998) found that increased coverage of guard hairs decreases the number of underfur hairs in the hair bundles, which leads to a negative correlation between the traits.

Genetic correlations of the size traits with the pelt quality traits pGC and pDE were mostly low and favourable. This is in line with the results of Peura *et al.* (2005), where pSI and gSI had mainly low or moderate favourable genetic correlations with corresponding grading and pelt quality traits.

Conclusions

Eye infections among farmed Finnish blue foxes are fairly common in the winter months. We therefore recommend that breeding value estimation for EYE should be based on actual observations recorded on the farms during this period. Overall, the etiology of EYE needs to be studied further, as multiple eyelid problems such as entropion, ectropion and ectopic cilia may be inheritable. Although these defects can be diagnosed without instruments, the diagnosing itself may be challenging for fur farmers. Nonetheless, it is essential to develop tools for the reliable diagnosis of such eye defect in order to determine the primary causes of eye infection and to plan breeding measures against it.

The genetic parameters estimated for EYE indicate that the trait has a genetic background and that genetic improvement through selection might be effective in reducing its prevalence in the Finnish blue fox population. However, single-trait selection for better eye health may impair some of the pelt quality traits such as woolliness, and also decrease animal and pelt size due to a correlated response to selection. Indeed, the results imply that a relatively large amount of pedigree data is needed to increase the accuracy of genetic correlations in variance component estimation. We suggest that the detected antagonistic connections between EYE, size traits and pelt quality traits be taken into account in the national blue fox breeding programme to improve the animals' eye health.

References

Berg, P., 1993. Present knowledge about heritability of different traits in mink. NJF Utredning/Rapport, 90, NJF workshop, Viborg, Denmark. 10 pp.

Blomstedt, L., 1998. Pelage cycle in blue fox (*Alopex lagopus*): A comparison between animals born early and late in the season. Acta Agriculturae Scandinavica 48: 122-128.

Kempe, R., Koskinen, N., Mäntysaari, E. and Strandén, I., 2010. The genetics of body condition and leg weakness in the blue fox (*Alopex lagopus*). Acta Agriculturae Scandinavica 60: 141-150.

Kempe, R., Koskinen, N., Peura, J., Koivula, M. and Strandén, I., 2009. Body condition scoring method for blue fox (*Alopex lagopus*). Acta Agriculturae Scandinavica 59: 85-92.

Kempe, R., Strandén, I., Koivula, M., Rekilä, T., Koskinen, N. and Mäntysaari, E., 2008. Genetic parameters of feed efficiency and its relationships with feed intake, daily gain and animal size traits in Finnish blue fox (*Alopex lagopus*). Scientifur 32: 47-52.

Lagerkvist, G., Johansson, K. and Lundeheim, N., 1994. Selection for litter size, body weight and pelt quality in mink (*Mustela vison*): Correlated responses. Journal of Animal Science 72: 1126-1137.

Linton, L. L. and Collins, B. K. 1993. Entropion repair in a Vietnamese pot-bellied pig. Journal of Small Exotic Animal Medicine 2: 124-127.

Madsen, P. and Jensen, J., 2000. A user's guide to DMU. A package for analysing multivariate mixed models. Danish Institute of Agricultural Sciences, Tjele, Denmark. 22 p.

Nikula, S., Smeds, K., Hietanen, H., Kenttämies, H. and Ojala, M., 2000. Confident behaviour and production traits – results from a field study of foxes. In: Murphy, B. D. and Lohi, O. (eds.), Proceedings of the VIIth Congress in Fur Animal Production, Kastoria, Greece. Scientifur 24, pp. 99-102.

OMIA 2012. Online Mendelian Inheritance in Animals, OMIA. Faculty of Veterinary Science, University of Sydney, 28.3.2012. World Wide Web URL: http://omia.angis.org.au/

Peura, J., Strandén, I. and Smeds, K., 2004. Genetics of litter size, age at first insemination and animal size in blue fox (*Alopex lagopus*). Scientifur 28: 206-210.

Peura, J., Strandén, I. and Mäntysaari, E. A., 2005. Genetic parameters in Finnish blue fox population: Pelt character and live animal grading traits. Acta Agriculturae Scandinavica 55: 137-144.

Rekilä, T., Korhonen, H., Pölönen, I. and Harri, M. 2000. Relationships between feed intake, body mass and skin length in blue fox. In: Murphy, B. D. and Lohi, O. (eds.), Proceedings of the VIIth International Scientific Congress in Fur Animal Production, Kastoria, Greece. Scientifur 24: 155-158.

Sakul, H. and Kellom, T. R., 1996. Heritability of entropion in several US sheep breeds. Small Ruminant Research 23: 187-190.

Strandén, I. and Vuori, K., 2006. RelaX2: pedigree analysis program. In: Proceedings of the 8th World Congress on Genetics Applied to Livestock Production, August 13-18, 2006, Belo Horizonte, MG, Brazil, CD-ROM communication No. 27-30.

The size and constitution of the body of dark-brown mink under different breeding schemes

L.V. Gerasimova
Bashkir State Agrarian University, 34, 50 let Oktyabrya str., Ufa, 450001, Russia; lysigera@mail.ru

Abstract

Body size of young animals of two generations in a Russian mink farm, depending on the scheme of selection of parents, was studied. Traits 'Live weight', 'Body length', 'Quality of downiness', 'Color', 'Reproductive ability' taken into account when designing the scheme of combinations of crossing. The best crossing schemes were revealed using homogeneous and heterogeneous selection.

Keywords: *Neovison vison*, selection, live weight, body length, quality of fur

Introduction

American mink (*Neovison vison*) is largely used as a fur species and its world-wide population has rapidly grown due to increased commercial farming. But the livestock in Russia decreases due to economical difficulties in past several decades. In 2012 the country produced about 410 thousands animals or only 3-3.5% of the world production (Balakirev, 2011). It is known (Plotnikov, 2010), that in many areas livestock sector there is a tendency of the transition from breeds to lines and crossings between them. This is not only due to the desire to use heterosis, but because of the impossibility combine many of the economically important traits in one breed or line. Therefore it is important to increase profitability of the industry and new more efficient breeding schemes should be undertaken. We studied several traits of body size of the American mink, depending on various homogenous and heterogeneous schemes of selection of crossed pairs.

Material and methods

The study was carried out by using a population of standard dark-brown mink of Iglinskiy fur farm (the Republic of Bashkortostan, Russia). Live weight, body length (from tip of nose to base of tail), fur quality and fertility of the experimental animals were measured and estimated. The information about the shape of their distribution was obtained by performing the Kolmogorov-Smirnov test to verify the assumption of normality of variables before using methods of parametric statistics. Only mink with extreme values of the measured traits were included to further analysis. Two different generations were utilized to study assessing parameters. First of them consisted of 14 types of pairs according to a combination of traits (Table 1) of males ('Heavy': ≥2.8 kg; 'Lightweight': ≤2.4 kg; 'Long': ≥58.0 cm; 'Short': ≤52.0 cm; 'Large': ≥2.8 kg and ≥58.0 cm; 'Small': ≤2.4 kg and ≤ 52.0 cm; 'Excellent fur': fifth of five maximal points in the density and intensity of black color of the fur; 'Very fertile': ≥8 live and stillborn puppies per a fertilized female) and females ('Heavy':≥ 1.8 kg; 'Lightweight': ≤1.4 kg; 'Long': ≥48.0 cm; 'Short': ≤42.0 cm; 'Large': ≥1.8 kg and ≥48.0 cm; 'Small': ≤1.4 kg and ≤42.0 cm; 'Excellent fur': fifth point; 'Very fertile': ≥8

Table 1. Scheme of the mink pairs selection (1st generation).

Group	Father	Mother
1×1	Heavy(10)[1]	Heavy (16)
1×2	Heavy (8)	Lightweight (17)
2×1	Lightweight (12)	Heavy (29)
2×2	Lightweight (9)	Lightweight (11)
3×3	Long (9)	Long (21)
3×4	Long (11)	Short (23)
4×3	Short (6)	Long (14)
4×4	Short (10)	Short (20)
5×5	Large (18)	Large (29)
5×6	Large (18)	Small (35)
6×5	Small (12)	Large (21)
6×6	Small (10)	Small (16)
Fu×Fu	Excellent fur (10)	Excellent fur (41)
Re×Re	Very fertile (5)	Very fertile (16)

[1] Number of fathers and mothers who participated in the pairing in the corresponding group is indicated in parentheses.

Table 2. Scheme of the mink pairs selection (2nd generation).

Group	Father			Mother		
	n[1]	Father of father	Mother of father	n[1]	Father of mother	Mother of mother
5/5×5/5	5	Large	Large	15	Large	Large
5/5×5/6	5	Large	Large	12	Large	Small
5/6×5/5	5	Large	Large	13	Large	Large
5/6×5/6	5	Large	Small	12	Large	Small
5/5×3/4	5	Large	Large	15	Long	Short
3/4×5/5	6	Long	Short	16	Large	Large
1/2×4/3	4	Heavy	Lightweight	13	Short	Long
4/3×1/2	5	Short	Long	12	Heavy	Lightweight
Fu/Fu×Fu/Fu	6	Excellent fur	Excellent fur	16	Excellent fur	Excellent fur
Re/Re×Re/Re	4	Very fertile	Very fertile	15	Very fertile	Very fertile

[1] Number of fathers and mothers who participated in the pairing in the corresponding group.

live and stillborn puppies). Second population was created by using the best part of the offspring of first generation and consisted of 10 types of pairs (Table 2). Inbreeding was absent in both generations. The offspring of both generations were composed of unrelated individuals, as well as half and full sibs. The measurements of unrelated animals were recorded in October at their age of 5 months. The computer program 'STATISTICA 6' (StatSoft) was used to perform descriptive statistical analysis.

Results

An analysis of the results of the breeding of first generation (Table 3) revealed that homogeneous selection of parents by using only criterion 'Live weight' without consideration of the trait 'Body length' leads to a negative result in the offspring. Live weight of males of the group 1×1 was lower by 6.8% in comparison to the group 2×2 (the differences are statistically significant at the level $P<0.05$). Heterogeneous selection for the breeding of heavy and lightweight animals also leads to decrease of the trait: -4.6% (differences are not significant) and -7.3% ($P<0.05$) for the groups 1×2 and 2×1, respectively. Lengths of bodies of animals were larger under heterogenic choice of parents (groups 1×2, 2×1, 51.7-51.8 cm) in comparison to the homogeneous type (samples 1×1 and 2×2, 50.7-51.0 cm). A selection of mink by using only the parameter 'Length of body' without recording of the trait 'Body weight' does not increase the live weight of the offspring. Males which were derived from parents with long bodies (3×3) had lower mass (-15.4%, $P<0.001$) than the

Table 3. The average live weight and body length of the mink of the 1st generation.

Group	Males			Females		
	n[1]	Live weight, g	Body length, cm	n[1]	Live weight, g	Body length, cm
1×1	36	2,173.3±83.56	50.7±0.33	32	1,569.0±38.38	45.4±0.84
1×2	24	2,225.4±73.78	51.7±0.50	26	1,611.8±31.63	45.0±0.58
2×1	15	2,162.8±57.88	51.8±0.43	14	1,557.6±46.25	47.3±0.52
2×2	25	2,332.5±112.83	51.0±0.84	25	1,483.3±67.82	46.4±0.43
3×3	27	1,894.2±63.78	52.4±1.36	29	1,380.8±36.90	45.1±0.52
3×4	25	1,974.6±41.62	51.8±0.41	23	1,397.4±40.85	44.9±0.42
4×3	22	2,218.4±81.54	51.2±0.85	22	1,533.0±64.55	46.3±0.45
4×4	19	2,240.0±96.71	50.8±0.60	22	1,673.1±37.90	44.9±0.44
5×5	48	2,252.1±132.26	51.1±0.36	37	1,676.7±86.31	46.6±0.38
5×6	35	2,123.8±58.09	50.6±0.61	39	1,390.9±50.58	47.2±0.49
6×5	27	2,247.7±68.05	50.2±0.55	20	1,706.7±52.88	45.8±0.47
6×6	17	2,128.2±71.01	50.1±0.41	21	1,555.9±40.28	44.5±0.30
Fu×Fu	42	1,854.5±51.82	49.4±0.48	44	1,420.6±36.74	44.8±0.34
Re×Re	36	1,858.1±50.63	49.5±0.42	35	1,442.4±29.36	44.6±0.35

[1] Number of the animals.

offspring of relatively short mink (4×4), but they developed better body length (+3.2%, $P<0.05$). Heterogeneous selection leads to an intermediate position in these parameters.

A selection of pairs was the most efficient in using both traits simultaneously. The males of the group 5×5 revealed more live weight (+3.2%, $P<0.05$) and body length (+2.0%, $P<0.05$) than the sample 6×6.

We found a maternal effect of the heterogeneous selection on live weight of animals. Males of 'Large' mothers and 'Small' fathers (6×5) had almost the same live weights (2,247.7 g) as males of the group 5×5 (2,252.1 g). 'Small' mothers and 'Large' fathers (5×6) formed the offspring on the level of the sample 6×6 (2,123.8 and 2,128.2 g, respectively). Body length reveals a tendency of a paternal effect. Males of the group 5×6 were longer in comparison to the sample 6×5.

Measurements of females revealed that there is a tendency of parental effect as a result of crossing of relatively heavy and lightweight parents. Live weights and body lengths of animals in groups 1×1 and 1×2 were larger than in 2×1 and 2×2. Selection of homogeneous parents of females by using the only criterion 'Body length' did not increase their body weight. Offspring of 'Long' parents had less mass (-17.4%, $P<0.001$) than puppies of 'Short' pairs. Females of the groups 3×3 and 4×3 show larger body lengths (45.1 and 46.3 cm, respectively) in comparison to the samples 3×4 and 4×4 (44.9 cm). The differences of the sample 4×3 with 3×4 and 4×4 are statistically significant at the level $P<0.001$. Simultaneous selection of pairs by using parameters 'Live weight' and 'Body length' was relatively effective for females. Mink of the group 5×5 had better values of these parameters (+7.8%, $P<0.01$ and +4.7%, $P<0.001$) than animals of the group 6×6.

Heterogeneous selection by body size also revealed a maternal effect on the live weight and paternal one on the length of the body. Females from 'Large' mothers and 'Small' fathers (6×5) had a mass of 1,706.7 g, which is even higher (1.8%) than in the offspring of larger parents (5×5). Mink of 'Small' mothers and 'Large' fathers (5×6) had less (-10.6%, $P<0.001$) mass than females from small parents (6×6). Again, as in males, we observed some paternal influence and heterosis effect, when females of 'Large' fathers and 'Small' mothers were even longer (+1.3%) in comparison to mink of larger parents. Values of body lengths of animals in 6×5 were in an intermediate position between parameters of the groups 5×5 and 6×6.

Selection of mink by quality of fur and color, as well as by the level of fertility, without consideration of body size, negatively affected to the traits 'Live weight' and 'Body length'. These traits were here even less than in offspring of homogenous selection of 'Lightweight' (2×2), 'Short' (4×4) and 'Small' (6×6) parents. All differences are statistically significant at different levels, excluding differences of 6×6 with Fu×Fu (males, 'Body length'), 2×2 (females, 'Live weight'), 4×4 and 6×6 (females, 'Body length') with Fu×Fu and Re×Re.

The results of measurements of mink of the second generation (table 4) revealed that the best development of body sizes and their constitution in male and female mink was achieved in the sample 5/6 x 5/5 (father: heterogeneous selection by body size; mother: homogeneous selection by 'Large' parents). The males reached an average live weight of2699.5 g, body length of 52.5 cm (1794.0 g and 46.1 cm for females, respectively). A good development of these traits was obtained

Table 4. The average live weight and body length of the mink of the 2nd generation.

Group	Males			Females		
	n[1]	Live weight, g	Body length, cm	n[1]	Live weight, g	Body length, cm
5/5×5/5	30	2,409.1±48.52	52.1±0.32	34	1,648.0±24.08	45.9±0.22
5/5×5/6	28	2,163.7±72.06	50.2±0.57	24	1,627.1±54.84	45.8±0.48
5/6×5/5	29	2,699.5±110.04	52.5±0.56	26	1,794,0±37.43	46.1±0.37
5/6×5/6	25	2,316.0±58.79	51.3±0.47	27	1,568.0±40.59	45.0±0.30
5/5×3/4	33	2,370.8±73.57	51.0±0.46	32	1,678.4±30.44	45.1±0.31
3/4×5/5	30	2,129.7±49.65	50.6±0.39	35	1,693.6±33.80	45.7±0.32
1/2×4/3	27	2,180.3±67.98	50.4±0.47	23	1,594.0±45.86	44.8±0.31
4/3×1/2	27	2,409.6±113.19	50.4±0.82	27	1,780.8±33.61	46.3±0.23
Fu/Fu×Fu/Fu	29	2,258.2±119.86	51.3±1.0	34	1,766.3±80.19	46.6±0.99
Re/Re×Re/Re	36	2,131.1±111.95	50.9±0.75	39	1,650.8±44.9	45.1±0.41

[1] Number of the animals.

for males under a homogeneous selection of progenitors (the group 5/5 x 5/5). Females under such conditions had a good developed body length but not good live weight. Offspring of mink of both sexes in 4/3 x 1/2 (father: heterogeneous selection on body length; mother: heterogeneous selection on live weight) was relatively large, excluding relatively small length of males.

A homogeneous selection by fur quality and fertility rate led to form puppies with relatively small body size in both generations (the group Fu/Fu×Fu/Fu and Re/Re×Re/Re).

Discussion

Traditionally, selection in mink breeding in Russia has been focused on mass selection for fur quality and body composition traits. Breeding programs in Russia included predominantly mating of the best animal pairs (mainly with respect to live weight or body length) (Kuznetsov *et al.*, 1974). The role of heterogeneous selection here has often been underestimated (Karyukhina *et al.*, 2011). We used the possibilities of the heterogeneous selection, which is known (Ovsyannikov, 1976) to undermine the genetic stability, increasing the possibility of development of symptoms in many intercrossing ways. The data analysis shows that the selection of such prime indicator for the selection and matching, as the live weight, without the body development has not been effective enough. Selection and matching of mink according to the range of indicators of body size – live weight and body length at the same time is more effective. This confirms the opinion of M.V. Mina and G.A. Klevezal (1976), who believe that the size of the animal can not be regarded in isolation, without reference to other properties of the organism, because they are formed in the process of ontogenesis.

Thus, we conclude that for good development of the trait 'Body size' in mink of first generation the combinations 'large father and large mother' or 'small father and large mother' are suitable options. In the second generation the father may be obtained under a homogeneous selection of large animals, or under heterogeneous selection, when a male is large and a female is small. The mother can be obtained by homogeneous selection of large parents. Homogeneous selection of the mink (on the quality of downiness and color of pelts, on the reproductive ability, on the length of the body without the development of body weight) negatively affects the body size of the young generation.

We plan further to use the results to create an Animal model and to develop indices of productivity based on body weight, body length, quality of fur and level of fertility as an alternative to the use of line crossing in the breeding work. We are going as well to use the results of the study obtained at the mink in the selection of sables (*Martes zibellina* L.). Such investigations are initiated by us in 2012 at the fur farm Oktyabrskoye Ltd. (The Republic of Bashkortostan, Russia). The research is important for the animal, which are used for a long-term (10-12 years) period. Sables have got a late sexual maturity and relatively low fertility, making difficult their breeding work.

References

Balakirev, N.A., 2011. Condition and prospects of the cellular fur farming in Russia. Journal of Rabbit Breeding and Fur Farming 3: 5-8 (in Russian).

Gerasimova, L.V., and Fenchenko, N.G., 2008. Biological and economic useful features of ferrets in terms of cell breeding. Gilem, Ufa, Russia, 385 pp. (In Russian).

Karyuhina, S.S., Budyanu, I.V., and Steshenko A.V., 2011. Characteristic features of the mink pups growth at age of 45 days. Journal Rabbit Breeding and Fur Farming 2: 18-19 (in Russian).

Kuznetsov, G.A., Tsepkov, N.M., Evreinov, A.G., Kulichkov, B.A., and Polunina, M.M., 1974. Mink breeding to increase the size. In: Proceedings of the 1st All-Union. scientific. conference: Biology and pathology of fur-bearing animals. Petrozavodsk, Russia, pp. 14-15 (in Russian).

Mina, M.V., and Klevezal, G.A., 1976. Growth of animals. In: Problems of Developmental Biology, Moscow, 291 p. (in Russian).

Ovsyannikov, A.I., 1972. Maternal inheritance. In: Pigs genetics and the theory of selection in pig breeding. Kolos, Moscow, pp. 5-10 (In Russian).

Plotnikov, V.G., 2010. The evolution of technology in rabbit breeding. Journal Rabbit Breeding and Fur Farming 1: 17-21 (in Russian).

Reproductive efficiency of scanbrown and mahogany mink females, selected for litter size

M. Brzozowski, B. Maćkowiak-Kondrak and R. Głogowski
Warsaw University of Life Sciences, Ciszewskiego 8, 02-786, Warsaw, Poland; marian_brzozowski@sggw.pl

Abstract

Reproduction results of one-year-old mink females were compared with the data obtained from older mothers, selected for the number of kits weaned in the first reproductive season. Two color types: Scanbrown and Mahogany were studied. The size of the litter in three-year-old females was significantly higher (P>0.05) than that in one-year-old (6.15 vs. 5.01, respectively) in Scanbrown mink, while in Mahogany significantly more kits per female were born (P>0.05) in the second reproduction year than those in the first (6.18 vs. 5.34, respectively). There were no statistically significant differences in reproduction results between color types.

Keywords: scanbrown and mahogany color types, primiparous and multiparous females

Introduction

The main criterion of the evaluation in animal breeding is the effectiveness of reproduction. The profitability increases when larger liters are weaned, therefore the improvement of mating results are of special importance in mink production (Hansen and Berg, 2008; Koivula *et al.*, 2008, Lefèvre and Murphy, 2008).

The aim of the study was to compare the reproduction results of mink females in different age and color type as a result of selection for the number of kits weaned from female.

Material and methods

On the mink farm in north central Poland, breeding records of females in various reproductive seasons (1st, 2nd and 3rd) were collected from three following years and analyzed. The main continuous selection criterion on the farm (commenced before the present experiment was conducted), was the number of kits weaned per female in the first reproductive season, therefore we assumed, that all females used in reproduction on that farm originated from litters that were born under this selection. Scanbrown and Mahogany females were evaluated (n=278 and n=320, respectively) (Table 1).

The average litter size (number of kits born) and the number of weaned kits were evaluated. Results are presented in tables as means±SD. The statistical significance of differences was assessed with t-Student's test (Statistica 6, StatSoft, Cracov-Poland).

Table 1. The number of females evaluated in the study.

Reproductive season	Color type		Total
	Scanbrown	Mahogany	
1	162	173	335
2	78	83	161
3	38	64	102

Results and discussion

The results concerning litter size at birth and weaning are presented in tables (Tables 2 and 3).

The evaluated reproductive parameters differed between continuous seasons in both estimated color types (Tables 2 and 3). Primiparous females had smaller litters than multiparous mink.

Table 2. The litter size at birth.

Reproductive season	Color type	
	Scanbrown	Mahogany
1	5.41 ± 2.56^{A}	5.79 ± 2.42^{a}
2	5.85 ± 2.71	6.54 ± 2.21^{b}
3	6.88 ± 2.18^{B}	5.96 ± 2.41

$^{A, B}$ Values in columns with different superscript letters differ significantly at $P<0.01$.
$^{a, b}$ Values in columns with different superscript letters differ significantly at $P<0.05$.

Table 3. The number of kits weaned per female.

Reproductive season	Color type	
	Scanbrown	Mahogany
1	5.01 ± 2.65^{a}	5.34 ± 2.59^{c}
2	5.28 ± 2.87	6.18 ± 2.31^{d}
3	6.15 ± 2.33^{b}	5.44 ± 2.53

$^{a, b, c, d}$ Values in columns with different superscript letters differ significantly at $P<0.05$.

There were no observed statistical significantly differences between Scanbrown and Mahogany females in the same reproduction season.

Similar tendency was shown for the number of kits weaned per female, as it was presented previously in numerous papers (Sulik and Felska, 2000; Rozempolska-Rucińska *et al.*, 2000; Rozempolska-Rucińska *et al.*, 2002; Socha and Markiewicz, 2001; Święcicka, 2006).

However, reproduction results obtained in this study were poorer than those observed by Sulik *et al.* (2007) and Ślaska *et al.* (2009), likely reflecting the effect of other factors like nutrition or husbandry.

The body size of female mink was proved to affect the litter size (Koivula *et al.*, 2010) but it seems probable that the color type and/or the age should also be taken under consideration in assessing the breeding value of mink.

Nevertheless it may be concluded, that significantly different number of weaned kits observed in various seasons, may indicate that the highest reproductive efficiency in Scanbrown females appears later than that in Mahogany. However, these tendencies require further elucidation.

References

Hansen, B.K. and Berg, P., 2008. Reduced litter size and percent kits alive is a consequence of selecting for high body weight. Scientifur 32: 15.

Koivula, M., Strandén, I. and Mäntysaari, E.A., 2008. Genetic parameters for litter size and grading traits in Finnish mink population. Scientifur 32: 53-58.

Koivula, M., Strandén, I. and Mäntysaari, E.A., 2010. Genetic and phenotypic parameters of age at first mating, litter size and animal size in Finnish mink. Animal 4(2): 183-188.

Lefèvre, P. and Murphy, B.D., 2008. Physiological constrains on litter size in mink. Scientifur 32: 13-14.

Lisiecki, H. and Sławoń, J., 1980. Hodowla norek, PWRiL, Warszawa (in Polish).

Rozempolska-Rucińska, I., Jeżewska, G., Tarkowski, J., Socha, S. and Zięba, G., 2000. Charakterystyka cech użytkowych norek odmiany standard. Scientific Annals of Polish Society of Animal Production 53: 45-53 (in Polish).

Rozempolska-Rucińska, I., Jeżewska, G., Tarkowski, J. and Zięba, G., 2002. Charakterystyka poziomu cech funkcjonalnych norek pastelowych. Scientific Annals of Polish Society of Animal Production 64: 57-62 (in Polish).

Socha, S. and Markiewicz, D., 2001. Analiza czynników wpływających na plenność norek. Medycyna Weterynaryjna 11: 840-843 (in Polish).

Sulik, M. and Felska, L., 2000. Ocena wpływu samca i terminu krycia na plenność i długość ciąży u norek. Scientific Annals of Polish Society of Animal Production 53: 115-121 (in Polish).

Sulik, M., Seremak, B. and Matyja, A., 2007. Analyse des Einflusses der gewählten Faktoren auf die Wurfgröße bei Nerzen mit Berücksichtigung verschiedener Farbschläge. Archive Tierzucht 50: 214-219 (in German).

Ślaska, B., Rozempolska-Rucińska, I. and Jeżewska-Witkowska, G., 2009. Variation in some reproductive traits of mink (*Neovison vison*) according to their coat colour. Annals of Animal Science 9: 287-297.

Święcicka N., 2006. Wpływ wieku samic oraz terminu wykotu na wyniki rozrodu u wybranych odmian norek. Komunikaty Naukowe Polskiego Towarzystwa Zootechnicznego w Bydgoszczy 71: 18-21 (in Polish).

Investigation of placental scars in mink females

L. Tinggaard[1], *T. N. Clausen*[2] *and P.F. Larsen*[2]
[1]*Kopenhagen Farm, Herningvej 112C, Holstebro, 7500, Denmark;* [2]*Kopenhagen Research, Agro Food Park 15, Aarhus N, 8200, Denmark; lti@kopenhagenfur.com*

Abstract

After the nursing period 2010 the uteri from 245 females with kits were investigated for placental scars (implantation sites). The litter size varied from 1 to 14, the number of placental scars varied from 3 to 18. The high number of placental scars showed that the ovulation rate was high. There were on average 3.5 foetuses more than the number of kits counted the day after birth. It is expected that some foetus are lost in the pregnancy period and some during birth, where the female often eat the dead kits. Females in different body conditions at birth had the same number of scars but there were fewer kits if the female was fat. There was a tendency to get more live born kits at birth, if the feeding and management was optimised in the pregnancy period and the females were in the right body condition at parturition.

Keywords: placental scars, body condition, litter size.

Introduction

The gestation period is very important for the females' reproductive performance in the reproduction period. Many factors have influence on the gestation, such as feeding, management and disease. Investigations have shown that staining (Turnbull blue method) of placental scars (implantation sites on uterus) is a useful tool to estimate how many foetuses a female has been pregnant within the latest breeding period (Hammer *et al.*, 2008).

In order to investigate the presence of placental scars in the females it is important to do it as soon as possible after parturition, because the scars disappear after some time. If the investigation takes place in the autumn (4-7 months after parturition) too many females will be estimated as barren and the number of kits will be evaluated too low compared to the number of kits the females actually have given birth to. In primiparous females it is not possible to detect placental scars in the uterus 12 month after the birth (Hammer *et al.*, 2008). Hammer *et al.* (2008) found that 39% of primiparous females considered barren had placental scars, which indicates that the females had been pregnant, but the kits died late in pregnancy or during birth. Similar data from 2005 showed that 27% of the barren females had placental scars.

After the nursing period 2010 the uteri from females with kits were analysed for placental scars, to study the effect of female body condition, the counted number of kits after birth and the number of scars in the uterus.

Materials and methods

The trial was conducted with 245 2-year-old females of the brown/glow mink type. All females had given birth to between 1 and 14 kits. The number of live and stillborn kits was counted the day after birth. Females' body condition was evaluated at birth and half of the females were weighed on day 28 and 42 after parturition. All females were euthanized July 2, with PentobarbitalNa.

The uterus was removed, washed with isotonic sodium chloride solution and frozen in isotonic sodium chloride for later staining of placental scars after the Turnbull blue method. Turnbull blue method was modified after the staining method, which first was developed by Salewski (1964) for rats and then used for European hares (Bray *et al.*, 2003).

The statistical calculations were performed using the statistical program SAS (SAS Institute Inc., 1996). For weight differences the procedures GLM (SS4), LSMEANS/PDIFF were used and for calculations of litter size and placental scars a non-parametric test, GENMOD was used. A 5% significance level was applied.

Results and discussion

The females were euthanized in early July (2/7), to obtain the best possible estimate of the number of scars. It is possible that foetuses which died early in the gestation period can be completely reabsorbed without leaving any visible scar tissue.

The investigation showed that in 12 females (4.9%) fewer scars were observed than the number of kits counted the day after birth. In 17 females (6.9%) the number of scars and counted kits were identical, but for the majority of females we found more scars than kits. For the investigated females the average litter size was 7.78 (live + dead), and there were on average 11.32 placental scars (Table 1). This means that, on average, there were 3.5 foetuses more than the counted numbers of

Table 1. Body condition at parturition, weight of the females, live born and stillborn kits at birth, and number of placental scars.

Body condition at birth[1]	Live born kits per litter[2]	Stillborn kits per litter[2]	Placental scars, no.	Weight day 28, gram[2,3]	Weight day 42, gram[2,3]
Thin	7.2 (2.7)^{ab}	1.2 (2.6)^{a}	11.4 (1.9)	1,341 (210)^{b}	1,189 (200)^{b}
Normal	7.7 (2.3)^{a}	0.4 (1.1)^{b}	11.3 (2.9)	1,389 (153)^{b}	1,210 (153)^{b}
Fat	6.9 (2.5)^{b}	0.3 (0.7)^{b}	11.3 (2.4)	1,523 (154)^{a}	1,329 (169)^{a}
P-value	0.02	0.002	NS	<0.001	0.001

[1] Estimated according to Hynes *et al.*, 2004.

[2] Numbers in brackets are standard deviation; NS indicate not significant, different superscripts (^{a,b,c}) indicate differ significantly ($P<0.05$).

[3] About half of the females were weighed.

kits. Hammer *et al.* (2008) found on average 2 placental scars more than the number of kits, when counting 2-3 month after parturition. At counting 4-5 months and 6-7 months after parturition, the number of kits in the litter was higher than the number of scars. Females investigated more than 12 month after birth retained no placental scars (Hammer *et al.*, 2008). In dogs' similar studies have shown that 11-13% of the foetuses die and are reabsorbed (Andersen and Simpson, 1973). In pigs 30-40% of potential piglets are lost before farrowing (Vonnahme *et al.*, 2002).

Foetal death occurs for various reasons such as diseases and insufficient feeding, but the general strategy is that once a female has become pregnant the foetus has priority, and the female will utilise her own body reserves to nourish the foetus, that most often survives, but may get permanent effects of the foetal malnutrition.

An earlier investigation of low protein diet to females in the gestation period has shown an increase in number of barren females and in number of barren females with placental scars (Clausen, 2008). Up to 50% of the barren females in the study by Clausen (2008) had been pregnant versus about a third on normal farms (Hammer *et al.*, 2008; unpublished data). The result corresponds with similar studies in pigs, where low protein levels in the feed increases the frequency of embryonic dead and foetal mummification (Rasbech, 1981).

Foetal death is most often seen in females with many embryos in the uterus. In the investigation we found up to 18 scars, but did not count more than 14 kits (Figure 1) Clausen (2008) found

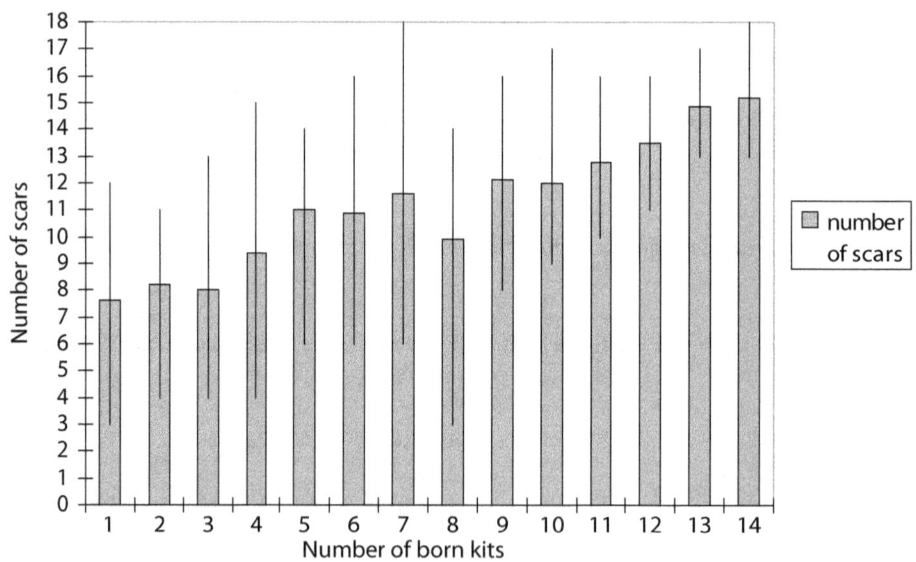

Figure 1. Number of live born and dead kits counted the day after birth, compared with the average number of scars on uterus. The lines go from the minimum counted number of scars to the maximum counted. At the smallest and largest litter size there are 5-7 females in each group, at the middle there are 24-27. The correlation between the counted number of kits and the average number of scars (r=0.958, P<0.05).

that approx. 50% of the barren females had 10 or more foetuses. These studies suggest that the load on the uterus may be too severe (high number of kits), and the embryos are then aborted or reabsorbed after foetal death. In pigs the capacity of the uterus to nourish the foetuses is, by some, considered to be the limiting factor for survival of the foetus (Rhodes *et al.*, 1991). Vonnahme *et al.* (2002) showed a positive correlation between the number of foetuses in the end of the pregnancy and the length of the uterus horns.

Thus, foetal death is also recorded in females with few embryos in the uterus. In pigs, 4 embryos are necessary to maintain pregnancy. In a recent study of barren mink females approximately 25% had been pregnant with only one embryo (Clausen, 2008). In the present study we did not find less than 3 placental scars in females giving birth (Figure 1), suggesting this being a lower limit to maintain pregnancy in mink females.

Because the females eat some of the stillborn kits, before they are counted, there will be a difference between counted kits and number of scars. According to Gade and Malmkvist (2005) the average female loses about 2 embryos pr litter. However higher numbers of dead / weak kits are seen if the females are fat at the birth (Clausen *et al.*, 2007; Bækgaard *et al.*, 2007).

An evaluation of the females' body condition at birth, number of counted kits and number of scars (Table 1) showed that females with different body condition at parturition had the same number of placental scars but fewer kits were born if the female was fat. This suggests that fat females loose more kits during and shortly after birth.

In general, the more placental scars we observed the larger the litter size (Figure 1), and also in the smallest litters there were many scars. Elofson *et al.* (1989) found on average 7.4 foetuses in the uterus of females euthanized in the gestation period, but counted 13.1 (from 7 to 21) copora lutea (ovulations) in the same females. The potential to increase the number of kits at birth seems to be high.

Conclusion

In this study the litter size varied from 1 to 14 kits, the number of scars varied from 3 to 18. There were on average 3.5 foetuses more than the number of kits counted the day after birth. This suggests that some kits are lost in the pregnancy period, others during or shortly after birth, where the female have eaten the dead kits.

The high number of placental scars suggested a high ovulation rate, resulting in many eggs ovulated and implanted. Females with different body conditions at birth had the same number of placental scars but there were fewer kits in litters born by fat females. Therefore, to get more live born kits at birth the feeding and management in the pregnancy period should be optimised and the body condition of the females controlled until birth.

References

Andersen, A.C. and Simpson, M.E., 1973. The ovary and reproductive cycle of the dog (Beagle). Geron-X, Inc., Los Altos, California. 11-14.

Bray, Y., Marboutin, E., Péroux, R., and Ferron, J. 2003. Reliability of stained placental scars counts in European hares. Wildlife Society Bulletin 31: 237-246.

Bækgaard, H., Hansen, M. U., Sønderup, M. and Clausen, T. 2007. Connection between early kit mortality and the body condition of the females and feed comsumption from January to birth. Annual Report 2006, Danish Fur Breeders Research Center, Holstebro, Denmark, pp. 185-192. (In Danish).

Clausen, T.N., 2008. Feeding with low level of protein in the gestation period reduce the litter size and skin length. In Danish. Intern Report (DJF Husdyrbrug), nr. 10, September 2008, pp. 45-51.

Clausen, T. N., Sandbøl, P. and Hejlesen, C., 2007. Body score of females in the winter and breeding period. Annual Report 2006, Danish Fur Breeders Research Center, Holstebro, Denmark, pp. 181-184. (In Danish).

Elofson, L., Lagerkvist, G., Gustafsson, H. and Einarsson, S., 1989. Mating systems and reproduction in mink. Acta Agric Scand 39: 23-41.

Gade, M. and Malmkvist, J. 2005. Behaviour around birth in relation to early kit mortality in mink. Annual Report 2004, Danish Fur Breeders Research Center, Holstebro, Denmark. pp. 7-14. (In Danish).

Hammer, A. S. Hammer-Jensen, T., Sørensen, C. M. and Clausen, T. N. 2008. Evaluation of staining methods for quantification of placental scars in farm mink (*Mustela vison*). Annual Report 2007, Danish Fur Breeders Research Center, Holstebro, Denmark, pp. 77-80. (In Danish).

Hynes, A.M., Rouvinen-Watt, K. and Arm-strong, D., 2004. Body condition and glycemic control in mink females during reproduction and lactation. VIII International Scientific congress in Fur Animal Production, Den Bosch, the Netherlands, September 15-18. Scientifur, 28, 3, pp. 79-86.

Rasbech, N.O., 1981. Husdyrenes reproduktion 3, Gynækologi 4. 123pp. (In Danish).

Rhodes, M.T., D.L. Davis and J.S. Stevenson, 1991. Flushing and Altrenogest affect litter traits in gilts. J. Anim. Sci. 69: 34-40.

Salewski, E.. 1964. Färbemethode zum makroskopischen Nachweis von Implantationsstellen am Uterus der Ratte. Naunyn-Schmiedebergs Archiv für experimentelle Pathologie und Pharmakologie: 247-367.

SAS Institute Inc., 1996. SAS – Systems for Mixed Models. SAS Institute Inc., Cary, NC, USA, 633 pp.

Vonnahme, K.A., Wilson, M.E., Foxcroft, G.R. and Ford, S.P., 2002. Impacts on conceptus survival in a commercial swine herd. J. Anim. Sci 80: 553-559.

Comparative analysis of morphometrics of wild and farm foxes (*Vulpes vulpes* L.) – preliminary results

M. Zatoń-Dobrowolska[1], M. Moska[1], H., Wierzbicki[1], P. Przysiecki[2] and A. Mucha[1]
[1]Wroclaw University of Environmental and Life Sciences, Department of Genetics, Kozuchowska 7, 51-631 Wroclaw, Poland, [2]Przysiecki P. Institute of Agriculture, State School of Higher Education in Leszno, Mickiewicza 5, 64-100 Leszno, Poland; magdalena.dobrowolska@up.wroc.pl

Abstract

Nowadays, an interesting aspect of animal farming is the possibility to analyze genetic, morphometric and behavioural changes in farm animals, which have taken place as a result of selective breeding. Then, the comparative analysis with wild animals can be done. Fur animals (fox, mink, raccoon dog) have relatively short history of domestication, so an interesting point is if this process affected body measurements of those species. The aim of the present study was to analyze and compare several body measurements of wild and farm foxes. The analyzed traits were: body weight, body length and circumference, tail length, ears height, length of front and hind limbs, length of front and hind feet. Farm animals were characterized by higher values of all traits (except tail length and ear height), than wild ones and these differences were statistically significant. Smaller differences between wild and captive populations were found for traits which were not crucial for species survival (e.g. tail length and ears height). Comparison between sexes showed that they differed in many size measurements, but measurements of females from the farm population and males from the wild population were comparable.

Keywords: body size measurements, selection pressure, sex differentiation

Introduction

The red fox (*Vulpes vulpes* L.) is an example of species, which lives both in wild and captive populations. Wild and captive animals are subject to different selection pressure and are affected by different environmental factors. Therefore, individuals living in wild and those kept on farms may differ in terms of morphology, anatomy, genetics and behaviour. Very often process of natural and artificial selection has different goals. In farmed animals, selection which is the main tool of genetic improvement focuses on traits which are of economic importance (e.g. body weight and length), whereas in wild animals natural selection favors those individuals which are better fitted. These differences exist also between populations and sexes living in similar conditions but different regions (Frafjord, 1993; Rosalino *et al.*, 2005; Yom-Tov *et al.*, 2007). Domestication which takes place in fur farming makes these differences more significant. Very strong selection (usually directional) in farm populations, which aims at genetic improvement of economically important traits (e.g. body size, litter size), results in loss of morphological and genetic variability. The traits which are important for breeders, for example pelt size (which is affected by body weight and length), coat quality and density are quantitative traits. It means that those traits are mainly determined by polygenes, and their genetic variation is usually high. Many studies performed in captive populations of silver fox, arctic fox, raccoon dog or mink have reported high estimates of heritability of economically important traits (Cholewa *et al.*, 2006; Lagerkvist *et al.*, 1994; Peura *et al.*, 2005; Peura, 2007; Zatoń-Dobrowolska *et al.*, 2011). This allows effective genetic improvement of animals resulting (among others) in changes in morphology.

Sometimes, as a result of selection, new variants of traits which do not exist or have been eliminated in wild populations are expected (e.g. coat colour). Many studies comparing the American mink populations have shown that farm and wild animals were genetically diversified, had different body size and skull morphometrics (Belliveau *et al.*, 1999; Hammershøj *et al.*, 2004; Kidd, 2008; Kidd *et al.*, 2009; Tamlin, 2009). Greater genetic diversity was found in the wild raccon dog than that kept on farms (Slaska *et al.*, 2010). Comparative analysis of wild and farm fox anatomy showed important differences in biometric characteristics of aorta (Nowicki, 2005).

The aim of the present study was to analyze and compare several body measurements of wild and farm foxes. The additional goal of the study was to compare these measurements between sexes.

Material and methods

Data on nine traits describing body size for 199 farm silver foxes and 69 wild red foxes were collected. The analyzed traits were: body weight (BW), body length (BL, measured from top of nose to base of tail), body circumference (BC, measured behind front limbs), tail length (TL), ears height (EH), length of front (LFL) and hind (LHL) limbs, length of front (LFF) and hind (LHF) feet. Normality of the traits distribution was tested in order to use either analysis of variance and Duncan test to statistically analyze normally distributed traits (only BW) or Wilcoxon rank sum test with continuity correction to study not normally distributed traits (BL, BC, TL, EH, LFL, LHL, LFF, LHF). Measurement differences between wild and captive populations as well as between sexes were tested.

Results

The statistical characteristics of the studied traits for both populations and sexes are shown in Table 1. Measurements of all traits, except tail length and ear height, had higher values in farm animals than in wild ones. This shows that living conditions – environment, availability and kind of food – have very important influence on animal morphology. Smaller differences between wild and captive populations were found for traits which are not crucial for species survival (e.g. tail length and ears height). An interesting finding is comparable limbs length in both populations, and at the same time significant differences in foot length (Table 2).

Statistical differences between populations and sexes are shown in Table 2. Significant differences between wild and captive populations were found for almost all studied traits. Only tail length and length of hind limbs had similar values in both populations. Significant differences for almost all traits were also found between sexes in both populations (except length of front limbs in case of farm animals and length of hind limbs for wild animals). The comparative analysis between sexes from wild and captive populations showed that measurements of females from the farm population and males from the wild population were similar. The comparison of males from both populations revealed higher differentiation in the studied traits. Most of the studied traits had

Table 1. Means and their standard deviations (in brackets) of the studied traits. Means of BW which differ significantly (P≤0.01, Duncan test) are marked with different capital letters.

Trait	Farm population			Wild population		
	Total (n=199)[1]	**Male (n=85)**[2]	**Female (n=114)**[3]	**Total (n=69)**[4]	**Male (n=36)**[5]	**Female (n=28)**
Body weight [g]	6,953.90 [A] (1,161.88)	7,779.76 (861.49)	6,332.65 (954.37)	5,744.30 [B] (929.56)	6,217.71 (893.00)	5,169.29 (654.31)
Body length [cm]	71.38 (4.33)	74.19 (2.83)	69.28 (4.05)	68.27 (3.59)	70.21 (3.06)	65.77 (2.55)
Body circumference [cm]	40.35 (4.86)	41.55 (2.33)	38.93 (2.44)	36.17 (4.41)	36.94 (2.78)	33.39 (3.13)
Tail length [cm]	39.66 (2.76)	40.82 (3.02)	38.79 (2.17)	39.09 (6.04)	41.12 (3.47)	37.11 (7.07)
Ears height [cm]	9.34 (0.64)	9.52 (0.54)	9.20 (0.69)	9.77 (1.01)	9.92 (1.06)	9.50 (0.96)
Length of front limbs [cm]	33.83 (2.90)	34.24 (2.98)	33.52 (2.81)	31.77 (3.24)	32.90 (3.13)	31.50 (1.90)
Length of hind limbs [cm]	36.11 (2.42)	36.82 (2.35)	35.58 (2.34)	35.58 (3.96)	36.76 (3.34)	35.34 (2.96)
Length of front feet [cm]	12.38 (0.95)	12.79 (0.91)	12.07 (0.86)	10.75 (1.41)	10.80 (0.76)	10.13 (0.86)
Length of hind feet [cm]	17.70 (0.95)	18.10 (0.94)	17.40 (0.84)	15.57 (1.18)	15.71 (0.99)	15.05 (1.00)

[1] 195 individuals for TL and 198 for BW.
[2] 84 individuals for TL.
[3] 113 individuals for BW and 111 for TL.
[4] 68 individuals for BW, LFL, LFF and EH; 64 individuals for BL.
[5] 35 individuals for BW, LFL, LFF, EH and 34 individuals for TL.

Table 2. Anova's P-values (for BW between populations) and Wilcoxon comparison test (for other traits) between wild (WP) and farm (FP) populations and sexes.

Trait	FP/WP	♂FP/♀FP	♂WP/♀WP	♂FP/♂WP	♂FP/♀WP	♀FP/♂WP	♀FP/♀WP
Body weight	<0.0001	$<2.2e^{-16}$	$1.712e^{-06}$	$4.695e^{-12}$	$5.107e^{-15}$	0.5512	$5.368e^{-08}$
Body length	$1.969e^{-08}$	$<2.2e^{-16}$	$8.017e^{-07}$	$7.715e^{-09}$	$6.883e^{-15}$	0.1434	$9.646e^{-09}$
Body circumference	$9.475e^{-14}$	$3.872e^{-12}$	$1.047e^{-05}$	$1.465e^{-12}$	$1.421e^{-13}$	0.0002547	$6.043e^{-12}$
Tail length	0.4629	$4.302e^{-06}$	0.004685	0.5649	0.003236	0.0001367	0.5464
Ears height	$4.033e^{-05}$	0.0004417	0.07788	0.004072	0.9785	$1.510e^{-05}$	0.02463
Length of front limbs	$2.09e^{-05}$	0.1187	0.006336	0.0808	$6.065e^{-05}$	0.3058	0.0002409
Length of hind limbs	0.9623	0.004019	0.1154	0.8174	0.07861	0.03371	0.8094
Length of front feet	$<2.2e^{-16}$	$6.072e^{-08}$	0.003071	$2.665e^{-15}$	$5.551e^{-15}$	$2.553e^{-11}$	$5.507e^{-14}$
Length of hind feet	$<2.2e^{-16}$	$5.352e^{-07}$	0.01368	$6.661e^{-16}$	$4.219e^{-15}$	$2.793e^{-13}$	$2.109e^{-14}$

significantly higher values in males from the captive population as compared to the wild one. The same situation was found when comparing females from both populations.

Discussion

The results of our study are comparable to those reported by other authors who studied morphometrics of other species. Kidd (2008) who investigated mink populations also reported, that body size measurements were higher in farm mink than in wild ones. Perhaps it is a consequence of selection for higher animal length, which is associated with larger pelt, as well as living conditions in captivity (especially regular feeding). The same tendency is observed for fish. The comparison of size measurements of individuals coming from fish farming and ocean populations showed that fish from aquaculture are significantly bigger (Blanchet *et al.*, 2008; Fleming *et al.*, 1994).

Acknowledgements

This study was supported by the National Science Centre (Project no N N311 229938).

References

Belliveau, A.M., Farid, A., O'Connell, M. and J.M. Wright, 1999. Assesment of genetic variability in captive and wild American mink (*Mustela vison*) using microsatellite markers. Canadian Journal of Animal Science 79: 7-16.

Blanchet, S., Páez, D.J., Bernatchez, L. and Dodson, J.J., 2008. An integrated comparison of captive-bred and wild Atlantic salmon (*Salmo salar*): Implications for supportive breeding programs. Biological Conservation 141: 1989-1999.

Cholewa, R., Pawliczak-Maj, K. and Szwaczkowski, T., 2006. Genetic and environmental effects on body weight and fur hight in nutria (*Myocastor coypus* L.). EJPAU 9 (2): http://www.ejpau.media.pl/volume9/issue2/art-24.html.

Fleming, I.A., Jonsson, B. and Gross, M.R., 1994. Phenotypic divergence of sea-ranched, farmed, and wild salmon. Canadian Journal of Fisheries and Aquatic Sciences 51: 2808-2824.

Frafjord, K., 1993. Circumpolar size variation in the skull of the arctic fox *Alopex lagopus*. Polar Biology 13: 235-238.

Hammershøj, M., Asferg, T. and Kristensen, N.B., 2004. Comparison of methods to separate wild American mink from fur farm escapees. Mammalian Biology, 69: 281-286.

Kidd, A.G., 2008. Mink gone wild: hybridization between escaped farm and wild American mink (*Neovision vision*) in a natural context. MSc Thesis, School of Graduate Studies Laurention University Sudbury, Ontario, Canada.

Kidd, A. G., Bowman, J., Lesbarrères, D. and Schulte-Hostedde, A. I., 2009. Hybridization between escaped domestic and wild American mink (*Neovison vison*). Molecular Ecology 18: 1175-1186.

Langerkvist, G., Johansson, K. and Lundeheim, N., 1994. Selection for litter size, body weight, and pelt quality in mink (*Mustela vision*): correlated responses. Journal of Animal Science 72: 1126-1137.

Nowicki, W., 2005. Comparison of biometric characters of aorta branches in farm and wild fox (*Vulpes vulpes* L.). Folia biologica (Kraków) 53 (Suppl.): 35-38.

Peura, J., Strandén, I. and Mäntysaari, E.A., 2005. Genetic parameters In Finnish blue fox population: Pelt charakter and live animal grading traits. Acta Agriculturae Scandinavica, Section A – Animal Science 55: 137-144.

Peura J., 2007. Genetic parameters for Finnish blue fox population: litter size, age At first insemination and pelt size. Agricultural and Food Science 16: 136-146

Rosalino, L. M., Santos, M. J., Domingos, S., Rodrigues, M. and Santos-Reis, M., 2005. Population structure and body size of sympatric Carnivores in a mediterranean landscape of sw Portugal. Revista Biol. 23: 135-146.

Slaska, B., Zieba, G., Rozempolska-Rucinska, I., Jezewska-Witkowska, G. and Jakubczak, A., 2010. Evaluation of genetic biodiversity in farm-bred and wild raccoon dogs in Poland. Folia biologica 58: 195-199.

Tamlin, A.L., Bowman, J. and Hackett, D.F., 2009. Separating wild from domestic American mink *Neovison vison* based on skull morphometrics. Wildlife Biology 15: 266-277.

Yom-Tov, Y., Yom-Tov, S., Barreiro, J. and Blanco, J.C., 2007. Body size of the red fox Vulpes vulpes in Spain: the effect of agriculture. Biological Journal of the Linnean Society 90: 729-734.

Zatoń-Dobrowolska, M., Mucha, A., Wierzbicki, H. and Przysiecki, P., 2011. Genetic parameters and factors affecting body size and fur traits in arctic foxes (*Vulpes lagopus*). Electronic Journal of Polish Agricultural Universities, Animal Husbandry 14 (4): http://www.ejpau.media.pl/volume14/issue4/art-05.html.

Benefit of high number of active mammary nipples in mink

J. De Rond[1] *and F.C. Kleyn van Willigen*[2]
[1]*Edelveen – Research Farm Dutch Fur Breeders' Federation, Postbox 488, 6600 AL, Wijchen, the Netherlands;* [2]*Animal Health Service, P.O. Box 9, 7400 AA, Deventer, j.derond@edelveen.com; f.kleyn.v.willigen@gddeventer.com*

Abstract

From 2009 till 2011 Edelveen has looked into the effect of the number of active mammary nipples on kit survival, kit growth and the heritability of mammary nipples. The projects showed the significant negative correlation in kit mortality between birth and 3 weeks of age. There is a clear variation in number of active mammary nipples in the total population. The repeatability of nipples is high. The correlation between mammary nipples and litter size is not strong. Females with 8 or 9 active mammary nipples had kits with a higher weight at 7 weeks of age and those females lost less weight between 3 and 7 weeks in lactation. The variation in number of nipple dots at kits of 2 weeks of age is high (6 to 11). The dots are strongly correlated with the number of mammary nipples of the mother and father. Nipple dots develop into active mammary nipples when the female has kits. The number of active mammary nipples is genetic correlated and showed a high heritability. Selection on number of active nipples is effective and helps the female mink in nursing their kits and results in a better welfare in mink.

Keywords: mammary nipples, nipple dots, kit losses, growth

Introduction

The relation between the number of active mammary nipples and various parameters has been studied previously. Pedersen (1978) reported in scientifur the works of Mohr and of Hoogenbrugge and a variation in the number of active nipples from 1 to 10, and they assessed a poor positive correlation between the number of nipples and litter size. Fiedler and Siler (1978) reported that the heritability for the number of nipples is high (h2 = 0.53). Møller and Brzozowski (1996) described how mammary glands develop with high intensity during the last weeks of gestation. Individual mammary glands can be activated up to 12 days post partum, and kits activate glands such that the nipples become available for the kits (Møller, 2008).

These results strongly suggest that it is worthwhile use the mammary nipple number as a selection criteria, as is done in pig and rabbit breeding. Still, on Dutch mink farms little attention is paid to mammary nipple numbers. The aim of this study was to collect recent Dutch data on active mammary nipple numbers, heritability and the impact of the number of nipples on litter size and kit development. We will argue that selection in favour of larger number of active mammary nipples is profitable for the farmers, and improves mink welfare.

Material and methods

Animals

Edelveen consisted of 1,500 brown/glow females which were introduced as young females short before 2009's mating season. The size of the stock was the same in 2010 and 2011, but the amount of old females raised.

Housing and food

The animals were housed in standard mink cages (30×85×45cm) provided with a nest box. They were fed with the food produced by food-kitchen Flevo, using an automatic feeding system (Farm Pilot).

Recorded data

These usual breeding data were recorded: date of birth, litter size one day after birth and at 3 and 6 weeks. Weight of mother and kits at 3 and 5 and 7 weeks. For this study we also recorded the numbers of nipple dots and active mammary nipples (MN). At the age of ±18 days, nipple dots were counted. Nipple dot can develop into mammary nipples during adulthood. Active nipples were counted also at ±18 days after whelping. Table 1 lists how many animals contributed to the various topics during the three years. Data were analysed with Anova.

Table 1. Materials and methods projects active mammary nipples.

Topic	Year(s)	Actions / selected mink[1]	n Litters	days in Lactation
Kit losses	2009/2010/2011	counting kits	± 1,450	d 1 and d 21
		counting active MN		± day 18
MN next lactation	2009/2010/2011	counting active MN	± 700	± day 18
Litter size	2009/2010/2011	counting kits	± 700	d 1 and d 21
		counting active MN		± day 18
Nipple dots kits	2010	OF with 6-7-8 or 9 active MN	100	± day 18
		and sons /daughters of OF	140	
Nipple dots kits	2011	OF with 6-7-8 or 9 active MN	100	± day 18
		sons /daughters 6-7-8 or 9 nipple dots	140	
Nipple dot -> MN	2010/2011	counting dots 2010 and MN 2011	100	± day 18
Kit growth	2011	whelping date: 28/4, 03/05, 09/05	±60/date	
		n kits 21 days: 5or6, 7or8, 9or10	±60/n kits	
		N active MN: 6, 7, 8 or 9	±45/n MN	
		weighting kits and OF	175	d 21, 35 and 49

[1] MN = mammary nipples; OF = old female mink.

Results

Active MN and kit losses during the first 21 days

The relation between number of active MN, litter size and kit survival was consistent over the years so that all data are lumped. Generally spoken, more kits survived the more active MN's were available, but Table 2 with lumped data from 2009-2011 shows that the exact relations between MN, litter size and survival rate are difficult to unravel. Reading the rows from left to right makes clear that the number of active MN is negatively correlated with mortality, in litters of all sizes from 10 to 6. Reading the columns from top to bottom reveals that mortality also depends on litter size. The smaller the litter in the range 6-10, the lower the mortality. The number of active MN seems to have more impact than litter size.

Table 2 does not show whether litter size and number of active MN depend on each other. Table 3 is composed to show their relation: the more active MN, the more live born kits per litter.

Active MN and kit growth from 3 to 7 weeks

De Rond (2007) already reported that kits gained more weight and mothers lost less, the more MN were active. In later studies (De Rond, 2011a,b), 175 mothers and kits were weighed 3 and 5 and 7 weeks after birth. The weight gain was not only influenced by the number of active MN, but more so by litter size. Kits from litter size 5/6 were the heaviest (615 g), litter size 7/8 weighed 585 g and litter size 9/10 only 515 g. Females lost most weight with litter size 5/6 (140 g). With litter size 9/10 they only lost 100 g. The birth date was less important, but not neglectable. Kits from 9/10 May weighed after 7 weeks 595 g and kits from 28/29 April 565 g.

Table 2. Litter size, number of mammary nipples, and mortality between 1 and 21 days after birth during 2009-2011.

Litter size	**Number of mammary nipples**				
	9	**8**	**7**	**6**	**5**
10 kits	3%^{a}	4%^{a}	6%^{a}	10%^{b}	
9 kits	2%^{a}	3%^{a}	7%^{a}	9%^{b}	17%^{b}
8 kits	3%^{a}	5%^{a}	5%^{a}	8%^{ab}	11%^{b}
7 kits	1%^{a}	4%^{a}	6%^{a}	9%^{a}	11%^{b}
6 kits	2%^{a}	3%^{a}	3%^{a}	8%^{ab}	10%^{b}
Females (n)	155	1,028	1,401	1,066	255
Liveborn kits (avg.)	8.4	8	7.7	7.1	6.4
Kit losses/litter	2%^{a}	4%^{a}	5%^{a}	7%^{ab}	11%^{b}

^{a,b} Groups with different letters are significantly different at $P<0.05$.

Table 3. Mammary nipples (MN) and litter size.

Mammary nipples at first lactation	Reproduction 2nd lactation	
	litters (n)	liveborn kits (n)
9 MN	25	7.5[a]
8 MN	193	7[a]
7 MN	231	6.7[b]
6 MN	162	6.5[b]

[a,b] Groups with different letters are significantly differents at $P<0.05$.

The relation between active MN and kit weight and Female weight loss is shown in table 4. Apparently, the relation between weight gain/loss and number of active MN during the second part of the lactation period is not strong. Females with 8/9 active MN scored the best.

Heritability and selection

Selection in favour of a large number of active MN only makes sense if the numbers vary. As a matter of fact they do. The number of active MN varied from 1 tot 10 and the number of dots varied from 6 to 11 per kit. Even within litters the numbers varied, in fact only in 25% of the litters (5 or more kits) the kits of the same sexe (male or female) had the same number of nipple dots.

If heritability plays an important role, then it is to be expected that the number of active MN of one female is more or less the same every year. Indeed, the numbers of active MN in subsequent years were strongly correlated with each other (Table 5).

Table 4. Active mammary nipples (MN), kit weight and female weight loss.

Active MN (n)	Litters (n)	kits/ litter at 21 days (n)	Kit (m/f) weight at 7 wks	Female weight loss 3-7 wks lact
6 MN	45	7.1	570[ab]	135
7 MN	45	7.2	560[a]	150
8 MN	45	7.3	585[ab]	105
9 MN	40	7.7	600[b]	105
Standard variation			55	115

[a,b] Groups with different letters are significantly differents at $P<0.05$.

Table 5. Number of active mammary nipples (MN) in the next lactation period.

Females (n)	MN 2009 (1st)	MN 2010 (2nd)	higher MN in 2nd (%)
20	9	8.9	0%
103	8	7.8	4%
143	7	6.9	9%
102	6	6.2	23%
25	5	5.8	60%

Table 5 also shows that the number of active MN tended to increase rather than decrease in the second year. Litter size dictated these numbers far less as shown earlier. Thus, an environmental factor like litter size, influenced the number of active nipples less than did the female itself.

The crucial test in assessment of heritability is looking for correlations between parents and offspring. Various correlations are worth to investigate: correlations between the mother's active MN en the kits' dots. And correlations between the mothers dots and the kit's dots. Only lactating females have active MN during part of their life, whereas kits and males of any age only have nipple dots. From the females introduced in 2009 we did not count the old female's dots, but only their active MN. Hence, the results of 2010 and 2011 are not lumped. Table 6 shows the average numbers of kit's dots and the numbers of their parents. In 2010 the only information of the females and males was the number of active MN of the old female. In 2011 the information of the parents was complete (n MN or n dots). Both in 2010 and 2011, the numbers in the kits were larger, the more dots the parents had. So, selection in favour of increased numbers of nipples must be effective.

Table 6. Nipple dots in kits and mammary nipples (MN) in parents in 2010 and 2011.

Year	Number	number of MN parents or nipple dots kits			
2010	150 (old) females	9	8	6	4.5
	1000 kits	8^a	7.6^b	7.3^c	7.3^c
2011	200 females	8.7	8	7.3	6.3
	1,130 kits	8.4^a	8^b	7.6^c	7.3^d

$^{a-d}$ Groups with different letters are significantly differents at $P<0.05$.

Discussion

The results confirm earlier reports on the number of active mammary nipples. The variation in the numbers of nipples is about the same as reported by Pedersen (1978). Our correlations between the numbers of parental nipple dots and those of their kits fits in the h2=0.53 reported by Fiedler and Siler (1977) and the development from dots to active nipples as described by Møller (1996, 2008) was seen as well.

The number of active nipples was positively correlated with some of the characteristics which determine the breeding results. Litter size and number of active nipples were positively correlated. The causal relationship is not fully clear, since suckling induces the transformation from nipple dots into nipples. Mortality was clearly less, the more nipples were available and active. If all females at Edelveen had a minimum of 7 active mammary nipples, 200 kits would have been saved per year. The weight of the kits slightly increased with the number of active nipples. Hence, it seems worthwhile to increase of the number of nipples.

Various information justify the conclusion that hereditary factors influence the number of nipples. Most convincing is of course the positive correlation of the mean number of the parental nipple dots with the number of the kits nipple dots. These correlations were already assessed by Fiedler and Siler (1977) and confirmed in this study. The fact that the number of active nipples of one female is rather consistent over the years supports the conclusion that the number of nipples is certainly not only determined by environmental factors.

All in all we conclude that it is worthwhile to select for more mammary nipples. In spite of the fact that some of our results are already known for decades, selection for nipples is hardly or not practised. Apparently, some practical problems have to be solved such as the variation in the number of dots of littermates. The next step at Edelveen is to solve the practical problems and to develop an effective way of working.

References

De Rond, J., 2007. Number of active mammary nipples affect kit growth. De pelsdierenhouder, pp. 121-125.

De Rond, J., 2011a. Selection on number of active mammary nipples is effective. NJF no 450, Knistva, Sweden, Finland, 8 p.

De Rond, J., 2011b. Selection Effect of active mammary nipples on kit growth and female condition in the second part of the lactation period. NJF no 450, Knistva, Sweden, Finland, 8 p.

Fiedler, J. and Siler, R., 1978. the heritability of the number of teats and the occurrence of white spots in standard mink. Scientifur no 4, November 1978, pp 20.

Møller, S.H. and Brzozowski M., 1996. Development of the mammary glands during gestation and lactation. VI IFASA congress, Warsaw, Poland, pp 103-112.

Møller, S.H., 1996. Development of the mammary glands in female mink from weaning through first lactation. VI IFASA congress, Warsaw, Poland, pp. 121-128.

Møller, S.H., 2008. Development of the mammary glands in mink. IX IFASA congress, Halifax, Canada, pp. 22-23.

Pedersen, H., 1978. Early kit losses related to number of teats in mink females. Scientifur no 4, November 1978, pp. 21-28.

Similarity of coat color aberrations in northern fur seal (*Callorhinus ursinus*), mink (*Neovison vison*), and sable (*Martes zibellina*)

O.V. Trapezov[1], S.V. Fomin[2] and N.S. Fomina[3]
[1]Institute of Cytology and Genetics, Siberian Branch, Russian Academy of Sciences, Novosibirsk, Russia; [2]Vyatka Agricultural Academy, Kirov, Russia; [3]Komandorski State Natural Biosphere Reserve, Commander Islands, Nikol'skoe, Russia; trapezov@bionet.nsc.ru

Abstract

The frequency of color aberrations in the Northern fur seal population of Commander Islands was studied. *Albino, acromelanism, nonspecific piebaldness, golden*, and *ferret type* color phases were recorded at the frequencies about 10^{-4}. They are phenotypically similar to the same phases in farm bred sables and minks. The evolutionary sense of color aberrations in phenotypic parallelism of color mutations in the Northern Fur Seal, Mink, and Sable is discussed.

Keywords: color, seal, mink, sable, evolution

Introduction

Within a species, a huge number of different mutations always occur that affect phenotypic characters and in the population many mutant genes that are the fund of variations (the gene pool in the language of genetics) may be present simultaneously. Such individuals in populations of wild animals that sharply deviate in color from the norm are commonly called aberrants (in latin: *aberration* – deviation) (Figure 1). In nature, aberrants are vary rare. At the same time, for both the genetic and population studies and for private genetics of fur animals, study of color aberrations is of great interest (Dunn, 1921; Haldane, 1927).

Counting color aberrations in the Northern fur seals of the Commander archipelago is assumed to have started in 1954, when S.V. Marakov made the first black-and-white photo of a seal with a nonspecific piebaldness on its head at the Southeast rookery of marine animals in Copper Island of the Commander archipelago (Northern-West part of the Pacific Ocean) (Marakov, 1974).

In this study, the following task were set: (1) to consider by the presence of color aberrations the presence of genes affecting the color of hair in the Commander population of Northern fur seal, (2) to compare color aberrations in Northern fur seal with similar color phases in representatives of the farm-bred mink and sable.

a) *Albinism* in fur seal, mink and sable.

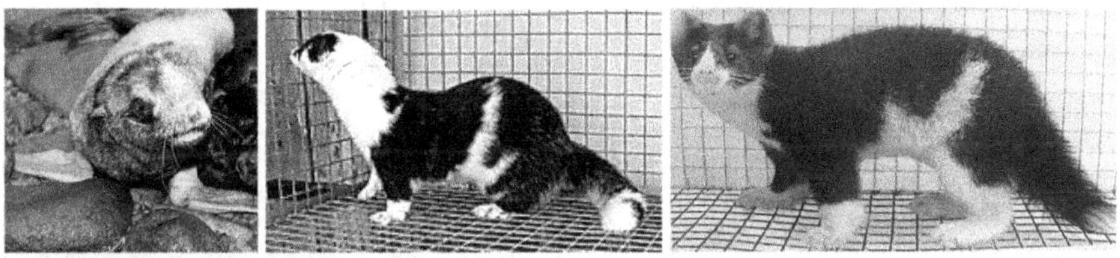

b) *Piebaldnes* in fur seal, mink and sable.

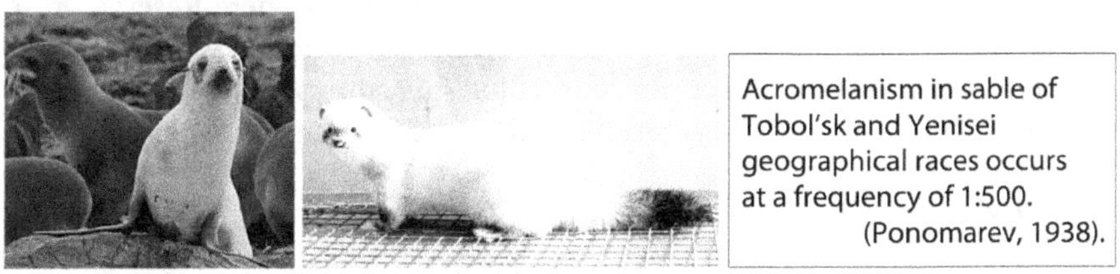

c) *Acromelanism* in fur seal, mink and sable.

Figure 1. Coat color parallelism in different species of fur bearing animals. (A) Albinsm in fur seal, mink and sable. (B) Piebaldness in fur seal, mink and sable. (C) Acromelanism in fur seal and mink.

Material and methods

Northern fur seal

Superfamily: *Pinnipedia* (pinnipeds). Family: *Otariidae* Gill, 1866 (Eared seals). Genus: *Callorhinus* Gray, 1859 (Northern fur seals). Species: *Callorhinus ursinus* Linnaeus, 1758 (the Northern fur seal). Chromosome set $2n$=36.

Procreation in Northern fur seals usually occurs from mid June to early August. Newborn pups have a body weight of about 5 kg and body length of 55-65 cm. By the end of the lactation period molting in cubs begins with a change of juvenile black hair to gray (Il'yina, 1940, 1951). The general coloration of adult males of the Northern fur seal ranges from silvery gray to blackish-brown. One can meet almost gray and pitchy individuals. The dorsal part of the body differs from the ventral one, it varies from dark gray to brown; ventral part is reddish brown. In adult females back is grayish-brown or gray, the breast is pale yellow, dark gray and light gray. Fur of cubs has a black or black-brown shiny guardian hair and rare light underfur. Young males and females are silvery gray, lighter on the ventral side (Marakov, 1974). The hair in fur seals consists of coarse guardian hairs and soft underfur. Hair grows in beams consisting of 1 guardian, and 30 underfur. There are 50 000 hairs per square centimeter of skin (Table 1).

Table 1. Characteristics of the hair in Northern fur seals in comparison with several other fur-bearing animals.

Fur	Wear on a 'Parker scale'	Density of hair, thousands per cm^2
See otter	120	154.1
River otter	100	58.6
Northen fue seal	85	50.0
Sable	80	12.86
American mink	70	33.72
Ermine	25	17.55

The work on the studying the presence of color aberrations among juveniles of Northern fur seal was carried out in the Commander Islands Nature and Biosphere Reserve on the North and Northwest rookeries on the Bering Island and on the Southeast and Uril'e rookeries on the Copper Island. The frequency of occurrence of color aberrations was detected in 1998, 2002, 2004, 2006, 2008, 2009, and 2010 during routine counts of the number of animals in August.

Farm-bred mink

Animals from the genetic collection of mutations of American mink coloration from the Experimental fur farm of the Institute of Cytology and Genetics SB RAS were used for a phenotypic comparison of the similarity of color aberrations in the Northern fur seal with those of mink.

Farm-bred sable

Taxonomists assign the sable to the most numerous species of the predatory mammalian superfamily (*Carnivora*). Like the American mink, the sable belongs to the musteline family (*Mustelidae*), marten genus (*Martes*) and marten subgenus (*Martes*). The intensity of pigmentation of the hair in the sable is variable: Ural and Kamchatkan sable are the lightest ones; dark specimens are more common among Yakut and Barguzin sable. It should be noted that the dark coloration forms of sable are rare in natural conditions, even among the famous Barguzin sable, which are obtained by hunting; about half of all individuals have an average or light color tone (Sabaneev, 1875; Bakeev *et al.*, 2003).

The great value of sable fur led to predatory and uncontrolled hunting that caused a deep depression in the sable population as a species. By the 1930s, the sable range was divided into separate island pockets that were left mainly in the southern part of Eastern Siberia (Altai-Sayan highlands and Lake Baikal). In some regions of Western Siberia, it disappeared completely (Monakhov and Bakeev, 1981). One of the measures to protect the sable as a species was the creation of specialized sable farms (Portnova, 1966).

The modern gene pool of farm-bred sable was completed in 1929-1932 from the Amur, Yenisei, Barguzin, and Altai geographic races (ridges). Animals taken from nature did not have the right color for selection, as to selectively catch a sable with the desirable dark color in the taiga was technically impossible. Therefore, in the generation of the 1932 source for cage farming, only 4% of the animals had the necessary dark-colored fur for further matching (Il'ina, 1935; Portnova, 1941, 1966). N.A. Ponomarev, along with livecatching sable conducted research on color aberrations seen in natural conditions. He highlighted the following phenotypes: *acromelanists*, *chromists*, *piebald*, *white* (with pale creamy coating on the sacrum, neck and legs, as well as brownish-orange colors on the abdomen and back along the spine) (Ponomarev, 1938). Later researchers added sable of *smoky*, *tannic*, and other colors to the list (Monakhov and Cheglakov, 1973). Regarding the concentration of aberrant forms of coloration in the population of sable, Bakeev *et al.* wrote: '...on average, for 25,000-30,000 sable of standard colors, there are animals of mutant forms with an unusual colors: *cream*, *orange*, *light ocher*, *blue-gray*, *piebald*, *white*, and other colors' (Bakeev *et al.*, 2003).

Over the last 25 generations of sable farming, the appearance *de novo* animals with nonspecific piebaldnesses on the feet, tail tip, muzzle, and torso was recorded. The size and color of the spots vary greatly. Spot colors can be not only white, but also yellow of varying intensity. In spotted sable, the eyes are often blue or dark blue and the nasal mirror is partially or completely depigmented. According to the nature of the inheritance of this coloration novelties, the presence of two mutations in sable that determine the manifestation of white spotting was suggested. Homozygous individuals die in the prenatal or postnatal stage of development. There has been only a single case of the birth of a white sable; it lived 45 days (Kudina, 1969).

Since the 1970's, the Research Institute of hunting and fur bearing animal farming has performed brief experiments to elucidate the inheritance of the aberrant coloration in sable (Monakhov and Cheglakov, 1973; Cheglakov, 1980).

Beginning from 1981, in the State fur farm 'Pushkinsky', the appearance of *de novo* original coat color aberrations similar to the well-described mutant colorations in mink started to occur in sable: *palomino*, *pastel*, *lavender*, and *silverblue* (Kuznetsov *et al.*, 2007). The coloration innovations that emerged in the sable farm were used for phenotypic comparison with the color aberrations that are found in the Commander population of Northern fur seals.

Results

In all the rookeries of the Commander archipelago islands, about 240,000 Northern fur seals occur annually. Of these, the share of juveniles that are born and survive at the time of observation is 56,000 on average. Taking deaths into account, the total annual production of juveniles in the Commander fur seal population is 58,500 on average (from the results of studies in 1998-2010). Standard colored animals make up the bulk of the counted population of seals. Aberrant color forms are represented by single individuals (usually by young animals-underyearlings). The frequency of their presence on average is on the order of 10^{-4} (Table 2).

Table 2. Frequency of color aberrations in the Commander population of the Northern fur seal on North, Northwest, and Uril'e rookeries in August (adults and juveniles at the age of 35-40 days) from 1998 to 2010.

Phenotype of color aberrations	Quantity of color aberrations among the total quantity of counted juveniles together with dead animals on a rookery in the year of investigation						
	1998	**2002**	**2004**	**2006**	**2008**	**2009**	**2010**
	26,738 (North); 9,532 (North-west)	**28,745 (North); 9,014 (North-west)**	**23,074 (North); 7,803 (North-west)**	**24,598 (North); 10,346 (North-west)**	**23,220 (North); 8,288 (North-west)**	**21,402 (North); 10,181 (North-west)**	**26,130 (North); 9,406 (North-west); 7,300 (Uril'e)**
Black and white				North 1 August 1 pup (dead)			
Piebaldness + blue eyes						North 3 August 1 pup	
Piebaldness					North 2 August 1 pup		
Nonspecific piebaldness							Uril'e 6 August 1 pup
Golden coloration	North 2 August 1 pup						North 5 August 1 pup
Acromelanism					North August adult female		
Albinism	North 2 August adult female	North 15 August adult female	North 2 August adult female	North August 1 pup			
Brown			Northwest 3 August 1 pup		Northwest 2 August 1 pup		

Of all the aberrations of hair coloration in the Commander population of the Northern fur seal, *albinism* is the most common one. Under good conditions, an albino has a chance to survive to sexual maturity and to create offspring. For example, in the North rookery of Behring Island, an adult female albino was observed for several years from 2002. In 2006, in the Colonic section of the North rookery, this female fed normally colored black pup, which involves coloration recessiveness of albino mother and heterozygosity for the same color of her cub. Experts who study the behavior of the fur seal know that a female feeds only her cub, finding it among thousands of others (possibly with the help of smell and hearing) (Marakov, 1972, 1974; Kruchenkova, 2009). In this case we can assume that this black colored wild-type pup was exactly her cub. But we must bear in mind that published data on maternal behavior in Northern fur seals with aberrant coloration are lacking and it is possible that mutations that affect the color of hair can cause a disruption in maternal behavior such that a female that is aberrant in color is able to raise any pup, including a foreign normally colored one. Therefore, a final answer on the genotypes of the mother and offspring may be possible only with a complete genetic analysis.

The hair of all categories of *golden* animals are golden in color. The eyes are usually light blue and the skin of the flippers is pink. For *piebald* fur seals, it is characteristic to have depigmented or weakly pigmented parts of the body to a greater or lesser extent. Piebaldness is often localized on the head; in rare cases it is nonspecific and covers the torso. Piebaldnesses often correlates with light-blue eyes. The spots are often a white and creamy color. In *the acromelanist type of fur seal*, only the ears, flippers, tip of the muzzle, and eyes are pigmented.

For aberrants of the *brown type*, it is characteristic to have weakening of coloration of the underfur against the background of a brown color of the guardian hair. The eyes are pale blue and the skin of the flippers is pink. It is impossible to say anything about the phenogenetics of the above-mentioned color aberrations of the Commander population of Northern fur seals, as the organization of hybridological experiments with them at this point is not possible due to extreme technical complexity. One can only assume that these aberrations are inherited monogenically and have Mendelian segregation during crossing.

Discussion

The vast majority of Northern fur seals of the Commander archipelago are animals with a standard type of coloration. At the same time, the observed color aberrations in the Northern fur seal are not a unique phenomenon that is peculiar only to this species. Aberrant variation in color is a universal process. The study of hair coloration in different animals has a rather long history in genetics (Sturm, 2006) but there is still no definitive explanation of the surprising phenomenon called *phenotypic parallelism in variability*, which has been observed for many years (Duval-Jouve, 1865; Cope, 1896; Osborn, 1917).

How can one interpret the phenotypic similarity in color aberrations in the representatives of three completely different taxa, viz., the Northern fur seal, American mink, and sable? At first glance, this visible similarity leads one to appeal to the law of homologous series in autogenous variation (Vavilov, 1922). Based on its predictive abilities with respect to parallel variability in different taxa, this law was perceived as the 'periodic table in biology' (Hammer and Schubert,

1994; Popov, 2002; Sommer, 2008). Despite the strong heuristic side of the law, it does not explain the cause and the evolutionary and genetic significance of the phenotypic similarity of the color aberrations in these three species.

Is it possible to interpret the phenomenon of parallelism in color aberrations in representatives of three different taxa (seals and cage farmed mink and sable) without analyzing their adaptive value? It is unlikely, because any manifestation of such a characteristic of variation (just like any other variation) undergoes evaluation for its adaptivity via natural selection (Mednikov, 1981).

Proceeding from the striking similarity in the phenotypic variability in color aberrations of the three types considered here, which are fairly distant in origin, this may mean that they have a specific genetic component that falls under the pressure of a similar selection channel.

What is common to the conditions of Northern fur seals in comparison with mink and sable under conditions of confinement in a cage? In their history, suddenly, in evolutionary terms, all three species found themselves under human influence and were subjected to unprecedented pressure from anthropogenic stress. In fact, they were involved in the same channel of selection, and more specifically, on the ability not only to survive but also to create offspring under the conditions of a stressing anthropogenic environment. We are talking about the similar hereditarily determined canalizable variability of phenotypes, the variability in a specific channel of selection, where in a number of generations hormonal regulatory mechanisms and neurochemical mechanisms of the brain are reorganized in the same direction and the same morphological and physiological changes emerges. One can say that in different species, along with their inherent specificity, there is a common symptom, 'stress tolerance genes' marked by similar aberrant colorations. One of such color deviations that associated with *tyrosinase (TYR)* gene mutations (the albino phenotypes) documented in various species including the American mink for example (Anistoroaei *et al.*, 2008).

References

Anistoroaei, R.M., Fredholm, M., Knud Christensen, K. and Leeb T., 2008. Albinism in the american mink (*Neovison vison*) is associated with a *tyrosinase* nonsense mutation. Animal Genetics 39(6): 645-648.

Bakeev, N.N., Monakhov, G.I. and Sinitsin, A.A., 2003. Sable. 2nd ed. (ext. revis.). Kirov. Vyatka. 336 pp. (in Russian).

Cheglakov, V.N., 1980. First Results of Work with Color Sable. In: Biology, Breeding, and Housing of Sables in Cages. Moscow. pp. 115-117. (in Russian).

Cope, E., 1896. The Primary Factors of Organic Evolution. Chicago, pp. 475-484.

Dunn, L.C., 1921. Unit Character Variation in Rodents. Journal Mammalian 2: 125-140.

Duval-Jouve, M.J., 1865. Variations Paralleles des Types Congeneres. Bulletin Societe Botanique France 12: 196-211.

Haldane, J.B.S., 1927. The Comparative Genetics of Colour in Rodents and Carnivora. Biological Review 11: 199.

Hammer, K. and Schubert, I., 1994. Are Vavilov's Law of Homologous Series and Synteny Related? Genetic Research Crop Evolution 41: 123-124.

Il'ina, E.D., 1935. Basics of Genetics and Breeding of Fur-Bearing Animals. Glavpushnina. NKVT. (in Russian).

Il'ina, E.D., 1940. Sea Furs on Commander Islands. Moscow. Rossel'khozizdat. (in Russian).

Il'ina, E.D., 1951. Island Animal Breeding. Moscow. Mezhdunar. Kniga. (in Russian).

Kruchenkova, E.P., 2009. Maternal Behavior of Mammals, Moscow. MGU. (in Russian).

Kudina, N.I., 1969. Proceedings of the MVA Conference on Veterinary. Breeding and Commodity. Moscow. (in Russian).

Kuznetsov, G.A., Sergeev, E.G. and Ruchkina, Z.S., 2007. The White Sable. Int. Conf. 75th Anniversary of NIIPZK: Problems of Rehabilitation and Further Development of Breeding Fur-Bearing Animals in Cages. Rodniki. NIIPZK. 2007. pp. 92-97. (in Russian).

Marakov, S.V., 1972. Nature and Fauna of the Commander Islands. Moscow. Nauka.

Marakov, S.V., 1974. Northern Fur Seal. Moscow. Nauka. (in Russian).

Mednikov, B.M., 1981. Current Status and Development of the Law of Homologous Series in Hereditary Variation. In: Problems of the Modern History of the Evolutionary Theory. Leningrad, 1981, pp. 127-135. (in Russian).

Monakhov, G.I. and Cheglakov, V.N., 1973. To the Problem of Breeding of Color Sables. In: Sb. NTI VNIIOZ. pp. 109-112. (in Russian).

Monakhov, G.I. and Bakeev, N.N., 1981. Sable. Moscow: Lesn. Prom. 240 pp. (in Russian).

Osborn, H.F., 1917. The Origin and Evolution of Life. New York: Charles Scribner's Sons.

Ponomarev, A.L., 1938. The Variability and Inheritance of Color in Sables. Zoological Journal. V. 17 (3): 482-503. (in Russian).

Popov, I.Y., 2002. Periodical Systems in Biology. In: Die Entstehung Biologisher Disciplinen. Hossfeld, U. and Junker, T. Eds., Berlin: VWB. pp. 55-69.

Portnova, N.T., 1941. Experience of the Sable Breeding in the State Fur Farm 'Pushkinskij'. Krolikovodstvo i Zverovodstvo 6: 7-9. (in Russian).

Portnova, N.T., 1966. Our Experience of Sable Breeding. Krolikovodstvo i Zverovodstvo 4: 15-16. (in Russian).

Sabaneev, L.P., 1875. The Sable and Harvesting of Sables. Moscow: Tip. V. Got'e. (in Russian).

Sommer, R.J., 2008. Homology and the Hierarchy of Biological Systems. Bio Essays 30: 653-658.

Sturm, R.A., 2006. A Golden Age of Human Pigmentation Genetics. Trends in Genetics 20: 464-468.

Vavilov, N.I., 1922. The Law of Homologous Series in Variation. Journal of Genetics 12: 47-89.

Coat color mutations and the monoamine content in the brain of the farm-bred mink (*Neovison vison*)

O.V. Trapezov, L.I. Trapezova, T.A. Alekhina, D.V. Klochkov and Yu.N. Ivanov
Institute of Cytology and Genetics, Siberian Branch, Russian Academy of Sciences, Novosibirsk, 630090 Russia; trapezov@bionet.nsc.ru

Abstract

The effects of mutations affecting the coat color on the dopamine, noradrenaline, and serotonin contents of the hypothalamus and brainstem of the farm-bred mink have been studied. The sample comprised *standard* (+/+) and mutant minks, including the monorecessive *pastel* (*b/b*), *silver-blue* (*p/p*), and *white hedlund* (*h/h*) and the combination double recessive *sapphire* (*a/a p/p*) and *pearl* (*k/k p/p*) ones. The dopamine content of the brainstem of the monorecessive *pastel* (*b/b*) and *silver-blue* (*p/p*) minks has been found to be higher than in *standard* (+/+) minks. Conversely, the homozygosity for two coat color loci in double recessive *pearl* minks (*k/k p/p*) significantly decreases the noradrenaline and serotonin contents of the hypothalamus. In addition, monorecessive and double recessive minks differ from each other in the serotonin contents of the midbrain and medulla.

Keywords: color mutations, serotin, dopamine, noradrenaline, brain

Introduction

A total of 35 dominant, semidominant, and recessive mutations affecting the coat color have been found in farm-bred American minks (Il'ina and Kuznetsov, 1983; Ness *et al.*, 1988; Trapezov, 1997; Koldaeva, 2004). Analysis of the pleiotropic effects of coat color mutations and their combinations has demonstrated that they affect serotonin metabolism in the striatum by modulating the activity of A type monoamine oxidase, which deaminates serotonin, in the midbrain (Popova *et al.*, 1994), and change the dopamine content of the striatum by affecting the activity of B type monoamine oxidase, which deaminates dopamine, in the midbrain, striatum, and hippocampus (Popova *et al.*, 1996). Comparative analysis has shown the A and B type monoamine oxidase activities in the brainstem and brain hemispheres of various mutant minks differ from those in *standard* (+/+) minks (Voitenko and Trapezov, 2001).

The pleiotropic effects of the genes affecting the coat color in the double recessive *sapphire* minks (*a/a p/p*) on the penetrance and expressivity of domestication behavior may be correlated with changes in the brain monoamine metabolism, which, in turn, suggests that the more pronounced domestication behavior of *sapphire* minks towards humans is related to the functions of monoamine systems (Trapezov *et al.*, 2012).

The purpose of this study was to estimate the effects of mutations affecting the coat color of the American mink on the monoamine contents of the hypothalamus (an evolutionarily conserved brain division modulating behavioral and neuroendocrine responses to environmental factors)

and the brainstem (a brain division most strongly protected against environmental conditions), whose nuclei are characterized by the largest accumulation of monoaminergic neurons (Felten and Sladek, 1983).

Material and methods

The genotypes and phenotypes of the mink.

We studied six color genotypes of American minks raised at an experimental fur farm of the Institute of Cytology and Genetics of the Siberian Branch of the Russian Academy of Sciences. The color of *standard* minks (+/+) (Il'ina and Kuznetsov, 1983; Ness *et al.*, 1988). *Pastel monorecessive brown minks* (*b/b*) are also called *royal pastel, brown eyed pastel, chocolate*, and *blond* minks (in addition, they are known under the trademark 'Autumn Haze') (Smith *et al.*, 1941; Shackelford and Cole, 1947). *Silver-blue monorecessive minks* (*p/p*) are also called *platinum* and *maltese* and are known under the trademark 'Argenta' (Shackelford, 1948). *White hedlund monorecessive minks* (*h/h*) are purely white all over the trunk, without pigmented spots or even single hairs. They are also known under the trademark 'Jasmine' (Shackelford and Moore, 1954). *Sapphire double recessive minks* (*a/a p/p*) have a trademark 'Cerulean'. *Pearl double recessive minks* (*k/k p/p*) have a trademark 'Tourmaline' (Il'ina and Kuznetsov, 1983; Ness *et al.*, 1988).

Feeding and other conditions were the same for all genotypes of the mink. They were kept under the conditions that are accepted at specialized fur farms. Minks from all genetic groups were sacrificed on the same day. The animals were about six months old. They were sacrificed by instantaneous cervical dislocation. After quick decapitation in the cold, the brain was extracted, and the hypothalamus and brainstem were isolated. In the given case, we assumed the brainstem to comprise the inferior colliculi of the midbrain tectum and 3 to 4 mm portions of the medulla adjacent to the midbrain. The isolated parts of the brain were kept in a freezer at -80 °C until monoamine assay. The dopamine, noradrenaline, and serotonin contents of the hypothalamus and brainstem were assayed fluorometrically (Schlumpf *et al.*, 1974). The monoamine fluorescence was measured by means of a Hitachi device (Hungary) at wavelengths 330/370 nm (for dopamine), 390/485 nm (for noradrenaline), and 360/470 nm (for serotonin). Statistical treatment of the data included the estimation of the mean values, which were compared using Student's *t*-test; the significance of the effects of the mutant genotypes on the monoamine levels was estimated using one-way ANOVA.

Results

For treatment of the data, we divided all animals into three genetic groups (standard, monorecessive, and double recessive) and carried out the corresponding comparisons (Tables 1 and 2). As evident from Table 1, the concentrations of noradrenaline and serotonin, neurotransmitters of the two main signal modulating systems, in the hypothalamus of the double recessive *pearl*(*k/k p/p*) minks were significantly lower ($P<0.05$) than in *standard* (+/+) minks. Additional ANOVA confirmed that the noradrenaline and serotonin contents of the hypothalamus depended on the number of mutant alleles affecting the coat color (Table 3).

Table 1. The dopamine, noradrenaline, and serotonin contents of the hypothalamus of standard minks and minks that are monorecessive and double recessive for mutations affecting coat color (mean±SE).

Phenotype	Genotype	Number of animals	Levels in the hypothalamus (µg/g)[1]		
			dopamine	noradrenaline	serotonin
Standard	+/+	10	1.05±0.12	1.39±0.25	0.73±0.21
Monorecessive					
pastel	*b/b*	6	0.60±0.11	1.61±0.31	0.55±0.07
silver-blue	p/p	10	1.03±0.07	1.57±0.18	0.53±0.03
white hedlund	h/h	8	1.16±0.14	1.15±0.28	0.55±0.11
Double recessive					
sapphire	a/a p/p	9	1.04±0.15	0.90±0.16	0.45±0.13
pearl	k/k p/p	8	0.86±0.10	0.58±0.16*	0.19±0.08*

[1] * Significant difference from standard minks according to Student's t-test (*P*<0.05).

Table 2. The dopamine, noradrenaline, and serotonin contents of the midbrain-medulla region of the brain of standard minks and minks that are monorecessive and double recessive for mutations affecting coat color (mean±SE).

Phenotype	Genotype	Number of animals	Levels in the midbrain-medulla region (µg/g)		
			dopamine[a]	noradrenaline[a]	serotonin
Standard	+/+	10	0.74±0.04	0.34±0.03	0.71±0.16
Monorecessive					
pastel	*b/b*	6	0.22±0.03***	0.24±0.03*	0.36±0.07[b]
silver-blue	p/p	10	0.47±0.07**	0.34±0.04	0.36±0.07[b]
white hedlund	h/h	8	0.61±0.09	0.25±0.04*	0.31±0.09[b]
Double recessive					
sapphire	a/a p/p	9	0.82±0.07	0.37±0.03	0.48±0.02
pearl	k/k p/p	8	1.13± 0.15*	0.37±0.03	0.66±0.09

[a] Significant difference from standard minks: * (*P*<0.05); ** (*P*<0.01), and *** (*P*<0.001).
[b] Significant difference from double recessive minks (*P*<0.05) (Student's t-test).

Regarding the brainstem, which contains the largest aggregations of monoaminergic neurons (the substantia nigra, locus coeruleus, and nucleus raphe), we obtained the following data. The level of dopamine, the main neurotransmitter regulating locomotor activity, in the monorecessive *pastel* (*b/b*) and *silver-blue* (*p/p*) minks was significantly lower than in *standard* (+/+) minks (*P*<0.001) and *P*<0.01, respectively). In the double recessive *pearl* (*k/k p/p*) minks, the dopamine content

Table 3. The significance levels of the effects the genotype factor and the number of mutant alleles on the monoamine contents of the mink brain calculated using the ANOVA ***F****-test.*

Factor	Target of the effect					
	hypothalamus			**medulla**		
	dopamine	**noradrenaline**	**serotonin**	**dopamine**	**noradrenaline**	**serotonin**
Genotype	-	0.05	-	0.001	-	0.05
Number of mutant alleles	-	0.01	0.05	0.001	0.05	0.01

was significantly higher (P<0.05) than in *standard* (+/+) minks, which suggests that the former can endure more intense motor activity than the latter (Table 2).

The noradrenaline concentration in the hypothalamus of the monorecessive *pastel* (*b/b*) and *white hedlund* (*h/h*) minks was significantly lower (P<0.05) than in *standard* (+/+) minks. In addition, the brainstem of all monorecessive varieties studied (*pastel* (*b/b*) *silverblue* (*p/p*), and *white hedlund* (*h/h*) contained significantly less serotonin than not only the *standard* (+/+) color form, but also the double recessive *sapphire* (*a/a p/p*) and *pearl* (*k/k p/p*) ones (P<0.05).

ANOVA showed a greater number of significant influences of mutant alleles on the monoamine content of the brainstem compared to the hypothalamus (five versus three). The genotype factor affected the variation of the levels of two monoamines (dopamine and serotonin), and the monorecessive/double recessive state affected the variation of the levels of all the three monoamine neurotransmitters.

Data on the significance levels calculated using the ANOVA F-test as measures used for comparing the effects of the genotype and the number of mutant alleles on the monoamine contents of the brain showed the following. The factor of the number of mutant alleles (an additive effect) had a stronger influence than the genotype factor: its effect was significant in a greater number of cases (five versus three out of six parameters) and stronger (P<0.01 versus P<0.05 in two cases) (Table 3).

Discussion

Our experimental data have demonstrated that domestication-related transformations of fur animals, such the appearance of new coat color variants (Trut, 2000) and changes in estrous cyclicality (Klochkov *et al.*, 2005), affect brain monoamines, which are intermediate links in the control of these changes. Special studies in model objects (sables and minks) on estimation of the effects of coat color alleles on the animals' defensive behaviors towards humans have shown that both the proportion of individuals with domestication behavior and the expressivity of tame behavior in animals with aberrant coat colors are significantly higher than in animals with the standard coat color (Popova *et al.*, 1994, 1996; Voitenko and Trapezov, 2001; Trapezov *et al.*, 2012). The modulating effects of the genotype on the levels of noradrenaline and serotonin, which regulate stress reactivity, have been also detected in other models, including various mouse and rat strains (Szot *et al.*, 1999; Alekhina *et al.*, 2007).

We found a significantly lower dopamine level in monorecessive minks and a significantly higher one in double recessive minks compared to minks with the standard coat color. The pleiotropic effects of alleles affecting the mink coat color modulate the activities of monoamine neurotransmitter systems in different divisions of the brain. According to the results of ANOVA, the coat color genes of mutant minks have stronger effects on the variation of the monoamine concentrations in the brainstem than in the hypothalamus. This is probably because the main nuclei synthesizing and monoamines and forming their storage pools are located in the midbrain-medulla region; in addition, this region is the most protected against external influences, which is necessary because it contains the vitally important respiratory and cardiac rhythm centers. In the hypothalamus, the effects of coat color alleles on the neurotransmitter levels are weaker.

What is the evolutionary genetic role of the pleiotropic effect of the genes affecting the coat color on the animal neurochemical characteristics? It is conceivable that coat color genes affect not only the neurotransmitter metabolism in different divisions of the brain, but also ensure the tolerance to the psychoemotional stress under the conditions of commercial domestication, primarily, the state of captivity, by modulating the rate of domestication-related transformations in successive generations of fur animals in the course of long-term breeding under anthropogenic conditions. Earlier, the pleiotropic effects of mutations affecting the coat color on the penetrance and expressivity of domestication behavior was found in animals with aberrant coat color. For example, both the proportion of animals with domestication behavior and the expressivity of this behavior were found to be significantly higher in *sapphire* minks (*a/a p/p*) compared to *standard* minks (+/+) (Trapezov *et al.*, 2012). Keeler (1947) noted similar effects of coat color aberrations in fur animals (foxes and minks) as early as the mid-20th century; however, these were only observations not confirmed by statistical data.

References

Alekhina, T.A., Prokudina, O.I. and Ryazanova, M.A., 2007. Typological Patterns of Behavior in Rats Selected for Elevating and Reducing of Pathological Moving, Linkage with Monoamines of Brain, Pavlov Journal of Higher Nervous Activity 57(3): 336-343.

Felten, D.L. and Sladek, J.R., 1983. Monoamine Distribution in Primate Brain: V. Monoaminergic Nuclei: Anatomy, Pathways and Local Organization. Brain Research Bulletin 58(2): 171-284.

Il'ina, E.D. and Kuznetsov, G.A., 1983. Principles of Fur Animal Genetics and Breeding. Moscow. Kolos. 280 pp. (in Russian).

Keeler, C.E., 1947. Coat Color, Physique, and Temperament. Journal of Heredity 9: 271-277.

Klochkov, D.V., Alekhina, T.A., Trapezov, O.V. and Petrenko, O.I., 2005. Estrus Circle, Folliculogenesis, Catecholamines after Horionic Gonadotropin Stimulation in Mink Selected for Behavior. Journal of Evolution Biochemistry and Physiology 41(3): 266-271. (in Russian).

Koldaeva, E.M. 2004. Farm Fur Animals: Genetics and Selection. 1st book. Moscow. Izvestiya. 296 pp. (in Russian).

Ness, N.N., Einarson, E.J., Lohi, O. and Jorgensen, G., with contributions from Jarosz, S. and Scheelje, R., 1988. Beautiful Fur Animals and Their Color Genetics. Glostrup, Denmark. Scientifur. 271 pp.

Popova, N.K., Voitenko, N.N., Kulikov, A.V. and Trapezov, O.V., 1994. Pleiotropic Effect of One-Locus (*p/p*) and Two-Locus (*a/a p/p*) Fur Color Mutations on Serotonin Metabolism in the Mink Brain. Russian Journal of Genetics 30(11): 1308-1310.

Popova, N.K., Voitenko, N.N., Kulikov, A.V. and Trapezov, O.V., 1996. The One-Locus Silver-Blue Fur Color Mutation (*p/p*) in Minks Affects the Dopamine System of the Brain. Russian Journal of Genetics 32(3): 452-454.

Schlumpf, M., Lichensteiger, W. and Langemann, H., 1974. A Fluorometric Microtechnique for Simultaneous Assay of 5-Hydroxytryptamine, Noradrenaline and Dopamine in Milligrams of Brain Tissue. Journal of Biochemistry and Pharmacology 23(6): 2337-2346.

Shackelford, R.M. and Cole, L.J., 1947. Screw-Neck in the Pastel Color Phase of Ranch-Bred Mink. Journal of Heredity 38: 203-209.

Shackelford, R.M., 1948. The Nature of Coat Color Difference in Mink and Foxes. Genetics. 33:311-336.

Shackelford, R.M. and Moore, W.L., 1954. Genetics Basis of Some White Phenotypes in the Ranch Mink. Journal of Heredity 45: 173-176.

Smith, S.E., Whitaker, C.H., Davis, L.F. and Nobel, P.V., 1941. The Inheritance of Three Coat Color Mutations in Ranch-Raised Mink. Journal of Heredity 32: 173-176.

Szot, P., Weinshenker, D. and White, S.S., 1999. Norepinephrine-Deficient Mice Have Increased Susceptibility to Seizure-Induced Stimuli. Journal of Neuroscience 19(24): 10985-10992.

Trapezov, O.V., 1997. Black Crystal: A Novel Coat Color Mutant in the American Mink. Journal of Heredity 88: 164-166.

Trapezov, O.V., Trapezova, L.I. and Sergeev, E.G. 2012. Coat color mutations and defensive reaction towards man in farm-bred minks and sables. In: Proceedings from the X International Scientific Congress in Fur Animal Production. August 21-24, 2012 Copenhagen, Denmark.

Trut, L.N., 2000. The Problem of Destabilizing Selection in Development, in Modern Concepts of Evolutionary Genetics. Novosibirsk. pp. 7-21. (in Russian).

Voitenko, N.N., and Trapezov, O.V., 2001. Effect of Four Coat Color Mutations (C_R, S, S^H, and h) on Brain Monoamine Oxidase in Mink. Russian Journal of Genetics 37(1): 519-522.

A single nucleotide polymorphism in the promoter region of the mink tumor necrosis factor-alpha gene

N.S. Yudin[1], R.B. Aitnazarov[1], I.V. Kulikov[2], E.V. Ignatieva[1] and O.V. Trapezov[1]
[1]Institute of Cytology and Genetics, Siberian Branch, Russian Academy of Sciences, 630090 Novosibirsk, Russia; [2]Institute of Internal Medicine, Siberian Branch, Russian Academy of Medical Sciences, 630039 Novosibirsk, Russia; trapezov@bionet.nsc.ru

Abstract

The aim of the study was screening for SNPs of promoter region of the *TNFalpha* gene in American mink (*Neovison vison*). Nucleotide sequence of 145 bp fragment of the promoter region of the mink *TNFalpha* gene was established. One SNP was found – a substitution C>T in position 111 from the beginning of the sequence. C allele contains potential LBP1/UBP1 transcription factor binding site, and T allele includes potential binding sites for NF-1 and C/EBP transcription factors.

Keywords: SNP, TNFapha, promotor region, mink

Introduction

Cytokines play a central role in the regulation of immune response and its integration with endocrine and hematopoietic systems. Tumor necrosis factor alpha (TNFalpha) is a multifunctional proinflammatory cytokine. TNFalpha is produced by macrophages, activated T-cells, natural killer cells and mast cells. TNFalpa initiates of different mechanisms that lead to an effective immune response. TNFalpha also is able to induce normal mammary gland development, apoptotic cell death, and anti-tumoral response (Cantaert *et al.*, 2010).

TNFalpha provides a rapid form of host defense against infection but is harmful in excess. Because TNFalpha is employed against a variety of pathogens, each involving a different pattern of risks and benefits, it might be expected that this would favor diversity in the genetic elements that control TNFalpha production. Single-nucleotide polymorphisms (SNPs) in promoter region of *TNFalpha* gene are associated with susceptibility to a number of infectious diseases. SNPs in the human *TNFalpha* promoter region had been found to be associated with susceptibility to human T-lymphotropic virus type I-induced uveitis (Seki *et al.*, 1999), with susceptibility to cerebral malaria (McGuire *et al.*, 1994), with mucocutaneous leishmaniasis (Cabrera *et al.*, 1995), with death from meningococcal disease (Nadel *et al.*, 1996), with lepromatous leprosy (Roy *et al.*, 1997), with scarring trachoma (Conway *et al.*, 1997), etc. SNP in the promoter region of cattle *TNFalpha* gene could at least in part contribute to the progression of lymphoma in bovine leukemia virus infection (Konnai *et al.*, 2006). Two single-nucleotide polymorphism genotypes within the *TNFalpha* promoter were more frequent in bank voles (*Myodes glareolus*) from areas with low prevalence of Puumala hantavirus infection (Guivier *et al.*, 2010).

Development of canine, feline, and panda marker genome maps has brought an opportunity for extensive studies on genome organization in other Carnivora, including species kept on fur-bearing animal farms. The aim of this report was screening for SNPs of promoter region of the *TNFalpha* gene in American mink (*Neovison vison*). The promoter regions contain specific DNA sequences (10-15 bp long) that are recognized by regulatory proteins known as transcription factors. Specific transcription factor binding to the promoter sequence controls the rate of gene transcription. To investigate the regulatory potential of the mink *TNFalpha* promoter region, the *in silico* recognition of potential transcription factor binding sites was performed.

Material and methods

The work was conducted on 4 American mink males from Experimental Farm of the Institute of Cytology and Genetics SB RAS (Novosibirsk, Russia). Blood samples (3-5 ml) were collected from tail vein, stored on ice, brought to the Institute of Cytology and Genetics SB RAS and further stored at -40 °C until DNA was extracted. Genomic DNA was prepared from peripheral blood leukocytes by digestion of cells with proteinase K in the presence of EDTA and SDS, followed by extractions with phenol-chloroform.

The primer set TNFalpha_d (5'-CCGAGAAATGGGACAACCT-3') and TNFalpha_r (5'-GCCATGTATCCCCAAAGAAT-3') was used to amplify a fragment of the promoter region of the mink *TNFalpha* gene. PCR primers were designed on the basis of the *TNFalpha* gene sequence of cattle deposited in the Genbank (accession no. Z14137, positions 3090-3108 and 3215-3234, respectively). The reaction mixture (25 μl) contained 200 ng DNA template, 10 pM each oligonucleotide, 1 unit of Taq DNA polymerase (Lyteh, Russia) in a buffer containing 75 mM Tris-HCl (pH 9.0), 200 mM $(NH_4)_2SO_4$, 2.5 mM $MgCl_2$, 200 μM each dNTP, 6% dimethyl sulfoxide, and 0.01% Tween-20. Samples were amplified for 30 cycles (94 °C for 50 s, 53 °C for 70 s and 72 °C for 2.5 min). The reaction products were purified by ethanol precipitation and sequenced by dye-terminator cycle sequencing on an ABI 3130xl genetic analyzer (Applied Biosystems, USA).

A multiple alignment of the sequences and phylogenetic tree reconstruction were made by AlignX module of Vector NTI software (Invitrogen, USA). The *TNFalpha* gene sequences of cow, cat, dog, and panda were obtained from UCSC Genome Browser website (http://genome.ucsc.edu/). The potential transcription factor binding sites were predicted using Transcription Element Search Software (TESS) (http://www.cbil.upenn.edu/cgi-bin/tess/tess/).

Results

Nucleotide sequence of the fragment of the promoter region of the mink *TNFalpha* gene was established (Figure 1). The length of studied sequence was 145 bp. The nucleotide sequence of the mink *TNFalpha* gene shared the highest identity with the cow (*Bos taurus*) sequence. Identity of the promoter region of the mink *TNFalpha* gene with three other Carnivora species was significantly weaker. A comparative analysis of the nucleotide sequences showed the smallest genetic distance between the mink and the cow, and the same genetic distance between this two species and cat, dog and panda (Figure 2). One SNP was found in the fragment of the promoter region of the mink *TNFalpha* gene: a substitution C>T in position 111 from the beginning of the sequence. All four animals were heterozygotes C/T.

```
                1                                                  50
Mink    (1)    CCGAGAAATGGGACAACCTCCAAGGCTGGGGAC-TAGAGAACACCAGAGG
Cow     (1)    .................................-................
Cat     (1)    .T...G.G.....G.G..C.......G....GG---.G.C..C.AG.G.A
Dog     (1)    .TA..G.......G.--.C..C...T.A...GGG.G.G.....TGG...A
Panda   (1)    GAAT.G.G.....G.-..C............GGA.G.G...T..AG...A
                51                                                100
Mink    (50)   CATTTCAGGAGACCTGGTCAC----ACACACAGGAGCTCTCAAGGGCAGT
Cow     (50)   .....................----.........................
Cat     (48)   ..C..T...G...AA...G.T------...G...G.....G.G..CT..G
Dog     (49)   .....T.......A.......GGTC.....G...G.....G.C..CT..G
Panda   (50)   .G........A..A......T------G..G...G.....G.GA.CT..G
                101            ↓                                  150
Mink    (96)   GCTGTCTCCAGGAAANTGGAGGGAGAAGGGATTCTTTGGGGATACATGGC
Cow     (96)   ...............C..................................
Cat     (92)   ...AC......A..TC.........TT..AG.........-.........
Dog     (99)   ...A.......A..TC...G..C.--T..AG.........-.C...A...
Panda   (94)   ...A.......A..TC.......T.CT..AG.........-.........
```

Figure 1. Alignment of the TNFalpha *promoter fragments from different species. Identical nucleotides are represented by a dot; dashes (-) indicate gaps. The position of the SNP is represented by the arrow.*

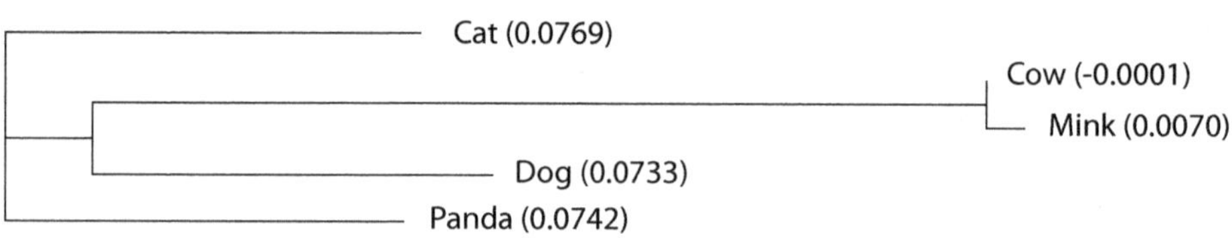

Figure 2. Phylogenetic tree of fragments of promoter of the TNFalpha gene.

It was found that the *TNFalpha* promoter region C allele contains the potential LBP1/UBP1 binding site which is co-localized with C/T polymorphic site (Figure 3). But the DNA region containing thymidine in the polymorphic site (T allele) does not reveal any similarity with LBP motif. LBP motif includes five nucleotides CCTGG and the second cytidine (underlined) matches the polymorphic site. On the other hand the T allele contains the potential NF-1 and C/EBP binding sites which are co-localized with SNP. NF-1 and C/EBP motifs contain five nucleotides (TTGGC and ATTGG) where SNP variant T coincides with the first nucleotide in the NF-1 motif and with the second nucleotide in the C/EBP motif.

The nucleotides C and T corresponding to SNP are indicated in bold character. The potential LBP1 (C allele) and NF-1 (T allele) transcription factor binding sites are underlined. The location of potential binding site for C/EBP (T allele) is indicated by rectangle. LBP1, NF-1 and C/EBP motifs and model identifiers used for prediction are shown above and below the *TNFalpha* promoter sequences.

```
        LBP1/UBP1 motif       CCTGG    Model_ID: I00191
C allele 5'- GCTGTCTCCAGGAAACTGGAGGGAGAAGGGA -3'
T allele 5'- GCTGTCTCCAGGAAATTGGAGGGAGAAGGGA -3'
        NF-1 motif             TTGGC    Model_ID: Q00112
        C/EBP motif           ATTGG    Model_ID: I00135
```

Figure 3. Potential LBP1, NF-1 and C/EBP transcription factor binding sites predicted in the mink TNFalpha *promoter region containing SNP.*

Discussion

Nucleotide sequence of the promoter region of the mink *TNFalpha* gene in our study was completely identical to the *Bos taurus TNFalpha* promoter sequence. Mink sequence was found to be far from other Carnivora species (cat, dog, and panda). The causes of such similarity are not clear yet, but it is known that there are a lot of highly conserved elements from hundreds to thousands of bases long in vertebrate genomes (Siepel *et al.*, 2005). We found a new SNP in the promoter region of the mink *TNFalpha* gene. Mink is insufficiently studied in relation to SNP so far. We didn't find any reference SNP cluster report in dbSNP (http://www.ncbi.nlm.nih.gov/snp) using search terms 'mink' or '*Neovison vison*'. It is notable that there is SNP in the same position in *TNFalpha* gene of cow (Konnai *et al.*, 2006). The existence of the same SNPs in distantly related species is a very rare phenomenon (Sechi *et al.*, 2010).

It was found that a substitution of C to T in the *TNFalpha* promoter region changes the pattern of the potential binding sites for transcription factors. C allele contains potential LBP1/UBP1 binding site, and T allele includes potential binding sites for NF-1 and C/EBP. Taking into account that LBP1/UBP1, NF-1 and C/EBP are expressed in a wide variety of cells and tissues, including cells of monocytic origin (Friedman, 2007; Kato *et al.*, 1991), we can assume that these proteins may regulate the transcription of the *TNFalpha* gene. The different regulatory mechanisms which may be mediated by two alleles mean potential advantages for heterozygous individuals

Acknowledgements

This work was supported by a grant (12-04-00917-a) from the Russian Foundation for Basic Research, and Programs of Russian Academy of Sciences (B.27, A.II.6).

References

Cabrera, M., Shaw, M.A., Sharples, C., Williams, H., Castes, M., Convit, J. and Blackwell, J.M., 1995. Polymorphism in tumor necrosis factor genes associated with mucocutaneous leishmaniasis. Journal of Experimental Medicine 182: 1259-1264.

Cantaert, T., Baeten, D., Tak, P.P. and van Baarsen, L.G., 2010. Type I IFN and TNFα cross-regulation in immune-mediated inflammatory disease: basic concepts and clinical relevance. Arthritis Research and Therapy 12: 219.

Conway, D.J., Holland, M.J., Bailey, R.L., Campbell, A.E., Mahdi, O.S., Jennings, R., Mbena, E. and Mabey, D.C., 1997. Scarring trachoma is associated with polymorphism in the tumor necrosis factor alpha (TNF-alpha) gene promoter and with elevated TNF-alpha levels in tear fluid. Infection and Immunity 65: 1003-1006.

Friedman, A.D., 2007. Transcriptional control of granulocyte and monocyte development. Oncogene 26: 6816-6828.

Guivier, E., Galan, M., Salvador, A.R., Xuereb, A., Chaval, Y., Olsson, G.E., Essbauer, S., Henttonen, H., Voutilainen, L., Cosson, J.F. and Charbonnel, N., 2010. Tnf-α expression and promoter sequences reflect the balance of tolerance/resistance to Puumala hantavirus infection in European bank vole populations. Infection, Genetics and Evolution 10: 1208-1217.

Kato, H., Horikoshi, M. and Roeder, R.G., 1991. Repression of HIV-1 transcription by a cellular protein. Science 251: 1476-1479.

Konnai, S., Usui, T., Ikeda, M., Kohara, J., Hirata, T., Okada, K., Ohashi, K. and Onuma, M., 2006. Tumor necrosis factor-alpha genetic polymorphism may contribute to progression of bovine leukemia virus-infection. Microbes and Infection 8: 2163-2171.

McGuire, W., Hill, A. V. S., Allsopp, C. E. M., Greenwood, B. M. and Kwiatkowski, D., 1994. Variation in the TNF-alpha promoter region associated with susceptibility to cerebral malaria. Nature 371: 508-511.

Nadel, S., Newport, M. J., Booy, R. and Levin, M., 1996. Variation in the tumor necrosis factor-alpha gene promoter region may be associated with death from meningococcal disease. Journal of Infectious Diseases 174: 878-880.

Roy, S., McGuire, W., Mascie-Taylor, C.G., Saha, B., Hazra, S.K., Hill, A.V. and Kwiatkowski, D., 1997. Tumor necrosis factor promoter polymorphism and susceptibility to lepromatous leprosy. Journal of Infectious Diseases 176: 530-532.

Sechi, T., Coltman, D.W. and Kijas, J.W., 2010. Evaluation of 16 loci to examine the cross-species utility of single nucleotide polymorphism arrays. Animal Genetics 41: 199-202.

Seki, N., Yamaguchi, K., Yamada, A., Kamizono, S., Sugita, S., Taguchi, C., Matsuoka, M., Matsumoto, H., Nishizaka, S., Itoh, K. and Mochizuki, M., 1999. Polymorphism of the 5'-flanking region of the tumor necrosis factor (TNF)-alpha gene and susceptibility to human T-cell lymphotropic virus type I (HTLV-I) uveitis. Journal of Infectious Diseases 180: 880e.

Siepel, A., Bejerano, G., Pedersen, J.S., Hinrichs, A.S., Hou, M., Rosenbloom, K., Clawson, H., Spieth, J., Hillier, L.W., Richards, S., Weinstock, G.M., Wilson, R.K., Gibbs, R.A., Kent, W.J., Miller, W. and Haussler, D., 2005. Evolutionarily conserved elements in vertebrate, insect, worm, and yeast genomes. Genome Research 15: 1034-1050.

Detection of SNP markers and practical applications in mink (*Neovison vison*)

P.F. Larsen[1,4], D. Demontis[2], V.H. Nielsen[3], J.P. Thirstrup[3] and C. Pertoldi[4]
[1]Kopenhagen Research, Agro Food Park 15, 8200 Aarhus N, Denmark; [2]Institute of Biomedicine, Aarhus University, Wilhelm Meyrs Allé 4, 8000 Århus C, Denmark; [3]Department of Genetics and Biotechnology, Aarhus University, Blichers Allé, 8830 Tjele, Denmark; [4]Department of Biological Sciences, Aarhus University, Ny Munkegade, 8000 Århus C, Denmark; pfl@kopenhagenfur.com

Abstract

Single nucleotide polymorphisms (SNPs) are widespread sources of sequence variation in genomes and are currently applied in fields such as forensic and animal breeding genetics. Here we present how it is possible to quickly identify informative SNPs in mink based on a SNP marker panel created for the domestic dog which is phylogenetic far away from the mink. Applying the commercial available Canine SNP Genotyping BeadChip (Illumina®) to genotype 12 American mink from 4 different color strains we identified 111 SNPs in mink. Based on the 111 SNPs genetic population structure analysis showed similar levels of relatedness and genetic differentiation than previously observed using microsatellite markers in mink. We suggest strategies for applying SNP markers in practical farm management, by genotyping individuals with a few very informative genetic markers, considering genetic information that would increase the heterozygosity and have positive effects on the breeding result in mink production. Alternative applications of SNP markers for parentage analysis, individual identification and forensic identification in mink are discussed. Finally, identifying markers associated with important production traits and combined with neutral markers can combined be applied to improve genetic screening on farm levels and for developing new breeding strategies on Danish mink farms.

Keywords: genetic distance, relatedness, inbreeding, mink farm

Introduction

In recent years a large number of microsatellite markers have been published for mink (*Neovison vison*) (e.g. Anistoroaei *et al.*, 2006, 2007). However, microsatellite markers are expensive in genotyping and difficult to genotype in a large scale approach, furthermore standardization among different laboratories is a big problem. Therefore new markers are desired and here SNPs represent the most widespread source of sequence variation within genomes (Brumfield *et al.*, 2003) and they have emerged as valuable genetic markers also in animal and conservation genetics. SNPs are biallelic markers, and are inherently less informative if compared to the multiallelic microsatellites, when used for individual identification, parentage analysis and population genetics. However, SNPs compared to microsatellites has several potential advantages like a better amplification power, lower genotyping errors and unambiguous allele comparisons. However, their simpler mutational dynamics strongly reduce risks of homoplasy (Chen and Sullivan, 2003). Furthermore, fast and inexpensive methods are now being developed to screen hundreds or thousands of SNPs per sample per population (Wang *et al.*, 2009).

From the few genetic studies conducted on mink so far, it has been shown that there are large genetic differences between different strains (color types) of mink (e.g. Michalska-Parda *et al.*, 2009; Larsen *et al.*, 2009). Demontis *et al.* (2011) have recently shown a very strong and highly significant negative correlation between the level of relatedness (R) and fecundity ($r = 0.536$, $P<0.001$). The high correlation between the level of R and breeding result, indicate that inbreeding decreases fecundity in farmed mink, demonstrating therefore, the importance of avoiding severe inbreeding in order to keep the fecundity high. This information could potentially be applied in practical farm management. By genotyping individuals with a few very informative markers, the degree of genetic differentiation (F_{ST}) estimates between farms within strains, and the degree of R within farms and strains etc., can be estimated and used to evaluate which animals would be the best to purchase in practice in order to keep the fecundity as high as possible (Larsen *et al.*, 2011).

The main aims of this study, were to genotype American mink on the Illumina Canine BeadChip in order: (1) to identify Single Nucleotide Polymorphisms (SNPs) in the mink; (2) to use these SNPs to estimate the level of R within strains as a measure of the inbreeding level (as inbreeding is produced by mating among relatives); (3) to elucidate the genetic divergence between strain, and finally (4) to estimate the probability of identity (PI) for an increasing number of loci, i.e. the probability that different individuals by chance share an identical genotype (potential useful for individual analysis and forensic genetics in mink).

Material and methods

DNA extraction and SNPs genotyping

12 American mink representing four different strains were included in this study; three Mahogany (consisting of a trio: female, male and their male offspring, included to follow segregation of alleles from parent to offspring and ensure SNPs are 'real SNPs'), three Sapphire, three Black and three Brown/Glow, from different Danish mink farms. Except for the mink in the Mahogany strain, all collected mink were unrelated. The unrelated mink did not have the same mother or father in the breeding season 2010. Blood or liver samples were kept in a freezer at -18 °C until analyses were performed. DNA was extracted the using the QIAGEN DNeasy Blood and Tissue Extraction Kit (QIAGEN Ltd). Genotyping of SNPs (single nucleotide polymorphisms) was performed using Canine SNP BeadChip (Illumina®) according to the manufacturer's protocol (Infinium II Multi-Sample), followed by analysis using the software GenomeStudio (Illumina®).

Statistical analyses

Variability analysis: Unbiased Expected Heterozygosity (UH_E), percent of polymorphic loci (P%), was performed using the software GENALEX v. 6.1 (Peakall and Smouse, 2006). GENALEX was also used for estimating the level of relatedness (R) within each strain (Ritland, 1996), and for estimating the probability of identity (PI) for an increasing number of loci, i.e. the probability that different individuals by chance share an identical genotype (the male Mahogany offspring was excluded from this test).

The genetic divergence between strains and within strains among farms was estimated by calculating the pair-wise F_{ST} (Weir and Cockerham, 1984) values using the software GENALEX and the significance of the F_{ST}s and linkage disequilibrium between loci (within strain) were estimated using the software GENEPOP 3.4 (Raymond and Rousset, 1995), (the male Mahogany offspring was excluded from this test).

Results

The CanineHD genotyping BeadChip SNP chip contains more than 170,000 markers placed on the CanFam2.0 reference sequence. Very stringent quality control was performed in order to identify SNPs with reliable calls. Only SNPs with a call rate above 91% (called in all but one individual) having a minor allele frequency above 0.25 and below 0.45 were included. Only 440 SNPs passed these criteria. After manually checking all clusters 111 SNPs remained for further analysis all demonstrated nice clusters with a call rate of 100%.

Among the known mother-father-offspring trio, no errors were detected for the 111 SNPs. Although the power to detect errors at individual loci was low, the fact that no errors were detected among the loci investigated suggests that SNPs can be typed reliably and that the true error rate may be lower than the 1% error rate assumed in the paternity simulations. If the error rate was assumed to be zero, a slightly smaller panel of SNPs would be sufficient for successful paternity analysis (data not shown).

The unbiased expected heterozygosity (UH_E) varied between 0.110 in Mahogany to 0.570 in Brown/Glow and the percent of polymorphic loci (P%) varied between 19.82% in Mahogany and 96.40% in Black and Brown/Glow (see Table 1). No linkage disequilibrium was detected between the 111 loci investigated, which were therefore all included in the statistic analyses.

Table 1. Id-number (N) of individuals genotyped from each farm and breeding line. The unbiased expected heterozygosity (UH_E), Standard Error (SE), % of polymorphic loci (P %) for each strain.

N	Strain	UH_E	SE	P %
1	SAPH	0.491	0.013	93.69%
2	SAPH			
3	SAPH			
4	BLACK	0.475	0.012	96.40%
5	BLACK			
6	BLACK			
7	WILD	0.570	0.012	96.40%
8	WILD			
9	WILD			
10 (son)	MAH	0.110	0.021	19.82%
11 (female)	MAH			
12 (male)	MAH			

When comparing the Black, Sapphire, Brown/Glow and Mahogany mink we found the highest mean R for the Mahogany strain (0.478) (bringing however in mind that the Mahogany strain is composed by a trio) followed by the Sapphire strain (0.238) and the lowest R in the Brown/Glow strain (0.067) and lastly, Black (-0.099) (see Table 2).

The genetic divergence (F_{ST}) between strains ranged from 0.074 (between the Black and Brown/Glow strains) to 0.449 (between the Sapphire and Mahogany strains) and were all highly significant (*P*<0.0001) (Table 3).

Probability of identity based on the 111 SNPs was estimated to be between 9.4E-09 for the Mahogany strain and 5.4E-45 for the Brown/Glow (see Figure 1).

Discussion

The rank of the R-values is mostly in agreements with the rank obtained by Demontis *et al.* (2011). It was not surprising that the highest degree of R was found in the Mahogany as the individuals representing this strain was a trio. However the Mahogany strain has been maintained isolated at very low population size for five generations and that is also reflected in the extremely

Table 2. Mean relatedness R (Ritland 1996) Upper (Uv error) and lower (Lv error) error bars bound the 95% confidence interval about the mean values as determined by bootstrap resampling (99 bootstrap) for SAPH, BLACK and BROWN/GLOW (B/G) without MAHOGANY (MAHO) (a) and with MAHO (b) (TRIO, mother, father and son).

		SAPH	BLACK	B/G	MAHO
a. Mean Bootstrap Value		0.163	-0.073	0.032	
	Uv error	0.058	0.08	0.015	
	Lv error	0.1	0.137	0.023	
b. Mean Bootstrap Value		0.238	-0.099	0.067	0.478
	Uv error	0.078	0.084	0.023	0.032
	Lv error	0.138	0.119	0.017	0.02

Table 3. Pairwise F_{ST} between strains. All the F_{ST} values were highly significant (P<0.0001).

	Sapphire	Black	Brown/glow
Sapphire			
Black	0.161		
Brown/glow	0.041	0.074	
Mahogany	0.449	0.169	0.301

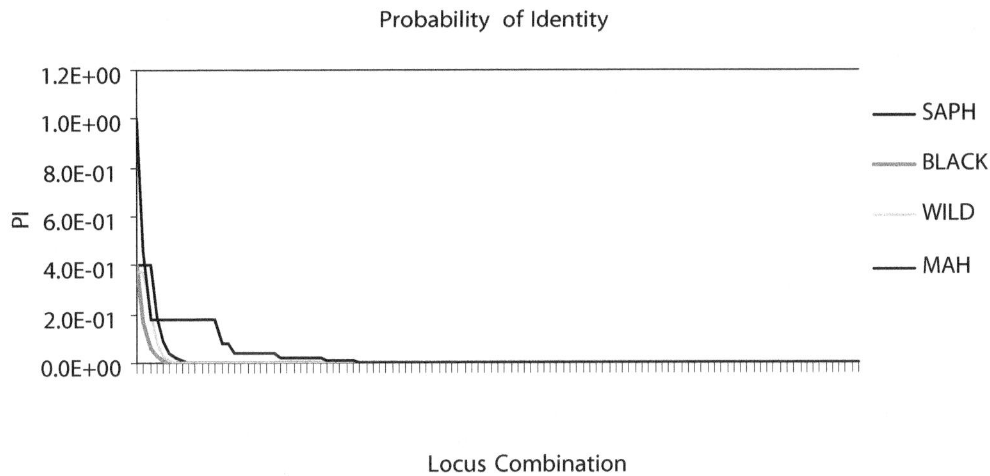

Figure 1. Probability of identity with increasing locus combination (excluding the son in Mahogany).

low level of polymorphism observed in this strain compared to the others. The relatively low level of R found in the Black strain in this study is the result of a large production group of Black mink at the Research Farm in Foulum. The observed degree of genetic differentiation between strains was also in agreement with the results shown by Demontis *et al.* (2011).

The use of SNPs as a genetic marker has proven advantageous although some disadvantages are experienced. Thus SNP identification is a challenge due to major laboratory efforts and cost associated with the findings of SNPs (Ryynänen *et al.*, 2007). Identification of polymorphic and informative SNPs in small or differentiated populations is difficult, even when polymorphic sequences for closely related species are available. Furthermore, not all SNPs detected in the sequences are in specific positions suitable for primer and probe design. Thus in this study we have shown that it is possible to use a chip developed for another species and detect useful SNP, this allow for detecting and applying SNPs in species where genomic resources are still limited. To improve genotyping success and reliability, it is preferable to amplify DNA fragments as short as possible. The use of single nucleotide polymorphisms (SNPs) requires only the amplification of very short fragments and this makes them particularly suitable for non-invasive genetic monitoring projects (Seddon *et al.*, 2005).

In mink SNPs could be used for selection of breeding animals or by farmers when buying in new males to ensure high fecundity in their breeding stock. Knowing the genetic distance between pure lines will make it possible to suggest better breeding strategies. Thus it is expected to obtain heterosis by mating genetically distant strains/lines. Qualified choices of breeding animals can be made when knowing both the 'genetic profile' of the animals on the farm as well as the profile of the potential breeding animals on top farms. Thus heterosis may be obtained for several traits such as fecundity, body size, skin length, weight at birth within color types. Knowing, the genetic distance among breeding lines from different farms within color types, it may be possible to observe heterosis within color types, which is often more attractive in practical farm management. QTL studies or association mapping of quantitative traits would help to identify markers associated with important production traits and these important markers combined with neutral markers can combined be applied to improve genetic screening for optimal animals for breeding on Danish mink farms.

A subset of very informative SNPs (48 or 96) could be genotyped using relatively cheap technologies like, for example, VeraCode or Fluidigm EP1 System. The costs of SNP microchip beads application is gradually decreasing, making it available to a wider range of users. These new techniques for high-throughput SNP genotyping could be used in non-invasive DNA monitoring projects that usually require a huge number of samples to be analyzed.

Based on the present study, a panel of 50-60 SNP-based genetic markers that are highly informative in mink could be designed and would be sufficient to conduct identity analyses in mink (Figure 1). This SNP panel would be conveniently genotyped using VeraCode (Illumina Inc.) or SNPlex (ABI) technology at a much reduced cost relative to typical microsatellite genotyping. Usage of an SNP panel as proposed here would allow not only reliable parentage and identity analysis but also would provide a standardized panel for exchange of genotype data between laboratories.

Conclusions

In mink, genotyping of individuals could help overcome problems with reduced fecundity, by mating genetically divergent breeding animals to optimize fitness or to obtain heterosis effects among strains within color types. Genotyping could also be employed to check identity of samples for diagnostic purposes and to detect identity of recovered stolen skins. We therefore see a huge advantage in applying molecular genetic information in practical mink farming in the future.

References

Anistoroaei R., Farid A., Benkel B., Cirera S. and Christensen K., 2006. Isolation and characterization of 79 microsatellite markers from the American mink (*Mustela vison*). Animal Genetics 37: 185-188.

Anistoroaei R., Menzorov A., Serov O., Farid A. and Christensen K., 2007. The first linkage map of the American mink (*Mustela vison*). Animal Genetics 38: 384-388.

Bonin A., Bellemain E., Eidesen P.B., Pompanon F., Brochmann C. and Taberlet P., 2004. How to track and assess genotyping errors in population genetics studies. Mol Ecol, 13: 3261-3173.

Broquet T. and Petit E., 2004. Quantifying genotyping errors in noninvasive population genetics. Mol Eco 13: 3601-3608.

Brumfield R.T., Beerli P., Nickerson D.A. and Edwards S.V., 2003. The utility of single nucleotide polymorphisms in inferences of population history. Trends Ecol Evol 18: 249-256.

Chen X. and Sullivan P.F., 2003. Single nucleotide polymorphism genotyping: biochemistry, protocol, cost and throughput. Pharmacogenom J 3: 77-96.

Demontis D., Larsen P.F., Baekgaard H., Soenderup M., Hansen B.K., Nielsen V.H., Loeschcke V., Zalewski A., Zalewska H., and Pertoldi C., 2011. Inbreeding affects fecundity of American mink (*Neovison vison*) at Danish farm mink. Animal Genetics (in press).

Larsen, P. F., Bækgaard, H., Sønderup, M., Møller, H. H., Hansen, B. K., Nielsen, V. H., Demontis, D., Loeschcke, V. and Perfoldi, C., 2010. Level of inbreeding and genetic distance in Danish farm mink (*Neovison vison*). Annual report 2009, 27-31, Danish Fur Breeders Research Center, Holstebro, Denmark

Larsen, P.F., 2011. The effect of inbreeding on the breeding result in Danish mink (In Danish: Indavls betydning for avlsresultat i danske mink). Internal report Foulum September 2011.

Michalska-Parda A., BrzeziŇski M., Zalewski A., and Kozakiewicz M., 2009. Genetic variability of feral and ranch American mink *Neovison vison* in Poland. Acta Theriologica, 54: 1-10.

Morin P.A. and McCarthy M., 2007. Highly accurate SNP genotyping from historical and low-quality samples. Molecular Ecology Notes, 7: 937-046.

Peakall R. and Smouse P.E., 2005. genalex 6: genetic analysis in Excel. Population genetic software for teaching and research. Mol. Ecol. Notes, 6: 288-295.

Quintáns B., Alvarez-Iglesias V., Phillips C., Lareu M.V. and Carracedo A., 2004. Typing of mitochondrial DNA coding region SNPs of forensic and anthropological interest using SNaPshot minisequencing. Forensic Sci Int, 140: 251-257.

Raymond M. and Rousset F., 1995. GENEPOP (version 1.2): Population genetic software for exact tests and ecumenicism. Journal of Heredity, 86: 248-249.

Ritland K., 1996. Marker-based method for inferences about quantitative inheritance in natural populations. Evolution; 50: 1062-1073.

Ryynänen H.J., Tonteri A., Vasemägi A. and Primmer C.R., 2007. A comparison of biallelic markers and microsatellites for the estimation of population and conservation genetic parameters in Atlantic Salmon (*Salmo salar*). J Hered, 7: 692-704.

Seddon J.M., Parker H.G., Ostrander E.A. and Ellegren H., 2005. SNPs in ecological and conservation studies: a test in the Scandinavian wolf population. Mol Ecol, 14: 503-511.

Sobrino B., Brión M. and Carracedo A., 2005. SNPs in forensic genetics: a review on SNP typing methodologies. Forensic Sci. Int. 154: 181-94.

Troell K., Mattsson J.G., Alderborn A. and Höglund J., 2003. Pyrosequencing analysis identifies discrete populations of *Haemonchus contortus* from small ruminants. Int J Parasitol, 33: 765-771.

Tschentscher F., Frey U.H. and Bajanowski T., 2008. Amelogenin sex determination by pyrosequencing of short PCR products. Int J Legal Med. 122: 333-335.

Vignal A., Milan D., SanCristobal M. and Eggen A., 2002. A review on SNP and other types of molecular markers and their use in animal genetics. Genet Sel Evol, 34: 275-305.

Wang J., Lin M., Crenshaw A., Hutchinson A., Hicks B., Yeager M., Berndt S., Huang W-Y., Hayes R.B., Chanock S.J., Jones R.C. and Ramakrishnan R., 2009. High-throughput single nucleotide polymorphism genotyping using nanofluidic Dynamic Arrays. BMC Genomics, 10: 561-574.

Weir B.S. and Cockerham C.C., 1984. Estimating F-statistics for the analysis of population structure. Evolution, 38: 1358-1370.

Part 4. Behaviour and welfare

Stress influencing production and welfare in farmed mink

J. Malmkvist
Aarhus University, Department of Animal Science, Research Centre Foulum, Blichers Allé 20, 8830 Tjele, Denmark; jens.malmkvist@agrsci.dk

Abstract

Results linking stress, production and welfare in farmed mink, with special focus on the hypothalamic-pituitary-adrenal (HPA)-axis, are presented. Interpretation of stress responses are discussed, combining context, neuroendocrine and behavioural indicators. Results from the development and validation of a novel non-invasive sampling of faecal cortisol metabolites (FCM) are presented. Both genes and environment affect the state of stress in farmed mink, and an overview of the major experimental findings in mink are given. This includes results from (1) selection for and against fearfulness with effects on the HPA-axis and the serotonin system, (2) studies of parturition/early kit mortality, and (3) cage enrichment studies affecting stress responses. Finally, recent results exploring the link between stress and abnormal behaviour are presented, investigating the HPA-axis response of stereotypic animals and whether fur-chewing/ stereotypic behaviour is associated with hippocampal neurogenesis in mink. Perspectives and suggested future research aiming to improve production and welfare in farmed mink are given.

Keywords: behaviour, hypothalamic-pituitary-adrenal (HPA)-axis, *Neovison vison, Mustela vison*, stress.

Introduction

The autonomic nervous system (ANS) and the hypothalamic-pituitary-adrenocortical (HPA) axis are activated when animals anticipate or are faced with stressors. Measuring HPA activity is a standard approach to the study of stress and welfare in farm animals, usually integrated with data from behavioural observation, production, and pathology records (Morméde *et al.*, 2007). In addition to be involved in the stress response, hormones from the HPA-axis play a role in normal body functions, illustrated by increased concentration in circulation before waking-up in the late stages of sleep; a situation not usually considered as stress.

The main hormone of HPA-axis in mink is cortisol secreted from the adrenal cortex, under control of the pituitary hormone ACTH (adrenocorticotropic hormone) from the anterior pituitary gland. Cortisol exerts a negative feedback on the axis at several levels, including within the brain. The feed-back ensures that plasma cortisol usually decline after the acute response. Chronic stress, certain diseases, and aging impair this feed-back action in a range of mammals (Sapolsky, 2002). In general, factors like predictability, control, and possibility of behavioural outlet have been found to modify the HPA-axis response in a range of animals when faced with stressors (e.g. classical papers by Weiss, 1968, 1971); these factors modulate the animals' perception of stress. Also the behavioural reactivity of the animals, shaped by the interplay between genes and environment, affects the HPA-axis response. Stress can arise both as a consequence of adverse external stimulation and as a consequence of internal causal factors driving the animal to attempt to carry out species-specific behaviour, without having the right possibility or sufficient feedback (Jensen and Toates, 1997).

Detrimental effects of prolonged stress responses

Although responses towards acute stressors are adaptive in nature, prolonged or repeatedly exposure to stress and stress hormones can have negative consequences, due to e.g. mobilization of energy at the cost of reserves, suppression of digestion, growth, reproduction, immunity and the inflammatory response (reviewed in Sapolsky, 2002). Additionally, high concentrations of glucocorticoids impair both memory and learning skills as shown in several mammalian species, and the link between stress and these functional deficits is believed to relate to changes in the hippocampal formation (Lupien *et al.*, 1998; McEwen and Sapolsky, 1995). One major finding across mammals is that stress or stress hormones can reduce the neuron number (Sapolsky *et al.*, 1985; McEwen, 1999; Jayatissa *et al.*, 2008) and decrease neurogenesis which occurs in the dentate gyrus of the hippocampal formation (Gould *et al.*, 1997; Sapolsky, 2004; Morris, 2007). Additionally, the hippocampus is involved as the central part of the feed-back control of the HPA-axis, with a high concentration of receptors for the corticoid hormones (McEwen, 1999). The negative effect on the hippocampus is one of the destructive effects of chronic stress.

Use of HPA-axis function to evaluate stress

Care should be taken when comparing animals of different genotypes and across studies in an attempt to evaluate consequences of e.g. different management and housing systems. However, within studies HPA-axis function has successfully been used as an indicator of stress, when sampled carefully. Baseline levels of glucocorticoid hormones have been sampled in blood, urine, and faeces in mink (an overview in Table 1). In addition, enlargement of adrenals – weight relative to individual body or brain size – has been taken as indicative of prolonged stress in some studies. Challenge tests are occasionally included to further evaluate the HPA-axis function, by following the change in cortisol concentration after standardised handling/immobilisation or after stimulation by ACTH injection. These HPA-challenges are used in order to detect effects – after chronic or repeated exposure to stressors – which potentially may change the set-point of the HPA-axis, making the animal less sensitive to secrete cortisol. There is no simple correlation between any single physiological/behavioural indicator and perceived stress. Farmed mink are seasonal breeders, with the photoperiod as the main regulator of reproduction, fur moulting, and hormonal rhythms. Therefore seasonality in the secretion of glucocortioids is observed, and the concentration may differ between sexes (summarized in Mormede *et al.*, 2007; also during gestation/around parturition in females: Malmkvist and Palme, 2008). Besides the influence of natural photoperiods on the hormonal levels, the normal farm routine with changing feeding strategies during the reproductive cycle may contribute to the variation. Feed restriction may for example increase both the baseline cortisol and the activity of mink (Bildsøe *et al.*, 1991).

Sampling of HPA-axis responses

The majority of blood cortisol reports in mink (Table 1) are on females, with the mean concentration from 25 to 95 nmol/L in plasma. These values cover a range of sampling procedures, such as nail

Table 1. HPA-function used to evaluate stress in relation to production and welfare in mink, examples from the literature. The listed indicators are typically linked to behavioural observations/other physiological measures. The table is not exhaustive.

Indicator	Focus variable/treatment	Reference
Weight of adrenals	Cage environment	Hansen 1988
	Housing system	Hanninen *et al.*, 2008a;b
	Fuel-oil feed contamination	Mohr *et al.*, 2008, 2010
Size of adrenals	Stereotyping mink	Mason 1992
Morphology of adrenals	Fuel-oil feed contamination	Mohr *et al.*, 2008
Cortisol in plasma	Infection, immobilisation	Jeppesen 1988
	Feed restriction, immobilisation in high and low stereotyping females	Bildsøe *et al.*, 1991
	Cage environment, immobilisation	Hansen and Damgaard 1991a
	Social environment	Hansen and Damgaard 1991b
	Early handling	Hansen 1993
	Selection for tame or fearful behaviour	Damgaard and Hansen 1996; Malmkvist and Hansen 2001
	Selection for inactivity or stereotypic activity	Zanella *et al.*, 1998
	Cage and social environment in two colour types	Pedersen and Jeppesen 2001
	Time of moving bitches from litter	Sørensen *et al.*, 2001
	Serotonin receptor agonist to mink selected tame or fearful behaviour	Malmkvist *et al.*, 2003
	Stereotypies and access to running wheel	Hansen and Damgaard 2009
Cortisol in serum	Pesticides in feed	Beard and Rawlings 1998
	Housing system, ACTH challenge	Hanninen *et al.*, 2008a;b
	Fuel-oil feed contamination, ACTH challenge	Mohr *et al.*, 2008, 2010
Cortisol in urine	PCB in food to pregnant females	Madej *et al.*, 1992
	Temperature, water availability in lactating mink	Tauson 1998
	Deprivation of food and resources in cage environment	Mason *et al.*, 2001
Cortisol in faeces	Offspring from high and low stereotyping parents	Tauchi *et al.*, 1999
Cortisol metabolites in faeces (FCM)	ACTH challenge, handling	Malmkvist *et al.*, 2011
	Cage size and environment	Hansen *et al.*, 2007
	Low and high stereotyping breeding lines	Svendsen *et al.*, 2007
	Low and high stereotypic mink	Malmkvist *et al.*, 2011
	Nest building possibilities, parturition	Malmkvist and Palme 2008
	Type of nesting materials, gestation-parturition	Lund and Malmkvist 2009
	Fuel-oil contamination	Mohr *et al.*, 2010
	Mink with/without abnormal behaviour	Malmkvist *et al.*, 2012
Cortisol binding protein -transcortin	Behavioural selection	Gulevich *et al.*, 2000
ACTH in plasma	Serotonin receptor agonist to mink selected for tame or fearful behaviour	Malmkvist *et al.*, 2003

clipping (Hansen and Damgaard, 1991a; Pedersen and Jeppesen, 2001), vein puncture (Malmkvist *et al.*, 2003) and catherization (Børsting and Damgaard, 1992) across seasons in different breeds. A duration of 2-2.5 min from the first capture until sample obtained is generally regarded as the upper time limit for detecting baseline values. Otherwise, the baseline concentration of cortisol hormones in the blood is not easily obtained as the capture of one animal may arouse and activate the secretion in other not yet sampled animals housed nearby (Pedersen, 1996), and the handling and blood sampling itself act as a stressor eliciting secretion of corticosteroid hormones into circulation. Therefore, examination of HPA-axis hormones in other biological excreta, such as urine and faeces sampled non-invasively, has been introduced. Sampling for saliva cortisol has to my knowledge not been used in mink. Repeated blood sampling by insertion of permanent catheters is possible (Bonneford *et al.*, 1988; Børsting and Damgaard, 1992), but this method is not stress-free (involves surgery and some movement restrictions) and infeasible when larger groups are to be sampled.

In particular, the measurement of faecal 11,17-dioxoandrostanes, a group of cortisol metabolites, has been proven useful to evaluate adrenocortical activity. In mink, these cortisol metabolites in faeces increase after injection of adrenocorticotropic hormone (ACTH) (Malmkvist *et al.*, 2004). Other studies have evaluated HPA-axis responses in mink based on non-invasive measurement of urine cortisol (e.g. Bildsøe *et al.*, 1991; Madej *et al.*, 1992; Mason *et al.*, 2001).

Mammalian species differ in the main excretory route of glucocorticoids; e.g. being predominately faeces for cats, and urine for dogs. The excretory route has only recently been studied in mink (Malmkvist *et al.*, 2011). Following an injection of radioactive cortisol, faeces were found to be the predominating excretory route in mink, with the ratio of cortisol in faeces:urine being about 5:1. The time lag until maximum concentration of cortisol was 3-7 hours in faeces and 2-5 hours in urine after a single dose injection (Figure 1).

Both faeces and urine integrate circulating cortisol levels over time, but are several steps away from the active hormone in the blood, which may interfere with validity/reliability. The cortisol to creatine ratio can be used to control for the amount of urine voided. Fresh faeces samples are more easily obtained than urine in the farm situation; besides faeces is the predominant excretion route for cortisol in mink. The concentration of cortisol itself is very low compared to that for cortisol metabolites (FCM), which are more detectable (Malmkvist *et al.*, 2011). FCM has mostly been used when mink are housed singly, and to my knowledge the effect of pooling faeces within/between animals has not been studied in detail. The calculation of concentration per unit wet weight versus dry weight of collected sample should be investigated in order to improve reliability. The interaction with steroid hormones (e.g. the sex hormone testosterone, sharing the precursor 17-a-hydroxyprogesterone with cortisol) on FCM should be studied to increase validity. Diurnal as well as seasonal effects should be taken into account when the sampling procedure is prepared. Knowledge of these and other sampling effects is to some extent described in the available literature using FCM. Adrenal evaluation may be an indicator of prolonged/severe stress, but is an invasive post mortem measure. This indicator is suspected to be insensitive to transient, milder types of stress. At pelting adrenal weight could with some effort be used, in case of post-mortem examination. The size of adrenals is dependent on sex/size of the animal, so the relative weight to e.g. body and brain size may be considered.

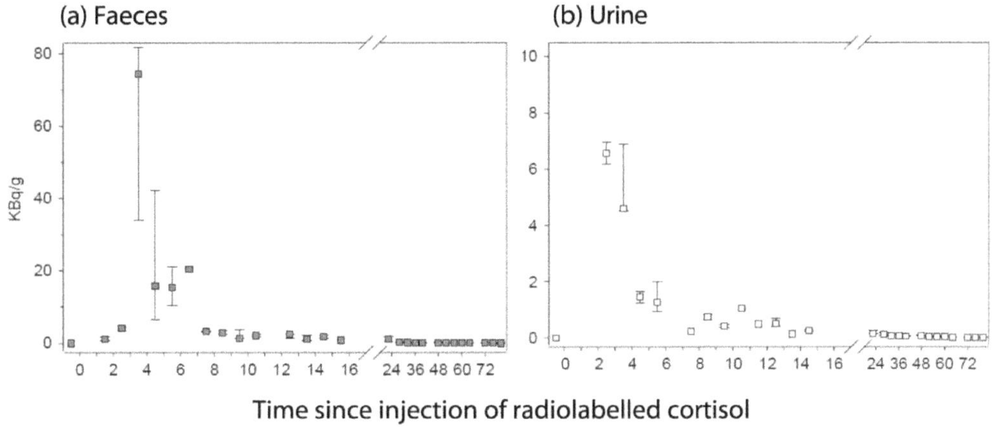

Figure 1. Time course of excretion of radioactive metabolites in (a) faeces (kBq/g) and (b) urine (kBq/ml) in mink females (n=8) after injection of ^{3}H-cortisol at 0 h. Data are given as hourly median with 25 and 75% quartiles for the first 16 h and per every 4th hour for the rest of the period. Note the different scaling of the y-axis for radioactivity, as faeces was the major route of excretion (83% vs. urine 17%). The proportion of injected radioactivity recovered was 84%. Modified from Malmkvist et al. *(2011).*

Stress in the production of mink

In this section, I have selected examples from studies into both genetic and environmental factors modulating stress responses, being of relevance for mink welfare and production.

Selection for behaviour

Selection for behaviour can influence the HPA-function in mink. Long-term selection based on reaction towards humans in the stick test have created lines differing markedly and consistently in their fearfulness (Malmkvist *et al.*, 2002) and in the HPA-axis response (e.g. secretion of cortisol) (Malmkvist and Hansen, 2001; Malmkvist *et al.*, 2003). Administration of an anxiolytic drug (Buspirone, a serotonin receptor agonist), neutralized the HPA-axis differences between fearful and explorative mink, illustrating the regulating role of the brain serotonin (5-HT) system (Figure 2). Consequently, the possibility to render mink more explorative, by increasing the relative amount of the essential serotonin precursor – the amino acid tryptophan – in mink feed has been investigated (Malmkvist *et al.*, 2011); the rationale was to reduce stress responses e.g. during periods of inevitable handling such as evaluation of fur and transport of mink. Following the selection for more tame/confident mink, other production relevant difference emerged, such as an increased mating willingness/earlier onset of oestrus (Klotchkov and Trapezov, 1991; Malmkvist *et al.*, 1997), and signs of quicker ontogeny (postnatal kit development) including an earlier onset of drinking (Malmkvist *et al.*, 2007a). In another behavioural selection experiment, mink females from a highly stereotypic breeding line had a higher cortisol baseline (Svendsen *et al.*, 2007).

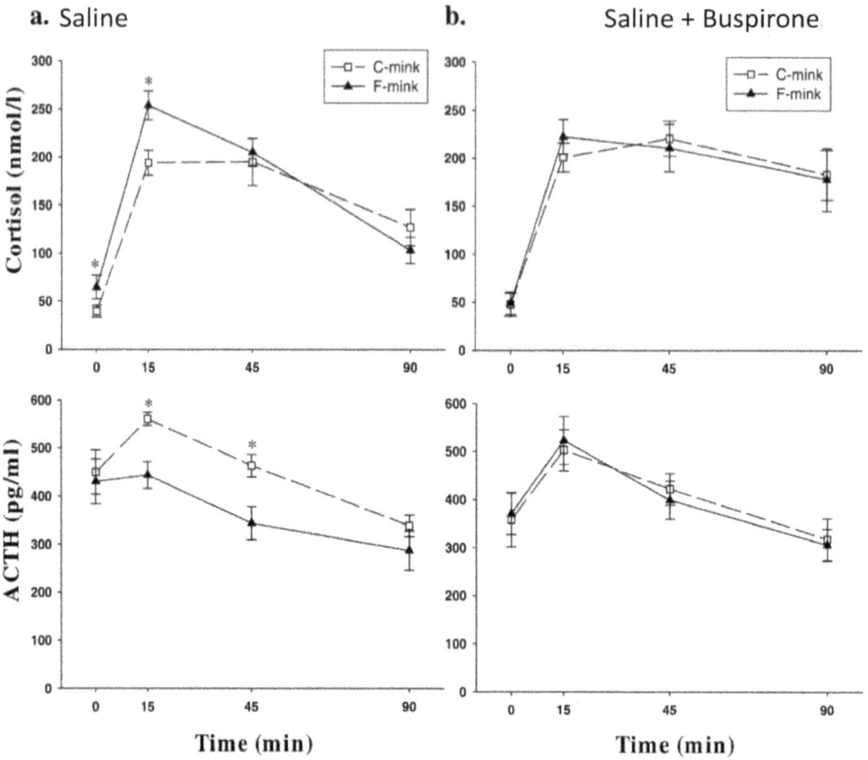

*Figure 2. Cortisol and ACTH concentration (mean±SE) in plasma vs. time after capture, in confident (C mink) and fearful mink (F mink) treated with (a) saline or (b) the anxiolytic serotonin (5-HT$_{1A}$) receptor agonist, Buspirone (1.25 mg/kg/day), for 28-29 days, administered continuously via osmotic minipumps placed s.c. in the neck. *Significant difference between the genetic lines. From Malmkvist* et al. *(2003).*

Cage enrichment

Changes in the cage environment can influence the stress responses markedly. Including tube platforms and chewing-ropes in cages resulted in significantly reduced abnormal behaviour (tail-chewing, stereotypies) and the baseline cortisol level (FCM). The mean concentration of FCM was approximately 25% lower in females kept in enriched compared to barren cages during the winter season. Thus, the cage additions reduced HPA-axis output significantly (P<0.01). In contrast, there was no effect whether females were housed in single or double cages (P=0.53) and no interaction between cage type (barren, enriched) and cage size (one, double) (P=0.28) (Hansen *et al.*, 2007). Hansen *et al.* (2007) concluded that increased cage complexity in form of occupational materials improve the welfare of the mink, whereas doubling the cage size without additional enrichment had little or no effect in relation to mink welfare.

Additionally, a nest box is important for the welfare of farmed mink as documented in studies of demand elasticity (longer stays more valued; Hansen and Jensen, 2001), HPA-axis responses (50%

higher plasma cortisol in mink without access to a nest box; Hansen and Damgaard, 1991), and behaviour (more stereotypic behaviour in mink without access to a nest box; Hansen *et al.*, 1994).

Parturition, maternal behaviour and early kit survival

As illustrated above, fulfilment of behavioural needs by adding relevant resources to the cage environment can reduce stress in mink. However, their stress responses can also vary with age and reproductive state; one example being during the time of parturition. Birth problems and prolonged parturition characterized mink females losing a high proportion of their litter (Malmkvist *et al.*, 2007b). Stressors may influence the labour controlling hormone oxytocin and prolong the delivery in other polytocous species, such as rats (*R. norvegicus*) and pigs (*S. scrofa*). For example, moving to a new environment during mid-parturition prolonged the birth process in rats (Leng *et al.*, 1987, 1988). Similarly, moving of sows to a farrowing crate during mid-parturition resulted in depressed oxytocin release through opioid inhibition and prolonged the birth process (Lawrence *et al.*, 1992). Stress can arise both as a consequence of adverse external stimulation and as a consequence of internal causal factors driving the animal to attempt to carry out species-specific behaviour, such as nest building, in case of no or insufficient feedback (Jensen and Toates, 1997). In one study, we compared different possibilities for female mink to perform nest building behaviour prior to and during the delivery season. Access to straw significantly reduced the variation in inter-birth intervals by 31-40% (P=0.044, during the delivery of in average 8.7-9.3 kits), indicating that access to straw for nest building is beneficial for the progress of parturition (Malmkvist and Palme, 2008).

Hypothermia is involved in early kit mortality, and may be the reason of the higher mortality and less growth in neonates raised by females with access to wood shavings only for nesting material. Thus, in spite of its worldwide use at mink farms, this nesting environment appears insufficient, which the higher basal cortisol level (faecal 11,17-dioxoandrostanes) of dams raising kits in this environment also indicate (at P=0.064). The nesting environment also affected the maternal behaviour of mink. Females with access to straw in combination with an artificial nest were more attentive and quicker to retrieve one of their 5-day-old progeny placed away from the safe nest. An artificial nest in combination with *ad libitum* access to straw significantly enhanced maternal kit-retrieval, and an artificial nest alone or in combination with straw tended to reduce maternal stress postpartum. For the kit vitality and survival, an artificial nest appeared as good as a nest of straw created by the dam, due to an improved in-nest climate postpartum. However, access to straw for nest building resulted in a less variable parturition, whereas the feedback from an artificial nest had no such effect (Malmkvist and Palme, 2008).

Abnormal behaviour and stress

The most common type of abnormal behaviour in farmed mink is stereotypies and fur-chewing, however, the link between abnormal and stress in mink has been debated. Although results linking individual stereotypic performance and welfare are inconsistent (cf. mink: Svendsen *et al.* (2007); review: Mason and Latham, 2004), a recent study reported that stereotypic behaviour is concurrent with higher baseline concentrations of faecal cortisol metabolites (FCM), and highly stereotypic mink react with more adrenocortical activity towards stressors than non-stereotypers (Malmkvist *et al.*, 2011) (Figure 3).

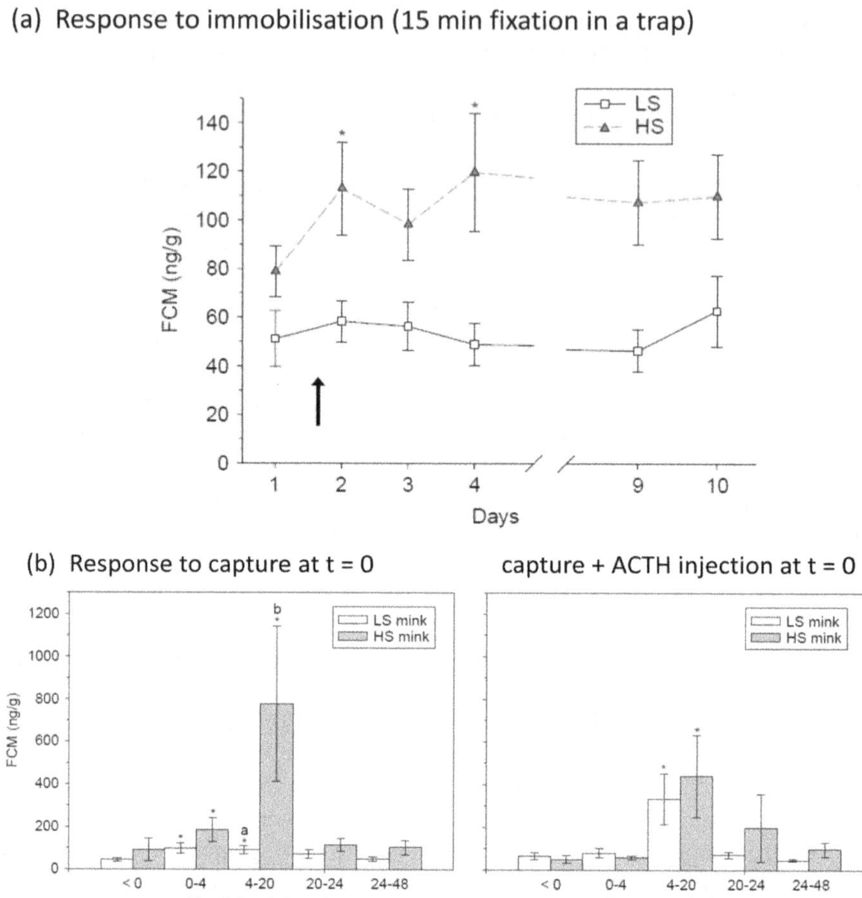

Figure 3. (a) FCMs (ng/g) in non-stereotypic (LS, n=71-80 per day) and high stereotypic (HS, n=69-76 per day) mink females sampled once daily at 14:00 to 18:00, given as mean±SEm. All mink were exposed to an immobilisation challenge for 15 min at Day 2 between 09:00 and 10:00 h (indicated by the arrow). The concentration of FCM was higher in HS than in LS mink (F-test, P<0.001), also at Day 1 (pairwise post test P = 0.002). Significant differences from within-group baseline (Day1) are marked with an asterisk. (b) Response to handing without and with ACTH injection at time t = 0 given as mean±SEM concentration of FCM in low (LS, n=12) and high (HS, ns=12) stereotypic mink females. Stereotypic mink reacted more during 4-20 h (F-test, P=0.001) after handling, but not after ACTH injection (F-test, P=0.74). Concentrations significantly greater than baseline are marked with asterisk, difference between LS and HS mink is marked by the letter 'a'. Modified from Malmkvist et al. *(2011).*

One major finding across mammals is that stress or stress hormones decrease neurogenesis which occurs in the dentate gyrus of the hippocampal formation (Gould *et al.*, 1997; Sapolsky, 2004; Morris, 2007). Adult neurogenesis – the generation of new neurons – has been demonstrated in the hippocampal formation in a range of mammals (e.g. rat, mouse, rabbit, guinea pig, cat, tree shrew and in primates, reviewed in Balu and Lucki, 2009), and recently also in mink (Malmkvist *et al.*, 2012). Comparing groups of mink females with and without abnormal behaviour, self-inflicted fur-chewing and stereotypic behaviour, we fail to demonstrate significant difference in number of new hippocampal cells (P=0.18) using 12 adult mink females per group. Repeating this study with more experimental animals could be beneficial to increase statistical power. In contrast to our expectations, there was positive correlation between the amount of stereotypic behaviour performed and the number of new hippocampal cells formed (P=0.016). Thus, the study did do not support that stereotypies in mink is concurrent with reduced hippocampal neurogenesis, typically taken as a sign of chronic stress and increased activation of the HPA axis. On the contrary, cell proliferation increased with increasing performance of stereotypic behaviour, being of an active/locomotory nature in mink (Figure 4). Further studies are needed to conclude whether stereotypy is an adaptive response rather than only reflecting pathology in mink. Additionally, it would be relevant to further understand effects of activity level, in order to investigate whether the positive effect on hippocampal cell proliferation relate to increased motor activity rather than to reduced stress in mink. In my opinion, these studies are needed for adding bricks to the puzzle of understanding different type of activity and stereotypies in mink, and how these relates to stress and welfare of relevance under production conditions.

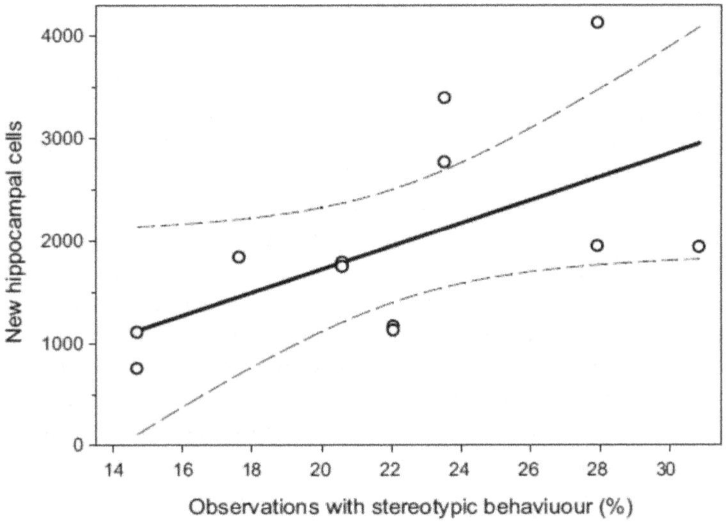

Figure 4. The number of new hippocampal cells (in the granule cell layer in the dentate gyrus) versus the amount of stereotypic behaviour in stereotypic mink (n=12). The line represents the linear regression, with 95% confidence intervals indicated by the broken lines. The number of new cells increase with the amount of stereotypic behaviour. From Malmkvist et al. *(2012).*

Conclusion and perspectives

Stress influence both welfare and production, and several studies have successfully used indicators of HPA-axis function to evaluate stress in farmed mink. Different sampling options are possible for determination of both baseline and response concentrations. The selected method should depend on the study hypothesis, remembering that handling and capture of the individual are acute stressors, eliciting secretion of HPA-axis hormones. When using the HPA function to evaluate stress one should particularly consider the influence of the fixed annual cycle of mink.

Recently, non-invasive FCM measurement reflecting circulating cortisol has been developed and validated for mink females. Faeces are the predominant excretory route of cortisol in mink, and nearly all cortisol present in faeces and urine are in metabolized forms. Consequently, assays measuring cortisol in mink urine may be less feasible to reflect the concentrations in circulation. Due to a relative high variation in non-invasive FCM, this method are suggested mainly in larger controlled experiments, e.g. when comparing different breeding lines, or different type of cage resources, with a sufficient number of animals per group. Further methodological developments are suggested, and include additional validation for FCM in males. Enlargement of adrenals could perhaps be developed into a measure of repeated/chronic stress, in case post-mortem examinations are considered; however, this indicator needs further investigation in order to increase validity.

We are in many cases interested in long-term effects of stress responses, due to its greater implications for both production and welfare. Therefore some focus in research should be devoted into understanding long-term effects, including e.g. dose-response relationships between stressors (including stress alleviation/fulfilment of needs) and the HPA-axis function in mink, and an improved understanding of how several indicators (e.g. abnormal behaviour, affective states, positive and negative effects on the brain) interact in mink. It would be naïve to rely on only one indicator as the ultimate stress measure; most study performed in mink include other indicators as well, and consider the influence of previous experiences, emotional states on perceived stress in mink. Future studies could address the border between adaptive and maladaptive stress responses. For example, what levels of prolonged stress/circulating cortisol is detrimental to the mink and their brain? Can we distinguish between detrimental vs. positive type of arousal/activity in mink? These may not be easy questions to answer, cf. studies of stereotypic behaviour. However, as the understanding of behavioural needs and welfare depend on reliable stress indicators, answering these questions are valuable for future progress in the production of farmed mink.

References

Balu, D. T. and Lucki, I., 2009. Adult hippocampal neurogenesis: regulation, functional implications, and contribution to disease pathology. Neurosci. Biobehav. Rev. 33: 232-252.

Beard, A.P. and Rawlings N.C., 1998. Reproductive effects in mink (*Mustela vison*) exposed to the pesticides Lindane, Carbofuran and Pentachlorophenol in a multigeneration study. J. Reprod. Fertil. 113: 95-104.

Bonneford, C., Fouche, L.G. and Martinet, L.A., 1984. A chronic jugular catheterization for remote blood sampling in freely moving mink. Physiol. Behav. 44: 141-146.

Bildsøe, M., Heller, K.E. and Jeppesen, L.L., 1991. Effects of immobility stress and food restriction on stereotypies in low and high stereotyping female ranch mink. Behav. Proc. 25: 179-189.

Børsting, C. and Damgaard, B., 1992. Use of chronic jugular catheterization for repeated blood sampling to measure diurnal variations in blood parameters in mink. Nor. J. Agric. Sci. 9: 550-556.

Damgaard, B.M. and Hansen, S.W., 1996. Radioimmunoassay of cortisol in mink plasma and the effect of adrenocorticotrophic hormone. Applied Science reports 29: 227-231.

Gould, E., McEwen, B.S., Tanapat, P., Galea, L.A.M. and Fuchs, E., 1997. Neurogenesis in the dentate gyrus of the adult tree shrew is regulated by psychosocial stress and NMDA receptor activitation. J. Neurosci. 17: 2492-2498.

Gulevich, R.G., Oskina, I.N., Kharlamova, A.V. and Trapezov, O.V., 2000. The cortisol and transcortin blood levels in the mink *Mustela vison* selected for behaviour after a long-term maintenance in pairs. J. Evol. Biochem. Physiol. 36: 531-536.

Hansen, S.W., 1988. Effect of variable cage size and lack of admission to nest box on the behaviour, physiology and production of mink kits. In: Murphy, B.D., Hunter, D.B. (eds.). Biology, pathology and genetics of fur-bearing animals, pp. 153-163. Proceedings of the IVth International Scientific Congress on Fur Animal Production, Toronto, Canada.

Hansen, S.W., 1993. Effects of early handling on later behaviour and stress response in farm mink. NJF seminar no. 239, subsection on fur animals, 23-24 September, Norway. 2pp.

Hansen, S.W. and Damgaard, B.M., 1991a. Effect of environmental stress and immobilization on stress physiological variables in farmed mink. Behavioural Processes 25: 191-204.

Hansen, S.W. and Damgaard, B.M., 1991b. Stress Physiological, Haematological and Clinical-chemical Status of Farm Mink Placed in Groups or Singly. Acta Agric. Scand. 41: 355-366.

Hansen, S.W. and Damggard, B.M., 2009. Running in a running wheel substitutes for stereotypies in mink (*Neovison vison*) but does it improve their welfare? Appl. Anim. Behav. Sci. 118: 76-83.

Hansen, S.W. and Jensen, M.B., 2002. Reward duration – a matter of concern in relation to constrction of demand curves. Proceedings of the 36th international congress of the ISAE, p. 213.

Hansen, S.W., Hansen, B. K. and Berg, P., 1994. The effect of cage environment and *ad libitum* feeding on the circadian rhythm, behaviour and feed-intake of farm mink. Acta Agriculturae Scandinavica. Section A: Animal Science 44: 120-127.

Hansen, S.W., Malmkvist, J., Palme, R. and Damgaard, B.M., 2007. Do double cages and access to occupational materials improve the welfare of farmed mink? Anim. Welfare 16: 63-76.

Hänninen, S., Ahola, L., Pyykonen, T., Korhonen, H.T. and Mononen, J., 2008a. Group housing in row cages: an alternative housing system for juvenile mink. Animal 2: 1809-1817.

Hänninen, S., Mononen, J., Harjunpaa, S., Pyykonen, T., Sepponen, J. and Ahola, L., 2008b. Effects of family housing on some behavioural and physiological parameters of juvenile farmed mink (*Mustela vision*). Appl. Anim. Behav. Sci. 109: 384-395.

Jayatissa, M.N., Bisgaard, C., West, M. and Wiborg, O., 2008. The number of granula cells in rat hippocampus is reduced after chronic mild stress and re-established after chronic escitalopram treatment. Neuropharmacol. 54: 530-541.

Jensen, P. and Toates, F. M., 1997. Stress as a state of motivational systems. Appl. Anim. Behav. Sci. 53: 145-156.

Klotchkov, D.V. and Trapezov, O.V., 1991. [Effects of domestication on variation in the reproductive function in mink] in [Evolutionary-genetic and genetic-physiological aspects of fur animal domestication] L.N. Trut, L.V. Osadchuk, P.M. Borodin (eds.), Institute of Cytology and Genetics, Siberian division, Novosibirsk. Translated from Russian, pp. 232-244.

Lawrence, A. B., Petherick, J.C., McLean, K., Gilbert, C., Chapman, C. and Russell, J. A., 1992. Naloxone prevents interruption of parturition and increases oxytocin following environmental disturbances in parturient sows. Physiol. Behav. 52: 917-923.

Leng, G., Mansfield, S., Bicknell, R. Brown, D., Chapman, C., Hollingsworth, S., Ingram, C.D., Marsh, M. I., Yates, J.O. and Dyer, R.G., 1987. Stress-induced disruption of parturition in the rat may be mediated by endogenous opioids. J. Endocrinol. 114: 247-252.

Leng, G., Mansfield, S., Bicknell, R. J., Blackburn, R. E., Brown, D., Chapmand, C., Dyer, R. G., Hollingsworth, S., Shibuki, K. Yates, J. O. and Way, S., 1988. Endogenous opioid actions and effects of environmental disturbance on parturition and oxytocin secretion in rats. J. Reprod. Fertil. 84: 345-356.

Lupien, S.J., de Leon, M., de Santi, S., Convit, A., Tarshish, C., Nair, N.P.V., Thakur, M., McEwen, B.S., Hauger, R.L. and Meaney, M.J., 1998. Cortisol levels during human aging predict hippocampal atrophy and memory deficits. Nat. Neurosci. 1: 69-73.

Lund, V. and Malmkvist, J., 2009. Periparturient nest building: quality of the nest, kit survival, maternal stress and behaviour in farmed mink (*Mustela vison*). In proceedings of the 21st Nordic Symposium of the International Society for Applied Ethology, 19-21 January 2009, Bjerringbro, Denmark, Aarhus University report in animal science 13: 27.

Madej, A., Forsberg, M. and Edquist, L., 1992. Urinary-excretion of cortisol and estrone sulfate in pregnant mink fed PCB and fractions of PCB. Ambio 21: 582-585.

Malmkvist, J., 2011. Påvirker foderets tryptofan indhold minks temperament? In annual report 2010, Danish fur breeders research center, Kopehagen Fur, Holstebor, p. 17-24.

Malmkvist, J., Brix, B., Henningsen, K. and Wiborg, O., 2012. Hippocampal neurogenesis increase with stereotypic behaviour in mink (*Neovison vison*). Behav. Brain Res. 229: 359-364.

Malmkvist, J. and Hansen, S.W., 2001. The welfare of farmed mink (*Mustela vison*) in relation to behavioural selection: a review. Anim. Welfare 10: 41-52.

Malmkvist J., Lassen, A.R.W., Clausen, K.T., Sauer, C.D., Houbak, B. and Decker, E., 2007a. Ny viden om tæve-hvalp samspillet. [New knowlegde regarding the bitch-kit interaction]. In DJF report 5, September 2007, pp. 77-95. (In Danish, with English abstract).

Malmkvist, J., Gade, M. and Damm, B. I., 2007b. Parturient behaviuor in farmed mink (*Mustela vison*) in relation to early kit mortality. Appl. Anim. Behav. Sci. 107: 120-132.

Malmkvist, J., Houbak, B. and Hansen, S.W., 1997. Mating time and litter size in farm mink selected for confident or timid behaviour. Anim. Sci. 65: 521-525.

Malmkvist, J. and Palme, R., 2008. Periparturient nest building: implications for parturition, kit survival, maternal stress and behaviour in farmed mink (*Mustela vison*). Appl. Anim. Behav. Sci. 114: 270-283.

Malmkvist, J., Hansen, S.W. and Damgaard, B.M., 2003. Effect of the serotonin agonist buspirone on behaviour and hypothalamic-pituitary-adrenal axis in confident and fearful mink. Physiol. Behav. 78: 229-240.

Malmkvist, J., Jeppesen, L.L. and Palme, R., 2011. Stress and stereotypic behaviour in mink (*Mustela vison*): A focus on adrenocortical activity. Stress 14: 312-323.

Malmkvist, J., Palme, R., Hansen, S.W. and Damgaard, B.M., 2004. Cortisol og corticoide nedbrydningsprodukter i minkfæces. Faglig årsberetning 2003. Kopenhagen Fur, Holstebro, Denmark, pp. 7-15 [In Danish, with abstract in English].

Mason, G.J., 1992. Stereotypy and welfare in mink. J. Anim. Sci. (suppl.) 70: 158.

Mason, G.J., Cooper, J. and Clarebrough, C., 2001. Frustrations of fur-farmed mink. Nature 410: 35-36.

Mason, G. J. and Latham, N. R., 2004. Can't stop, won't stop: is stereotypy a reliable animal welfare indicator? Anim. Welf. 13: S57-69.

McEwen, B.S., 1999. Stress and hippocampal plasticity. Annu. Rev. Neurosci., 22: 105-122.

McEwen, B.S. and Sapolsky, R.M., 1995. Stress and cognitive function. Curr. Opin. Neurobiol. 5: 205-216.

Mohr, F.C., Lasley, B. and Bursian, S., 2008. Chronic oral exposure to bunker C fuel oil cases adrenal insufficiency in ranch mink (*Mustela vison*). Arch. Environ. Contamin. Toxicol. 54: 337-347.

Mohr, F.C., Lasley, B. and Bursian, S., 2010. Fuel oil-induced adrenal hypertrophy in ranch mink (*Mustela vison*): Effects of sex, fuel oil weathering, and response to adrenocorticotropic hormone. J. Wildlife Dis. 46: 103-110.

Mormede, P., Andanson, S., Auperin, B., Beerda, B., Guemene, D., Malmkvist, J., Manteca, X., Manteuffel, G., Prunet, P., van Reenen, C.G., Richard, S. and Vessier, I., 2007. Review: Exploration of the hypothalamic-pituitary-adrenal function as a tool to evaluate animal welfare. Physiol. Behav. 92: 317-399.

Morris, R., 2007. Stress and the hippocampus. In: P. Andersen, R. Morris, D. Amaral, T. Bliss and J. O'Keefe (eds.), The hippocampus book. Oxford University Press, New York, USA. pp. 751-768.

Pedersen, V., 1996 Combined behavioural and physiological measurements as a basis of the assessment of animal welfare. Acta Agric. Scand., Anim. Sci., Suppl. 27: 69-75.

Pedersen, V. and Jeppesen, L.L., 2001. Effects of family housing on behaviour, plasma cortisol and performance in adult female mink (*Mustela vison*). Acta Agric. Scand. Sect. A-Anim. Sci. 51: 77-88.

Sapolsky, R.M., 2002. Endocrinology of the stress response. In: J.B. Becker, S.M. Breedlove, D. Crews and M.M. McCarthy (eds.), Behavioral endocrinology, MIT press, London, UK, pp. 409-450.

Sapolsky, R.M., 2004. Is impaired neurogenesis relevant to the affective symptoms of depression? Biol. Psychiatry 56: 137-139.

Sapolsky, R.M., Krey, L.C. and McEwen, B.S., 1985. Prolonged glucocorticoid exposure reduces hippocampal neuron number: implication for aging. J. Neurosci. 5: 1222-1227.

Svendsen, P.M., Hansen, B.K., Malmkvist, J., Hansen, S.W., Palme, R. and Jeppesen, L.L., 2007. Selection against stereotyped behaviour may have contradictory consequences for the welfare of farm mink (*Mustela vison*). Appl. Anim. Behav. Sci. 107: 110-119.

Sørensen, B., Clausen, T.N., Wamberg, S. and Hansen, O., 2001. Physiological changes in mink (*Mustela vison*) dams subjected to weaning at different times during lactation. Acta Agr. Scand. Sect. A Anim. Sci. 51: 148-154.

Tauchi, M., Yuan, Y., Aulerich, R.J. and Zanella, A.J., 1999. The association between gentoype, stereotypies and fecal steroids in mink. In: KE Bøe, M Bakken and BO Braastad (eds.), Proceedings of the 33rd International Congress of the ISAE, Lillehammer, Norway, August 17-21. Agricultural University of Norway: Ås, Norway, p. 87

Tauson, A.H., 1998. Water intake and excretion, urinary solute excretion and some stress indicators in mink (*Mustela vison*): effect of ambient temperature and quantitative water supply to lactating females. Br. J. Nutr. 80: 555-564.

Weiss, J.M., 1968. Effects of coping responses on stress. J. Comp. Physiol. Psychol. 65: 251-260.

Weiss, J.M., 1971. Effects of coping behavior in different warning signal conditions on stress pathology in rats. J. Comp. Phsyiol. Psychol. 77: 1-13.

Zanella, A.J., Mason, G. and Schmidt, P., 1998. Selective breeding can affect the performance of abnormal behavior and neurophysiological indicators in mink. In: I Veissier and A Boissy (eds.), Proceedings of the 32nd International Congress of the ISAE, Clermont-Ferrand, France, July 21-25 p 89. Institut National de la Recherche Agronomique: Clermont-Ferrand, France.

The effects of climbing cages on behaviour of female mink during the lactation period

L. Lidfors[1], H. Axelsson[1], J. Loberg[1] and S.W. Hansen[2]
[1]Swedish University of Agricultural Sciences, Department of Animal Environment and Health, P. O. Box 234, 532 23 Skara, Sweden; [2]University of Aarhus, Department of Animal Health, Welfare and Nutrition, Blichers Allé 20, P.O. Box 50, 8830 Tjele, Denmark; lena.lidfors@slu.se

Abstract

The aim was to investigate if there were differences in behaviour of female mink when kept in a climbing cage compared with a standard cage during the lactation period. The study was carried out on 90 mink of the colour type 'black cross'. Females were housed in either climbing cages (4,350 cm^2, n=46) or standard cages (2,550 cm^2, n=44). Behavioural observations were made four weeks before birth until kits were nine weeks old with one-zero sampling during two hours before feeding. During week 1-4 after females had given birth they were mostly recorded in the nest box and there were no differences between cage types. Week 5-8 after giving birth females in climbing cages were less in the nest box, less active out in the cage and had fewer abnormal behaviours, but were more on the platforms and more inactive out in the cage. In the bottom cage females were more often walking, grooming and inactive out in the cage and tended to be more often on the platform compared to in the upper cage. It is concluded that the climbing cage worked well for females with kits and that the upper cage can offer a refuge for females when kits get older.

Keywords: fur animals, cage types, behaviours, placement

Introduction

Farmed mink kits are usually weaned, i.e. separated from their mother, at 6-8 weeks of age in June-July (Joergensen, 1985; Vinke, 2004). On Swedish mink farms weaning often occur at 8-9 weeks of age. At 5-6 weeks the kits are nutritionally independent, meaning that they can survive without the mother´s milk. In the wild the kits are highly mobile at the age of 6 weeks and they may emerge from the den only to be dragged back to the safety of the nest by their mother (Dunstone, 1993). At the age of 10 weeks violent rejections towards the kits by the mother increases (Dunstone, 1993). Mother-young separation is normally a gradual process, which involves short-lasting and intermittent episodes of conflict between the adult and the young until mutual independence is fully achieved (Dunstone, 1993). In farmed mink the female can get away from her kits by leaving them in the nest when they are young, but when they get some weeks they emerge from the nest and follow the female around. For the female it may be stressful that she cannot get away from the kits and a cage that could give her that possibility might be better for the welfare of the female.

During the lactation period the females are fed ad libitum and this decreases stereotypies (Damgaard *et al.*, 2004; Hansen and Møller, 2008). The body condition of female mink can be affected by different factors such as stress, litter size and activity level (Joergensen, 1985).

Joergensen (1985) describes anecdotal observations that a poor mother is usually unsettled, dashes in and out of the nest, has no milk for her kits, and will not let them have necessary peace and quiet in which to develop, and these aspects can result in kit mortality.

In previous studies giving mink access to larger cages seems to have had no or little effect on abnormal behaviour (Vinke, 2004; Hansen *et al.*, 2007). Hansen *et al.* (2007) did a study on the welfare of farmed mink in double cages in the months from half July till the start of March. Access to double cages compared to standard cages had no effect on stereotypies, fur-chewing and physiology linked to welfare, but mink with access to a double cage used the nest box less and had a lower consumption of straw. Vinke (2004) doubled the cages as well and compared it to the standard cage, but there were no clear differences between the different housing systems in abnormal behaviour. In another study of Vinke (2004) there were no differences in stereotypic and in anticipatory behaviour of juvenile mink between the standard cage and the larger cage (two-floor family housing). However, these studies were not done on female mink before and after giving birth. In one study female mink and her litter were housed together in a larger cage (three connected standard cages) until pelting, and there were some concerns about the welfare of the females, but stereotypic behaviour was not observed during the lactation period (De Jonge and Iwaarden, 1995). Group housing has been found to lead to reduced stereotypies also in other studies (Pedersen and Jeppesen, 2001; Pedersen *et al.*, 2004). Cages where the females can climb up and get away from their kits more effectively than in connected cages have to our knowledge not been evaluated before.

The aim of this study was to investigate if there are any differences in the behaviour of female mink when kept in a climbing cage compared with a standard cage during the lactation period, and how they use a climbing cage when they have kits.

Material and methods

The study was carried out from 30 April until 28 June 2008 at a private farm in South-west Sweden. In total 90 unrelated mink of the colour type 'black cross' was used; 44 females were housed in standard cages and 46 females were housed in climbing cages since they were weaned. The females were one year old and had been born in May 2007. After weaning in July 2007 these females were housed with a male mink until pelting of the male in November. In March the females were mated with several male mink. The females and their kits were housed together until weaning at 9 weeks of age. Twelve females were not pregnant (8 in standard cage, 4 in climbing cage), but as this was not known until sometime after birth their behaviour was recorded. All management was done by the farm personal according to the farms normal routines during the period. No handling of female or kits have been taken place during the study period. The animals had free access to drinking water. The mink were housed in either a standard cage (Figure 1A) or in a climbing cage (Figure 1B). The climbing cages were constructed by a smaller cage being placed on top of a standard cage and a hole being made between the two cages so that the mink could climb up to the upper cage. Climbing cages were placed on one side of the shed and standard cages on the other side of the same shed.

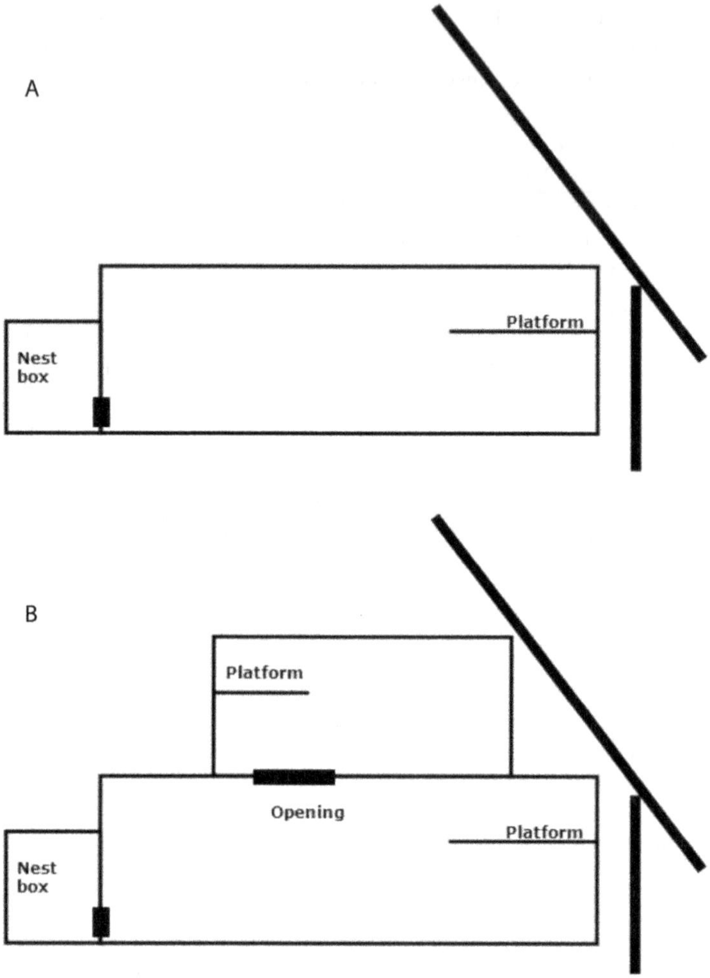

Figure 1. Drawing of (A) the standard cage and (B) the climbing cage. The total floor size of the standard cage was 2,550 cm^2 (length: 90 cm, width: 30 cm, height: 45 cm) and the total floor size of the climbing cage was 4,350 cm^2. The bottom cage had the same dimensions as the standard cage and the top cage was 1,800 cm^2 (length: 55 cm, width: 30 cm, height: 45 cm).

The female mink were observed daily four weeks before giving birth until weaning at a kit age of 9 weeks. The observations were done two hours before feeding in the morning, due to that we wanted to observe if they performed any stereotypies before feeding which is the aim of a larger project. Two observers made the behavioural recordings from the walking alley inside the nearby shed-house to minimize the observer's impact on the animals. The observers made observations an equal amount of time and on both cage types. The observation method used was one-zero sampling with 10 s intervals on focal animals. All 90 females were observed during the two hours of observation, resulting in 8 observations à 10 s per animal and day. Observations were first made on one cage type and then on the other cage type in an alternating way for the 8 observations within the same day. The following observation day the other cage type was observed first to equal out time before feeding between cage types.

Observations of the following behavioural categories and individual behaviours were done: In nest box, active out in cage (walk, stand, look-out, feed-seek, climb, drink, groom, urinate/ defecate, carry straw, bite cage), inactive out in cage (lie on cage floor), use of platform (both bottom and top cage), abnormal behaviour (pendling at least three times repeated, other stereotypies at least three times repeated, eating faeces) and maternal behaviour (wrestling (mother mouth and front legs on pup), carrying and interaction with at least one pup).

Statistical analysis was done in Statistical Analysis System (SAS) version 9.1 and in SPSS version 17. The 12 females that did not get any kits were removed from data analysis. Behavioural data were not normally distributed and therefore non-parametric statistical analysis was used. Median values of the main behaviours for the two cage types and 95% Confidence Intervals (CI) or percentages were calculated. Differences between the two cage types were tested with the Mann Whitney U-test (independent two samples). The Wilcoxon Signed ranks test was used to test if the female mink in the climbing cage performed behaviours more often in the lower or the upper part of the cage.

Results

The most common behaviour of the females was being in the nest box, during the first four weeks after birth of their kits and this did not differ between the cage types (Table 1). During week 5-8 after birth of kits the females spent almost half of the observations in the nest box, but at this time females in climbing cages were recorded less often in the nest box compared to females in the standard cages (Table 1). Instead they were recorded on the platforms both in the bottom and upper sections of the climbing cage (Table 1). Female mink in standard cages were recorded to

Table 1. Median recordings with 95% confidence intervals (CI) of different behaviours observed in female mink in standard (n=36) and climbing (n=42) cages from 1-4 and 5-8 weeks after birth of kits. Tested for significant differences with Mann-Whitney U-test.

Behaviours	Week	Standard	Climbing	CI	*P*<
In nest box	1-4	0.9098	0.9196	-0.0170; 0.0246	ns
In nest box	5-8	0.5738	0.4821	0.0045; 0.1741	0.05
Active in cage	1-4	0.0069	0.0050	-0.0026; -0.0000	0.05
Active in cage	5-8	0.0248	0.0177	0.0023; 0.0092	0.01
Use of platforms	1-4	0	0.0056	-0.0078; -0.0033	0.0001
Use of platforms	5-8	0	0.0770	0.0647; 0.1027	0.0001
Abnormal behav.	1-4	0.0002	0.0002	-0.0000; 0.0004	ns
Abnormal behav.	5-8	0.0092	0.0041	0.0000; 0.0102	0.1

be active out in the cage more often than female mink in climbing cages both week 1-4 and week 5-8 (Table 1). Abnormal behaviour, which mainly consisted of stereotyped movements, was rare during the first four weeks after giving birth, and more common during week 5-8 where it tended to be more often recorded in the standard cages than the climbing cages (Table 1). Inactive out in the cage and maternal behaviour was rarely recorded.

The recordings of the female mink being in the nest box before giving birth, during birth and during the kits development until weaning at 9 weeks of age is shown in Figure 2. When the kits were about four weeks old the female mink in the climbing cage started to spend more time outside the nest box than the females in the standard cage did. However, at 8 weeks there were no differences anymore, and at 9 weeks it seems to be the opposite. We have not tested difference between cage types for each week statistically.

There was a tendency that the mink in the climbing cages used the platform in the bottom cage more than the platform in the upper cage (Table 2). They were walking, grooming and inactive out in the cage more in the bottom cage than in the upper cage (Table 2). Standing and looking-out was performed as much in the upper cage as in the bottom cage (Table 2).

Discussion

The results of this study show agreements with previous studies performed in other seasons on the effect of larger cages on the behaviour of mink. Juvenile mink used the nest box less when kept in a climbing cage (Lidfors *et al.*, 2009) or a double cage (Hansen *et al.*, 2007) compared to

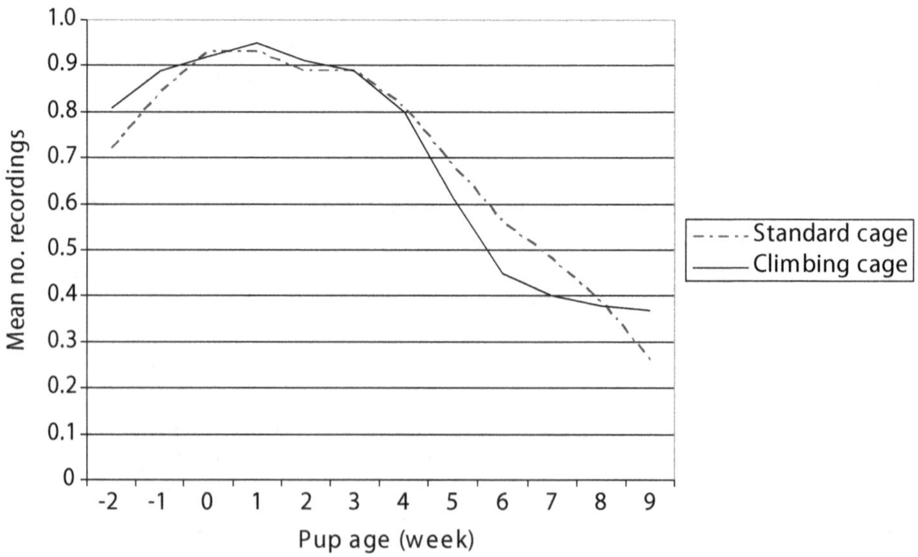

Figure 2. Mean number of recordings of female mink being in the nest box before giving birth and with increasing kit age (n=36 standard cage, n=42 climbing cage).

Table 2. Percentages of behaviours performed by female mink in the bottom cage compared to the upper cage when kept in climbing cages from four weeks before until nine weeks after birth of kits. Analysed for significant differences with Wilcoxon signed ranks test for paired data (n=45).

Behaviour	Bottom cage	Upper cage	Z-value	*P*-value
Use of platform	54.8	45.2	-1.921	0.055
Walking	66.9	33.1	-5.832	0.000
Standing	51.7	48.3	-0.818	0.413
Look-out	56.3	43.7	-1.166	0.244
Grooming	77.8	22.2	-5.668	0.000
Inactive out in cage	85.0	15.0	-5.735	0.000

when they were kept in a standard cage. Hansen *et al.* (2007) also found that juvenile mink in double cages used the shelves more than mink in standard cages.

The female mink in this study spent more time outside the nest box in the standard and climbing cage than wild mink would do (Dunstone, 1993). However, it is difficult to compare observations on wild mink during other times of the day with those of caged mink. In this study the female mink spent more time in the nest box in the standard cage than in the climbing cage. If older kits try to suckle from the mother when she is in the nest it may be demanding on her to be so close to her kits. The mink kits grow very fast and this put great demands on the female mink. The females will go out of the den more often to forage in this period in order to produce enough milk for their kits (Dunstone, 1993). This can explain why the female mink in the standard and climbing cage is more outside the nest box after a few weeks. However, there seems to be a relation between the date and the recordings of behaviours in the upper cage. This can be explained by the fact that the kits were becoming older and that the female mink took some distance from them. Being on the platform in the bottom cage increased as well, but one can question why the female mink in the standard cage did not use the platform if the purpose was to take a rest from the kits? Maybe the platform was too low so that the kits could still reach the mother when she was on the platform in the lower cage. This should be investigated further.

In this study the litter size is not taken into account and can be a confounder for stereotypic behaviour. Also the birth dates of the kits could be a confounder. The kits were born from mid-April to mid-May. The females in the standard cages gave birth to their kits approximately two weeks earlier than the females in the climbing cages, but we do not know if this was due to earlier mating or an effect of housing. As the females in both cage types were recorded equally much in the nest box during the first four weeks after birth we do not have any behavioural differences that can explain this. However, the female mink in the climbing cages were recorded more often outside the nest box from 5-8 weeks after birth than the females in the standard cages. It is possible that the female mink in the climbing cages placed itself at a longer distance from their pups than the female mink in the standard cages could do. A comparison of the development of behaviour in the bottom and in the upper cage has to be made to see if there is a difference of development of those behaviours after the fourth week. In this study it is not clear if the mink were in the upper cage to look-out or just to escape from the kits. Hansen *et al.* (2011) showed that especially female mink use a platform as a look out especially just before expected feeding time.

The recordings of the use of the platform in the climbing cages are more than the recordings in the standard cages. In fact there was only one mink in the standard cage which used the platform. The platforms are regarded as enrichments in the cages and have been shown to be used more than other enrichments by female mink during the winter time (Axelsson *et al.*, 2009). Access to platforms has been found to decrease stereotypic behaviour in mink (Vinke, 2004; Axelsson *et al.*, 2009). In this study, the females in the climbing cages were less active out in the cage, but used the platforms more.

It is concluded that the climbing cage was used for different activities by the females and that it did not cause any problems for the survival and welfare of the kits. The upper cage in a climbing cage can offer a refuge for females when kits get older and a preferred look out place before expected feeding time.

Acknowledgements

We want to thank the mink farmer for letting us test the climbing cages on his females, Eva Aldén for her advices on mink farming, Elin Kortes for summarizing the first draft of this study and Swedish Fur Breeders Association (SPR) for financing the study.

References

Axelsson, H.M.K., Aldén, S. and Lidfors, L., 2009. Behaviour in female mink housed in enriched standard cages during winter. Applied Animal Behaviour Science 121: 222-229.

Damgaard, B.M., Hansen, S.W., Børsting, C.F. and Møller, S.H., 2004. Effects of different feeding strategies during the winter period on behaviour and performance in mink females (*Mustela vison).* Applied Animal Behaviour Science 89: 163-180.

De Jonge, G. and van Iwaarden, M., 1995. Een nieuw huisvestingssysteem voor nertsen. De pelsdierhouder, pp. 19-24.

Dunstone, D., 1993. The mink. T and AD Poyser Ltd: London, UK.

Hansen, S.W., Malmkvist, J., Palme, R. and Damgaard, B.M., 2007. Do double cages and access to occupational materials improve the welfare of farmed mink? Animal Welfare 16: 63-76.

Hansen, S.W., 2008. Diurnal activity pattern of farm (*Mustela vison*) subjected to different feeding routines. Applied Animal Behaviour Science 111:146-157.

Hansen, S.W. and Møller, S.H., 2008. Diurnal activity patterns of farm mink (Mustela vision) subjected to different feeding routines. Applied Animal Behaviour Science, 111: 146-157.

Hansen, S.W., Møller, S.H. and Damgaard, B.M., 2011. Feed restriction and tubes for environmental enrichment in growing mink – consequences for behaviour and welfare. Applied Animal Behaviour Science 134: 193-200.

Lidfors, L., Lindberg, H., Aldén, E. and Hansen, S., 2009. How do larger enriched cages and lager group size affect juvenile mink behaviour? Process of the 43rd Society for applied ethology, Cairns Australia, pp. 62.

Joergensen, G., 1985. Mink Production. The care of mink. 164 pages

Pedersen, V. and Jeppesen, L.L., 2001. Effects of family housing on behaviour, plasma cortisol and performance in adult female mink (*Mustela vison*). Acta Agric Scand., Sect. A, Animal Sci. 51: 77-88.

Pedersen, V., Jeppesen, L.L. and Jeppesen, N., 2004. Effects of group housing systems on behaviour and production performance in farmed juvenile mink (*Mustela vison*). Applied Animal Behaviour Science 88: 89-100.

Vinke, C.M., 2004. Cage enrichments and welfare of farmed mink. PhD thesis.

Impact of weaning age on kit performance

T.N. Clausen and P.F. Larsen
Kopenhagen Research, Agro Food Park 15, 8200 Aarhus N, Denmark; tcl@kopenhagenfur.com

Abstract

To get more knowledge on the influence of weaning age on health and wellbeing among mink kits, autopsy was made and bite damages registered in the nursing period, in two groups of each 400 litters weaned either at 49 or 56 days of age. The influence of the females on whether the kits bite each other in the period from day 49 to day 56 seems to be of lesser importance, since there was no effect of weaning age on number of bite damages. Biting is seen from 6 weeks of age (day 42) and especially in the period from day 42 to day 49. In that period we found 0.4% of the kits dying from bite, from day 49 to 56 only 0.1% died or was euthanized. It is mainly in the big litters and among female kits we observe bites, 80 percent of kits with bites were female kits. To prevent bites it would probably be a good idea to separate big litters after 6 weeks of age.

Keywords: age of weaning, autopsy, bites

Introduction

Bites among kits are seen between four to eight weeks of age (Clausen, 2010a). Hansen *et al.* (2008) found bites to be the most common cause of death among kits in June. In the same period Clausen (2010a) found diarrhoea and emaciation as the most common causes of death among mink kits. The difference between the two investigations was that Hansen *et al.* (2008) included all kits with bites, whereas Clausen (2010a) divided kits into two groups, one group with kits obviously killed by siblings or by the female, and the second group with kits bitten/eaten after they died from other causes. Kits bitten after death from other courses is most often seen from the beginning of June until July (from 30 to 60 days of age), kits killed by siblings (0.3 percent of live borne kits) are especially seen from the middle of June to the beginning of July (from 44 to 68 days of age) (Clausen, 2010a).

Therefore it seems that when kits start to eat solid feed (around 4 weeks of age) they will naturally eat one of their siblings if they die. When the kits eat a lot of food and the milk production is decreasing and stopping (around 6 weeks after birth) there can be fights between siblings probably due to hunger or lack of water. The frequency of bites varies a lot between farms and from year to year depending on weather conditions, feed, water supply and management. A low milk production, insufficient water system, dry feed, feed with a bad taste or not sufficient feed to fulfil the need of the kits is important, but also bad weather or similar preventing the kits from starting to drink from the water nipple. Kits with broken skull (0.09% of live borne kits) indicating they had been killed by the female was in the investigation of Clausen (2010a) observed five times out of 828 litters.

To get a better understanding of the importance of weaning time on the health of the kits, dead kits from 800 litters weaned day 49 or day 56 after birth were autopsied and bites registered in the period May 24th of to June 28th 2011.

Materials and methods

To the investigation we used 800 brown litters born in the period April 25th to May 4th. Litter size varied from 5-10 kits at birth. Feed kitchen feed was used in the whole period. Litters were split in two equal groups, one group was weaned day 49 after birth (WEA49) and the other group was weaned at day 56 (WEA56). From day 28 to day 56 all dead kits were collected twice daily and autopsied. The number of kits in the litters was counted day 49 and 56. If kits with bite were observed at the daily control, the sex was registered and they were moved to another cage together with a sibling. Kits with small bites were treated with antibiotics kits with big bites were euthanized.

The statistical calculations were performed using the statistical program SAS (SAS Institute Inc., 1996). The PROBIT procedure was used with 5% significance level.

Results

The average litter size in both groups was equal, 7.37 in WEA49 and 7.33 in WEA56 day 49 after birth and 7.30 in WEA49 and 7.20 in WEA56 day 56.

A total of 2.5% of live borne kits died in the period May 24th to June 28th. Autopsy of dead kits showed that the main courses of death were diarrhoea (0.2%), emaciation (0.8%) and bite between siblings (0.5%) or from the female (0.25%). Bladder and kidney infections started up later in the period, when the kits ate a lot of food, the earliest bladder infection was observed at 44 days of age. Diarrhoea was seen from 25 to 45 day of age. Emaciation was seen in the same period. Kits eaten after death from other courses was seen from day 28, these kits were omitted from the investigation. Bites between siblings were seen from day 41, most of them from day 42 to 49 (0.4%), in the period from day 49 to 56 only 0.1% of the kits died due to bites from siblings (P=0.002 between weeks), there were no difference between weaning groups. Females killing their kits were seen during the whole period, before day 49 two kits in group WEA49 and one kit in WEA56 was killed by the female, in the period day 49 to 56 eight kits in WEA56 was killed by the female (P=NS).

Registration of bites on kits that did not die, showed like with the dead kits, that most bite damages were in the period from day 42 to 49 (1.3% of live borne kits). From day 49 to 56 the frequency of kits with bites damages was 0.47% in WEA49 vs. 0.51% in WEA56 (P<0.0001 between weeks; P=NS between groups).

There were on average one kit more in litters with bite damages than in litters without (P<0.0001). About 80% of kits with bite damages were female kits. In litters with bite damages on average one or two of the kits had bite damages the others had not.

Discussion

The autopsy results showed that the main causes of death from day 28 in the nursing period were diarrhoea, emaciation and bites between siblings corresponding to other investigations (Clausen, 2010a; Hansen *et al.*, 2008). Diarrhoea was seen when the kits started to eat solid feed, emaciation was seen in the same period most often probably as a consequence of the diarrhoea. The first bite damages were seen at day 41 this corresponds to earlier studies where bite damages was seen from day 44 (Clausen, 2010a).

Biting was mainly seen in big litters, this was also observed by Risager (1989). Not all of the kits in the litter are bitten, but especially the female kits (80% of bitten kits). Likewise Hansen *et al.* (2008) found that 65% of dead kits due to bites were females. Since female kits are smaller than mails our results indicates that big litters can be stressful and that fighting among kits will affect the smallest kits the most. But whether this is true and whether females are bitten because they are smallest and it is the smallest of the females that is bitten or perhaps they become smallest because they are bitten will be investigated further.

Biting is most likely due to competition on food, milk from the female and/or water. Body weight of kits is reduced with increasing litter size due to less milk for each kit, but perhaps it is also difficult for the smallest kits to be allowed to eat in the big litters. Increasing feeding frequency from day 28 to day 56 seems to be an advantage, since the number of kits killed by siblings was reduced (Clausen, 2011b).

Most of the bite damages were seen from day 42 to 49, before any of the kits was weaned in this investigation. However weaning at 42 days of age would not solve the problem, Clausen (2011a) used weaning at 42, 49 or 56 days and found no difference in the number of kits with bite damages between groups, and kits weaned at 42 days were smaller that the other kits. About 6 weeks after birth the female stops her production of milk, but with the Danish water nipple system it takes around 9 days from the kits start to eat until they start to drink water (Steffensen *et al.*, 2007), in that period they depend totally on milk from the female and water added to the feed. If they have access to an open water surface (Steffensen *et al.*, 2007) or drip water system (Møller, 1991) they will start to drink earlier, also placement of the water nipple is important (Brink *et al.*, 2004). Another way to improve the water uptake by the kits in that critical period is to increase the feed water content. A reduction of feed dry matter from 37 to 25% by adding water was of benefit for both female and kits (Risager *et al.*, 1992). Addition of fibres to the feed to bind more water is also a possibility however it is important which fibre is used (Møller, 1985). Different fibres and feed binders were tested from day 28 to 56 the results showed that the lowest number of kits dying from bite was seen in the group with the highest water uptake per 100 kcal (Clausen, 2011b,c).

In the period from day 49 to 56 bite damages were reduced compared to the pervious period, indicating that the kits learned to drink water. There were no difference in bite damages between weaning groups, so the effect of the female on whether the kit bite each other from day 49 to 56 seems to be of lesser importance.

Throughout the whole period there are some females that kill their kits. In WEA56 from day 49 to 56 eight kits were found with broken skull, indicating that the female get stressed by the kits. This was also seen in an earlier investigation by Clausen (2010). Jeppesen *et al.* (2004) suggest that weaning at 8 weeks is probably best for the kits whereas it can be too late for some females who increase stereotypic behaviour during the nursing period and sometimes kill their kits.

When bite damages is seen in the nursing period it is most often in big litters, it is not all kits that are bitten in the litter but most often female kits who are smallest. Most bite damages are seen in the period day 42 to 49 probably due to hunger or a lack of water. The females have only minor influence on whether kits bite each other in the period from day 49 to 56 in the nursing period. Some females kill their kits probably due to increased stress from being with the kits longer than necessary. To avoid biting it might be a good idea to wean the biggest kits in the big litters already at 42 days of age, further possibilities include optimised drinking water systems and increase feed water content.

References

Brink, A.-L., Jeppesen, L.L. and Heller, K.E., 2004. Behaviour in suckling mink kits under farm conditions: effects of accessibility of drinking water. Applied Animal Behaviour Science 89: 131-137.

Clausen, T.N., 2011a. Weaning of black mink kits day 42, day 49 ore day 56. (In Danish: Fravænning af standard hvalpe dag 42, dag 49 og dag 56). Faglig Årsberetning 2010, Pelsdyrerhvervets Forsøgs- og ForskningsCenter, Holstebro, Danmark, pp. 119-121.

Clausen, T. N., 2011b. Strong feeding of mink kits in June – July (In Danish: Kraftig fodring af hvalpe i juni - juli). Faglig Årsberetning 2010, Pelsdyrerhvervets Forsøgs og ForskningsCenter, Holstebro, Danmark, pp. 55-58.

Clausen, T. N., 2011c. Feeding mink kits day 28 to day 56 (In Danish: Fodring af hvalpe med fibre i perioden dag 28 til dag 56). Faglig Årsberetning 2010, Pelsdyrerhvervets Forsøgs- og ForskningsCenter, Holstebro, Danmark, pp. 49-53.

Clausen, T. N., 2010a. Kit death from birth to August the first (In Danish: Dødsårsager fra fødsel til 1. august). Årsberetning 2009, Pelsdyrerhvervets Forsøgs- og ForskningsCenter, Holstebro, Danmark, pp. 97-103.

Clausen, T.N., 2010b. Water balance in 8 week old mink kits. NJF Seminar no. 440, September 29 – October 1, 2010, Oslo, Norway, Poster presentation.

Hansen, M.U., Weiss, V., Clausen, T.N., Mundbjerg, B. and Lassén, M. 2008. Investigations in causes of death among mink kits from June to October (In Danish: Årsager til dødsfald hos minkhvalpe fra juni til oktober). Faglig Årsberetning 2007, Pelsdyrerhvervets Forsøgs- og Forskningscenter, Holstebro, Danmark, pp. 99-107.

Jeppesen, L.L., Simonsen, T. and Pedersen, V., 2004. Project welfare in practice (In Danish: Projekt velfærd i praksis). Faglig Årsberetning 2003, Pelsdyrerhvervets Forsøgs- og ForskningsCenter, Holstebro, Danmark, pp. 35-44.

Møller, S., 1991. Supplementary watering system. In: The influence of various management, environment and nutritional elements on behaviour, physiology and production in mink, 688 Report from the National Institute of Animal Science, Denmark. Møller, S.H., Hansen, S.W., Lohi, O., Brandt, A., Rasmussen, P.V. and Jensen, L.V. (eds.), pp. 33-39.

Møller, S.H., 1985. Importance of fibres on nutrient digestibility and content of mineral and water in faces. (In Danish: Fiberholdige stoffers indflydelse på næringsstoffernes fordøjelighed, og minkens udskillelse af vand og salte i gødningen). Hovedopgave i pelsdyrproduktion, KVL.

Risager, H.J., Clausen, T.N. and Olesen, C.R., 1992. Variation in feed dry matter and its importance for the frequency of nursing sickness. (In Danish: Variationer i foderets tørstofprocent og dets betydning for fremkomsten af diegivningssyge hos minktæver). Dansk Pelsdyravlerforening. Faglig Årsberetning 1991, pp. 190-197.

Risager, H.J., 1989. Optimal nursing period reduces mink kits fighting. Dansk Pelsdyr Avl, 5, pp. 315-316.

SAS Institute Inc., 1996. SAS® Systems for Mixed Models. SAS Institute Inc., Cary, NC, USA, 633 pp.

Steffensen, L.K., Hansen, S.W. and Jeppesen, L.L., 2007. Introducing an open water surface as an alternative to the traditional valve drinker for ranch mink (*Mustela vison*) in the lactation period. Scientifur, 31, 1, pp. 7-18.

High need for drinking water in young mink kits between 30 and 50 days of age

J. De Rond[1] *and F.C. Kleyn van Willigen*[2]
[1]*Edelveen – Research Farm Dutch Fur Breeders' Federation, P.O Box 488, 6600 AL, Wijchen;* [2]*Animal Health Service, P.O. Box 9, 7400 AA, Deventer; j.derond@edelveen.com; f.kleyn.v.willigen@gddeventer.com*

Abstract

From 2008 till 2011 the effect of drinking water in or near the nest box when the kits start eating was studied at Edelveen. In 2008 and 2009 the water consumption from an extra drinking nipple inside of the nest box was high (between 35 and 50 days, 365 ml/litter/day). This resulted in 2010 and 2011 in a behaviour study of the litters (30 to 50 days post whelping). The test group of mink received an extra drinking nipple inside of the nest box or near the entrance. Kits with water in or near the box started drinking 4 days earlier than controls. Between 35 and 45 days the frequency of consuming water was 3 times higher, of saliva licking 3 times lower and kits were 2 times calmer. Kit growth and kit losses tended to be better in the testing group in all 4 years. In 2010 5% of the control females suffered from nursing sickness, none in testing group. Mothers in the testing groups lost 30% less weight between 30 and 50 days, than did the controls. Females with the lowest weight loss had the heaviest kits at 50 days of age and a better reproduction next year. Drinking water close to the kits apparently improved their own and their mothers welfare.

Keywords: drinking water, young mink kits, saliva licking, restless kits, nursing sickness

Introduction

The moment mink kits start to eat farm food, they need water, but in the traditional housing systems the drinking nipple is difficult to approach. Ahlstrom (2010) described the importance of drinking water for little kits in the njf workshop energy supply. Møller (1991) and Brink *et al.* (2004) found that mink kits start to drink water at the drinking nipple/valve 2 weeks after they start to eat food. In this 2 week period the need for water was covered with water from the food, the milk and the saliva of the mother. Brink also found that easy availability of drinking water reduced saliva licking. Moreover he described that an increasing drinking activity is correlated with reduced aggression. A system with drip water (Møller and Lohi, 1989) or water on open plates (Steffensen *et al.*, 2007) stimulated the kits to start drinking 3 days earlier and improved their growth. De Rond (2009) showed that a drinking nipple inside the nest box resulted in high water intake in the nest box by mink kits and dam, which lowered with increasing age. The availability of drinking water also improved growth.

The period between the start of feed intake (4 weeks of age) and drinking at the nipple in the pen (5-6 weeks of age) is also the period were problems of nursing sickness and cannibalism occur. Clausen *et al.* (1992) found dehydration of the female as the main course for nursing sickness. When the kits receive farm food at 30 days of age, their weight is±0.3 kg and the wire floor is not easy to pass. The favourite place for the kits is the nest box. That changes in the next 3 weeks, but then the weight is ±0.6 kg (at 50 days).

This study started with looking at the consumption of water in the nest box and the effect on growth and kit losses in various whelping dates and litter sizes (2008). The next year (2009) the results were proved to be consistent. In 2010 the aim was to study also the behaviour of the kits in the age of 4 till 7 weeks, when drinking water was available inside the nest box or not. The effect on the dams' weight was also taken into account. This resulted in a new study in which more practical situations were tested (2011) like a drinking nipple near the entrance of the nest box. Is a drinking nipple inside or close to the nest box a practical solution on mink farms to help the kits and the dam?

Material and methods

In 2008 an individually water supply per pen was placed in the corner of the nest box, near the entrance. It was refilled twice a day and the amount was weighed. This system was also used in 2009. In 2010 a new system was developed together with Columbus Aqua A/S. On Edelveen, the mink could drink through a drinking nipple at the end of the pen and the drinking water circulated in a closed system for the whole farm. To provide a drinking nipple inside the nest box, a water tube was constructed outside the box. Through a little hole in the backside of the box, the adapted drinking nipple was placed inside the box at 4 cm above the floor and opposite to the entrance of the nest box. The nest box floor consisted of 2 floors of wire with straw and wood cuttings in between. In 2011 an extra drinking system was developed (like drinking nipple for rabbits) and placed near the entrance of the nest box on the site of the pen so that water did not drip into the nest box.

Table 1 shows the materials and methods in the 4 years: per year the period during which the data have been collected, which groups, the number of litters involved and the litter size (equally divided), and the recorded parameters. The groups were weighed at 3, 5 and 7 weeks of age or at 32 and 58 days. On average they were weighed as mentioned in Table 1.

In 2010 there were 2 experimental groups observed: drinking water in the nest box and controls.

In 2011 these 5 experimental groups were observed:
1. drinking water in the nest box;
2. drinking water near the entrance at the pen site;
3. three times per day feeding, and the nipple at the end of the pen;
4. at 4 weeks of age placing a clean kit-wire floor in the pen which stays till 7 weeks of age, and the nipple at the end of the pen;
5. controls, traditional housing and management and nipple at the end of the pen.

For behavioural observations 2 groups of 111 litters were selected in 2010 and 5 groups of 50 litters in 2011. The groups were composed on the basis of birth dates and litter size. In 2010 a whelping date group covered 4 days, in 2011 it was 2 days. A scan sampling method was used to record behaviour. Each litter was observed during 5 to 10 seconds at these times: 4-5 hours and

Table 1. Materials and methods of the drinking water projects 2008-2011.

Year	Period	Groups	Litters (n)	Litter size	Actions
2008	26-49 days in lactation	water in nest box, control	2×60	6, 7, 8, 9 and 10 kits (young & old females)	water consumption in the nestbox, weigts gain kits, kit losses
2009	32-58 days in lactation	water in nest box, control	2×60	6, 7, 8, 9 and 10 kits (young females)	water consumption in the nestbox, weigts gain kits, kit losses
2010	32-56 days in lactation	water in nest box, control	2×45 2×111	6, 7, 8, 9 and 10 kits (young females)	weigts gain kits + female, kit losses, behavior study kits
2011	28-54 days in lactation	water in nest box (nb), water in entrance nb 3 times/day feeding extra floor till 7 weeks control	5×30 5×50	6, 7, 8, 9 and 10 kits (young females)	weigts gain kits + female, kit losses, behavior study kits

1-2 hours before feeding and 1-2 hours and 4-5 hours after feeding. They were fed once a day, from 13.00 onwards, on the nest box and on the pen with the automatic feeding system (Pilot Fur Feeding), such that the feeding places were 2 to 3 hours per day empty. During each observation period of 5 to 10 seconds it was noted whether one or more kits performed these behaviours:

- busy with water in or near the nest box;
- busy with water at the end of the pen;
- saliva licking;
- active and aggressive behaviour;

If more than one kit performed the same behaviour, we only recorded that the behaviour was seen. Per whelping date group the data were collected in excel and those data show the kit behaviour between 30 and 50 days of age. The data of weight gain (kits) or weight loss (female) are analysed with Anova (analyse of variance). The behaviour data were analysed with the F-test. We tested at the 0.05 level.

Results

To simplify the tables data of some groups were lumped. The groups 1 and 2 with water in respectively near the nest box are lumped because they had in common that the drinking nipple was near the nest box, and called the testing groups. The groups 3-5 are lumped because they have in common that the drinking nipples were far from the nest box.

Water consumption in nest box

The water consumption (including spoiled water) in the nest box was measured in 2008 and 2009. The frequency of drinking out of the water system in/near the nest box was spotted in 2010 and 2011 and Table 2 shows the results. The lactation period is divided in 5 days steps from 30 to 50 days. In 2009 and 2010 data were also collected after 50 days and these are included.

The amount of water consumed by kits and mother of the testing group was consistent in both years. The drinking frequency reflects the amount of consumed water. After 45 days the intake and the drinking frequency in/near the box decrease, for kits spent more time in the pen.

Weight gain kits

Table 3 shows the average weight gain of male/female kits in the water projects. Per year the period differed and the exact moments of weighting are printed in the table. The deviation of the number of male and female kits in the litters per group was equal in each year.

Table 2. Water consumption in the nest box.

	Water consumption (ml/d)		Frequency kits drink (%)	
Year	**2008**	**2009**	**2010**	**2011**
Period in lactation				
30-34 days	65	100	1%	1%
35-39 days	215	335	10%	10%
40-44 days	395	440	12%	15%
45-50 days	390	420	8%	11%
>50 days		245	4%	

Table 3. Average kit weight gain (male/female) kits in water projects.

Year	Period in lactation	Litters (n)	Water in nest box		Control	
			Kit (m/f) weigth gain (g)	variation (g)	Kit (m/f) weigth gain (g)	variation (g)
2008	26-49 days	2×60	420	42	415	53
2009	32-58 days	2×60	570	45	555	43
2010	32-56 days	2×45	480	53	445	48
2011	28-54 days	3×30	495	92	465	85

The testing group consistently tended to weigh more than the controls, but the differences were never significant (P>0.05). The higher variation in 2011 is due to the 2 periods of weighing (at 3 and 7 weeks or 1st and 28th of June). No effect of the whelping date is detected.

Weight loss mothers

Table 4 shows the mother's average weight losses per group in 2010 and 2011. At start of the project, the average weight of the females was 1.3 kg (0.1 kg var.) in all groups.

The mothers in the control group lost in 2010 significantly more weight than those in the testing groups (P<0.05); a similar, but not significant difference was found in 2011. The variation is enormous; some females gained a little weight and others lost over 0.5 kg between 30 and 50 days in lactation. All females were fed with the automatic feeding system such that the feeding place was 2 to3 hours empty. Both years the control and testing group consisted of the same number of litters and same litter size. Weight loss and litter size were not related.

The reproduction of the females which were followed in June 2010 showed an interesting fact. For Table 5 the weight losses in 2010 are divided into 3 groups and for each group the average weight loss of the females in 2010 is given. The average number of live born kits (LB) per litter in 2010 was high (litter size 6-10). Table 5 also gives the reproduction in 2011 and the number of

Table 4. Average weight loss in mothers in June.

Year	Period in lactation	Litters (n)	Water in nest box		Control	
			weight loss (g)	variation (g)	weight loss (g)	variation (g)
2010	32-56 days	2×45	-110	90	-160	130
2011	28-54 days	3×30	-80	105	-125	90

Table 5. Weight loss in females in 2010 and reproduction in 2010 and 2011.

Weight loss 2010		Litter 2010		Litter 2011		
groups (g)	loss (g)	females (n)	live born kits	females (n)	live born kits	dead born kits
>200	265	12	8.5	11	5.1^{a}	1.5
100-200	155	21	8.1	20	7.3^{b}	0.8
<100	30	17	7.9	16	8.3^{b}	0.6

^{a,b} Groups with different letters are significantly different.

females, the live born kits and the dead born kits (DB) per litter. The reproduction in 2011 was significant lower for the females with the high weight loss in June 2010.

Kit and female losses

The kit losses per group are shown in Table 6. Kit losses seemed lowest each year in the testing group, but the differences were not significant (*P*=0.08). The kit losses in the testing group never exceeded 1 kit per litter, whereas many control group litters lost 2 or more kits.

Females were lost only in 2010. The feed supplier reported a too low sodium level between the 10th and 16th of June. Several females which were at 35 days in lactation at the 10th of June suffered from nursing sickness when they were at 40 days. Those females stopped food consumption as followed from the untouched feed portions on top of the pen (150 gram). The kits had no problems in consuming the feed portion on the nest box. In the control group, 3 females died and 3 were weaned from the kits and in total 10 females were treated with vitamins. In the testing group with drinking water in the nest box none of the females died or had to be weaned and only 1 was treated with vitamins.

Table 6. Kit losses during the water projects.

Year	Litters (n)	WNB/EB[1]	Control
2008	2×60	0.5%	2.2%
2009	2×60	0.2%	0.4%
2010	2×110	0.5%	1.2%
2011	3×50	1.2%	2.8%
Average 2008-2011		0.6%	1.7%

[1] WNB/EB = water in the nest box / near the nest box.

Behaviour observed in the litter

Table 7 shows the average frequencies of drinking, saliva licking and restless behaviour per day between 30 and 50 days of age and per group. For the table we lumped the groups 1 and 2 (water in or near the box, WNB/EB), and 3-5 (water at the end of the pen only). The levels mentioned in Table 7 are the average frequency of pens with kits showing a behaviour per observation round (4 times per day).

The differences in kit behaviour between the groups were clear in every spotted behaviour and significant over the total period from 30 till 50 days of age. Drinking was spotted all day but mostly after feeding. Kits with water in the nest box started drinking on average 4-5 days earlier (32 vs. 36 days) and kits with water near the entrance 2-3 days earlier. That system started in a little lower frequency, but after 45 days the frequency became even higher than with water in

Table 7. Frequency of kit behaviour between 30 and 50 days of age.

Period	Drinking water		Saliva licking		Restless kits	
	WNB/EB[1]	Control[2]	WNB/EB[1]	Control[2]	WNB/EB[1]	Control[2]
30-34 days	0.4%	0.0%	3.0%	3.7%	0.6%	1.0%
35-39 days	7.1%	0.4%	8.3%	18.4%	3.0%	5.9%
40-44 days	12.2%	3.9%	3.8%	13.3%	4.0%	9.4%
45-50 days	9.9%	4.9%	0.7%	2.4%	2.0%	4.9%
Av 30-50 days[3]	7.4%[a]	2.3%[b]	4.0%[a]	9.5%[b]	2.4%[a]	5.3%[b]

[1] WNB/EB = water in the nest box / near the nest box.

[2] No direct water access in or near the nest box.

[3] Groups with different superscript letters are significantly different at $P<0.05$.

the nest box. Both systems resulted in much more water drinking than done by the kits with the drinking nipple at the other side of the pen. After 40 days of age, those kits more often drank in the pen. The kits from the testing group and the controls started drinking in the pen at the same age, but the controls did so twice as much.

Saliva licking was spotted mostly after feeding. Kits looked for the lips of the mother when she was sleeping or awake. Kits seemed to prefer that over drinking milk from the mammary nipples from time to time. Saliva licking is most often spotted from several kits at the same time at the female and mostly between 35 and 45 days of age. The incidence of saliva licking is much lower in the testing group kits. When the control kits were spotted saliva licking, the kits with water were spotted drinking water. The sum of the frequencies of drinking water and saliva licking by the testing group approximates the frequency of saliva licking by the controls.

Active and aggressive behaviour of kits showed mainly between 35 and 45 days and prior to feeding, but they also did so when feed was available. The testing group showed less active behaviour than did the control group.

Discussion

The title 'high need for drinking water' is clear after the behaviour study. Kits of 5-6 weeks of age have one favourite place to be: the nest box. When kits start eating farm food at 4-5 weeks of age they need also water. The water in the food seems not enough and therefore the kits look for it through the saliva and milk of the mother (Brink *et al.*, 2004). When drinking water is available in or near the nest box, kits drink water instead of licking saliva, especially between 35 and 45 days of age. This saves the female while she looses less weight. This makes also the kits more calm when drinking water is near the kits. These results confirm the results of Brink *et al.* (2004) and show the steps towards a practical system that can be implemented on farms.

The amount of water which is really drank by the kits is unknown, but the mothers drank 250 ml/day at the 5th week of lactation from the drinking system at the end of the pen (De Rond 2008). Since the milk production declines after 35 days, the female's water intake most likely decreases. If the mother drinks all the water in the nest box (which is not the case), the kits still drink a lot of water. The drinking frequency reflects the daily amount of water.

The kit growth and kit losses with water in or near the nest box were consistent for 4 year. The female however showed the best benefit for the drinking water while the weight loss is lower and the incidence in saliva licking is much lower. The effect of calmer kits on the mothers is unknown, they did not seem to bother and never interfered when their kits fight.

In 2010 the sodium level in the food was too low and resulted in nursing sickness in the control females. The feeding system putted in June 150 gram food on top of the pen and the rest, the big volume on the nest box. The females who suffered from nursing sickness didn't touch the feed on top of the pen for a few days prior to the occurrence of the sickness and probably stopped eating at all. The feeding system will be used as a system to signal the females which come into a problem. Not eating is not right and asks for controlling the female.

Female mink are good mothers while they give their self away. That is not totally true, the females with the most weight loss in June have kits with the lowest growth in June (<100 gram weight loss -540 gram kit growth versus >200 gram weight loss -490 gram kit growth).

The weight loss in June comes on top of the weight loss in May. Projects on Edelveen in 2004 and 2006 showed an average weight loss of 50 till 150 gram per female between whelping and 30 days in lactation. That makes the total weight loss enormous in several cases (>30% of BW). What is known of the reproduction of females which suffered from and recovered of nursing sickness the year before? This item will be followed in 2012 also while the impression lives that this is animal bound. Also females with water in the nest box can loose a lot of weight and few females in control group loose no gram. If this is all true, in selection of females for reproduction the body condition at the end of June should be scored and used.

Drinking water in or close to the nest box helps the kits while they have a high need for drinking water when they prefer to stay in the nest box. It helps the female in nursing her kits and saving own weight, which helps the growth of the kits and her reproduction next year. Drinking water in or close to the nest box improves welfare of mink and farmer.

References

Ahlstrom A., 2010. Mink and foxes requirement for water. NJF workshop; energy supply of mink and foxes, Malmo, Sweden.

Brink A-L., Jeppesen L.L. and Heller K.E., 2004. Behaviour in suckling mink kits under farm conditions: effect of accessibility of drinking water. Applied Animal Behaviour Science 89: 131-137.

Clausen T.N., Olesen C.R., Hansen O. and Wamberg S., 1992. Nursing sickness in lactating mink, Epidemiological and pathological observations. Can. Journal of Vet Res. 56: 53-56.

De Rond J., 2008. Water intake female affects growth kits. De Pelsdierenhouder 2008: 144-145.

De Rond J and Kleyn van Willigen F., 2009. Drinking water in nest box when kits start eating farm food supports kits and dam. NJF no 427, Vaasa, Finland, 1 p.

De Rond J and Kleyn van Willigen F., 2011. Kit performance and behaviour between 30 an 50 days of age when easy access to drinking water. NJF no 450, Knivsta, Sweden, 10 p.

Moller S., 1991. Drinking behaviour of mink in relation to watering system and water temperature. NJF-seminar no 192, Uppsala, Sweden, 11 p.

Moller S. and Lohi O., 1989. Drikke adfaerd og vaegtutvikling hos mink med drypvandingssystem. Faglig Arsbereting 1988 Dansk Pelsdyravlerforening, Glostrup Danmark, pp 41-52.

Steffensen L.K., Hansen S.W. and Jeppesen L.L., 2007. Introducing an open water surface as an alternative to the traditional valve drinker for ranch min in the lactation period. Scientifur 2007, 31, 1: 7-18.

Mink's adaptation to group housing in practice

S.W. Hansen and S.H. Møller
Department of Animal Science, Aarhus University, Blichers Allé 20, P.O. 50, 8830 Tjele, Denmark; steffenw.hansen@agrsci.dk

Abstract

In this project, we test the hypothesis whether mink populations with a higher social tolerance have been developed in practice and thus are better adapted to housing in stacked cages. The hypothesis has been tested by comparing the level of bite damages and bite marks in mink kept in pairs and in mink kept in social groups. In total, the project included 1,191 brown mink housed on four farms in up to five different types of social groups. The result showed that mink in groups had significantly more bite marks than mink kept in pairs, and that mink with bite wounds had significantly more bite marks than mink that did not have bite wounds. Thus, based upon bite mark results, we cannot confirm the hypothesis that mink populations adapted to group housing have been developed. The number of mink that had died or had been removed during the test period, or that had bite wounds at the time of observation was low on farm A, B, and C, and significantly lower than on farm D. This partly supports the hypothesis about having developed mink populations and or management adapted to group housing. The investigation has confirmed that the level of bite marks is a useful, objective measuring parameter for the social tolerance in the groups and for the level of aggression.

Keywords: bite wounds, bite marks, fur chewing, group housing, welfare

Introduction

Several mink of the same sex are kept in the same cage when housing mink in groups during the growth season. This kind of housing is different from the mink's solitaire and territorial way of living, and this difference makes new claims to both mink and management. Thus, group housing increases the risk of aggression compared to traditional housing of pairs of male + female, which is also confirmed in the studies that have been carried through. The increased aggression documented as increased aggressive behaviour (Pedersen *et al.*, 2004; Hänninen *et al.*, 2008), increased occurrence of bites, wounds, and damages (Pedersen *et al.*, 2004; Møller, 2011), and more dead mink (Pedersen *et al.*, 2004). Several studies have shown that the number of mink with black spots on the leather side of the skin is higher when mink are kept in groups than when they are kept in pairs in standard cages (Hansen, 1997; Møller *et al.*, 2003; Pedersen *et al.*, 2004; Hansen and Houbak, 2005). It is assumed that black spots occur by disturbing or destroying hair follicles during maturing of the winter fur, for instance due to bites, but not necessarily by perforation of the skin. The black spots are therefore taken as indicative of bite marks and are only visible after pelting. Bite marks can be quantified and thus make an objective parameter for the level of aggression in the cage, which is easier to observe than aggressive behaviour and more sensitive than the occasions of aggression leading to bites, which luckily are rather rare. Bite marks in mink kept in pairs (male + female) most often occur in the neck area and are probably due to play or sexual activity and are most often seen in females. Bite marks on the body and tail, however, seem to be connected to aggression (Hansen and Jeppesen, 2008). The coherence between bite marks and reduced welfare is based on the perception that being subjected to aggressive bites is connected to pain and malaise.

It can be difficult to acknowledge aggression problems in mink before the effects are visible as wounds and bite damages. In practice, experience with aggression in group mink, housed in stacked cages, varies (cf. personal communication with several breeders). Differences in the aggression level can be due to difference in the management (Hansen, 2011) combined with difference in the mink's genetic basis of social tolerance (Berg and Møller, 2010). On this background, we wish to examine the hypothesis stating that mink populations have reached a higher social tolerance in practice via housing, care, and selection and thus better adapt to housing in stacked cages. The hypothesis has been tested by examining the level of bite wounds and bite marks on mink farms which are assumed to be adapted for group housing of mink in practice, and by examining whether there is coherence between bite marks on the leather side and bite wounds on the hairy side of the skin. Furthermore, we have registered the occurrence of fur chewing, not related to aggression, but an abnormal behaviour in farm mink.

Material and methods

In cooperation with Kopenhagen Forskning, four farms were selected. In several years, the four farms had practiced group housing of mink in stacked cages (i.e. double cages placed above each other) without the expected increased level of aggression, which potentially could be ascribed a higher social tolerance in mink on these farms and/or a changed management. These farms, thus, are not necessarily representatives for all mink farms in Denmark, however, selected exactly because of their good experience with group housing.

The breeders were encouraged to use their preferred group composition in the stacked cages. Furthermore, the breeders were encouraged to include ontest groupconsisting of 2 males + 2 females, as this is the best examined combination in scientific studies. In addition to that, the breeders were requested to place a group of mink housed in pairs (male + female) that was to work as control group. In order to avoid a difference in colour type between farms and groups, it was requested that each of the social groups should consist of approx. 100 brown mink.

After killing, the mink bodies were examined for bite wounds and fur chewing, and after skinning and scraping, they were examined for bite marks on the leather side of the skin. Chewing, bite wounds, and bite marks were registered on three areas of the body: *Neck*: from tip of the nose to the shoulder/foreleg; *Body*: from foreleg to 10 cm above tail head; *Tail*: from 10 cm above tail head to tip of the tail. Besides, we distinguished between whether the bite wounds were placed on the tip of the tail or at the tail head. The sizes of bite wounds as well as fur chewing were graduated from 0 to 9 with the score = 0 for none and the score = 9 for bite wounds or chewing with a diameter of more than 50 mm. The indication of bite wounds included open and healed wounds and also smaller swellings primarily on the tail (categorized as healed wounds), which can only be registered by feeling, carried out together with the visual examination. Fur chewing is the indication of areas on mink where the guard hairs are missing or where both guard hair and underfur are missing.

Likewise, the number of bite marks was graduated with the score = 0 for no bite marks and the score = 9 for more than 45 bite marks (see Berg and Møller, 2010). The sum of scores for bite marks in the neck, on the body and tail is characterized in the following as total bite marks, and likewise, a total score for bite wounds and fur chewing was calculated.

The number of mink that had died or had been killed, and mink that had been moved to an infirmary because of bite wounds or illness before pelting was noted

Results

The number of mink that totally had died or had been moved during the test period, from the kits were housed in test groups and until pelting, was significantly higher on farm D (14.4%) than on farm A (0.9%), B (3.7%), and C (3.0%). All the dead/moved mink on farm D were, according to the breeder's statement, removed because of bite wounds.

One thousand one hundred and ninety-one mink were examined for bite wounds and fur chewing. During pelting, some of the mink lost their nose mark, which identified each mink to the group, which is why bite marks could only be related to group in 1,135 mink.

When examining the killed mink, we found that 7.7% had at least one bite wound and 5.4% had bite wounds on the tip of the tail and 2.0% at the tail head. On farm D, there were more mink with bite wounds (12.9%) than on the other farms (farm A=6.0, farm B=3.2, and farm C=8.1%). Bite wounds were found in 3.5% of the mink kept in pairs, a number which was significantly lower than in mink kept in groups (6.8-11.0%; $P<0.001$).

Fur chewing most often occurred on the tail (90% of the mink with fur chewing) but opposite to bite wounds, there were fewer mink with fur chewing on farm D ($P<0.0001$). There were more mink with fur chewing when kept in pairs (31%) and 4 females together (31.9%), than in mink kept in groups of mixed sexes (17.1-21.1%, $P<0.027$).

Bite marks occurred in 93% of the mink. The number of mink with bite marks was highest on farm D (96.8%) and A (95.3%), and lower on farm B (88.9%) and C (87.8%). There were fewer mink with bite marks among mink kept in pairs (84.4%) than among mink housed in groups (95.4-97.9%; $P<0.0001$). As for mink kept in pairs, there were more females with bite marks (5.2%) than males (1.6%; $P<0.0001$), but there was no difference in sexes when mink were kept in groups (male: 9.3% and females: 10.1%). However, in grouped mink, more females (28.1%) than males (17.6%; $P<0.002$) had fur chewing, but not when they were kept in pairs (male: 30.7% and females: 31.6%).

In the following section, the average value of the score for bite marks is presented per test group in tabular form for each farm. The occurrence and the score of bite wounds and fur chewing were in general low within farm and are not shown in the tables, however, the main results of bite wounds and fur chewing appear in the text.

Farm A

Mink kept in pairs had significantly fewer bite marks (in total, in the neck, and on the body and tail) than mink kept in groups of 2 males + 1 female, and 4 females together (Table 1).

Females had more bite marks in the neck (3.1) than males (1.2; $P<0.0001$), but there was no significantly difference in sexes in the number of bite marks on the body and tail irrespective of social group (Table 1). Males, that were kept in pairs with females, had fewer bite marks in the neck (0.5), on the body (1.4), and on the tail (1.0) than when they were kept in groups with another male (neck: 1.6; $P<0.0015$; body: 3.9; $P<0.0001$; tail: 2.3; $P<0.0032$). The score for bite wounds was not different between groups or between sexes. Also, there was no difference in fur chewing in the neck or on the body between groups or between sexes. Mink kept as 1 female + 2 males had less tale chewing (0.3) than mink kept in pairs (1.1; $P<0.0001$), and 4 females together (1.1; $P<0.0023$).

Table 4. Bite marks for mink kept in groups for farm A.

Farm A	Social groups[1]		
	1♂+1♀	2♂+1♀	4♀
Total	4.5±4.9 b***	8.6±6.3 a	11.6±7.3 a
Neck	1.3±1.8 b***	2.4±2.3 a	3.3±2.4 a
Body	1.9±2.3 b***	3.8±2.9 a	4.6±2.9 a
Tail	1.3±2.0 b***	2.5±2.5 a	3.7±2.5 a

[1] Values in the same row with different letters are significantly different.***: $P<0.001$; **: $P<0.01$; *: $P<0.05$.

Farm B

Mink kept in pairs had significantly fewer bite marks (in total, in the neck, and on the body and tail) than mink kept in groups of 2 males + 2 females but not different from 4 females together. The group of 4 females also had fewer bite marks in the neck and on the body than mink kept in groups of 2 males + 2 females (Table 2).

Females had more bite marks than males (neck: 3.0 vs. 1.3; $P<0.001$; body: 2.6 vs. 1.5; $P=0.101$; tail: 3.2 vs. 1.4; $P<0.010$). For mink kept in pairs, the sexual difference in bite marks on the body was not significant. Males in the group of 2 males + 2 females had more bite marks in the neck=2.0, body=3.0, and tail=2.2 than males kept in pairs with a female (neck=0.8; $P<0.007$, body=0.7; $P<0.001$; tail=0.6; $P<0.007$). The score for bite wounds was not different between groups or sexes. Likewise, significant difference in fur chewing between groups or between sexes could not be shown.

Table 2. Bite marks for mink kept in groups for farm B.

Farm B	Social groups[1]		
	1♂+1♀	2♂+2♀	4♀
Total bite marks	3.6±4.3 b	11.0±8.1 a***	6.7±6.5 ab
Neck	1.3±1.6 b	3.7±3.0 a***	1.7±2.0 b*
Body	0.9±1.5 b	3.7±3.2 a***	2.3±2.7 b*
Tail	1.4±2.2 b	3.5±3.3 a***	2.7±3.3 ab

[1]Values in the same row with different letters are significantly different. ***: $P<0.001$; **: $P<0.01$; *: $P<0.05$.

Farm C

Mink kept in pairs had fewer bite marks (in total, in the neck, and on the body and tail) than 4 females together (Table 3). For mink kept in pairs, females had more bite marks than males (neck: 2.7 vs. 0.8; $P=0.001$; body: 1.8 vs. 0.7; $P=0.024$; tail: 1.8 vs. 0.3; $P=0.005$), and females that were housed with a male had fewer bite marks than when 4 females were housed together (neck: 2.7 vs. 3.7; $P=0.010$; body: 1.8 vs. 4.2; $P=0.001$; tail: 1.8 vs. 3.3; $P=0.001$).

Mink kept in pairs had smaller bite marks on the tip of the tail (0.2) than 4 females together (0.3; $P=0.036$). There was no difference in bites/wounds in total, in the neck, on the body, or by the tail head. No difference was registered in chewing between groups or between sexes.

Table 3. Bite marks for mink kept in groups for farm C.

Farm C	Social groups[1]	
	1♂+1♀	4♀
Total bite marks	4.3±4.6 b***	11.2±7.5 a
Neck	1.9±2.1 b***	3.7±2.7 a
Body	1.3±1.8 b***	4.2±3.1 a
Tail	1.1±1.7 b***	3.3±3.1 a

[1] Values in the same row with different letters are significantly different. ***: $P<0.001$; **: $P<0.01$; *: $P<0.05$.

Farm D

Mink kept in pairs had significantly fewer bite marks (in total, in the neck, and on the body and tail) than mink kept in groups of 4 females, 2 males + 2 females, and 1 male + 2 females. The group of 1 male + 2 females had significantly fewer bite marks than groups of 2 males + 2 females and 4 females. There was no difference between mink kept in groups of 4 females and mink kept in groups of 2 males + 2 females (Table 4).

Females had significantly more bite marks than males in total, in the neck, and on the body and tail ($P<0.01$). For the group of 2 males + 2 females, the sexual difference in bite marks on the body was not significant. Besides, females kept in pairs with a male had fewer bite marks in the neck, on the body and tail than those in groups of 4 females ($P<0.01$).

The score for bite wounds on the tip of the tail, tail head, body, and neck was different between groups ($P=0.033$). The group of 2 males + 2 females had more bite wounds in total than groups of male + female and more than groups of 1 male + 2 females, but not more than the groups of 4 females. The difference in bite wounds between groups was also significant for females but not for males. No significant difference in bite wounds between sexes could be shown. Fur chewing was not different between groups or between sexes.

Table 4. Bite marks for mink kept in groups for farm D.

Farm D	Social groups[1]			
	1♂+1♀	1♂+2♀	2♂+2♀	4♀
Total bite marks	6.9±7.4 c	10.5±8.6 b	14.9±7.7 a***	17.0±8.1 a***
Neck	2.4±2.6 b	3.4±3.0 b	5.0±2.8 a***	5.9±2.9 a**
Body	2.4±2.8 c	3.8±3.2 b**	5.5±3.0 a***	5.9±3.0 a**
Tail	2.1±2.8 c	3.4±3.3 b*	4.3±3.1 a***	4.3±3.3 a*

[1]Values in the same row with different letters are significantly different. ***: $P<0.001$; **: $P<0.01$; *: $P<0.05$.

Comparison of mink kept in pairs and four females together across all four farms

The group of 4 females had significantly more bite marks on farm D than on farm A, B, and C both in total and spread on the neck, body, and tail (Figure 1). Group housing with four females on farm B had, in total, fewer bite marks than those on farm A and C.

Mink kept in pairs had, in total, more bite marks on farm D than on farm A, B, and C. Farm A, B, and C were not significantly different. Mink kept in pairs on farm A, C, and D had fewer bite marks than mink kept in groups of 4 females, both in total and in the neck, body, and tail. However, there was no difference in bite marks between mink kept in pairs and groups of 4 females on farm B due to a general low level.

The score for bite wounds was not different between farms or between sexes. Between groups, there was a tendency to more bites/wounds, in total, in groups of 4 females (0.2) than in pairs (0.1;

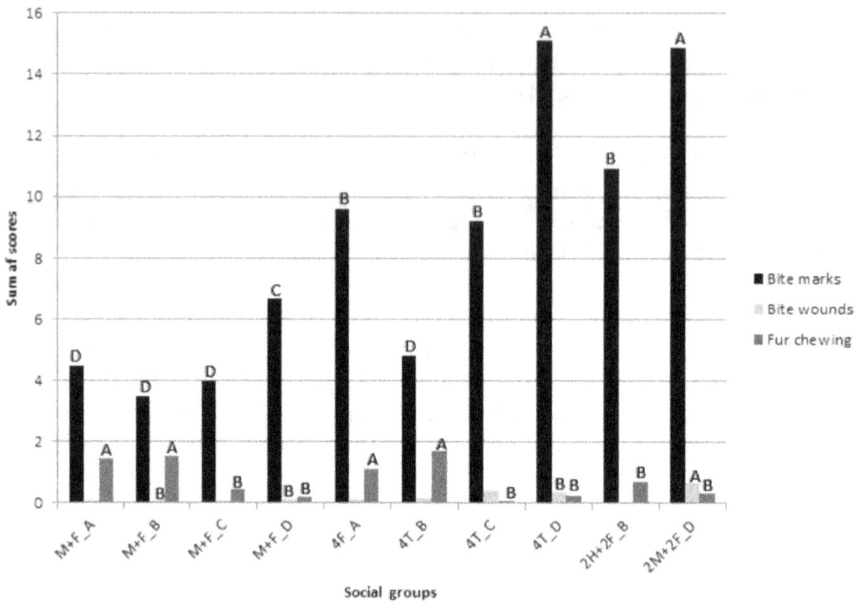

Figure 1. Comparison of the total score of bite marks, bite wounds, and fur chewing between social groups: mink kept in pairs (pair), 4 females (4T) on farm A, B, C, and D and between mink in pairs and 2 males + 2 females (2H + 2T) on farm B and D. Significant difference (P<0.01) in bite marks, bite wounds, or fur chewing between farm and social group is indicated by different letters (a, b, c, d).

P<0.054). Likewise, the group of 4 females had more bites/wounds on the tip of the tail than pairs (*P*<0.050). There was no difference in bites/wounds on the body and at the tail head. However, mink kept in pairs had more bites/wounds in the neck than the groups of 4 females (*P*<0.038), indicating that this behaviour may be intra-sexual. The score of fur chewing was higher on farm A and B than on farm C and D (total: *P*<0.001). No difference in fur chewing between groups or sexes could be shown.

Comparison of mink kept in pairs and 2 males + 2 females on farm B and D

Farm D had, as previously shown, significantly more bite marks than farm B. Mink kept in groups of 2 males + 2 females had, in total, significantly more bite marks than mink kept in pairs (figure 1). Females had more bite marks than males (*P*<0.001).

For bite wounds, in total, there was a significant interaction between farm and group. On farm D, groups of 2 males + 2 females had more bite wounds than mink kept in pairs, but not on farm B. The group of 2 males + 2 females on farm D had more bite wounds than on farm B, but there was no difference between farms for mink kept in pairs.

For fur chewing, there was also a significant interaction between farm and group. On farm B, mink kept in pairs had more chewing than groups of 2 males + 2 females, but not on farm D. On farm B, mink kept in pairs had, in total, more fur chewing than on farm D, but this difference between farms did not apply to groups of 2 males + 2 females. For tail chewing, which was the most common kind of fur chewing, there was the same significant interaction between farm and group as for chewing in total.

Comparison of bite marks in mink with and without bites/wounds

Mink with bite wounds had significantly higher score of bite marks in total (14.0±9.4) than mink without bite wounds (8.6±7.6; $P<0.001$). This difference in bite marks between mink with or without bite wounds was also significant within sexes, farm, and the social groups of 2 males + 2 females and 4 females, but not for mink kept in pairs, 1 male + 2 females, and 2 males + 1 female.

Discussion

On all four farms, the study has shown that mink kept in groups have more bite marks than mink kept in male+female pairs. Moreover, it has been shown that mink with bite wounds have more bite marks than mink without bite wounds. Thus, the status of bite marks cannot confirm the hypothesis that mink populations, via housing, care, and selection, have been developed to have a higher social tolerance and thus have adapted the housing in stacked cages.

On farm A, B, and C, the number of mink having been removed from the experiment, or that had been subjected to bite wounds at the time of observation was low and significantly lower than on farm D. Therefore, it was not possible to show a difference between social groups in the number of dead/removed mink, or in the number of mink with bite wounds on the three farms. This result partly supports that a variation may exist on farms, e.g. in the social tolerance of mink and/or in management factors, improving the performance in stacked houses in practice. However, a confirmation of the hypothesis will require a test on a bigger animal material because of the low frequency of dead/removed mink and mink with bite wounds.

Farm D differentiated from the other three farms by having significantly more mink with bite wounds, more mink that had died/had been removed and more bite marks in the neck, on the body and tail irrespective of social group housing. However, the number of bite marks in the neck, on the body and tail was lowest when mink were kept in pairs and significantly bigger in mink kept in groups. Thus, the pattern in bite marks on farm D was not different from the overall picture, namely that mink kept in groups had more bite marks than mink kept in pairs.

Mink that had been subjected to bite wounds had significantly more bite marks than mink without bite wounds. This difference in number of bite marks between mink with and without bite wounds was statistically significant within farm and sexes and within social groups of 4 females and 2 males + 2 females across farms.

A high occurrence of bite marks on farm with high mortality and many removed mink, and the fact that mink with bite wounds had more bite marks than mink without bite wounds, related to the fact that mink kept in groups have more bite marks, clearly indicate that aggression or other social contact, involving bites, is causing the black spots on the leather side of the skin. It is the first time that a direct link between bites causing marks on the leather side of the skin and bites causing wounds, removal or killing of mink has been shown.

The fact that the bite marks are caused by social interactions is also supported by the characteristic difference between sexes. More females than males had bite wounds, and more females than males had bite marks in the neck, on the tail and body, when being kept in pairs (male + female). This sexual difference is probably due to the male dominating the female by virtue of the bigger size of his body. The fact that mink kept in pairs at the same time had fewer bite marks than mink kept in groups is supported by studies that have shown a higher social tolerance between sexes than within sexes (Berg and Møller, 2010). The sexual difference in bite marks in mink kept in groups was less clear than in mink kept in pairs, but in the social groups of 2 males + 2 females and 2 males + 1 female, we found more females with bite marks on the tail and, for the latter group, also more females with bite marks in the neck. However, we never found that males had more bite marks than females.

Opposite from the number of mink with bite wounds and bite marks, the number of mink with fur chewing was larger among mink kept in pairs than among mink kept in groups of mixed sexes. Fur chewing is not considered linked to aggression but more likely to boredom/under-stimulation (Hansen *et al.*, 1998; Malmkvist and Hansen, 2001). For mink kept in pairs and in a group of four females, more fur chewing could be shown on farm A and B than on farm C and D. At the same time, farm A and B had smaller bites/wounds on the tail than farm C and D. When comparing fur chewing on farm B and D, that both had the groups of male + female and 2 males + 2 females, we found more fur chewing on farm B than D. At the same time, farm B had smaller bites/wounds and fewer bite marks than farm D. The result shows that fur chewing is reduced in mink in socially unstable groups measured as increased occurrence of bite wounds and bite marks.

The low level of dead/removed mink, the number of mink with bite wounds and the number of bite marks on farm A, B, and C compared to farm D indicate that also other factors than the size of the group influence the social stability in mink. The biggest direct difference between the four selected farms, that all had good experience with group housing previously, was the time and the mink kits' age at release into the test groups. On farm D, the mink were housed in the test groups July 17, which was approx. 3 weeks later both compared to the other three farms and compared to the normal procedure on farm D. The fact that the late time for establishment of groups and thus mixing of mink kits can influence the social stability in the groups, so markedly as the result implies, has not previously been shown. However, anecdotally, it seems to be good practice to have the kits put in place on the farm around July 1st. In Holland, the weaning time is not, as in Denmark, decided on the basis of the kits' age, but on the contrary on the basis of a firm date. In Holland, the kits must be weaned on July 1st. If the kits' age has a markedly influence on the social stability when establishing the groups, an early establishment of social groups in Holland can contribute to explaining the different experiences to be found in Denmark and Holland regarding bite marks and social tolerance in mink kept in groups.

The results from this study show that the number of bite marks on the leather side of the skin is influenced by whether mink are kept in pairs or in groups. Almost all mink with social contact are subject to one or more bite marks. Therefore, the level of bite marks is a useful and objective measure parameter revealing the social stability in the groups and the level of aggression. At the same time, it is, however, important to emphasize that knowledge and documentation are missing about the anatomic and physiologic mechanisms in the skin, which transform the mechanical influence of bites to black spots in the skin as well as about for how long the black bite marks are visible after having been bitten.

Acknowledgement

The project is financed by Kopenhagen Forskning/Fur production levy foundation. We would like to thank the four private breeders for having placed their farm at our disposal for the study as well as for their engagement in the project.

References

Berg, P. and Møller S.H., 2010. Mulighed for at selektere for reduceret aggression i gruppeindhusning. Pelsdyrerhvervets Forsøgs- og ForskningsCenter. Faglig Årsberetning. 2009. 6 pp.

Hansen, S. W., Houbak, B. and Malmkvist, J., 1998. Development and possible causes of fur damage in farm mink – significance of social environment. Acta Agric. Scand. Sect. A, Animal Sci. 48: 58-64.

Hansen, S.W., Houbak, B. and Malmkvist, J., 1997. Does the 'solitary' mink benefit from having company. NJF seminarium nr. 280 /NJF Utredning/Rapport nr. 116, Helsingfors, Finland, 6-8 october.

Hansen, S.W. and Houbak.B., 2005. To skridt frem og tre tilbage – gruppeindhusning af mink. reproduktion Pelsdyrerhvervets Forsøgs- og Forsknings Center, Faglig Årsberetning 2004. 8 p.

Hansen, S.W. and Jeppesen, L.L., 2008. Bidmærker som velfærdsindikator hos mink. Pelsdyrerhvervets Forsøgs- og ForskningsCenter. Faglig Årsberetning. 2007, pp. 13-23.

Hansen, S.W. and Malmkvist, J., 2011. Fodringsmæssige tiltag til begrænsning af bidmærker hos mink holdt i grupper – foreløbige resultater. Intern rapport nr. 109. Temadag om aktuel minkforskning (ed.) P. Berg, Aarhus Universitet. pp. 19-34.

Hännien, S., Mononen,J., Harjunpää, S., Pyykönen, T., Sepponen, J. and Ahola, L., 2008. Effects of family housing on some behavioural and physiological parameters of juvenile farmed mink (*Mustela vison*). Appl. Anim. Behav:Sci. 109: 384-395.

Malmkvist, J. and Hansen, S.W., 2001. The welfare of farmed mink (*Mustela vison*) in relation to behavioural selection: A review. Animal Welfare 10: 41-52.

Møller, S.H., 2003. Information value and applicability of health and welfare indicators observed at pelting of mink. Proc. 10th International Symposium of Veterinary Epidemiology and Economics (on CD-rom), Vina del Mar, Chile, Abstract 409, 3 pp.

Møller, S.H., Hansen, S.W. and Sørensen, J.T., 2003. Assessing animal welfare in a strictly synchronous production system: The mink case. Animal Welfare 12: 699-703.

Møller, S.H., 2011. Presence of wounds and injuries in the mink production. Intern rapport nr. 109. Temadag om aktuel minkforskning. P. Berg (ed.), Aarhus Universitet, pp. 61-67.

Pedersen, V., Jeppesen, L.L. and Jeppesen, N., 2004. Effects of group housing systems on behaviour and production performance in farmed juvenile mink (*Mustela vison*). Appl. Anim. Behav. Sci. 88: 89-100.

The effect of handling and training on measures of the affective state of farmed mink (*Neovison vison*)

P.M. Svendsen
Group of Animal Behaviour and Stress biology, Department of Animal Science, Aarhus University, Blichers Allé 20, P.O. Box 50, 8830 Tjele, Denmark; pernille.svendsen@agrsci.dk

Abstract

The present study aimed to examine if handling and training female mink on a learning task, as preparation for a cognitive bias test, had an effect on measures of their affective state. Adult female farm mink were used based on measures of approach/avoidance behaviour, and categorised as explorative (E, n=10) and fearful (F, n=10). Affective state was assessed on the basis of approach/avoidance behaviour in two contexts (stick-tests, novel object tests) and adrenocortical activity (faecal cortisol metabolites, FCM). The results showed that the F mink changed to being more explorative after handling and training for the cognitive bias test. Furthermore, the handling and training also had an effect on the E mink, but in the opposite direction. This study showed that preparing mink for a cognitive bias test had a marked effect on measures of their affective state, which has to be considered when interpreting the results from experiments using training to assess welfare.

Keywords: American mink, assessment of welfare, behaviour, cognitive enrichment, FCM

Introduction

The assessment of positive experiences and affect in animals has become a growing focus of animal welfare research. Boissy *et al.* (2007) have pointed out that animal well-being is not only the absence of negative experiences and affective states (e.g. fear) but indeed also the presence of positive ones, such as pleasure. Affective or emotional states can be defined as subjective states that are characterised by valence (how positive or negative they are) and arousal (how activated the animal is) (Mendl *et al.*, 2010b). Whilst it is currently not possible to directly assess the subjective experiences of other species, or to know for certain whether they have such experiences, it is possible to measure other components of affective states including behavioural, physiological, and cognitive responses (Paul *et al.*, 2005). The avoidance of a novel object or stick may indicate a general 'fearfulness' because approach/avoidance behaviours are thought to be important correlates of affective valence (Paul *et al.*, 2005). The hypothalamic – pituitary – adrenal (HPA) axis responds and cortisol is released from the adrenal cortex when confronted with a stressor which may correlate to fear and anxiety, but the HPA axis is also responsive to other types of activity that are associated with emotional arousal (Touma and Palme, 2005).

Cognitive bias refers to the influence that affect has on a range of cognitive processes in animals, including attention, memory and judgement (Mendl *et al.*, 2009). It is known from humans that individuals in a negative affective state pay more attention to threatening stimuli, retrieve more negative memories and judge ambiguous stimuli more negatively (being 'pessimistic') than happier individuals (e.g. Mendl *et al.*, 2009; Paul *et al.*, 2005). Harding *et al.* (2004) have developed a cognitive bias method for determining affective state in animals, based on measuring emotionally induced biases in decision making. Their findings indicated that rats in unpredictable housing, inducing a mild negative affective state (Willner, 1997), were more pessimistic in response to ambiguous stimuli than controls. The method has been applied successfully to a number of other species, including starlings (Matheson *et al.*, 2008), sheep (Doyle *et al.*, 2010) and dogs (Mendl *et al.*, 2010a). The method is suggested to be a useful indicator of animal emotion and it is a promising indicator of animal welfare but it requires a lot of intensive training (Mendl *et al.*, 2009).

Cognitive enrichment is based on instrumental learning directed to a highly motivating consummatory act that may have the capacity to increase animal welfare and it may initiate positive emotions (Manteuffel *et al.*, 2009). On the farm, cognitive enrichment usually refers to the provision of a specific and short-term problem solving activity or device (Puppe *et al.*, 2007). The possibility that mastering a problem-solving task is rewarding is suggested by Hagen and Broom's (2004) finding that cattle show an affective reaction (e.g. increased heart rate and behavioural vigour (indicating increased arousal, though valence is not entirely clear)) when they increase performance on a simple instrumental task.

The study aimed at investigating the effect of preparation for a cognitive bias test in mink, i.e. handling, habituation to the test chamber and training on a positive instrumental learning task, on measures of the affective state of female farmed mink. The study used traditional indicators of affective state, being measures of approach/avoidance behaviour in two contexts (stick test, novel object test) and adrenocortical activity (faecal cortisol metabolites, FCM), assessed before and after handling and training, as the preparation for a cognitive bias test.

Materials and methods

Animals and housing

The start population comprised 200 adult female farm mink (*Neovison vison*) of the color type wild, born around May 1, 2009. Based on approach/avoidance categorization using a stick test (see below) in the home cage, 60 animals (30 explorative and 30 fearful) were selected for a novel object test also in the home cage. Ten explorative mink with very short latencies to make contact and 10 fearful mink with very long latencies to make contact or no contact at all in the novel object test were selected and allocated to two groups, explorative (E) and fearful (F) respectively.

The selected mink were moved to new housing facilities where they were habituated for 7 days before the first faecal sampling, followed by the onset of the handling and training procedure. The mink were individually housed in standard-sized wire cages (W: 30 cm, H: 45 cm, L: 90 cm; Hedensted-Gruppen, Hedensted, Denmark) connected to one wooden transportable nest box (W: 25 cm, H: 30 cm, L: 21 cm), at the mink farm at Aarhus University, Research Centre Foulum, Denmark. Food was provided once daily at 12:00±15 min in amounts to keep the animals at maintenance level, water and straw was provided *at libitum*, and all mink had access to either a shelf or a tube in their cage. The study was performed in September – December 2010.

Handling and training procedure

The animals were shut into their transportable nest box and moved 50 m to another building, where they were let into a standard-sized wire cage and trained on a positive instrumental learning setup. The mink had to press a lever on a sound cue to obtain a food reward (positive reinforcement). The mink were returned to their home cage after approx. 20 min. All animals were handled 2-3 times per week for 12 weeks continuously. Once every 2 weeks faecal samples were collected from the backside of the cages (first and last collection described here). Human movement behind the cages is unusual.

Stick test

The approach/avoidance behaviour of the mink was characterized by the stick test (Hansen and Jeppesen, 2006; Malmkvist and Hansen, 2002) prior to and after 12 weeks of handling and training. The mink were excluded from the nest box using a metal shutter and were tested individually. A tongue spatula was put through the wire netting in the upper part of the lid of the cage and the immediate reaction as well as the position (front or back of the cage) of the mink after 30 s was registered. The reaction was allocated to one of four categories: (1) explorative if the mink approached and sniffed the spatula persistently; (2) fearful if the mink withdrew and did not touch the stick; (3) aggressive if the mink attacked and bit the stick and (4) uncertain if the mink did not respond or showed a mixture of responses and so could not be placed in one of the first three categories.

Novel object test

The first test was performed on two testing days one week after the stick test and prior to the handling and training. The second test was performed on one single test day following the handling procedure and stick test. The mink were shut into the nest box using a metal shutter placed at the entrance. A wooden cube (10×10×10 cm) was placed centrally approx. 60 cm into the cage from the entrance to the nest box as a novel object. A new cube was used in all cages. The duration of the test was 180 s and began when the metal shutter was removed and the mink had full view and access to the cage. Latency to approach and to contact, as well as the duration of manipulation of the novel object was recorded using a handheld computer (Psion workabout Copyright Psion PLC 1995). The wooden cube was removed after each test. Two mink (one from each group) were excluded from the analysis due to human error.

Sampling and analysis of cortisol metabolites

Concentrations of FCM are a validated measure in mink, reflecting circulating cortisol in female mink (Malmkvist *et al.*, 2011). Fresh faecal samples were collected non-invasively following habituation and after the intensive handling period between 10:30 h and 14:30 h. Fresh faecal samples (0.5 g) were frozen within 20 min after delivery at -21 °C until analysis. A total of 0.5 g of each sample was extracted with 5 ml methanol (80%) and concentrations of FCM were measured in an aliquot of the supernatant with two different enzyme-immunoassays (EIAs) described by (Palme and Mostl, 1997) (11-oxoaetiocholanolone EIA) and (Frigerio *et al.*, 2004)

(11β-hydroxyaetiocholanolone EIA). One E mink did not defecate during the sample collection hours in the sampling after the intensive handling period and was excluded in the further analysis. Only FCM values measured by the 11β-hydroxyaetiocholanolone EIA are presented, as both measures were highly correlated.

Statistical analysis

Differences between each of the approach/avoidance behaviour categories and position in the stick-test before and after handling within the two groups (E and F) were tested using a paired t-test. A Wilcoxon test for matched pairs was used to analyze latencies to approach (leave the nest box) and to touch the novel object in the novel object tests, between the first and second test for both groups. The duration of contact with the novel object before and after handling in both groups was analyzed using a paired t-test. The difference in pre- and post-handling on FCM concentrations within groups was analyzed using a Wilcoxon test for matched pairs. All analyses were performed according to (Siegel and Castellan, 1988) using the statistical software R 2.13.0 (R Core Development Team, 2012). Differences were considered significant if $P<0.05$.

Results

After the handling period, significantly fewer of the E mink were classified as explorative in the stick test (40%, t=3.67, df=9, $P<0.01$) and more as uncertain (40%, t=-2.45, df=9, $P=0.04$) compared to all being explorative in the first test. There were no significant differences in aggressive ($P=0.34$) or fearful reactions ($P=0.34$) between the two tests. In contrast, the F mink were significantly more explorative after the handling and training (60%, t=-3.67, df=9, $P=0.005$) with a reduction in fearful mink (20%, t=6, df=9, $P<0.001$) compared to all mink being fearful in the first test. There were no differences in aggressive ($P=0.34$) or uncertain ($P=0.34$) between the two tests. The position of the mink in the back of the test cage immediately after the stick test indicates an active movement to avoid the stick and the testing person. The F mink showed a significant change in position, from 1 mink being at the front in the first test to 8 mink in the second test (t=-4.58, df=9, $P=0.001$). There was no difference in position before and after handling in the E mink ($P=0.17$) with 10 mink being at the front in the first test and 8 mink in the second.

Median latency to approach and make contact with the novel object in the novel object test changed in both groups, illustrated in Figure 1. The change in the median in the F mink was significant (approach, W=72.5, $P=0.03$; contact, W=80.5, $P<0.001$), but not in the E mink (approach, $P=0.23$; contact, $P=0.77$).

All mink touched the novel object placed in their home cage in both tests, except for one F mink that did not touch the novel object in the first test. Mink from both groups spent significantly more time in contact with the object after the handling procedure (F: t=-4.32, $P<0.01$; E: t=-3.06, $P=0.02$).

Figure 2 illustrates the median concentrations of the FCM from the samplings prior to and after the handling period for both groups. There was a significant rise in FCM concentrations in the E group (W=9, $P=0.004$) with higher concentrations in samples after the intensive handling period. No difference was found in the F group ($P=0.43$).

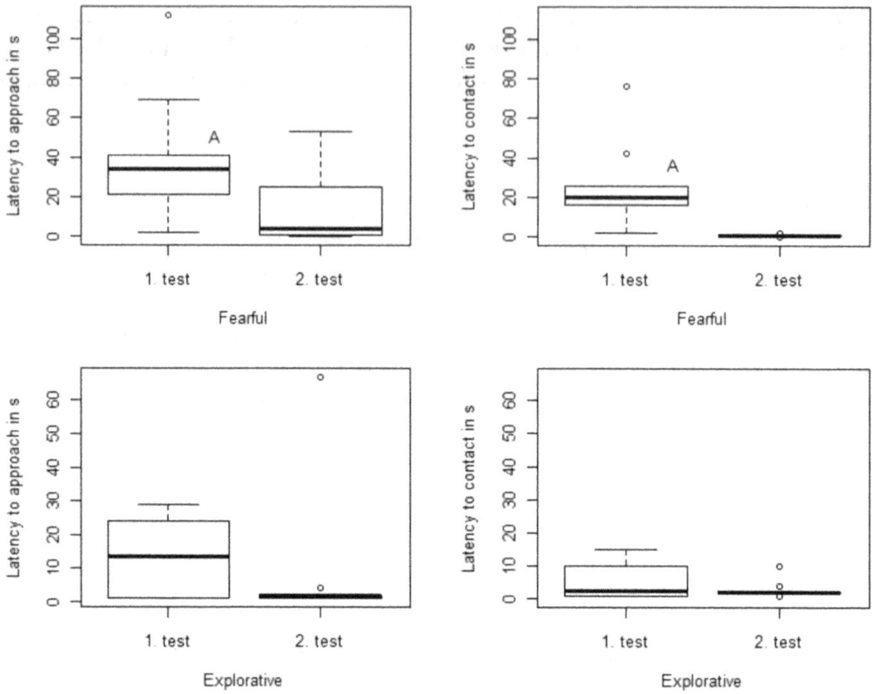

Figure 1. Median latencies to mink's reaction in two novel object tests in fearful and explorative mink, before and after handling. The horizontal lines show the median and the bottom and top of the box the 25th and 75th percentiles. Points are outliers and are plotted individually. When there are outliers the dashed lines show the interquantile range of the data and otherwise they show the max and min. An A indicates that the tests differ significantly (P<0.05).

Discussion

The purpose of handling and training the explorative (E) and fearful (F) mink was to prepare them for a cognitive bias test, a novel indicator of animal welfare. However, the handling and training had an effect on (1) measures of the affective state of the mink and (2) these effects were not in the same direction for the two types of mink. The F mink became less fearful and more explorative as measured in the stick test after a period of handling and training, and they were faster at approaching and touching the novel object as well as spending more time in contact with it. Contrary, even though the E mink spent more time in contact with the novel object in the second test and the fearfulness did not increase significantly, fewer were scored as explorative and more as uncertain in the stick test after the handling procedure. Additionally the E mink reacted to the handling procedure with an increased concentration of FCM. Thus the training and handling had an apparently positive effect on the affective state of F mink, becoming less fearful and an apparently negative effect on the E mink, becoming less explorative.

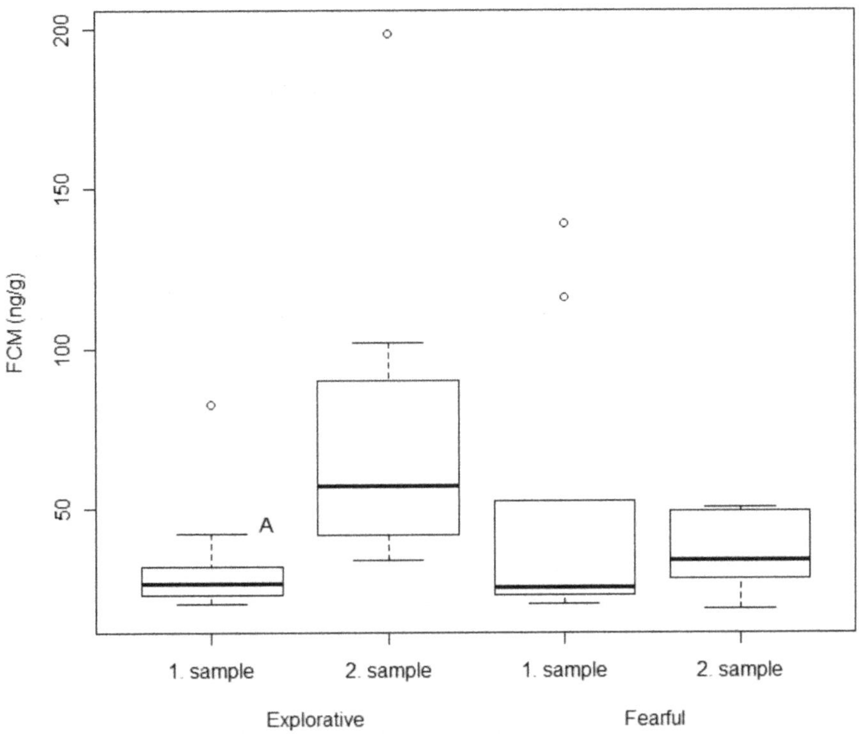

Figure 2. Median concentrations of FCM (ng/g) from one sampling before and one after handling in explorative and fearful mink. The horizontal lines show the median and the bottom and top of the box the 25th and 75th percentiles. Points are outliers and are plotted individually. When there are outliers the dashed lines show the interquantile range of the data and otherwise they show the max and min. An A indicates that the samples within groups differ significantly (P<0.05).

The handling in this study included training on an instrumental learning setup using positive reinforcement (food), i.e. a type of cognitive enrichment, likely to influence the affective state (Manteuffel *et al.*, 2009). The F mink showed reduced fearfulness, possibly due to a combination of the learning setup and habituation i.e. experience with the handling and training procedure as well as the ability to generalize across tests. Malmkvist and Hansen (2002) have showed that within two groups of mink, genetically selected for being either explorative or fearful, the animals generalized their fear responses across an exposure of six different tests becoming more explorative.

The elevated level of FCM concentration in the E mink suggests that they (1) perceived the handling and training as aversive or (2) that they were frustrated as an anticipated reward (in the following learning task) was unexpectedly absent (Burman *et al.*, 2008). The shift of some E mink from being categorized as explorative in the first test to uncertain in the second supports an aversive reaction. Classified as 'uncertain' in the stick test reflect that the observer is not able to classify the response unambiguously, but not necessarily that the mink experience an aversive situation. An aversive reaction would probably involve increased fear, whereas frustration more likely would be categorized by different motivational reactions in the stick test (uncertain). It would be interesting to investigate the behaviour of the animals prior to each handling and training session to determine whether they anticipate a reward (in the following learning task). Expectation of a positive reward has been found to increase the activity of mink in the cage (Hansen and Jeppesen, 2006) possibly causing inattentiveness towards the stick.

Fear and anxiety are negative affective states and in this study, F mink became less fearful thus moving towards a more positive affective state after handling and training. E mink showed the opposite, being less explorative. The handling and training procedure thus had a different effect on the animals depending on their initial affective state. Even though the two groups were subjected to the exact same treatment, they responded differently. There are some factors that need to be investigated further such as the inclusion of a non-handled control group and an investigation of anticipatory behaviour prior to each handling session.

Taken together, studies that involve induced affective states and a lot of training of the animals to assess their welfare, such as the cognitive bias method, need to be interpreted carefully as the handling and training has a different effect on animals in different affective states. The addition of established welfare indicators would support the interpretation.

References

Boissy, A., Manteuffel, G., Jensen, M.B., Moe, R.O., Spruijt, B., Keeling, L.J., Winckler, C., Forkman, B., Dimitrov, I., Langbein, J., Bakken, M., Veissier, I. and Aubert, A., 2007. Assessment of positive emotions in animals to improve their welfare. Physiology and Behavior 92: 375-397.

Burman, O.H.P., Parker, R.M.A., Paul, E.S. and Mendl, M., 2008. Sensitivity to reward loss as an indicator of animal emotion and welfare. Biology Letters 4: 330-333.

Doyle, R.E., Fisher, A.D., Hinch, G.N., Boissy, A. and Lee, C., 2010. Release from restraint generates a positive judgement bias in sheep. Applied Animal Behaviour Science 122: 28-34.

Frigerio, D., Dittami, J., Mostl, E. and Kotrschal, K., 2004. Excreted corticosterone metabolites co-vary with ambient temperature and air pressure in male Greylag geese (*Anser anser*). General and Comparative Endocrinology 137: 29-36.

Hagen, K. and Broom, D.M., 2004. Emotional reactions to learning in cattle. Applied Animal Behaviour Science 85: 203-213.

Hansen, S.W. and Jeppesen, L.L., 2006. Temperament, stereotypies and anticipatory behaviour as measures of welfare in mink. Applied Animal Behaviour Science 99: 172-182.

Harding, E.J., Paul, E.S. and Mendl, M., 2004. Animal behavior – Cognitive bias and affective state. Nature 427: 312.

Malmkvist, J. and Hansen, S.W., 2002. Generalization of fear in farm mink, *Mustela vison*, genetically selected for behaviour towards humans. Animal Behaviour 64: 487-501.

Malmkvist, J., Jeppesen, L. and Palme, R., 2011. Stress and stereotypic behaviour in mink (*Mustela vison):* A focus on adrenocortical activity. Stress-the International Journal on the Biology of Stress 14: 312-323.

Manteuffel, G., Langbein, J. and Puppe, B., 2009. Increasing farm animal welfare by positively motivated instrumental behaviour. Applied Animal Behaviour Science 118: 191-198.

Matheson, S.M., Asher, L. and Bateson, M., 2008. Larger, enriched cages are associated with 'optimistic' response biases in captive European starlings (*Sturnus vulgaris*). Applied Animal Behaviour Science 109: 374-383.

Mendl, M., Brooks, J., Basse, C., Burman, O., Paul, E., Blackwell, E. and Casey, R., 2010a. Dogs showing separation-related behaviour exhibit a 'pessimistic' cognitive bias. Current Biology 20: R839-R840.

Mendl, M., Burman, O.H.P. and Paul, E.S., 2010b. An integrative and functional framework for the study of animal emotion and mood. Proceedings of the Royal Society B-Biological Sciences 277: 2895-2904.

Mendl, M., Burman, O.H.P., Parker, R.M.A. and Paul, E.S., 2009. Cognitive bias as an indicator of animal emotion and welfare: Emerging evidence and underlying mechanisms. Applied Animal Behaviour Science 118: 161-181.

Palme, R. and Mostl, E., 1997. Measurement of cortisol metabolites in faeces of sheep as a parameter of cortisol concentration in blood. Zeitschrift fur Saugetierkunde-International Journal of Mammalian Biology 62: 192-197.

Paul, E.S., Harding, E.J. and Mendl, M., 2005. Measuring emotional processes in animals: the utility of a cognitive approach. Neuroscience and Biobehavioral Reviews 29: 469-491.

Puppe, B., Ernst, K., Schon, P.C. and Manteuffel, G., 2007. Cognitive enrichment affects behavioural reactivity in domestic pigs. Applied Animal Behaviour Science 105: 75-86.

R Core Development Team, 2012. R version 2.13.0. In: www.r-project.org.

Siegel, S. and Castellan, N.J., 1988. Nonparametric statistics for the behavioral sciences. McGraw-Hill, New York, N.Y.

Touma, C. and Palme, R., 2005. Measuring fecal glucocorticoid metabolites in mammals and birds: The importance of validation. Ann N Y Acad Sci. 1046: 54-74.

Willner, P., 1997. Validity, reliability and utility of the chronic mild stress model of depression: a 10-year review and evaluation. Psychopharmacology 134: 319-329.

Discrimination between the domesticated and the feral American mink in Poland based on the left track of a forelimb

M. Sulik[1], A. Radek[1] and P. Bilski[2]
[1]West Pomeranian University of Technology Szczecin, WBiHZ, Department of Fur Animal Breeding,Poland; [2]Drawa National Park, Poland; msulik@zut.edu.pl

Abstract

In Poland, American mink lives in a natural environment as a non-indigenous species which occupied the ecological niche left by European mink (*Mustela lutreola*). The first signals of American minks appearing in the wild date back to 1962-1963. According to some authors in the 1980's there were introductions of minks from fur-farms what resulted in modifications of genetic structures in the population of wild minks. They also claim that up to this day it is not known how the process of local population's augmentation ran or in what ways farm minks influenced the population of the wild ones. The aim of study was defining the differences between feral and currently bred in farms American minks. They were based on the track of a forelimb left by both types of minks. The study included minks caught in their natural habitat and taken from fur-farm. Both places where the biological material was obtained were in the same area, i.e. Western Pomerania in Poland. Modelling clay was used to show differences in track's size between the feral and the farm mink. Artificial imprints of left front mink's paw was made in it and photographed. As a result of studies found that basing on the metacarpal pad imprint left by a mink, it is possible to assess its origin. The feature for the farm minks ranged between 128.28-167.98 mm^2 and for the feral minks 75.99-106.1 mm^2. Therefore, it can be assumed that finding an imprint of a surface above 120 mm^2 can be a foundation to establish the track's origin as the farm mink's. A good indicator to establish a track's origin, if it is wild or farm mink's, is analyzing the proportions between the track's height and the height or the width of the second digital pad.

Keywords: paw imprint, American mink, farm, feral

Introduction

American mink is an animal which has precious and valuable fur that has been desired by people for many years. It used to occur in the wild in North America, but due to its attractive fur it became a farm animal. Breeding American minks dates back to the 19th century, but its real rise began in 20th century. The development of farming allowed its legal regulation what resulted in recognizing American mink as a domesticated species and livestock in Polish legislation.

Animal husbandry led to changes which helped to achieve features especially desired by men. Not only animals' body size, weight, features and colour of fur have undergone these changes, but also their psychological characteristics. All those transformations within the species are typical examples of changes which appear as a result of domestication (Filistowicz and Kuźniewicz, 1999).

In Poland, American mink lives in a natural environment as a non-indigenous species which occupied the ecological niche left by its European ancestor – European mink (*Mustela lutreola*). In the 1920's and 1930's American mink was brought to fur-farms in Europe, but it was not introduced in Poland until the 1950's, when the intensive growth of its breeding began in other European countries.

The first signals of American minks appearing in the wild date back to 1962-1963 (Ruprecht *et al.*, 1983). Most likely they were moving populations of minks from Russian fur-farms, where between 1933-1977 they were released into the wild (Birnbaum, 2006). In those years Russians introduced 21 300 minks, which developed into herds of wild American minks. The biggest population of minks was set free in Latvia in 1944. Thanks to favourable European climate this species adopted in our ecosystem and using the wildlife corridor it moved through east and middle parts of Poland to the west.

According to some authors (Brzeziński and Marzec 2003; Zalewski *et al.*, 2010, 2011) in the 1980's there were introductions of minks from fur-farms what resulted in modifications of genetic structures in the population of wild minks. They also claim that up to this day it is not known how the process of local population's augmentation ran or in what ways farm minks influenced the population of the wild ones.

In order to elucidate this issue, research defining the differences between feral and currently bred in farms American minks were undertaken. They were based on the track of a forelimb left by both types of minks.

Material and methods

The study included minks caught in their natural habitat (10 male and 22 female) and taken from fur-farm (10 male and 10 female). Both group were located in the same area, i.e. Western Pomerania in Poland.

The study material was divided into four groups, i.e. feral females, farm females, feral males and farm males. The animals were weighed with an accuracy of 1 g. Modelling clay was used to show differences in track's size between the feral and the farm mink. Artificial imprints of left front mink's paw was made in it and photographed. The images were specially processed and measured with a computer program Axio Version Rel. 4.8.2. by Zeiss.

Seven measurements were made for the tracks obtained this way (Figure 1):

1. vertical distance of the track – track's height;
2. horizontal distance of the track – track's width;
3. height of the second digital pad;
4. width of the second digital pad;
5. surface of the second digital pad;
6. height of the metacarpal pad;
7. surface of the metacarpal pad.

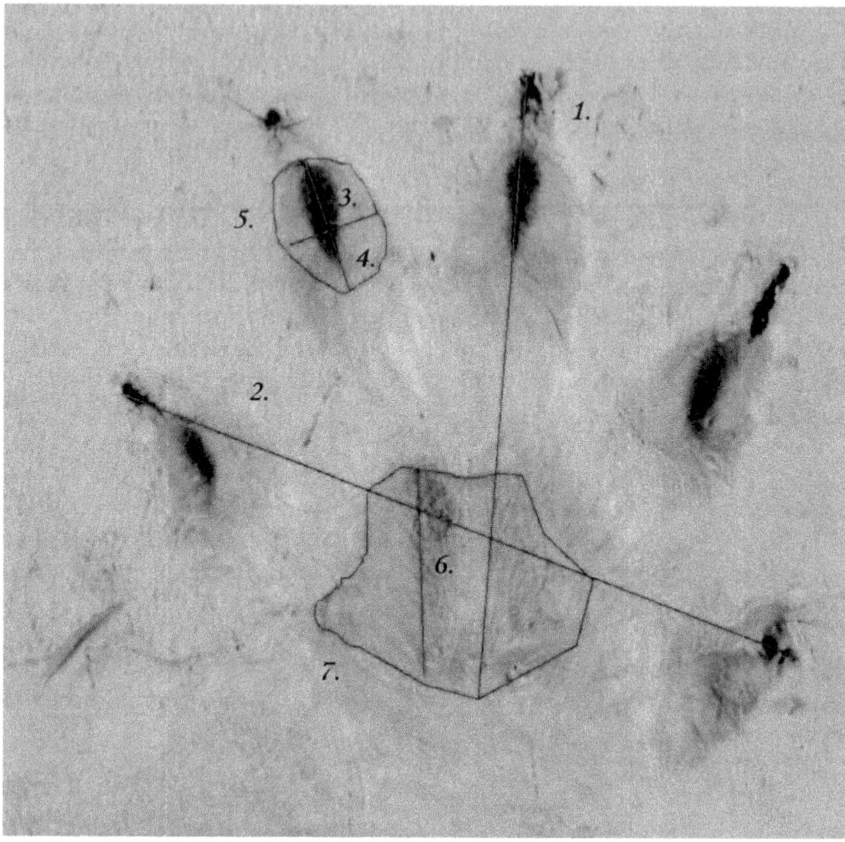

Figure 1. Places where the measurements of the mink's forelimb were made.

The obtained data were statistically assessed using the computer program Statistica 9.1 with the use of multi-factor analysis of variance and Duncan's range test.

Results

In Table 1 the results of mink's body weight were collated regarding the sex and origin. American minks are characterized by distinct sexual dimorphism, therefore, it was necessary to divide the groups additionally according to their sex. Comparison of body weight within sex showed that the farm population was twice as heavy as the one living in the feral. Maximum body weight of wild living minks was lower than the minimum value of that trait among farm animals. The diversity of the trait was on approximate level in all tested groups. This study shows that there are substantial differences in both examined populations of animals: feral and farm. The differences statistically significant with $P \geq 0.01$ were shown between all tested groups apart from the group of feral male and farm female.

Table 1. Feral and farm minks' body weight.[1]

	Female			Male		
	feral	farm	mean	feral	farm	mean
Average	752*	1,415*	1,101	1,388*	2,757*	1,865
Max	945	1,870	1,870	1,870	3,460	3,460
Min	599.5	1.110	599.5	960	2.130	960
Sd	121.5	258.3	394.4	271.8	406.4	723.9
V%	16.15	18.26	35.82	19.58	14.74	38.8

[1] * differences statistically significant with $P \leq 0.01$.

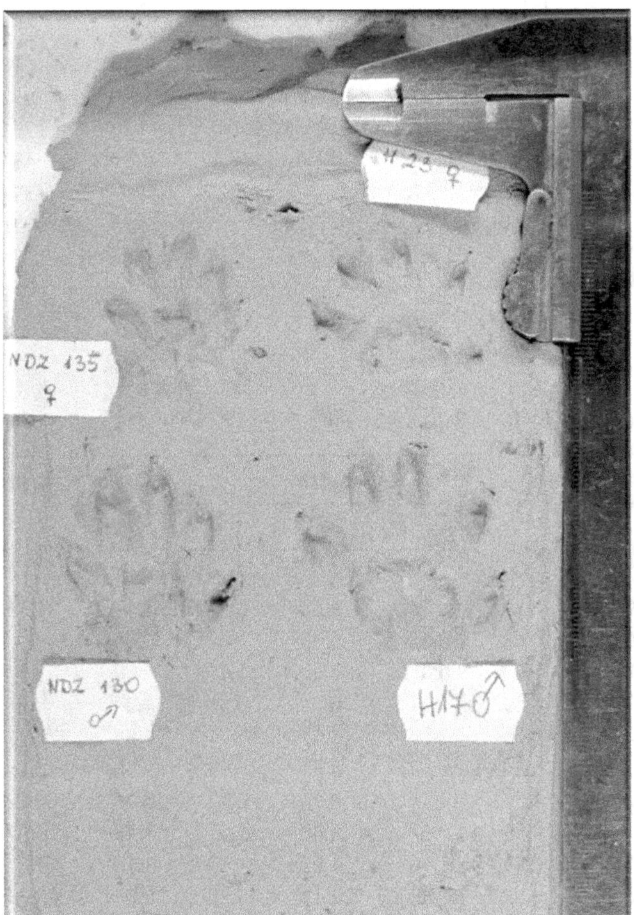

Figure 2. Minks' tracks imprinted in modelling clay: NDZ – wild minks, H – farm minks.

Seven measurements of the tracks imprinted in modelling clay were made and then set in tables with a division into:

- made on the whole track (Table 2);
- made on the second digital pad (Table 3);
- made on the metacarpal pad (Table 4).

Table 2. Measurements of the whole mink's forelimb track.[1]

Sex		Height			Width		
		feral	farm	mean	feral	farm	mean
Female	X	28.2*	35.74*	32.17	23.81	31.77	28
	Max	31.5	38.18	38.18	28.19	33.87	33.87
	Min	24.33	33.43	24.33	21.06	29.32	21.06
	Sd.	2.02	1.71	4.27	2.14	1.6	4.47
	V%	7.16	4.78	13.27	8.99	5.04	15.96
Male	X	32.58*	41.45*	35.75	28.78	36.79	31.64
	Max	35.71	44.2	44.2	31.59	43.28	43.28
	Min	28.43	37.9	28.43	24.6	33.83	24.6
	Sd.	2.46	2.14	4.9	2.25	2.77	4.58
	V%	7.55	5.16	13.7	7.82	7.53	14.48

[1] * statistically significant differences with $P \leq 0.01$.

Table 3. Measurements of the second digital pad of the mink's forelimb.[1]

sex		height			width			surface		
		feral	farm	mean	feral	farm	mean	feral	farm	mean
female	X	9.09^{A}	8.35^{B}	8.7	5.28	3.22	4.19	37.55	20.32	28.48
	Max	10.67	10.68	10.68	6.25	4.62	6.25	48.06	35.65	48.06
	Min	7.19	6.6	6.6	3.4	1.69	1.69	18.54	10.13	10.13
	Sd.	1.17	1.39	1.31	0.84	0.86	1.34	8.55	9.07	12.32
	V%	12.87	16.65	15.06	15.91	26.71	31.98	22.77	44.64	43.26
male	X	9.07	10.33^{AB}	9.52	5.51	5.65	5.56	40.26	43.38	41.38
	Max	12.13	11.51	12.13	8.19	7.13	8.19	69.22	68.88	69.22
	Min	5.67	8.69	5.67	3.43	4.07	3.43	16.07	25.48	16.07
	Sd.	1.68	0.88	1.56	1.22	1.09	1.16	17.01	12.12	15.28
	V%	18.52	8.52	16.39	22.14	19.29	20.86	42.25	27.94	36.93

[1] A, B, C – differences statistically significant with $P \leq 0.01$.

Table 4. Measurements of the metacarpal pad of the mink's forelimb.[1]

Sex		Height			Surface		
		feral	farm	mean	feral	farm	mean
Female	X	9.47*	10.58	10.06	75.99*	128.28*	103.51
	Max	10.95	13.72	13.72	87.52	171.64	171.64
	Min	7.44	7.35	7.35	61.58	94.13	61.58
	Sd.	1.17	1.97	1.69	7.25	27.19	33.36
	V%	12.35	18.62	16.8	9.54	21.2	32.23
Male3	X	10.73	12.04*	11.2	106.1*	167.98*	128.2
	Max	13.86	16.07	16.07	147.42	232.1	232.1
	Min	8	7.17	7.17	54.63	109.16	54.63
	Sd.	1.66	2.52	2.07	24.27	41.72	43.16
	V%	15.47	20.93	18.48	22.87	24.84	33.67

[1] * differences statistically significant with $P \leq 0.01$.

The height of the mink's track varied between 24.3-44.2 mm showing big diversity within the sex as well as the origin. Similar differences were observed in the width of the track which varied between 21.06-42.28 mm. Differences within sex, between feral and farm minks, were for the track's height 21.1% (females) and 21.4% (males), and for the track's width 25% (females) and 21.8% (males). The smallest difference was between the farm female's track and the feral male's track (height – 8.8%, width – 9.4%), where the farm female's track was higher and wider than the feral male's one. The measurements' differences were statistically verified between all the tested groups.

The measurements of the second digital pad showed that although the whole track was bigger, the digital pads of farm females were smaller in size (width and surface) than those of the feral females in most of the cases. However, both of the male groups had similar sizes of the digital pads. In the female group the differences were 8.1% for height, 39% for width and 45.9% for the track's surface. In males' group the biggest difference was in the height of the second digital pad and amounted 12.2% in farm males' favour.

The measurements of the metacarpal pad's height showed that it is lower over about 10% among wild animals within sex. The smallest height of the trait was observed among farm females (7.35 cm) and the biggest among farm males (16.07 cm). Great diversity of the trait among all tested groups (12.35-20.93%) indicates its instability in both tested populations. The differences between the feral females group and the farm males were statistically verified with $P \leq 0.01$.

The measurements of metacarpal pad's surface showed that the wild living animals have decidedly smaller surface of metacarpal pad and the differences between them reach 41.2% among males and 36.8% among females. Between all the tested groups the study showed highly statistically significant differences with $P \leq 0.01$. Interestingly, in the scope of the trait, the feral males as well as females got lower parameters than the farm animals. The difference between the farm females and the feral females was 22.18 cm^2 (i.e. 17.3%). Statistically significant differences (with $P \leq 0.01$) were shown between all the tested groups apart from the group of feral males and farm females where the differences were statistically insignificant with $P \leq 0.05$. Therefore, the obtained results can be a foundation for estimating the origin of a track.

It seemed important to draw proportions between the made measurements. They were collated in Table 5. Basing on these calculations, it can be estimated that the shape of the track left by a mink sustains its proportions irrespective of the origin and the sex. However, distinct changes occurred between the track's height and the width of the second digital pad, which mostly differed between feral and farm females (respectively 5.21 and 11.96). The reference of the track's height proportions to its elements appears to be a better parameter to describe its origin than the analysis of singular measurements.

Table 5. Proportions between chosen measurements of feral and farm minks' tracks.[1]

sex		HT/WT			HT/HFP2			HT/WFP2			HT/HMP			HFP2/WFP2		
		feral	**farm**	**mean**	**feral**	**farm**	**mean**	**feral**	**farm**	**mean**	**feral**	**farm**	**mean**	**feral**	**farm**	**mean**
Female	X	1.19	1.13	1.16	3.05*	4.38*	3.75	5.21*	11.96*	8.76	2.99*	3.5*	3.26	1.73*	2.71*	2.24
	Max	1.24	1.17	1.24	3.76	5.34	5.34	5.82	20.5	20.5	3.67	4.77	4.77	2.13	4.00	4.00
	Min	1.12	1.08	1.08	2.61	3.16	2.61	4.45	7.31	4.54	2.55	2.52	2.52	1.42	2.21	1.42
	Sd.	0.04	0.03	0.05	0.38	0.67	0.86	0.37	3.69	4.34	0.37	0.77	0.65	0.23	0.57	0.66
	V%	3.36	2.65	4.31	10.1	15.3	23.2	7.1	30.85	49.5	12.4	22.0	17.2	13.3	21.03	29.46
Male	X	1.13	1.13	1.13	3.74*	4.03*	3.85	6.27	7.56*	6.73*	3.09	3.59*	3.27	1.69	1.88	1.75
	Max	1.24	1.19	1.24	5.34	4.71	5.34	10.19	9.53	10.19	3.78	5.58	5.58	2.52	2.37	2.52
	Min	1.08	1.02	1.02	2.88	3.63	2.88	4.28	5.88	4.28	2.46	2.66	2.46	1.24	1.52	1.24
	Sd.	0.04	0.05	0.04	0.63	0.34	0.69	1.34	1.3	1.45	0.36	0.83	0.61	0.30	0.31	0.31
	V%	3.54	4.42	3.54	16.8	8.4	17.9	21.4	17.2	21.5	11.7	23.1	18.7	17.75	16.5	17.71

[1] * differences statistically significant with $P \leq 0.01$. HT = height of track; WT = width of track; HFP2 = height of second digital pad; WFP2 = width of second digital pad; HMP = height of metacarpal pad.

Conclusions

Based on the metacarpal pad imprint left by a mink, it is possible to assess its origin, i.e. if it is a wild living animal or a runaway from a farm. The feature for the farm minks ranged between 128.28-167.98 mm^2 and for the wild minks 75.99-106.1 mm^2. Therefore, it can be assumed that finding an imprint of a surface above 120 mm^2 can be a foundation to establish the track's origin as the farm mink's.

A good indicator to establish a track's origin, if it is wild or farm mink's, is analyzing the proportions between the track's height and the height or the width of the second digital pad. The relation between height and width of the foot print is a better measure for discrimination between feral and farmed mink than the single measurements. These differences were confirmed statistically significant with $P \leq 0.01$. If the proportions exceed the second digital pad's height four times or its width seven times, then the mink's track points to its farm origin. Lower results may be a foundation to describe a mink's origin as wild.

Acknowledgements

This study was financed by the National Centre for Research and Development, development project no 12-0140-10.

Welfare consequences of digging substrates in blue foxes

H. T. Korhonen, P. Eskeli and H. Huuki
MTT Agrifood Research Finland, Animal Production Research, Finland; hannu.t.korhonen@mtt.fi

Abstract

The aim of the present study was to evaluate welfare implications of digging substrates on blue foxes (*Vulpes lagopus*) in wire cages. Digging substrates compared here were sandbox and digging plates placed on cage floor and cage wall. Digging in sand was considered as a goal-directed behaviour. Plates provided foxes possible to perform digging behaviour *per se*. Each group comprised of 20 juveniles housed in pairs. The results showed that motivation to dig for a specific goal was no greater than to perform digging behaviour *per se* only. There was a significant difference in dirtiness of the fur between the groups (P<0.001). The pelts originating from animals in sandbox cages were dirtier than those from other groups. Plates on cage wall and floor did not negatively affect fur quality. Behavioural recordings did not reveal marked differences between the groups. The weight gain in animals with digging plates was similar to that in animals housed in standard cages. The variation in body weight was highest in animals with a sandbox. No statistical difference was found in blood picture between experimental groups. There was a slight tendency (P=0.07) for adrenal glands to be heaviest in animals from the sandbox group. Digging plates were considered to be more suitable as digging substrates than a sandbox.

Keywords: Digging, animal welfare, motivation, fur animal production

Introduction

A major factor contributing to the welfare of a farmed animal is its ability to fulfil its behavioural needs under given housing conditions (Dawkins, 2004). As not all environmental features occurring in the wild can be provided under farm environment, the animals must given enrichments that satisfy their most important needs (Dawkins, 1990). Clarification of farm foxes' need to dig is an essential and urgent issue at present. This is because of the implementation of new welfare legislation. Crucial is the question how to provide foxes possibility to dig? In the wild, foxes are digging for a certain specific purpose, i.e. make a den or cache and retrieve food. Such goal-directed behaviour is not essential for survival on farm conditions. However, digging as such may be rewarding, i.e. it provides sensory feedback from the feet and muscles. It would seem that it is not the purpose of digging behaviour that is important for wellbeing but rather the possibility to dig *per se* (Korhonen *et al.*, 2003, 2004). If this is the case, then the need to dig could be satisfied by providing a simple digging substrate such as a plate, not an entire earthen floor or even sandbox.

Our study sought to clarify farmed blue foxes (*Vulpes lagopus*) digging motivation on two different digging substrates, i.e. a sandbox and plates. Plates were placed on the cage floor and on the cage wall. They would permit foxes to dig without achieving a tangible goal such as a nest or finding food. A box with sand was provided to permit the animals to perform goal-directed digging. In addition, second aim was to establish the welfare implications of digging substrates studied.

Materials and methods

The study comprised four experimental groups housed in four different cage setups: (1) a standard cage (105 cm long×115 cm wide×70 cm high) without digging substrates, which housed the control group; (2) a standard cage with solid metal plate (210 mm×297 mm) on the wall for digging and scratching; (3) a standard cage with a solid metal plate (210 mm×297 mm) on the floor for digging and scratching; (4) a standard cage with a metal sandbox for digging and scratching (80 cm×40 cm×14 cm; L×W×H). The sandbox had a 10 cm layer of sand (ca. 25 kg, particle size 0-18 mm). The experimental groups were formed at weani9ng on 22nd July. There were 20 foxes in each group, one male and one female per cage at weaning, a platform (45 cm from the floor) and a wooden block (diameter 7 cm, length 35 cm) were also placed in the cages. Animals were fed equally large portions (750-1000 g daily) of fresh fox feed.

Animals were weighted four times during the experiment (July 22, Sept 3, Oct 1, Nov 25). Capture time as well behavioural reaction to capture were simultaneously evaluated. The dirtiness of the digging substrates was checked nine times during the study. The reactions of the foxes to humans were evaluated by a walking test. The behavior of the foxes was video recorded twice during the experiments (Aug 12, Sept 9). A ball test was conducted to evaluate the explorative behavior. Blood samples were used to determine haemoglobin (Hb) and haemotocrit (HCT) levels, the red blood cells (RBC) count and the white blood cell (WBC) count. At pelting, carcasses were prepared and the adrenal glands, liver and heart were weighed. Organ weights are presented as actual weights.

Statistical analyses were made with the SAS system for Windows 9.1 and SAS Enterprise Guide 3.0. The general linear model (GLM) and the Kruskal-Wallis ANOVA were employed for testing.

Results

Food intake was the same in all groups, i.e. 750-1000 g/animal daily. Basic data is given in Table 1. Weight development was normal in all groups. The foxes with a sandbox weighed the least and those with a plate on the wall most at the end of the study, but there were no statistical differences between the groups.

No statistical difference was found in organ weights between the groups, although a slight tendency (P=0.07) was noted for the adrenal glands to be heaviest in the sandbox group. The adrenal glands tended (P=0.07) to be lightest in the group with a digging plate on the wall. The animals remained healthy throughout the study. The blood picture parameters showed that general health was good in all groups. There was no statistical difference in Hb, HCT, WBC and RBC values between the groups.

The pelts from the foxes housed in cages with a sandbox were dirtier than those from the foxes housed in standard cages or cages with a digging plate on the wall. The pelts from the foxes in cages with digging plates on the floor were also dirty. The pelts from foxes housed in cages with a sandbox lost the most grading points. The sandbox was the dirtiest and the plate on the wall the cleanest.

Table 1. Basic data at pelting on Nov 25th. N=20 blue foxes per group.

Variable	Standard	Plate (wall)	Plate (floor)	Sandbox
Body weight, kg	13.8±0.7	14.4±0.6	14.0±0.2	13.3±1.3
Adrenal gland, mg	427±98	379±79	395±88	450±85
Heart, g	47.2±4.2	42.6± 4.7	46.3±4.3	42.6±4.5
Liver, g	348±95	388±95	383±92	358±111
Fur quality	5.9±1.8	6.8±1.2	6.3±1.6	6.2±1.7
Capture time, s	1.9±0.8	1.8±0.6	2.0±0.6	2.2±1.2

Only a small number of foxes (5-10%) in any group responded confidently to the feeding test. The foxes in cages with a digging plate on the wall tended to be the most confident. Only a couple of animal per group (10%) showed explorative behavior during the ball test. The foxes in cages with a sandbox seemed to be the least keen to explore a novel object. The amount of digging for sandbox and plate on the floor was 12 min/animal daily and for a plate on the wall 18 min/animal daily (not significant difference between groups). Any signs of digging behavior were not observed in animals of standard group.

Discussion

The behavioural repertoire of each species has evolved to be typical of that species and ensure survival under the prevailing living conditions. The blue fox may have two different digging motivations. First, digging for a certain survival-supporting goal. This is a typical pattern in the wild where foxes dig for a shelter or for catching/searching food. On farms, on the other hand, digging may have little to do with survival because farmer provides the foxes with shelter and regular food. Second, digging because digging *per se* may be rewarding, i.e. it provides sensory feedback from feet and muscles (Korhonen *et al.* 2003; 2004). The crucial question here is: do farmed foxes have a need to dig? (Korhonen *et al.*, 2001a). Previous debrivation study (Korhonen *et al.*, 2004) documented that performing digging behavior *per se* is rewarding under farm conditions. In the present study, digging plates permitted the foxes to dig without any visible goal. The sandbox, on the other hand, gave the foxes a chance to engage in goal-directed behaviours. Goal digging behavior here was defined as foxes digging in sand. Our previous study (Korhonen *et al.*, 2004) showed that the sandbox may serve as an actual digging substrate tracing an earthen floor.

The existence of a digging substrate in a cage allows the foxes to dig at will. Thus, it is their choice. Previous studies have shown that the time spent by farmed foxes on digging typically ranges from 4 to 15 min daily (Hovland and Bakken, 2000; Korhonen *et al.*, 2001a, 2003, 2004). In the present study, the amount of digging for sandbox and plate on the floor was 12 min and for a plate on the wall 18 min daily. So, these values fit well with the previous ones. They also reveal that the motivation to dig for a specific goal (sandbox group) is no greater than the motivation to perform digging behavior *per se* only (plate groups). Plates thus suffice to satisfy the foxes' need to dig under farm conditions. Furthermore, the amount of daily digging is actually rather low.

The motivation to eliminate was highest on sand. The digging plate on the wall, on the other hand, remained clean throughout the study. Its use for digging tended to be also highest. It would seem then that a plate on the wall could compensate for a sandbox as equipment for digging and scratching (Korhonen *et al.*, 2003). It can also be recommended in terms of hygiene. Evaluation of fur properties revealed that the coat was dirtiest in the foxes with a sandbox in their cage. The same result was found also in our previous study (Korhonen *et al.*, 2003). In other fur parameters, there were no significant differences between the groups. However, fur quality was most variable in the foxes with a sandbox. Thus, there were several foxes with lower quality pelts but also those with normal quality pelts in the sandbox group. Such variation is not acceptable from the farmer's point of view.

Prolonged stress is known to increase the size of the adrenal glands (Korhonen *et al.*, 2001b). In the present study, we found a certain tendency for the adrenals to be heaviest in animals with a sandbox and lightest in those with a plate on the wall. This finding coincides well with the dirtiness of the sandbox and plate and also with the dirtiness of the animals in these groups. We are thus tempted to conclude that dirtiness of both the digging substrate and the fur coat may in the long run be stressful to foxes. A dirty fur coat is undoubtedly unpleasant for foxes. Not only is it a problem from the hygienic point of view but it also may lower the thermophysical properties of the coat and thus prevents normal heat regulation (Korhonen, 1987). Our previous study (Korhonen *et al.*, 2001c) clearly showed that foxes kept in ground-floor pens had very dirty fur coats. These animals were also most stressed. In the present study, the digging substrates tested did not make any difference to the difficulty of catching the foxes in the cages. However, the reactions of the foxes at capture were not totally unambiguous.

No significant difference was found in body weights of animals between study groups. However, slight tendency was seen that final weights were lowest in animals from sandbox group. Animals with sandbox had dirtiest furs which may decline thermophysical properties of fur coat (Korhonen, 1987). However, this study was finished already in November. Until that the ambient air temperature in Finland typically is not very cold. Coldest part of winter normally is during January-March. Therefore, declined thermoregulation due to dirty fur coats cannot be seen here as a significant effect in body weights.

It can be concluded that a plate located on the cage wall can be considered the most suitable digging substrate for farmed blue foxes. A sandbox on the cage floor is least recommended because it makes the fur coat dirty and may therefore present problems for hygiene and thermoregulation.

Acknowledgements

This study was funded by the Finnish Fur Breeders' Association and MTT Agrifood Research Finland. The staff of MTT Kannus is acknowledged for valuable help in carrying out this experiment.

References

Dawkins, M.S., 1990. From animal's point of view: motivation, fitness and animal welfare. Behavioural Brain Science 13: 1-61.

Dawkins, M.S., 2004. Using behavior to assess animal welfare. Animal Welfare 13: 3-7.

Hovland, A-H. and Bakken, M., 2000. The welfare situation of farmed foxes in relation to domestication status and compared to other farmed species. Agricultural University of Norway (NLH), Department of Animal Science, 88 pp.

Korhonen, H., 1987. Significance of sleeping plate as a thermal protection for farmed raccoon dogs (*Nyctereutes procyonoides*). Comparative Biochemistry and Physiology 87A: 631-633.

Korhonen, H., Niemelä. P. and Wikman, I., 2001a. Extent of digging and its possible underlying causal factors in penned blue foxes. Acta Ethology 3: 127-133.

Korhonen, H.T., Jauhiainen, L., Niemelä, P., Harri, M. and Sauna-aho, R., 2001b. Effects of space allowance and behavioural responses in blue foxes (*Alopex lagopus*): Comparisons between space quantity and floor material. Physiology and Behaviour 69: 571-580.

Korhonen, H., Jauhianen, L., Niemelä, P., Harri, M. and Sauna-aho, R., 2001c. Physiological and behavioural responses in blue foxes (*Alopex lagopus*): comparisons between space quantity and floor material. Animal Science 72: 375-387.

Korhonen, H.T., Jauhiainen, L. and Rekilä, T., 2003. In-cage sandbox as a ground substitute for farmed blue foxes (*Alopex lagopus*): Effect of digging activity and welfare. Canadian Journal of Animal Science 83: 703-712.

Korhonen, H.T., Jauhiainen, L., Kokkonen, L. and Rekilä, T., 2004. Digging in farmed blue foxes: essential or not? Annals of Animal Science 4(2): 405-419.

Effect of slurry channels to collect urine and faeces in mink farming

K. Meldgaard and P.F. Larsen
Kopenhagen Farm, Herningvej 112C, 7500 Holstebro, Denmark; khv@kopenhagenfur.dk

Abstract

In practical mink farming slurry channels collect the majority of urine and faeces under the cages. However, no studies have previously focused on the exact ratio of urine and faeces collected by the slurry channels and the ratio that ends up on the area beneath the cages. Therefore the aim of this study was to develop a new method that could collect urine and faeces in specific areas under the cage. The second aim was to determine how much urine and faeces the slurry channel are able to collect, and third to study differences in the way mink depose the urine between young mink in pairs, single female, and single male mink and furthermore faeces for male mink in January and February and describe conceivable seasonal variations for urine deposition. Two different collections chutes were developed so urine and faeces from six cages could be registered in specific sections under the cages. Common for both setups were that both had 8 collecting bottles, i.e. three bottles representing the slurry channel and five bottles the soil area. Ones every 24 hours, urine and faeces was collected and weighted to quantify the amount. The preliminary findings show that the slurry channels collect the majority of the urine and faeces. Results from the three groups in October and November showed that young mink in pairs depose 93.0% of the urine in the slurry channels, whereas female and male mink depose 93.2% and 80.5%, respectively. In conclusion the collection chute can be used to define where the mink depose urine and faeces.

Keywords: urine, faeces, slurry chanel, seasonal variation

Introduction

Farm animals are frequently housed in systems where faeces and urine can be collected so the evaporation of nitrogen to the surrounding environment is kept on a low level. Mink are housed in open cage systems which makes it difficult to control the ammonia emission and drainage through the soil. Slurry channels collect the majority of urine and faeces under mink cages. However, no studies have previously focused on the exact ratio of urine and faeces collected by the slurry channels along with the ratio that ends on the ground beneath the cages. The aim of this study was to develop a method that could determine the effect of the slurry channels ability to collect urine and faeces. The second aim was to study the actual effect of slurry channels to collect urine and faeces. The final aim was to study differences in the way mink depose the urine between young mink in pairs, single female, male mink and describe seasonal variation.

Material and methods

To simulate the slurry channel and the soil area beneath the cages, two collections chutes were developed to collect urine from six cages in specific sections under the cages (see Figure 1A and 1B). The experiment included two different collecting methods to make sure that the deposing place did not change after placing a collecting system beneath the cages.

One collecting chute was composed by two plastic roof sheets witch were placed on a skeleton of wood. As can be seen in Figure 1A the collecting chute one was placed in the same level underneath the six cages from the slurry channel to the nest box. Collection chute two was divided in two sections. As can be seen in Figure 1B the section mimicking the slurry channel was placed in the same level as the slurry channel and the rest was placed close to the ground to mimic natural conditions. Both setups were equipped with 8 collecting bottles placed in a line per every 12 centimetres (see Figure 1). The slurry channels collecting area include 3 bottles.

The study included three groups of mink. 36 young mink in pairs, 12 single female, and 48 single male mink respectively. One registration period include both setups and lasted 4 days. 12 to 24 mink were stabled in each period. Every 24 hours urine was collected and weighed separately. All combinations of urine deposing between the two collecting chutes, sex and age were included and experiments were carried out 15 times. Subsequently registrations of faeces deposing from single male mink were also registered in January and February by placing netting on top of the collection chute to collect manure and ensure separation of the two fractions (5 experiments of 6 males).

To observe any change in faeces deposing after placed in the experiment, young mink in pairs were observed in the farm. To register faeces deposing in the farm netting was placed under the cages and the place of deposing was registered as inside or outside the slurry channel. Subsequently the deposing places in the farm were compared with the deposing spot in the experiment.

Urine experiments were performed in October and November for all three groups and furthermore in January and February for single male mink collecting both urine and faeces.

A B

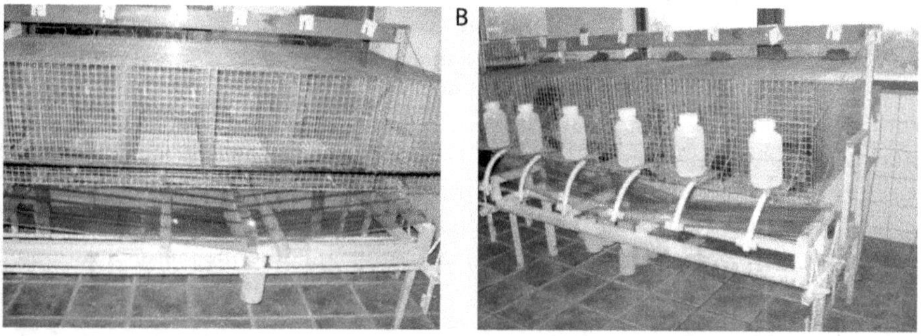

Figure 1. Two types of slurry chanels for collecting urine and faeces. (A) Collection chute; (B) divided collection chute.

To avoid water spillage in the urine bottles water was supplied in bottles with a lesser pressure than in the hose. Earlier studies (Clausen, 2010; Tinggard and Hvam 2010) used this system with drinking bottles in trails where water intake was examined (see Figure 1B). Water bottles were replaced every second day.

Results and discussion

Results from the three groups in October and November showed a significant differences in deposition among young mink in pairs and single female mink deposing 92.8% and 93.2% of the urine in the slurry channel, respectively, and single male mink deposing 80.5% ($P<0.0005$, Figure 2).

The observed difference between the groups could be ascribed to the older males need to mark their territory as they would do in nature. Single females and young mink in pairs are very similar in their deposition in the slurry channel and this most likely reflects that the young male mink is not sexual mature in the fall and therefore do not need to mark their territory in the same way as older males.

As seen in Figure 3, registrations of urine deposition in male mink during January and February showed that male mink in this period deposit more urine in the slurry channel compared to October and November ($P<0.0001$). A hypothesis theory could be the low temperature and dark period reduce activity and therefore less need for marking territory.

Results from the two collections chutes showed no difference in the deposition or quantity of the total amount of urine within the three groups of mink when the data sets were pooled ($P>0.91$). This shows that the different collection chutes do not affect the deposing place when object are placed underneath the cages. The same was observed in faeces deposing on the farm where young

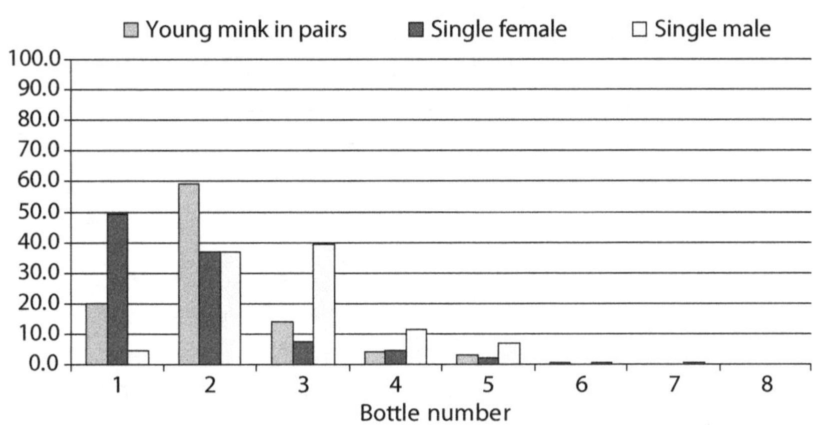

Figure 2. The distribution of urine in the eight collection bottles, October and November. Average is presented in percentages of total amount of urine in the different bottles. Bottle number 1-3 represent the slurry channel and bottle 4-8 the soil.

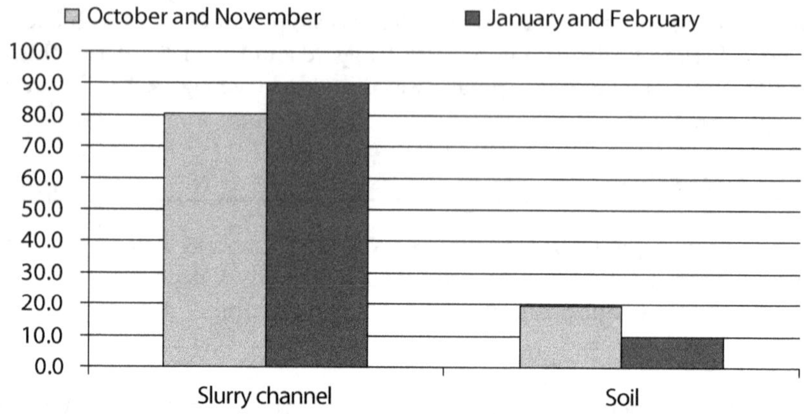

Figure 3. The distribution of urine fore male mink in the period October – November and January –February (in percentages) between slurry channels and the soil.

mink in pairs did not change deposing place after the experimental was started. Furthermore in the faeces collections from male mink registered after racks were placed underneath the cages – this did not change the deposing of faeces mainly in the slurry channel area (Figure 4). A possible cause can be search for ammonium more than a physical place to depose urine and faeces.

Faeces collecting from 30 male mink showed that the slurry channel collect 96.0% of the faeces (Figure 4). Registrations showed that this was due to one male consistently depositing faeces outside the slurry channel, whereas the rest of the males deposited close to 100% in the channels.

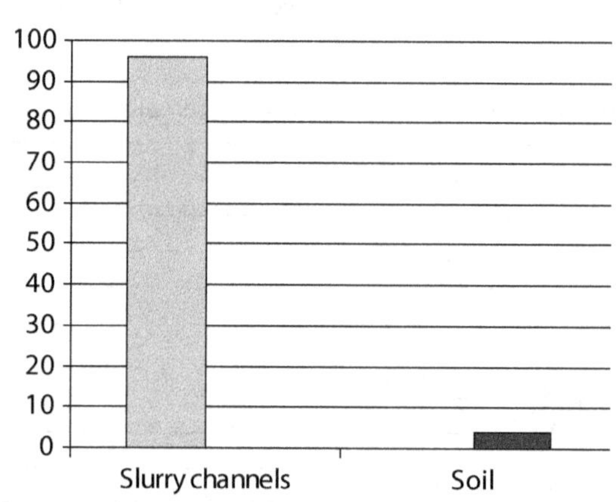

Figure 4. The distribution of faeces from single male mink in January and February (in percentages) between slurry channels and the soil.

Future studies will show to what extent seasonal variation exist in urine and faeces deposition in and outside the slurry channels. But the results so far demonstrate high efficiency of the channels for pairs of young mink and adult female mink.

The findings show that the addition for all groups of mink, the two collecting chutes and seasonal variation the slurry channel collects 89.7% of the urine. This corresponds to Sandbøl and Pedersen (2001) who added a part of the 11.4% they could not explain in a major study of nitrogen cycle in mink production, to be ascribed accumulation in the soil area. This new method shows that collection of urine and faeces can give a good estimate of where the mink depose urine and faeces dependent on sex, age and season. Future studies will provide more information on seasonal variation in deposition.

Conclusions

In conclusion the study shows that this new method had shown that it is possible to give a good estimate of the amount and location of urine and faeces deposited by mink in a standard production system. And the behavioural deposing place is not affected when objects are placed under the cage.

We found that single male mink place a higher amount of urine outside the slurry channel in October and November compared to young mink in pairs and single females. Furthermore the study of faeces deposition showed large individual variation among males.

References

Clausen, T.N. 2010. Water balance in 8 week old mink kits. Annual Report, Danish Fur Breeders Research Center.

Sandbøl P. and Pedersen S. 2001. Ammonia emission from mink farms (In Danish: Ammoniakfordampning fra minkhaller). Husdyrbrug nr. 22.

Tinggaard, L and Hvam K. 2010. Methodology to register water balance in mink. NJF Seminar no.440, September 29 – October 1, Oslo, Norway, Poster presentation.

Black spots in subcutis of mink pelts are no bite marks

F.C. Kleyn van Willigen[1], N.M. Meertens[1], J. De Rond[2] and L. Boekhorst[2]
[1]Animal Health Service, P.O Box 9, 7400 AA Deventer, the Netherlands; [2]Dutch Fur Breeders Organisation, P.O. Box 488, 6600 AL Wijchen, the Netherlands; f.kleyn.v.willigen@gddeventer.com

Abstract

Black spots in the meat side of mink pelts with dark coloured hairs have been found in the inguinal region, the back, the abdomen, the neck or spread over the whole body. The black spots have been explained by fur farmers and researchers as the remains of bite marks. Examination of the pelts learns that the distribution of the black spots did not concur with the places where bite marks usually develop. The density and distribution of the spots ware variable. When started with pelting in the first week of November of the experimental minks kept in different housing systems and finishing pelting the last animals of the same group in December, there was a reduction of occurrence and the density of the black spots was remarkably decreased. In breeder males were no black spots found. The males generally had less black spots than the females, but almost all the minks pelted in December were free of black spots. No real black spots were found in pelts from violet, pearl and white type minks. Histological examination of the black spots does not reveal indications of damage caused by biting, as haemorrhage, inflammation or fibrosis. No remnant iron products were found using the Perl's Prussian blue stain. During the development from summer coat to winter coat, anagen (grow) fase, the bulb portion of the hair follicles is deeply anchored in the superficial part of the subcutaneous fat tissue. If the maturation of the hairs is not yet finished or retarded at the moment that the animals are killed, the hair follicles are recognizable as blue pelts and black spots. This does not happen when the hair follicles are in telogen, hair-containing phase, in which the hair follicles have a more superficial location within the dermis and the bulb portion has a white color. The conclusions of the macroscopical and histological examinations demonstrated that black spots are normal hair follicles in anagen stage and are no bite marks.

Keywords: maturation hair, black spots, bite marks, mink (*Mustela vison*)

Introduction

Black spots are black-brown spots in the subcutis. The black spots can be seen single but also scattered over the body or linked up to groups. Black spots on the leather side of the pelts were always indicated as bite marks. Sometimes a few black spots and sometimes much more or in groups were seen. Black spots in groups were mostly seen above the tail between the hind legs and the neck. The single ones over the whole body as the backside, the breast, the belly or the sides. The spreading patron of the 'black bite marks' was not indicative for bite marks. Researchers in other countries claimed that the 'black bite marks' were seen more in group housing. There is a great difference between new or older bite marks and black spots.

Materials and methods

In the autumn of 2009, 2010 and 2011wild type females and males, kept in group (2 females and 2 males) housing in pairs (female and male) and individual, were selected for research. Also the breeder males in the spring of 2010 and 2011 have been examined. The minks were euthanatized in a mix of CO and fresh air gas. The dead bodies were pelted and scraped. After scraping the leather side of the minks were examined on black spots.

In the pelting period November/December 2009, 1,100 pelts from different pelting dates were judged on the black spots. The minks were kept till December 7th in single, in pairs or in group housing with 4 minks (2 females and 2 males) during the growing period till pelting. In the end of March 2010 were 400 male breeders examined. Sixty percent of these breeder males were raised in pair housing and 40 percent of the males were raised in group housing till December 2009.

In the pelting period November/December 2010 were 2350 pelts from different pelting dates judged on the black spots. The minks were kept till December 7th in single, in pairs or in group housing with 4 minks during the growing period till pelting. In the end of March 2011 were 100 breeder males examined. Sixty percent of these breeder males were raised in pair housing and 40 percent of the males were raised in group housing till December 8th 2010.

In the pelting period November / December 2011 were 1,750 pelts from different pelting dates judged on the black spots. The minks were kept till December 7th in single, in pairs or in group housing with 4 minks during the growing period till pelting. In November 2011 at 3 practical farms were judged 488 males and 398 females sapphires, violets and white ones, kept in group housing.

Macroscopic

After scraping, the leather side of all the pelts were brushed to clean them up from the wood dust and to have a clear view on the leather side of the pelts.

The leather sides of the pelts were with the naked eye controlled on red fresh bite marks and black spots. Black (brown) spots are seen in the subcutis and did not develop after scraping normal skins under increased scraping pressure several times.

Microscopic

The fresh bite marks and the black spots were in practise histological researched after Heamatoxyline-Eosine (HE) colouring and the Perl's Prussian Blue Strain colouring was used to prove rests of iron (Fe) from haemoglobin, possible in older bleeding after biting.

Classification

The back sides, the tail areas, the legs, the necks, the heads and the belly sides of the pelts were reviewed with the naked eye on the number of black spots or as real noticed bite marks. The black spots in leather side of the pelts were classified in 4 groups (Table 1).

Table 1. Classification of black spots in leather side of pelts of minks.

<5 spots	less than 5 black spots, totally clear pelt with or without a few black spots
>5 spots <10	some black spots at one side, after counting more than 5 black spots, more or less clear pelt
>10 spots <20	one or more concentrated places with spots, or black spots over the whole pelt but not more than 20 black spots
>20 spots	pelt with many black spots on the back, near the tail, around the legs or in the neck, always concentration of spots

Results for black spots on the subcutis of the pelts

Macroscopical results

The black spots are scattered over the whole body on the leather side of the pelts. Sometimes a few and some times more. The locations of the black spots could be the upper side and the underside of the head and the neck, the back, the lateral sides and underside of the belly, the lateral breast sides, between the for and hind legs and above the tail. The skin in areas with many black spots was very thin. On the fur side the hairs can stick together. If the hairs in a black spot were pulled out, the black spot disappeared, indicating that the black spots consisted of the hair and/or hair follicle itself. In the areas with black spots hairs were seen protruding from the subcutaneous side of the pelts after passage through the scraping machine. Potentially this could be caused by poor adjustment of the scrape machine. However, black spots did not develop after scraping normal skins under increased pressure several times. Very few real bite marks were seen. Identifiable real bite marks and surrounding are colouring reddish. The leather side of the skins from breeder males in the end of March or in the start of April showed an increase of the blue colour without black spots.

Histological results

Histological examination of the black spots similarly does not reveal indications of damage caused by biting, as haemorrhage, inflammation or fibrosis. No remnant iron products were found in the Perl's Prussian blue stain. In the examined animals with black spots the subcutis was often remarkably thin.

Classification results: males 2009

Table 2 shows the results of 2009. It was clear that pelts with black spots reduce firmly during the pelting period till the first week of December. The first animals were pelted in the first week of November. The leather side of early pelted minks showed in 36% of the cases blue colour, known as insufficient maturation.

Table 2. Black spots on the subcutis of males in the pelting periods 2009.

Male minks 2009	Black spots on subcutis					
	number	blue	<5	>5<10	>10<20	>20
Pelting date 4 Nov.	94	36%	70%	15%	9%	6%
Pelting date 11 Nov.	110	7%	81%	10%	5%	4%
Pelting date 19 Nov.	79	1%	77%	15%	6%	1%
Pelting date 7 Dec.	50	0%	90%	10%	0%	0%
Breeders March 2010	407		100%	0%	0%	0%

In the pelting period from November 4th till December 7th the most pelts of the males had a small amount of black spots (Table 2). During the pelting period the pelts with black spots firmly reduced in compare with the black spots from roommates in the same cage which were pelted earlier in the pelting period. In the last week of pelting in December 90% of the pelts were free of black spots. Pelts from breeders, kept in group housing till pelting time or in pairs, pelted at the end of March did not show black spots. Reduction of unripe blue pelts decrease firmly.

Classification results: males 2010

In 2010 the experimental minks were housed in groups composed of 2 young males and 2 young females and in pairs composed of one young male and one young female or one young male and one old female.

In 2010 the results are comparable with the results of 2009 (Table 3). In the first week of November high percentage of blue pelts and during the pelting period an evident reducing of blue pelts and black spots were seen. As expected no black spots were found in the breeder males.

Table 3. Black spots on the subcutis of males in the pelting period 2010.

Male minks 2010	Black spots on subcutis					
	number	blue	<5	>5<10	>10<20	>20
Pelting date 4 Nov.	130	54%	75%	16%	8%	2%
Pelting date 11 Nov.	133	26%	67%	20%	9%	5%
Pelting date 18 Nov.	96	3%	73%	14%	9%	5%
Pelting date 7 Dec.	110	0%	88%	7%	2%	3%
Breeders March 2011	100		100%	0%	0%	0%

Dutch male minks housed in pairs score lightly better than males housed in groups (Table 4). Pelts from males with 50% Danish blood, kept in groups and pelted in November score double in black spots than the Dutch population at the research farm. All the 50% Danish males were killed in November, so it was not possible to measure the reduction of black spots in December 2010.

Table 4. Black spots on the subcutis of males review 2010.

Male minks 2010	**Black spots on subcutis**				
	number	**<5**	**>5<10**	**>10<20**	**>20**
Pair Nov.	136	77%	13%	7%	2%
Pair Dec.	35	91%	6%	0%	3%
Group Nov.	223	66%	20%	9%	5%
Groep Dec.	75	86%	8%	3%	3%
Group DK (50%) Nov.[1]	74	45%	27%	8%	20%

[1] DK = mink with 50% Danish blood.

Classification results: males 2011

The pelts harvest from weeks of 11 and 18 was by circumstances out of control. So it was only possible to give review of the males pelts kept in groups in 2011 (Table 5). In November 58% of the Dutch pelts score (<5) almost free of black spots and the pelts from males with 50% Danish blood 53%. Four weeks later the Dutch males score 75% almost free of black spots (<5) and the males with 50% Danish blood 69%.

Table 5. Black spots in the subcutis of males review 2011.

Male minks 2011[a]	**Black spots on subcutis**				
	number	**<5**	**>5<10**	**>10<20**	**>20**
Group NL Nov.	294	58%	21%	13%	8%
Group NL Dec.	68	75%	18%	3%	4%
Group DK (50%) Nov.	138	53%	17%	13%	17%
Group DK (50%) Dec.	71	69%	18%	7%	6%

[a] NL = Dutch pelts; DK = mink with 50% Danish blood.

Classification results: females 2009

In 2009 the pelts of the females (Table 6) showed comparable results as the males (Table 2). In the first week of December 94% of the pelts were almost free of black spots. It is not clear why there was a rising to 10% of pelts with black spots in the category more than 20 in the 3rd week of the pelting period.

There was hardly any difference between pelts with black spots from males or females minks kept in group housing or in pairs. Pelts from single raised males and females from June-July till the pelting period showed black spots to.

Table 6. Black spots on the subcutis of females in the pelting period 2009.

Female minks 2009	Black spots on subcutis					
	number	blue	<5	>5<10	>10<20	>20
Pelting date 4 Nov.	97	21%	48%	24%	13%	14%
Pelting date 11 Nov.	98	11%	64%	22%	8%	5%
Pelting date 19 Nov.	99	1%	62%	19%	9%	10%
Pelting date 7 Dec.	50	0%	94%	2%	0%	4%

Classification results: females 2010

In 2010 the experimental minks were housed in groups composed of 2 young males and 2 young females and in pairs composed of one young male and one young female or one young male and one old female. In 2010 the results are comparable with the results of 2009 (Table 7). In the first week of November high percentages of blue pelts and during the pelting period an evident reducing of blue pelts and black spots were seen.

Table 7. Black spots on the subcutis of females in the pelting period 2010.

Female minks 2010	Black spots on subcutis					
	number	blue	<5	>5<10	>10<20	>20
Pelting date 4 Nov.	139	32%	63%	18%	7%	12%
Pelting date 11 Nov.	136	18%	59%	17%	12%	12%
Pelting date 18 Nov.	103	1%	64%	21%	8%	7%
Pelting date 7 Dec.	70	0%	86%	5%	9%	0%

In the review of 2010 (Table 8) scored the Dutch females kept in pairs better in (<5) than the females kept in groups. The females with 50% Danish blood scored in November the same as the Dutch females. All the females with 50% Danish blood are killed in the second week of November, so no reportage was possible 3 weeks after.

Classification results: females 2011

The results of the 383 pelts in 2011follows the same patron as the other years (Table 9). Starting in November with a high percentage of blue pelts and a clear reduction till the first week of December. The same is visible in the pelts without black spots (score<5).

The results agree with the results of females kept in pairs (Table 10). In the score >20 were less pelts found. Pelts from females housed in groups with 50% Danish blood had almost the same results as 100% Dutch female pelts at the research farm. In the score more than 20 black spots the Dutch pelts had better results after the second judgement in December. Pelts from minks with 50% Danish descent show in the groups >5< and >10<20 a small reduction in black spots. In the group >20 was no reduction remarkable in black spots. Females kept single after weaning show black spots on places where self-biting, chin or back of the head and the neck, not possible was.

Table 8. Black spots on the subcutis of females review 2010.

Female minks 2010	**Black spots on subcutis**				
	number	**<5**	**>5<10**	**>10<20**	**>20**
Pair Nov.	141	67%	21%	8%	5%
Group Nov.	237	58%	16%	10%	17%
Groep Dec.	70	87%	5%	9%	0%
Group DK (50%) Nov.[1]	36	58%	22%	11%	8%

[1] DK = mink with 50% Danish blood.

Table 9. Black spots on the subcutis of females in the pelting period 2011.

Female minks 2011	**Black spots on subcutis**					
	number	**blue**	**<5**	**>5<10**	**>10<20**	**>20**
Pelting date 7 Nov.	89	57%	56%	20%	9%	15%
Pelting date 14 Nov.	126	17%	56%	17%	11%	17%
Pelting date 21 Nov.	96	8%	72%	10%	8%	9%
Pelting date 6 Dec.	73	0%	78%	12%	8%	1%

Table 10. Black spots in the subcutis of females review 2011.

Female minks 2011[a]	Black spots on subcutis				
	number	<5	>5<10	>10<20	>20
Group NL Nov.	174	63%	17%	7%	13%
Group NL Dec.	73	78%	12%	8%	1%
Group DK (50%) Nov.	35	40%	29%	17%	14%
Group DK (50%) Dec.	71	59%	13%	13%	15%
Single from July-Nov.	30	83%	10%	0%	7%

[a] NL = Dutch pelts; DK = mink with 50% Danish blood.

Results practical farms 2011

The sapphires, the pearls and the violets were killed in the first week of November. The white minks were 2 weeks later killed and pelted. At the leather side of the pelts from white, pearl, sapphire and violet type (Table 11), from both sexes, no signs of black spots were noticed.

Thirty percent of the sapphire male pelts and 1% of the sapphire females pelts were blue. Real bite marc were noticed in the sapphire females (4), in the pearl males (5) and in de pearls females (8). In the sapphire males and in de violet males and females were not real bite marks noticeable. In the white males were 7 and in the females were 18 real bite marks noticed. Blue pelts and black spots or indicative signs for this were not noticed.

Table 11. Black spots in males and females on subcutis of mink pelts under field circumstances 2011.

Field farms	Black spots on subcutis					
	Number	Blue	<5	>5<10	>10<20	>20
Male minks						
Sapphire	106	30%	86%	4%	3%	8%
Violet	52	2%	100%	0%	0%	0%
Pearl	180	0%	100%	0%	0%	0%
White	150	0%	100%	0%	0%	0%
Female minks						
Sapphire	105	4%	85%	1%	5%	10%
Violet	43	0%	100%	0%	0%	0%
Pearl	100	0%	100%	0%	0%	0%
White	150	0%	100%	0%	0%	0%

Discussion

It appeared that the pelts of in the first week of November pelted minks showed high percentage of blue skins, more in the males than in the females. In 2009 the percentages of blue skins of the first pelted the skins were in 2009 lightly lower than in 2010. Blue skins were seen in pelts if the minks were pelted early and it looked as a sign of insufficient maturation.

In early pelted skins the black spots were evident more present than later in the pelting periodin 2009. The pelts from males in 2010 kept in pairs showed less black spots than the males kept in groups. In December 2010 were the results in the score less than 5 in the pairs 5% better in compare of 2009.

In 2011 were males and females judged with 50% Danish blood. Danish blood was imported at the research farm to improve the fur quality of the pelts. Perhaps there was a link between an increase of fur quality and later maturation of the pelts in the brown and mahogany types. Few black spots were seen in the sapphires in practical farms.

In the areas where black spots hairs were seen, was protruding of the hairs from the subcutaneous side of the pelts after passage through the scraping machine visible. The skins were often thin. Potentially this could be caused by poor adjustment of the scrape machine. However, black spots did not develop after scraping normal skins under increased pressure several times. Protruding of hairs and black spots at the leather side in the pearl, violet and white couloirs were not visible.

Very few identifiable real bite marks were seen in all colour types. Identifiable bite marks and the surrounding are purple blue or reddish. During the development of the fur from summer coat to winter coat, after a relatively long anagen (growth) phase in which the bulb portion of the hair follicles is deeply anchored in the superficial part of the subcutaneous fat tissue, the hair follicles progress towards the the telogen phase in which the bulb of the hair follicles is smaller and located more superficially, in the mid- to deep dermis. It appears that if the development of anagen to telogen is retarded or if the animals are killed before all follicles are in telogen phase, the hair follicles are recognizable as small black spots with the naked eye, and under the influence of the pelting machine, the bulb portion of the hair follicles is ruptured or even removed together with the fat tissue.

The hairs are then not firmly anchored in the ruptured follicle anymore and can easily be epilated. This does not happen when the hair follicles are in telogen, hair-containing phase, in which the hair follicles have a more superficial location within the dermis and therefore are not damaged by removal of the fat tissue by the pelting machine. This also explains why, when the animals are pelted later in the season, the black spots disappear spontaneously and the pelts have a normal morphology. The impression is that pelts of females are later matured than the pelts of males. The as real bite marks indicated spots did not show in the histology sings of bleeding or inflammation. Only fibrosis was to see. This means that the with naked eye as real bite marks indicated spots are old bite marks.

The as real noticed bite marks and the black spots were earlier judged in histological research after Heamatoxyline-Eosine (HE) colouring and the Perl's Prussian Blue Strain colouring (Kleyn van Willigen *et al.*, 2010) The last research was used to prove rests of iron (Fe) from haemoglobin, possible after older bleeding, in the black spots (Kleyn van Willigen *et al.*, 2010). In the follow up the histological judgement was not carried out in 2010 and 2011 (Kleyn van Willigen *et al.*, 2011).

Conclusion

In all the experiments from 2009 to 2011 was the reduction of black spots and blue pelts in pelted skins evident. The absence of the black spots in breeder male pelts, in combination with the results of the macroscopical and histological examination in 2009 (Kleyn van Willigen *et al.*, 2010) of the black spots, where the hair follicles were in the anagen stage, justify the conclusion that black spots cannot be pointed as bite marks.

References

Kleyn van Willigen F.C., De Rond J. and Boekhorst L., 2010. Black spots in subcutis of minks: no bite marks. NJF Seminar no.440, Oslo, Norway.

Kleyn van Willigen F.C., De Rond J. and Boekhorst L., 2011. The follow up of black spots in the subcutis of mink skins in The Netherlands. NJF Seminar no.102, Knivsta, Sweden.

Coat color mutations and defensive reaction towards man in farm-bred minks and sables

O.V. Trapezov[1], *L.I. Trapezova*[1] *and E.G. Sergeev*[2]
[1]*Institute of Cytology and Genetics, Siberian Branch, Russian Academy of Sciences, Novosibirsk, 630090 Russia;* [2]*Institute of Fur-Bearing Animals, Russian Academy of Agricultural Sciences, Rodniki, Moscow oblast, 140143 Russia; trapezov@bionet.nsc.ru*

Abstract

The role of coat color mutations on the defensive reaction towards man in farm-bred mink and sable was studied. 17,500 *standard dark brown* and 14,420 *sapphire* minks and 8,427 *standard dark brown* and 52 *black-and-white* sables had been involved into investigation. Both species were tested for defensive reaction towards man using the 'hand catch test'. Most animals (both minks and sables) from farm-bred populations displayed fear reaction towards humans. The domestic behavior was defined as absence of fear or defence reaction towards man. Coat color genes (both *sapphire* minks and *white-and-black* sables) have been found to have pleiotropic effects; the proportion of animals with domestic behavior and the expressivity of this behavior are significantly higher in coat color mutants ($P<0.01$ and $P<0.001$, respectively). The result is valid for both sexes.

Keywords: colour mutations, behaviour, fear, mink, sable

Introduction

Beginning from Darwin, many researchers of domestication concluded from paleontological, cultural, and historical data, as well as from comparison of domestic and wild animals, that domestication is often accompanied by an increase in reproductive parameters, changes in the proportions of organs and body parts, and development of complex behaviors (Darwin, 1868; Hale, 1960; Price, 1984; Zeder, 2006). Academician Belyaev used the results of a longterm domestication experiment using animals from family *Canidae*, namely, silver foxes (*Vulpes vulpes*) as a model object to enunciate the theory of destabilizing selection (Belyaev, 1969, 1979). Belyaev demonstrated that polymorphism of defensive reaction towards humans in the original animal population is a necessary prerequisite for successful domestication of fur animals (Trut, 1999). To test whether the theory of destabilizing selection was universal, a domestication experiment was started on representatives of another family. The purpose of this study was to determine the effect of coat color mutations on the defensive reaction towards man in farm-bred populations of mink (*Neovison vison*) and sable (*Martes zibellina*).

Material and Methods

The animals were housed on the same farm. There was no difference in housing and management between the two types of mink and two types of sables. Experiments were performed on *standard dark brown* (+/+) (n=17,500) and *sapphire* (*a*/*a p*/*p*) (n=14,420) minks. The *sapphire* color phase is often also called *aleutian silver* or *royal sapphire*. This combined color type results from additive interaction between the *aleutian* (*a*) and *silverblue* (*p*) recessive genes, which determine the new, sapphire, color type in the double homozygotes (*a/a p/p*) (Ness *et al.*, 1999).

Behavior of farm-bred sables was studied on the *standard dark brown* (n=8,427) animals (Figure 1A) and *black-and-white* (n=52) animals with white patches on the legs, tail tip, snout, and trunk (Figure 1B). The size and shade of the white patches considerably vary. Their color may be not only white, but also yellow of various intensities. The eyes of many *black-and-white* sables are blue (dark or light), and the nose is partly or entirely depigmented. Preliminary studies on the inheritance of white patches showed that this character is controlled by a semidominant gene with a recessive lethal effect. It determines white patches in the heterozygous state and white coat color in the homozygous state. Homozygous forms die at the prenatal or postnatal developmental stage. Only one white kit sable can survive until an age of 45 days (Koldaeva, 2004).

Behavioral testing was performed once at the age of 4.5-5 months (in September – October). Both males and females were tested for defensive reaction towards man during a day, with breaks for the morning and evening feeding. The 'hand catch test' was developed for this purpose. The experimenter opened the cage where the tested animal was kept, slowly reached for the animal, and tried to catch it with a hand protected with a special mitten.

The technique used allowed to distinguish three types of defensive behavior towards man: aggressive, fearful, and tame. Animals with fearful behavior tried to *avoid contact with a man* as much as possible; when the experimenter tried to catch them, they ran about the cage in a panic,

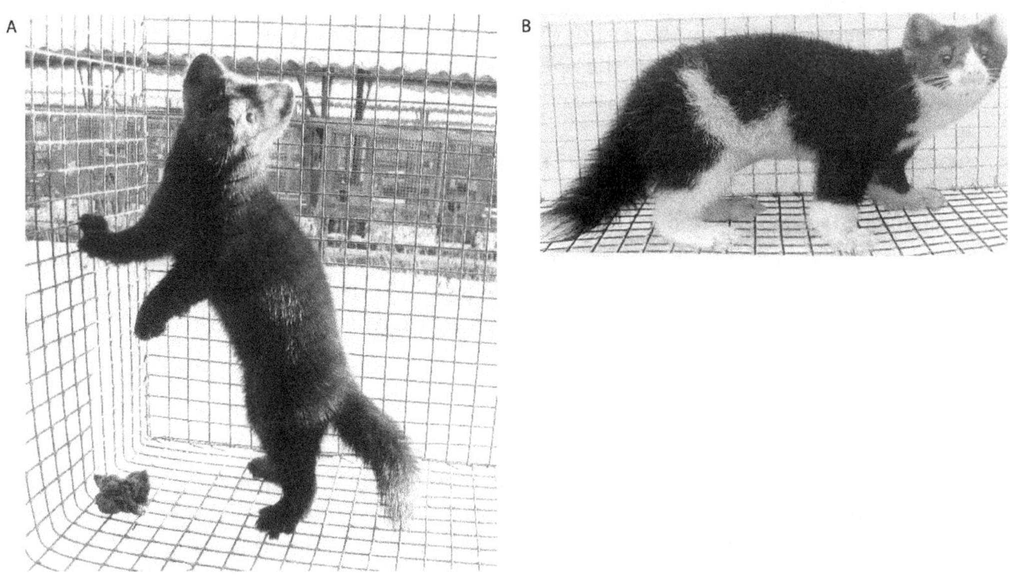

Figure 1. (A) Standard dark brown sable. (B) Black-and-white sable.

shrieking. This behavior was assigned score '0'. Aggressive behavior varied qualitatively, which allowed to estimate it in scores of a four-score scale as follows:

- Score -1. Fearful response towards humans. When attempts were made to catch the animal, it rapidly retreated, hid in its nest box, gaped and bared its teeth, cried shrilly or hissed; its posture was tense, showing severe emotional stress.
- Score -2. Attack from the nest box. When attempts were made to catch the animal, it jumped to the entrance to the nest box and hid there to attack the experimenter's gloved hand from the box and bite it fiercely.
- Score -3. Active attack outside the shelter. When attempts were made to catch the animal, it, instead of hiding, immediately attacked the experimenter's hand.
- Score -4. Onset of attack in response to approach of a human. Even before the test began, i.e. before the experimenter opened the cage, the animal loudly shrieked, ran about the cage, and gnawed the bars.

Animals with domesticated behavior did not show any signs of fear or aggression upon contact with man, which allowed the experimenters not to protect the hands with mitten. In the hole the domestic behavior was defined as absence of fear or defence reaction towards man.

The expression of domesticated behavior also varied qualitatively, making it possible to score it using a five-point scale as follows:

- Score +1. Exploratory responses. The animal calmly responded to the experimenter's stretched hand and showed an exploratory response, sniffing the hand and quivering the vibrissae.
- Score +2. A calm response to contact with the human hand. The animal accepted the forced contact with the experimenter's hand and did not retreat, allowing the hand to touch its face, chest, and paws.
- Score +3. Active contact on the part of the animal. When the experimenter opened the cage, the animal got up leaning against the open door and reached out with its snout to the human hand. Within the cage, the animal actively sniffed about the hand, examined it, and often leant against the hand with its paws.
- Score +4. The animal allowed any part of its body to be touched. The animal displayed an actively exploratory response, examined the experimenter's hands, played with them, but resisted all attempts to be handled.
- Score +5. The animal allowed humans to handle it. These animals were unique in the farm population. They displayed the most pronounced domesticated behavior, allowing the experimenters to handle them without showing any signs of fear or aggression.

Results

As can be seen in Tables 1 and 2, most minks and sables from farm-bred populations displayed fearful behavior in the 'hand catch test'. The proportions of animals with maximum expressivities of aggressive and domesticated behaviors (scores -4 and +5, respectively) were equally small (0.02-0.3%). The data shown in Table 2 demonstrate that, among tame animals (both sables and minks), the proportion of mutant individuals was twice as high as the proportion of those with the *standard* coat color ($P<0.01$ and $P<0.001$, respectively). Among animals displaying aggressive response towards humans, the difference between the proportions of color mutants

Table 1. Behavioral polymorphism of farm bred minks and sables with different color types.

Color phase	Sex		Number of animals	Behavioral phenotypes -4	-3	-2	-	0	+1	+2	+3	+4	+5
American mink													
Standard	♂	n	8,700	6	113	426	559	7262	188	88	41	15	2
		n%	100	0.07	1.3	4.9	6.4	83.5	2.1	1.0	0.5	0.2	0.02
	♀	n	8,800	14	214	643	844	6,917	91	41	28	5	3
		%	100	0.1	2.4	7.3	9.7	78.6	1.0	0.5	0.3	0.06	0.03
	♂+♀	n	17,500	20	327	1,069	1,403	14,179	279	129	69	20	5
		%	100	0.1	2.0	6.0	8.0	81.0	1.6	0.7	0.4	0.1	0.03
Sapphire	♂	n	7,200	7	81	318	413	5,782	261	138	120	57	23
		%	100	0.1	1.1	4.4	5.8	80.3	3.6	1.9	1.7	0.8	0.3
	♀	n	7,220	17	248	716	1	5,071	123	42	43	8	7
		%	100	0.2	3.4	9.9	13.2	70.2	1.7	0.6	0.6	0.1	0.1
	♂+♀	n	14,420	24	329	1,034	1,358	10,853	384	180	163	65	30
		%	100	0.2	2.3	7.2	9.4	75.3	2.7	1.2	1.1	0.4	0.2
Sable													
Standard	♂	n	4,304	1	19	2	222	2,974	512	532	32	8	2
		%	100	0.02	0.5	0.04	5.1	69.1	11.9	12.4	0.7	0.2	0.04
	♀	n	4,123	3	11	11	195	3,135	367	384	16	-	1
		%	100	0.07	0.3	0.3	4.7	76.0	9.0	9.3	0.4	-	0.02
	♂+♀	n	8,427	4	30	13	417	6,109	879	916	48	8	3
		%	100	0.05	0.3	0.4	4.9	72.5	10.4	10.8	0.6	0.1	0.03
White and black	♂	n	28	-	1	-	-	14	1	8	4	-	-
		%	100	-	3.6	-	-	50.0	3.6	28.6	14.3	-	-
	♀	n	24	-	1	-	-	13	1	6	3	-	-
		%	100	-	4.2	-	-	54.2	4.2	25.0	12.4	-	-
	♂+♀	n	52	-	2	-	-	27	2	14	7	-	-
		%	100	-	3.8	-	-	52.0	3.8	26.9	13.5	-	-

and *standard* animals was smaller. Among fearful animals, the ratio between color mutants and *standard* individuals was intermediate.

The score estimates of behavior (Table 3) show that the mean score of reaction towards humans in the entire sample of minks, including *standard* (+/+) and *sapphire* ones (*a/a p/p*) was slightly shifted towards aggressiveness. Conversely, in sables, both *standard* and *black-and-white*, the mean score was shifted towards domestic behavior, the shift being particularly marked in sables with the aberrant color type (*black-and-white*). These shifts were accounted for by the fact that the proportions of animals with tame behavior among both *sapphire* minks and *black-and-white* sables were twice as high as their proportions among the respective groups with the standard

Table 2. Percentages of behavioral phenotypes among animals with different color types in farm-bred populations of minks and sables.

Phenotype	Aggressive	Fearful	Tame[1]
American mink			
Standard	16.2	81	2.8**
Sapphire	19.1	75.3	5.6**
Sable			
Standard	5.5	72.5	22*
White and black	3.8	52	44.2*

[1] ** $P<0.001$; * $P<0.01$.

Table 3. Effects of coat color mutations on the expressivity of behavioral responses in farm-bred populations of minks and sables.[1]

Phenotype	Sex	Aggressive			Fearful		Tame			Total		
		no.	mean score	σ2	no.	mean score	no.	mean score	σ2	no.	mean score	σ2
American mink												
Standard	♂	1,104	-0.61±0.02	0.23	7,262	0.00	334	+1.67±0.05	0.81	8,700	-1.14±0.008	0.26
	♀	1,715	-1.65±0.02	0.28	6,917	0.00	168	+1.74±0.07	0.85	8,800	-0.29±0.008	0.38
	♂+♀	2,819	-1.63±0.02	0.26	14,179	0.00	502	+1.69±0.06**	0.64	17,500	-0.21±0.006	0.32
Sapphire	♂	819	-1.61±0.02	0.24	5,782	0.00	599	+2.07±0.05	1.69	7,200	-0.01±0.01	0.67
	♀	1,926	-1.66±0.02	0.29	5,071	0.00	223	+1.81±0.07	1.14	7,220	-0.39±0.01	0.74
	♂+♀	2,745	-1.64±0.01	0.27	10,853	0.00	822	+2.00±0.04**	1.69	14,420	-0.20±0.008	0.76
Sable												
Standard	♂	244	-1.18±0.04	0.33	2,974	0.00	1,086	+1.58±0.02	0.54	4,304	+0.33±0.01	0.78
	♀	220	-1.18±0.04	0.33	3,135	0.00	768	+1.55±0.02	0.64	4,123	+0.22±0.01	0.56
	♂+♀	464	-1.18±0.03	0.33	6,109	0.00	1,854	+1.56±0.01**	0.34	8,427	+0.28±0.009	0.63
White and black	♂	1	-3		14	0.00	13	+2.23±0.17	0.36	28	+0.93±0.23	1.58
	♀	1	-3		13	0.00	10	+2.20±0.20	0.4	24	+0.79±0.26	1.67
	♂+♀	2	-3		27	0.00	23	+2.23±0.12**	0.36	52	+0.86±0.19	1.96

[1] ** $P<0.001$.

color types (Table 2). In the sample of *black-and-white* sables, there were only a few aggressive animals.

As evident from Table 3, *sapphire* minks with aggressive behavior did not differ in the expressivity of this character from animals with the *standard* color type and aggressive behavior (mean scores -1.64±0.01 and -1.63±0.02, respectively, $P>0.01$). On the other hand, the expressivity of tame behavior among minks with a 'positive attitude' towards humans was obviously enhanced by the direcessive *sapphire* animals: the mean behavioral scores in *standard* and *sapphire* minks were +1.69±0.06 and +2.00±0.04, respectively ($P<0.001$). The same effect was observed in sables: animals with the aberrant color type (*black-and-white*) were, on average, significantly tamer than those with the *standard* coat color (+2.23±0.12 and +1.56±0.01), respectively, $P<0.001$). Since aggressive *black-and-white* sables were too few (only two animals in the whole sample), we could not estimate the effect of coat colour aberrations on the expressivity of aggressiveness. In addition, coat color mutations of both sables and minks modulated the variability of defensive reaction towards man: the variance of this character in animals with mutant color types was drastically increased: σ^2 was more than two times higher in *sapphire* minks than in *standard* ones (0.32 and 0.76, respectively) and three times significantly higher in *black-and-white* sables than in *standard* ones (0.63 and 1.96, respectively).

Discussion

One of the main conditions of 'stochastic' domestication (without special selection for behavior) of minks and sables bred at fur farms is the system of keeping them in cages, the animals being always at an arm's length from humans, which inevitably leads to selection for animals less prone to the stress caused by the presence of humans. Therefore, the number of animals tolerant to anthropogenic stress, which allow them to be handled during zootechnic manipulations, transportation, valuation, vaccination, etc., is constantly increasing. This is directional selection for domestic behavior, which is possible due to polymorphism of the expression of defensive reactions towards humans in farm-bred populations of both minks and sables, as evident from this experimental data.

The pleiotropic effect of coat color mutation on the penetrance and expressivity of domestic behavior was found: in animals with aberrant color types (*sapphire* minks and *black-and-white* sables), both the proportion of individuals with domesticated behavior and the expressivity of this behavior are significantly higher. Earlier, Keeler noted these effects of aberrant coat color and suggested that the so-called domestication genes in individual foxes and minks should be taken into account in the course of domestication in farm populations (Keeler, 1947). These, however, were only references to observations of breeders of these fur animals at specialized fur farms, which were not confirmed by statistical data.

This raises the question as to why coat color aberration so strongly affect animal behavior. Special experiments were performed to determine the monoamine content of the brain of minks differing in their behavior towards humans. The minks were sampled from a common farm population that had never undergone intentional selection for behavior.

These studies demonstrated that the serotonin contents of the hypothalamus and corpus striatum and the content of the serotonin metabolite 5-hydroxyindolacetic acid in the corpus striatum were decreased in minks with nondomesticated (aggressive and fearful) behaviors towards humans (Nikulina *et al.*, 1985).

Many studies have shown that the serotonergic system inhibits intraspecific and predatory aggressions (Popova *et al.*, 1978). An increased serotonin content has also been found in domesticated Norway rats. However, these changes were observed after 20 generations of selection, whereas in mink populations, the same phenomenon was found in animals that had never been specially selected for domesticated behavior but nevertheless displayed different response towards humans (Nikulina *et al.*, 1985). This polymorphism with respect to the activities of cerebral monoaminergic systems corresponds to the polymorphism of reactions towards humans: minks displaying negative responses towards humans have reduced serotonin contents of the hypothalamus and corpus striatum.

Experimental data have shown that coat color genes also affect neurotransmitter metabolism in various regions of the brain (Popova *et al.*, 1994; Voitenko and Trapezov, 2001). Changes in the brain serotonin metabolism have been observed in *sapphire* minks (*a*/*a p*/*p*). The pleiotropic effects of the *a*/*a* and *p*/*p* loci are expressed in changes in the activities of the main enzymes of serotonin metabolism: tryptophan hydroxilase (the key enzyme of its biosynthesis) and monoamine oxidase A (the enzyme of its catabolism). Serotonin is a phylogenetically old cerebral neurotransmitter with a wide spectrum of activity (Popova *et al.*, 1978); therefore, changes in the serotonin metabolism in the brain may explain some physiological characteristics of mutant animals. The observed changes in serotonin metabolism in sapphire minks (*a*/*a p*/*p*) suggest that the decreased aggressiveness of these minks towards humans is related to an increased serotonin synthesis rate in the brain.

Regarding aberrant color types of sables, we have no experimental data on serotonin metabolism and can only preliminarily assume that a mechanism similar to that found in minks modulates neurotransmitter metabolism in different regions of the brain of these sables, thereby determining the behavioral polymorphism of farm populations.

References

Darwin, C., 1868. Variation of Animals and Plants under Domestication. London: John Murray.

Belyaev, D.K., 1969. Domestication of Animals. Science 1: 47-52.

Belyaev, D.K., 1979. Destabilizing Selection as a Factor in Domestication. Journal of Heredity. 70: 301-308.

Hale, E.B., 1969. Domestication and Evolution of Behavior. In: The Behavior of Domestic Animals. Hafes, E.S.E. (ed.) Baltimore, pp. 22-24.

Koldaeva, E.M., 2004. Farm Fur Animals: Genetics and Selection. 1st editon. Moscow. Izvestiya. 296 pp. (in Russian).

Keeler, C.E., 1947. Coat Color, Physique, and Temperament. Journal of Heredity 9: 271-277.

Ness, N.N., Einarson, E.J., Lohi, O. and Jorgensen, G., with contributions from Jarosz, S. and Scheelje, R., 1988. Beautiful Fur Animals – and Their Color Genetics. Glostrup, Denmark. Scientifur. 271 pp.

Nikulina, E.M., Trapezov, O.V. and Popova, N.K., 1985. Monoamine Content in Mink Brain Differing in Defense Reaction against Humans. Zhurnal Vysshej Nervnoj Deyatel'nosti. 35(6): 1142-1146. (in Russian).

Popova, N.K., Naumenko, E.V. and Kolpakov, V.G., 1978. Serotonin and Behavior. Novosibirsk. 304 pp. (in Russian).

Popova, N.K., Voitenko, N.N., Kulikov, A.V. and Trapezov, O.V. 1994. Pleiotropic Effect of One-Locus (*p/p*) and Two-Locus (*a/a p/p*) Fur Color Mutations on Serotonin Metabolism in the Mink Brain. Russian Journal of Genetics 30(11): 1308-1310.

Price, E.O., 1984. Behavioural Aspects of Animal Domestication. Quarterly Review of Biology 59(1): 1-32.

Trut, L.N., 1999. Early Canid Domestication: The Farm-Fox Experiment. American Scientist 87: 160-169.

Voitenko, N.N., and Trapezov, O.V., 2001. Effect of Four Coat Color Mutations (C_R, S, S^H, and h) on Brain Monoamine Oxidase in Mink. Russian Journal of Genetics. 37(1): 519-522.

Zeder, M.A., 2006. Documenting Domestication: The Intersection of Genetics and Archaeology. Trends in Genetics 22(3): 139-155.

Diversity of selected features of integumentary system in mink

M. Piórkowska[1] *and A. Natanek*[2]
[1]*National Research Institute of Animal Production, Department of Animal Genetic Resources Conservation, Sarego 2, 31-047 Krakow, Poland; Agricultural University, Department of Animal Reproduction and Anatomy, Mickiewicza 24/26, 30-059 Krakow, Poland; mpiorko@izoo.krakow.pl*

Abstract

Domestication by humans was aimed to modify the morphological, physiological, developmental and mental traits of the wild animals so as to obtain desirable characteristics. As a result of long-term intensive farming, the productive traits of farmed mink differ considerably from those of feral mink. Clear differences are observed in reproductive traits, and animals are improved mainly for hair coat quality, coat colour, body size and temperament. The wild American mink, a non-indigenous species that took over the European mink's ecological niche, is found in Poland. The aim of the study was to determine differences in the integumentary system of mink depending on the animals' place of origin and habitat. The biological material was represented by skins of feral mink caught in Canada and Poland following the winter period and skins of farmed Polish mink obtained after winter fur priming. Physical parameters of all raw skins were measured, the degree of skin damage was evaluated, and the histological examination of the skins was performed during the study because hair coat is associated developmentally and biologically with cutaneous tissue. A total of 150 skins were analysed. In relation to Canadian skins, Polish skins were 22% heavier, 15% longer and had 31% greater area, while the skins of farmed mink were 130% heavier, 52% longer and had 88% greater area.

Keywords: mink, skins, hair coat, integumentary system, histology, defects

Introduction

The use of animals dates back to the Stone Age, when man probably learned how to utilize animals before taming and domesticating them. In the first stage of domestication, animals were tamed to make better use of wild species for human needs and to ensure that access to them was easier and not limited by external factors. Compared to wild animals, animals kept near home settlements were a more reliable source of food and raw material (wool, skin, fur, bones) for production of clothes and other objects such as ornaments, tools, parts of weapons, etc. The domestication of different species was easier and quicker for animals living in herds compared to solitary animals, which retain a degree of independence even after domestication. Breeding was conducive to a settled way of life. The importance of many domesticated species of animals varied according to region, age, and culture. The domestication process was aimed to modify the morphological, physiological, developmental and mental traits of the wild animals so as to obtain desirable characteristics. This occurred as a result of long-term effects of humans on animal populations, especially through human interferences in genetic and environmental factors, which led to the creation of many breeds and even specific types within the same breed.

The history of fur animal domestication is relatively short. Originally, American mink and common foxes were only farmed in North America. Fur-bearing carnivores came to European farms as breeding animals in the 1920s. In Europe, free-living populations of mink appeared when mink farming began. These animals emerged in the East European habitat as a result of deliberate introduction of the American mink and sporadic escapes of farmed animals. Farmed mink were also deliberately released, some of which survived in the wild.

The wide distribution area of fur-bearing carnivores, which easily adapt to various environmental conditions, gave rise to a large number of species and colour types, differing in coat colour, size and fur parameters of skins. In the relatively short process of fur animal domestication, rapid changes in the coat colour and quality of these animals are particularly noticeable.

The animal's skin is a highly complex organ that fulfils many physiological functions. For centuries, this raw material has been used for different purposes due to the basic advantages of skin such as the ability to control temperature and humidity, respiration, high strength and resistance to damage. The hair coat of every fur animal is different and has characteristic properties associated with the adaptation to a specific environment. Compared to wild animals, farmed animals are better nourished and larger, their skins are thicker with less damage and greater variation in colour. The hair coat of wild and farmed fur animals differs in length, thickness and microscopic structure of different hair types. Important factors during the determination of fur value and thus technological suitability are the amount of fur damage, as well as the location, depth, size and number of defects and damages. Essential among these factors is the topographic area of damage.

The aim of the study was to determine changes in phenotypic characteristics (physical parameters) of skins in the farmed and wild populations of mink and to evaluate the amount of damage to the analysed skins and their histological structure.

Materials and methods

The experiment used raw mink skins, which were divided into three groups, each consisting of 50 skins depending on the animal's habitat:

- Group I – wild mink living in Canada, not subjected to domestication;
- Group II – wild mink living in natural habitat in Poland;
- Group III – farmed mink from the Chorzelów Farm of the Experimental Station of the National Research Institute of Animal Production.

Skins in each group were representative of the different populations. Skins from wild mink were collected from animals caught in the autumn-winter period, and those from farmed mink were taken from animals born in a given calendar year after reaching winter fur maturity. Measurement was made of the physical parameters of all raw skins such as weight, length, width, area and lightness (weight of 1 dm^2). Cutaneous tissue and hair coat of the furs was evaluated and the damages found were treated as defects that reduce the functional value of furs to various degrees. Histological preparations were made by fixing the samples of cutaneous tissue in 6% buffered formalin. The paraffin blocks were cut by a Leica 2450 microtome into sections 6-8 μm thick, which were later stained differentially with Delafield's hematoxylin and eosin. The preparations were analysed and morphometric measurements were made using MultiScanBase v. 18.03 imaging software. The results obtained were analysed statistically by analysis of variance and Duncan's test using Statistica 9.

Results

The size of skins and the size of animals from which the furs were obtained varied between groups (Table 1). Wild Canadian mink were characterized by significantly lowest skin size parameters in terms of weight, length and area, averaging 67.02 g, 49.16 cm and 7.30 dm^2, respectively. In relation to Canadian skins, Polish skins of wild mink were 22% heavier, 15% longer and had 31% greater area, while the skins of farmed mink were 130% heavier, 52% longer and had 88% greater area. Canadian and farmed mink were characterized by similar tail length (about 18.6 cm) and skin width (±15 cm). The tails of wild Polish mink were shorter by about 4 cm. In terms of lightness, the skins of wild mink were lighter than those of farmed mink. The skins of wild Polish mink were the lightest, those of wild Canadian mink slightly heavier (by 0.54 g per dm^2 skin) and the skins of farmed Polish mink the heaviest (by 2.58 g per dm^2 skin).

Histomorphological examination showed highly significant differences between different skin layers, namely epidermis, dermis and hypodermis (Table 2). The thickest epidermis and hypodermis was characteristic of wild Polish mink. In comparison, Canadian mink were characterized by lower values, i.e. about 18% thinner epidermis, 41% thinner dermis and 68% thinner hypodermis. Farmed mink were characterized by 635 μm thick dermis, which was twice as thick as in Canadian mink and about 14% thicker than in wild Polish mink. Thickness of epidermis and hypodermis was higher than in group I and lower than in group III.

Detailed histological examination showed that the number of bundles per tuft was the smallest (1.4) in wild Canadian mink, intermediate in farmed mink, and the highest in wild Polish mink (Table 2). Despite the differences in the number of bundles per tuft, wild mink had a similar number of down hair (25-26) per tuft. Farmed animals had a higher number of down hair per

Table 1. Physical parameters of raw mink skins.[1]

Raw, dried skins	Wild mink		Farmed mink	SEM
	Canadian	Polish		
Weight of skin (g)	67.02 Aa	81.86 Ba	154.34 AB	4
Length of skin (cm)	49.16 CD	56.57 CE	74.87 DE	1
Length of tail (cm)	18.58 F	14.91 FG	18.70 G	0
Width of skin (cm)	15.54 H	13.56 HI	15.05 I	1
Area (dm^2)	7.295 JK	9.572 JL	13.753 KL	0
Lightness (weight of 1 dm^2)	8.879 M	8.339 N	10.914 MN	0

[1] Means in rows with the same letters differ significantly: small letters at $P \leq 0.05$; capital letters at $P \leq 0.01$.

Table 2. Results of histological examination of mink skins.[1]

Item	Wild mink		Farmed mink	SEM
	Canadian	Polish		
Thickness of skin layers (μm):				
Epidermis	6.010 A	7.304 Aa	6.400 a	0.150
Dermis	322.721 BC	545.602 Bb	634.732 Cb	18.336
Hypodermis	121.754 DE	380.715 DF	240.845 EF	13.870
No. of bundles per tuft	1.36 c	1.71 c	1.46	0.060
No. of down hairs per tuft	25.35	26.13	30.00	1.499
No. of down hairs per bundle	18.64 G	15.28 GH	20.55 H	0.427
Sebaceous gland				
Area (μm^2)	3502.11 IJ	6803.80 I	7018.14 J	350.513
Length	95.71 Kd	122.79 de	155.57 Ke	4.637
Diameter	54.76 L	77.02 L	62.71	2.715

[1] Means in rows with the same letters differ significantly: small letters at $P \leq 0.05$; capital letters at $P \leq 0.01$.

tuft compared to groups I and II, by about 15.5% and 12.9%, respectively. Farmed mink were characterized by the highest number of down hair per bundle (20.6) and wild Polish mink by the lowest (15.3), with significant differences. The analysed preparations contained many sebaceous glands, the largest of which occurred in farmed mink.

The analysed skins were characterized by a large number and area of hair coat blotches, especially in wild Canadian and Polish mink, and the presence of single and tuft white hairs in undercoat in both wild and farmed Polish mink.

Discussion

Fur-bearing carnivores have been farmed in Europe for around 90 years. Genetic studies of wild American mink found in the area of Poland showed that only a small percentage of feral mink originate from farmed ancestors (Michalska-Prada *et al.*, 2009). Various authors (Duda, 1992; Gugołek *et al.*, 2008) report that differences between the skins of farmed and wild mink were small only in the initial period of fur farming. Today, the differences in the size and quality of fur material are so large in favour of farmed animals that over 86% of the auction house sales are generated by farmed animals (Gugołek *et al.*, 2008), most of which are mink (80%). The skins of wild mink account for only 0.4% of all skins sold. This is due to their lower quality, the lack of colour differences, and the greater number of defects (Gugołek *et al.*, 2008). In a study investigating changes in the conformation of fur animals over 1975-2002, Brzozowski (2002) found that the size of skins from farmed Polish mink increased by two auction sizes, i.e. by about 12 cm.

The number of down hairs in the coat of mink varies according to season. Fur animals are characterized by a complex, tuft distribution of hair on skin, with each tuft consisting of several bundles located around guard hair. The number of developed and active hair roots determines the density and heat-insulating properties of the hair coat. It is important that as many hair roots become active as possible during the time when winter hair coat is formed (Natanek *et al.*, 2001). In mink, hair density on the back ranges, according to different authors, from 18,000 to 24,000/cm^2 (Kondo and Nishiumi, 1988, Natanek *et al.*, 2001). Blomstedt (1998) reports that in mink, the mean number of down hairs per bundle is 18, with 1-3 dormant hair roots per bundle. The same author found 11-12 down hairs in bundles with awn hair and up to 20 hairs in bundles without awn hair.

Histological examination of hair coat density in mink (Natanek *et al.*, 2001) showed that mean hair density in bundles was about 18,100 on the neck, 22,400 in the sacral region, and 19,700 on the belly. The fact that hairs fail to grow out of all roots and structural hair defects are due to genetic (Vanek and Hanzlova, 1992), climatic and environmental factors, the most important of which is a poor diet (Kerminen-Hakkio *et al.*, 2000).

In summary, it is concluded that the analysed skins and mink populations which provided the test material showed differences ($P \leq 0.01$) in the features of the integumentary system. Farmed mink are superior to free-living mink in terms of size parameters except the lightness of skins. Modern skin tanning and dyeing techniques and garment cutting and sewing methods make it possible to improve the lightness of ready-to-wear fur garments (Duda, 1992). Farmed animals are also characterized by the highest number of down hairs per bundle and low number of bundles per tuft.

References

Blomstedt L., 1998. Pelage development in mink, ferret and blue fox, and some influencing factors. Academic Disseration, University of Helsinki, Finland.

Brzozowski M., 2002. Zmiany pokroju mięsożernych zwierząt futerkowych jako efekt pracy hodowlanej (Fur animal exterior changes as a result of provided selection). Zeszyty Naukowe Przeglądu Hoddowlanego, 66: 81-87.

Duda I., 1992. Skóry surowe futrzarskie. Akademia Ekonomiczna w Krakowie, 240 pp.

Gugołek A., Lorek M.O., Kuśmierkiewicz M. 2008. Dominująca pozycja skór zwierząt hodowlanych na światowym rynku futrzarskich. Hodowca Zwierząt Futerkowych 33: 40-42.

Kerminen-Hakkio M., Dahlman T., Niemela P., Jalava T., Rekila T.and Syrjala-Qvist L., 2000. Effect of dietary protein level and quality on growth rate and fur parameters in mink. Journal of Mammalogy 51, 1: 27-34.

Kondo K. and Nishiumi T. 1988. The pelage development in young mink (*Mustela vision*). Proccedings of the IV International Congress in Fur Animal Production, Ontario (Canada), pp. 397-407.

Michalska-Prada A., Brzeziński M., Zalewski A. and Kozakiewicz M. 2009. Genetic variability of Farel and ranch Americanmink *Neovision Visio* in Poland. Acta Theriologica 54: 1-10.

Natanek A., Wojtysiak D., Barabasz B. and Langenfeld M. 2001. Badania nad gęstością okrywy włosowej u norek z uwzględnieniem obrazu histologicznego skóry (Hair coot in the mink with regard to histological appearance of skin). Roczniki Naukowe Zootechniki, Suplement 12: 209-214.

Vanek M. and Hanzlova J., 1992. Surface morphology and innervation of defective guard hairs of American mink (*Lutreola vison,* Schreb. 1974). Norwegian Journal of Agricultural Sciences, Supplement. 9: 629-636.

Part 5. Theme on 'Welfur for mink and foxes'

WelFur – mink: development of on-farm welfare assessment protocols for mink

S.H. Møller, S.W. Hansen, T. Rousing and J. Malmkvist
Department of Animal Sciences, Aarhus University, Blichers Allé 20, P.O. Box 50, 8830 Tjele, Denmark; steenh.moller@agrsci.dk

Abstract

European Fur Breeders' Association initiated the 'WelFur' project in 2009 in order to develop a welfare assessment protocol for mink and fox farms after the Welfare Quality® standards. The assessment is based on four welfare principles (*Good feeding, Good housing, Good health* and *Appropriate behaviour*) and 12 underlying criteria, to be measured on-farm. The major steps in the development of the WelFur mink protocols are described: (1) Writing literature reviews and listing potential measures. (2) Identifying valid, reliable and feasible welfare measures. (3) Developing registration protocols, descriptions, and schemes. (4) Testing preliminary protocols in relevant seasons of the annual production. This paper focus on the evaluation of validity, reliability and feasibility of the 22 measures that have been selected for the WelFur assessment protocols. These protocols have been tested in the three mink production seasons: Winter, spring, and autumn, in order to cover the life cycle of mink and proved feasible for a one-day visit.

Keywords: validity, reliability, feasibility of welfare measures, *Neovison vison*

Introduction

Society is showing a growing interest in the welfare of livestock and an increasing number of on-farm welfare assessment systems are being developed. 'Welfare Quality®' (WQ) is one of the latest and most ambitious animal welfare assessment approaches. It is based on a multidimensional animal welfare definition focusing on feeding, housing, health and behavioural aspects and aimed on certification. WQ have been developed for swine, cattle and poultry herds, as well as abattoirs (e.g. Welfare Quality®, 2009a,b,c). WQ was funded under the EU 6 Framework Programme 'Food quality and safety' and fur animals were therefore not included.

Existing welfare assessment systems, including WQ, are multi dimensional i.e. are based on numerous measures including different aspects of animal welfare (e.g. hunger, pain, fear etc.). Both resource-based (system and management) and animal-based measures (health and behaviour) have been considered in existing animal welfare systems (Figure 1). The focus has mainly been on indication of poor animal welfare – but recently also indicators of good welfare are included (Mononen *et al.*, 2012; Welfare Quality®, 2009a,b,c).

In order to assess the welfare in the strictly synchronous annual mink production (Møller *et al.*, 2003), this must include three seasons (Møller and Hansen, 2001) (Figure 2). The observations are limited to the last 6-8 weeks in each season, where the information value of the welfare

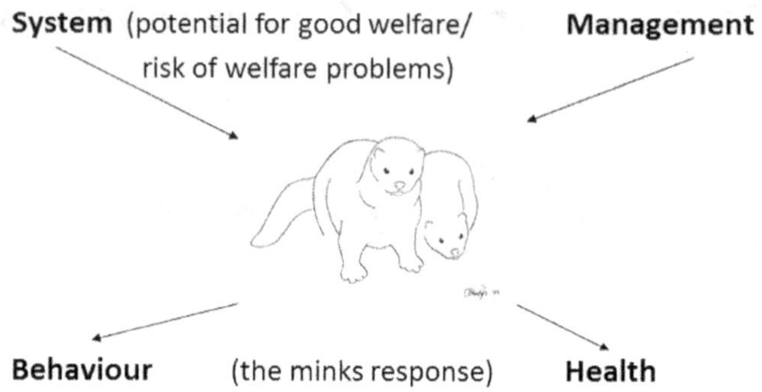

Figure 1. Resource- and animal-based based sources of information for assessing the welfare of farm mink.

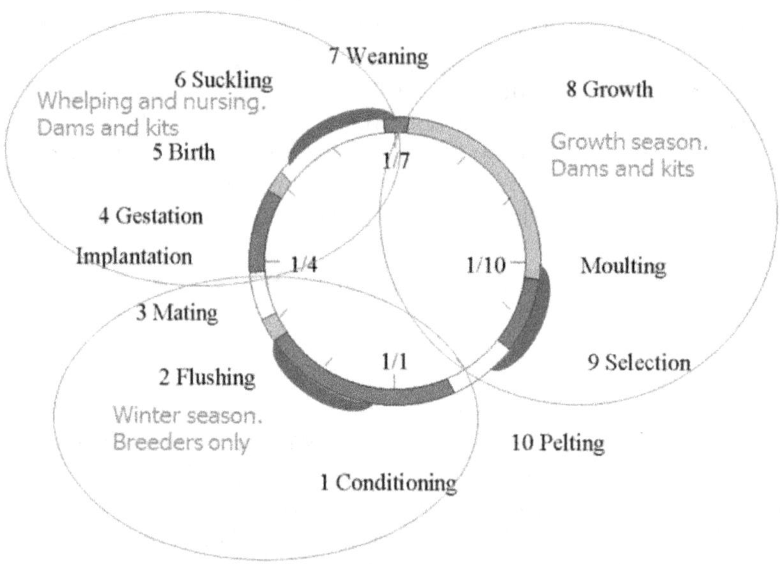

Figure 2. Three seasons of the year to be covered by WelFur Mink. Periods for assessment marked within each season.

measures is highest. These seasons are: (1) Winter. January 1st to February 20th, (2) Spring. May 5th to July 1st/weaning, and (3) Autumn. October 1st to November 30th/pelting (Figure 2).

'Welfare Quality®' seems to become the de facto standard of welfare assessment in the European Union. The European Fur Breeders' Association (EFBA) launched the project 'WelFur' in 2009 with the aim to develop a protocol for use in welfare assessment of mink and fox farms by the standards developed in Welfare Quality ®. This involved a number of steps, such as: getting familiarised with Welfare Quality® project and protocols; writing literature reviews and listing potential measures; identifying valid, reliable, and feasible welfare measures; developing registration protocols, descriptions, and schemes; testing the protocols in all three seasons. The WelFur protocols for fox and mink (Mononen *et al.*, 2012) are based on the four principles and 12 criteria developed in WQ as these are considered independent of animal species (Veissier *et al.*, 2009). Furthermore, WQ dictates, that wherever possible, measures should be animal based. The aims of this paper are to: (1) Present the development of the WelFur-mink protocols. (2) Explain evaluation of the measures included (i.e. the evaluation of validity, reliability, and feasibility of the welfare measures included), using WQ concepts to develop a WelFur protocol for mink

A large number of potential measures have been applied in research on mink welfare or was suggested during the development. Many of these require blood-sampling, laboratory tests, video recording or were considered not feasible for other reasons. The definition of 'a measure' is somewhat arbitrary as some involves several observations (e.g. *Quality of nest box* includes 4 observations) while others could be divided into a large number of individual measures (e.g. the many potential diseases included in the measure *Obviously sick mink*) (Table 1). By reviewing previous work and thorough discussions we came up with 54 potentially feasible welfare measures for mink. For each of the three seasons each measure was evaluated according to their:

- Validity: the measure reflect some aspect of the actual welfare of the mink relative to the criteria.
- Reliability: the measure has acceptable inter or intra observer reliability and robustness to external factors e.g. time of day or weather condition.
- Feasibility: the measure is possible and practical to apply in practice with reasonable costs.

The review written for each of the 12 criteria was concluded by an assessment of the validity, reliability and feasibility of the measures based on a three point scale (EFSA, 2008: 64) (Table 1):

1. Low certainty: Scarce or no data available; evidence provided in unpublished reports, or based on personal observations or communications; authors' conclusions vary considerably between the reports.
2. Medium certainty: Some or only incomplete data available; evidence provided in small number of references; authors' conclusions vary from one to the other; solid and complete data available from other species which can be extrapolated to the species considered.
3. High certainty: Solid and complete data available; strong evidence in multiple references with most authors coming to the same conclusion.

A total of 22 welfare measures passed this evaluation and are included in one or more of the three seasonal protocols (Table 1). This inclusion is discussed according to the WQ template.

Validity of measures included

Good feeding: absence of hunger and thirst

Body condition score (BCS) was chosen as an animal-based measure with high validity (e.g. Damgaard *et al.*, 2004) and acceptable reliability (Hansen *et al.*, 2009) in all seasons for the criterion 'Absence of prolonged hunger'. *Continuous water availability*, and *Function and cleanliness of the drinkers* were chosen as valid (Møller, 1991), and reliable resource-based measures for 'Absence of prolonged thirst', as it was not possible to find feasible animal-based measures for this criterion.

Table 1. Welfare measures included in one or more of the three seasonal assessment protocols for mink, under the four principles and 12 criteria. R/A=resource/animal-based measure. V=validity, R=reliability, and F=feasibility evaluation: 1=low, 2=medium, 3=high. Season of the measurement: 1=winter (breeding animals), 2=spring (breeding animals and pre-weaned kits), 3=autumn (growing juveniles and breeders).

Welfare principles and criteria	Welfare measures	R/A	V	R	F	Season
I. Good feeding						
1. Absence of prolonged hunger	body condition score	A	3	2-3	3	1, 2, 3
2. Absence of prolonged thirst	continuous water availability	R	3	3	3	1, 2, 3
	function and cleanliness of the drinkers	R	3	3	3	1, 2, 3
II. Good housing						
3. Comfort around resting	access to a nest box	R	3	3	3	1, 2, 3
	quality of nest box: dry, clean, sharp edges, fleas	R	3	2	3	1, 2, 3
	cleanliness of the fur	A	1	3	3	1, 2, 3
4. Thermal comfort	protection from exceptional weather conditions	R	2	2	3	1, 2, 3
	nest box and bedding material	R	2-3	3	3	1, 2, 3
5. Ease of movement	space available for moving	R	1-3	3	3	1, 2, 3
III. Good health						
6. Absence of injuries	lesions and injuries	A	3	2	3	1, 2, 3
7. Absence of disease	mortality	A	3	2	3	1, 2, 3
	diarrhoea (incl. sticky kits)	A	3	2	3	1, 2, 3
	lameness and impaired movement	A	3	3	3	1, 2, 3
	obviously sick mink	A	3	3	3	1, 2, 3
8. Absence of pain induced by management procedures	methods of killing:					
	individual mink	R	3	3	3	1, 2, 3
	mink at pelting	R	3	3	3	1, 2, 3
IV. Appropriate behaviour						
9. Expression of social behaviours	weaning procedures: age, distance, time between weaning and separation	R	2	3	3	2
	social housing	R	2	3	3	3
10. Expression of other behaviours	cage enrichment	R	2	3	3	1, 2, 3
	stereotypy	A	3	3	3	1, 2, 3
	fur-chewing	A	3	2-3	3	2, 3
11. Good human-animal relationship	temperament (stick test)	A	3	3	3	1, 3
12. Positive emotional state	frequency and duration of handling	R	3	3	3	1, 2, 3

Good housing: comfort around resting, thermal comfort and ease of movement

Most measures under this principle reached the highest validity scores (Table 1). The availability (Hansen *et al.*, 1994, 2000) and quality (Møller, 1990; Malmkvist, 2008) of a nest box are of great importance to mink, and are thus valid resource-based measures. Research is lacking or scarce on the measures *Cleanliness of the fur* and *Protection from exceptional weather conditions* and the inclusion of fur cleanliness is still under evaluation. The *Cage dimensions* are obviously important for the minks' ease of movement, but once a satisfactory cage size has been reached, the validity of welfare improvement with increased size is undocumented (see discussion in Mononen *et al.*, 2012).

Good health: absence of injuries, disease, pain

The prevalence of *Lesions and injuries*, which is generally low in mink production (Møller 2011) is a valid measure of *Absence of injuries*. European farmers are obliged to register the *Mortality* which is taken as a valid measure of *Absence of disease*. Due to the low prevalence (e.g. Rattenborg *et al.*, 1999) most potential diseases are pooled by the measure *Obviously sick mink*. Due to a higher expected prevalence *Lameness and impaired movement* and *Diarrhoea* (incl. *Sticky kits* in season 2) is observed separately. These are all animal-based measures with high validity in all three seasons (Table 1). Effective *Methods of killing* are scrutinised for pelting as well as for individual mink that need to be euthanized, are considered valid measures of '*Absence of pain induced by management procedures*'.

Appropriate behaviour: social behaviour, other behaviours, human-animal-relationship and positive emotional state

The majority of welfare research in mink is based on applied ethology research and most behavioural measures have been tested and documented prior to the WelFur project. The resource-based measure *Weaning procedures* needing further investigation and *Social housing* still under intense investigation are therefore considered having medium validity. Due to vast scientific documentation the resource-based *Cage enrichment* (e.g. Hansen *et al.*, 2007; Mason *et al.*, 2007), and the animal-based *Stereotypy* (e.g. Malmkvist *et al.*, 2011; Mason and Latham 2004), and *Fur-chewing* (e.g. Hansen *et al.*, 1998) are highly valid measures. *Stereotypy* and *Fur-chewing* are by definition abnormal behaviours mainly caused by inadequate environment and therefore often used signs of poor animal welfare although the relation between these measures and welfare at the individual level is not fully understood (e.g. Mason and Latham 2004). Stereotypic mink are more stress-sensitive (Malmkvist *et al.*, 2011), but the performance of stereotypic behaviour may relate to reduced stress at the brain level (Malmkvist *et al.*, 2012).

Numerous references render the test of *Temperament* in mink a very well validated indicator of the emotional state in mink (see e.g. Malmkvist and Hansen, 2002; Kirkden *et al.*, 2010). It can also be argued to reflect a human-animal relationship although it is different from social species that often relate more direct and long term towards humans. This animal-based measure is therefore taken as a highly valid score of the combined criteria of 'Good human-animal relationship' and 'Positive emotional state' in season 1 and 3 (Table 1). Testing the protocol proved the temperament test less valid in season 2, as the nursing dams often stay with or aggressively defend their litter. The *Frequency and duration of handling* are situations that inflict fear and frustration in the mink and the amount of research is sufficient to grade the validity with high certainty (Hansen and Damgaard, 1991).

Reliability of measures included

The reliability of most resource-based measures is high (Table 1) as they are measured without the involvement of subjective assessment. Furthermore the registrations are independent of external factors such as time of day or weather condition. The two exceptions from objective assessments are: (1) *Quality of nest box* (assessment of whether the nest box is *wet, dirty,* have *sharp edges,* or *fleas*) and (2) *Protection from exceptional weather conditions* (assessment of protection against wind, heat, cold, and rain from the shed and cage system). The reliability of these depends highly on skills and training of observers.

The reliability of most animal-based measures is also high (Table 1) as they are measured under standardized conditions making them robust to external factors. Their reliability have been tested for use in some seasons (see discussions in Mononen *et al.*, 2012) and are being tested further if included also in other seasons (Møller *et al.*, 2012). The three exceptions are: (1) *Lesions and injuries* for which the observations depend on how the mink present itself. (2) *Mortality* that depends on the farmers' observations and registrations. (3) *Diarrhoea* for which the observations depends to some degree on the frequency of mucking out the trough for the slurry system. These are all measures depending highly on observer skills and training, and measures for which robustness to external factors are not yet documented.

An important part of reliability of behavioural welfare measures is the robustness to external factors e.g. time of day. Stereotypic behaviour illustrates this challenge, as it is mainly displayed during active periods at dawn and dusk, and, if feeding is restricted, also in relation to feeding time (Hansen and Møller, 2008). Quantification of mink that display stereotypic behaviour as part of their repertoire, therefore, depends heavily on time and method of observation. In WelFur, the goal is to measure the proportion of stereotypers in a standardized way, by postponing the feeding and observing the unfed mink. This will prompt only the established stereotypers to display this behaviour. Robustness is obtained by calculation of stereotypy in percent of all active mink. E.g. the sound of a feeding machine during observation will tend to increase activity as well as stereotypy, rather than the proportion between them. This standardization allows for observation of stereotypy in most situations, while further validation is needed, e.g. for farms that feed at different times every day. In that case, the sound of a feeding machine has been suggested as a way to standardize the test situation – but this has to be validated.

Feasibility of measures and protocol

The literature rarely reports on feasibility so the evaluation is mainly based on practical experience. Many potential animal-based welfare measures are unfeasible due to cost, time or procedure involved e.g. video recordings, hormone analysis and catching the animals. Animal-based measures are often time-consuming because they vary, and thus have to be measured for each individual mink. Resource-based measures often have little (e.g. cage design) or no (e.g. watering system) variation within a farm and are therefore fast to sample.

As animal-based measures are favoured in WelFur, the observation procedures are planned to facilitate this. All measures have high feasibility and are taken on a stratified sample of 120 cages irrespective of farm size. This sample is taken to represent the farm according to welfare relevant variation in housing and animals and is described further in Rousing *et al.* (2012). Therefore most animal-based measures can be observed simultaneously, e.g. while scoring the body condition or the temperament.

Not only each measure but also the combination of measures in the welfare assessment protocol should be feasible to assess at a one-day farm visit. For this purpose the mink WelFur protocols have been tested in 9 farms in Denmark in each of the three seasons and testes have also been carried out in The Netherlands, Sweden and Norway. Based on these tests and the experience from different countries the protocols, description of measures, and registration schemes have been refined. Furthermore the sampling strategy and size (Rousing *et al.*, 2012), and the Inter Observer Reliability (Møller *et al.*, 2012) have been tested in season 2 and 3, while the robustness of the protocol to the age of the kits in season 2 is currently under investigation.

Concluding remarks

The present WelFur mink welfare assessment protocols, descriptions, schemes, and scoring calculation systems (see Botreau *et al.*, 2012) are almost ready for implementation. Despite the validation described in this paper, it is anticipated that testing and implementing the protocols in all EFBA member countries will reveal farm conditions and situations in which the protocols are not robust, where some measures will prove less valid or reliable than expected. Based on the results and experiences from these countries a refined Version 2 of the WelFur mink protocol can be build.

Acknowledgements

We are grateful to all WelFur partners who have generously contributed to the development, test, and refinement of the WelFur assessment protocols and schemes. Thanks to farmers for letting us in, to colleagues and advisors for making the testing possible, and to INRA for making our reasoning logic and consistent. EFBA is acknowledged for financial and practical support of the WelFur project.

References

Botreau R., Gaborit M. and Veissier, I., 2012. Applying Welfare Quality® strategy to design a welfare assessment tool for foxes and mink farms. Scientifur. 35, 4, pp. 6.

Damgaard, B.M., Hansen, S.W., Børsting, C.F. and Møller, S.H., 2004. Effects of different feeding strategies during the winter period on behaviour and performance in mink females (*Mustela vison*). Applied Animal Behaviour Science 89: 163-180.

European Food Safety Authority, 2008. Animal welfare aspects of husbandry systems for farmed Trout, Prepared by Working Group on Trout Welfare, Issued on 11 September 2008. Annex I to The EFSA Journal 796: 1-97.

Hansen, S.W. and Damgaard, B.M., 1991. Effect of environmental-stress and immobilization on stress physiological variables in farmed mink. Behav. Processes 25: 191-204.

Hansen, S.W., Hansen, B.K. and Berg, P., 1994. The effect of cage environment and *ad libitum* feeding on the ciradian rythm, behaviour and feed intake of farm mink. Acta Agric. Scand. sect. A 44: 120-127.

Hansen, S.W., Houbak, B. and Malmkvist, J., 1998. Development and possible causes of fur damage in farm mink – significance of social environment, Acta Agriculturae Scandinavica. Section A. Animal Science 48: 58-64.

Hansen, S.W., Jensen, M.B., Pedersen, L.J., Munksgaard, L., Ladewig, J. and Matthews, L., 2000. Methodological aspects of operant conditioning: do type of work and duration of access to reward affect demand functions in mink. In: Proceedings of the 12th Nordic Symposium of the International Society for Applied Ethology, pp. 29-30.

Hansen, S.W., Malmkvist, J, Palme, R and Damgaard, BM, 2007. Do double cages and access to occupational materials improve the welfare of farmed mink? Animal Welfare 16: 63-76.

Hansen, S.W. and Møller, S.H., 2008. Diurnal activity patterns of farm mink (*Mustela vison*) subjected to different feeding routines. Applied Animal Behaviour Science 111: 146-157.

Hansen, B.K, Møller, S.H., Berg, P. and Bækgaard, H., 2009. Validation of a method for body condition scoring. Annual report 2008, Danish Fur Breeders Research Center, Holstebro, Danmark. pp. 129-135. [In Danish].

Kirkden, R.D., Rochlitz, I., Broom, D.M. and Pearce, G.P., 2010. Assessment of on-farm methods to measure confidence in mink and foxes on Norwegian farms. Report prepared for Dyrevernalliansen (Norwegian Animal protection Alliance), Oslo, Norway. Cambridge University Animal Welfare Information Centre, Department of Veterinary Medicine, University of Cambridge, Cambridge, UK, 43 pp.

Malmkvist, J and Hansen, S.W., 2002. Generalization of fear in farm mink, *Mustela vison*, genetically selected for behaviour towards humans. Anim. Behav. 64, 487-501.

Malmkvist, J. and Palme, R., 2008. Periparturient nest building: Implications for parturition, kit survival, maternal stress and behaviour in farmed mink (*Mustela vison*). Appl. Anim. Behav. Sci. 114: 270-283.

Malmkvist, J., Jeppesen, L.L. and Palme, R., 2011. Stress and stereotypic behaviour in mink (*Mustela vison*): A focus on adrenocortical activity. Stress 14: 312-323.

Malmkvist, J., Brix, B., Henningsen, K. and Wiborg, O., 2012. Hippocampal neurogenesis increase with stereotypic behaviour in mink (*Neovison vison*). Behav. Brain Res. 229: 359-364.

Mason, G. and Latham, N.R, 2004. Can't stop, won't stop: is stereotypy a reliable animal welfare indicator? Animal Welfare 13: 57-69.

Mason, G., Clubb, R., Latham, N. and Vickery, S., 2007. Why and how should we use environmental enrichment to tackle stereotypic behaviour? Appl. Anim. Behav. Sci. 102: 163-188.

Mononen, J., Møller, S. H., Hansen S. W., Hovland A. L., Koistinen, T., Lidfors, L., Malmkvist, J., Vinke, C. and Ahola, L., 2012. The development of on-farm welfare assessment protocols for foxes and mink: the WelFur project. Animal Welfare (in press).

Møller, S., 1990. The need for nest boxes and drop-in bottoms in the whelping period of female mink. Scientifur 14, pp. 95-100.

Møller, S. 1991. Drinking behaviour of mink in relation to watering system and water temperature. NJF seminar no. 192. Drinking water for farm animals. Uppsala, Sweden. March 6-7. pp. 12.

Møller, S.H. and Hansen S.W. 2001. Assessment of mink welfare at farm level. NJF seminar nr. 331. Snekkersten, Danmark, 1/3/10. 8 pp.

Møller, S.H., Hansen, S.W. and Sørensen, J.T. 2003. Assessing Animal Welfare in a Strictly Synchronous Production System: The Mink Case. Animal Welfare 12: 699-703.

Møller, SH., 2011. The incidence of wounds and injuries in farm mink. Internal rapport no. 109, September 2011. Aarhus University, pp. 61-67. [In Danish]

Møller, S.H., Rousing,T. and Hansen, S.W., 2012. WelFur – Mink. Inter Observer Reliability of on-farm welfare assessment in the growing season. Scientifur 35, 4, pp 6.

Rousing, T., Møller, S.H., and Hansen, S.W., 2012. WelFur – Mink. Effect of sample size on test results of animal based measures. Scientifur 35, 4, pp 6.

Rattenborg E., Dietz H.H., Andersen T.H. and Møller S.H. 1999 Mortality in Farmed Mink: Systematic Collection versus Arbitrary Submissions for Diagnostic Investigation. *Acta Veterinaria Scandinavica 40*: 307-314

Veissier, I., Botreau, R. and Perny, P., 2009 Scoring animal welfare: difficulties and Welfare Quality® solutions. In: Keeling, L. (ed.), Welfare Quality Report No. 12. An overview of the development of the Welfare Quality® project assessment system, Cardiff University, Cardiff, UK, pp. 15-32.

Welfare Quality®, 2009a. Welfare Quality® assessment protocol for cattle. Welfare Quality® Consortium: Lelystad, the Netherlands, 182 pp.

Welfare Quality®, 2009b. Welfare Quality® assessment protocol for pigs. Welfare Quality® Consortium: Lelystad, the Netherlands, 122 pp.

Welfare Quality®, 2009c. Welfare Quality® assessment protocol for poultry. Welfare Quality® Consortium: Lelystad, the Netherlands, 114 pp.

WelFur-mink: on-farm welfare assessment of mink (*Neovision vision*) – effect of sample size on animal based measures

T. Rousing, S.H. Møller and S.W. Hansen
Dept. of Animal Science, Aarhus University, Blichers Allé 20, P.O. Box 50, 8830 Tjele, Denmark; tine.rousing@agrsci.dk

Abstract

European Fur Breeders' Association initiated the 'WelFur project' in 2009 which is aiming at developing an applicable on farm welfare assessment protocol for mink based on the Welfare Quality® principles. Such a welfare assessment system should possess the following qualities: It should be 'high' in validity, reliability as well as feasibility – the latter both as regards time and economy costs. This paper based on empiric data addressed the questions on needed sample size for a robust herd assessment of animal based measures. The animal based part of the full WelFur protocol including 9 animal based measures such as 'stereotypy', 'temperament' and 'body condition'. These measures were in each of 9 Danish mink farms on approximately 250 adult mink on each farm collected during the nursing season 2011 by a total of 8 observers working in pairs. Descriptive result showed a large between herd variations in herd prevalence of the mentioned parameters. Statistical analyses showed that a sample size of 125 adult mink was a robust estimate of the herd level of animal based measures.

Introduction

The present paper reports a part of a research activity 'WelFur' initiated by European Fur Breeders' Association (EFBA) focusing on developing a 'Welfare Quality®' inspired animal welfare system for mink and fox farms (Mononen *et al.*, 2012). Validity, reliability as well as feasibility – here among time and economy costs – are key words when developing any animal welfare systems for implementing on commercial farms. In total twenty two measures have been selected for the WelFur assessment protocols. These protocols are to be tested in the three mink production seasons: Winter, spring, and autumn, in order to cover the life cycle of mink and proved feasible for a one-day visit (for a general presentation of and discussion of these see Møller *et al.*, 2012 in the present proceedings).

Most mink farms in Europe range from 1000 to 10,000 breeder dams, and thus around 5 time as many kits during the growth season. It is therefore not possible to test all animals and a sample has to be taken for the WelFur on-farm welfare assessment. Focusing on mink farms, this paper will address the question: 'What sample size is needed for a robust estimate of the herd level of animal based measures?' The quality of such a sample should be to represent the mink on the farm at the time of assessment as good as possible under the restraint that the assessment should be possible to carry out in one day – e.g. preferably in about 5 hours – in order to allow also for transportation of the assessors. One approach could be to stratify a random sample large enough for the outcome of the welfare assessment to be the roughly same as if all mink on the farm were tested. This would also mean that the result will only change marginally if a different random stratified sample of the same size were tested. Most likely the sample size needed would vary somewhat between welfare measures – first of all because of differences in herd prevalence of different measure outcomes. However, for practical reasons, it would be a large benefit if all the measures can be taken from the same sample, not least because the stratification on housing system, sex, age, and colour type of mink etc. would only have to be done once. If some measures could reliably be taken in a smaller number, this could then be done by taking, e.g. every second or third cage in the defined sample.

It was decided in advance of each welfare assessment based on farmer interviews to define a sample to take the measures from. For feasibility reasons a section of typically 6-7 cages was defined as the smallest sampling unit. Based on prior knowledge (e.g. Møller, 1993; Møller and Berg, 2002) of mink farm layout and distribution of animal groups during the annual production cycle, it was suggested that a sample of 20 sections, i.e. 120-140 cages with mink would fulfil the needs of a representative sample.

Animal based welfare measures are in comparison with resource based measures often considered less reliable because collection of these often includes subjective assessments. Therefore, this paper focus on the animal based part of the WelFur assessment protocols. This paper tests the hypothesis that a random stratified samples of 20 cage sections (120-140 cages with mink) has a sufficient size to give a valid estimate of the farm. This is tested by comparing the result from two such samples by 9 animal based WelFur measures taken during the nursing season on 9 farms.

General experimental design for the exploration of effect of sample size

The measures were taken in the spring season – more precisely in the nursing period between 9 and 23 June 2011 by a total of 8 observers working in pairs at 9 Danish mink farms. Within herd two stratified samples of 20 sections each, i.e. two times 120-140 cages were randomly appointed. The stratification was based on housing system, sex, age, and colour type of mink. In practise the two samples were taken by appointing one random stratified sample and then double it to make two sample series: If for example 2 sections of brown adult females housed in two row sheds were included in a standard sample – the first two sections defines the first sample – the next two sections the second sample. The farmers were asked to postpone the feeding at the day of observation by approximately 1 hour. All other farming procedures were carried out as normal.

The 9 animal based measures

The methodologies of measurement of the 9 animal based measures relevant for the nursing period are presented in Table 1. For an overview of development of the protocols, selection of measures based on validity, reliability and feasibility assessments of these measures see Mononen *et al.* (2012) and Møller *et al.* (2012).

Results as regards the effect of sample size

The measures *cleanliness of the fur, diarrhoea, sticky kits,* and *obviously sick mink* had very low herd prevalence for what reason calculation of sample size effects were considered meaningless.

Table 1. WelFur measures of mink during the nursing period.

Measure	Categories and definition
Body condition score	cage wise individual (adult animals) body condition scoring of standing animals (allured by a stick or the like to stretch upwards) classification: 1= very lean, 2 = lean, 3= normal, 4= obese or 5 = very obese
Cleanliness of the fur	cage wise individual (adult animals) classification of fur cleanliness: 'clean' 'moderately dirty': faeces, feed remains in smaller parts of the body 'severely dirty': as above but for fur parts larger than in total $5cm^2$
Lesions and injuries	cage wise individual (adult animals and kits) classification of lesion and injury appearance: 'no lesions' 'unhealed wound or lesions with a diameter <10 mm or healed lesions e.g. missing an ear or less than half the tail 'unhealed wounds or lesions with a diameter <30 mm or healed lesions e.g. missing more than half the tail 'unhealed wounds or lesions with a diameter >30 mm or unhealed wound or lesion from e.g. missing more than half the tail
Diarrhoea	cage wise classification based on faeces classification in the cage or the slurry system: 'no evidence of diarrhoea' 'presence of fluid and/or discoloured faeces'
Sticky kits	cage wise counting of affected litters (dirty nails, sweaty, wet, or greasy kits)
Lameness and impaired movement	cage wise individual (adult animals and kits) classification of mobility: 'un-impaired mobility, 'impaired movement'
Obviously sick mink	cage wise individual (adult animals and kits) classification of general appearance: 'no signs of obviously sickness', 'obvious sickness'
Stereotypy	section wise 2 minute habituation to an observer standing nearby cage within section wise 2 minute one-zero individual (adult animals) sampling: 'pacing' and other distinctly repeated movements other than walking. categorisation: 'not active' 'active but nut stereotyping' 'stereotyping'
Temperament (stick-test)	cage wise individual (adult animals) presentation of wooden tongue spatula through the front wire mesh classification of the primary behavioural response: 'fearful': avoidance of the stick 'explorative': sniffing maybe licking the stick 'aggressive': attacking, biting the stick hard 'uncertain': mixture of the above or not 'reacting' to the stick – e.g. mink that stayed in the nest box

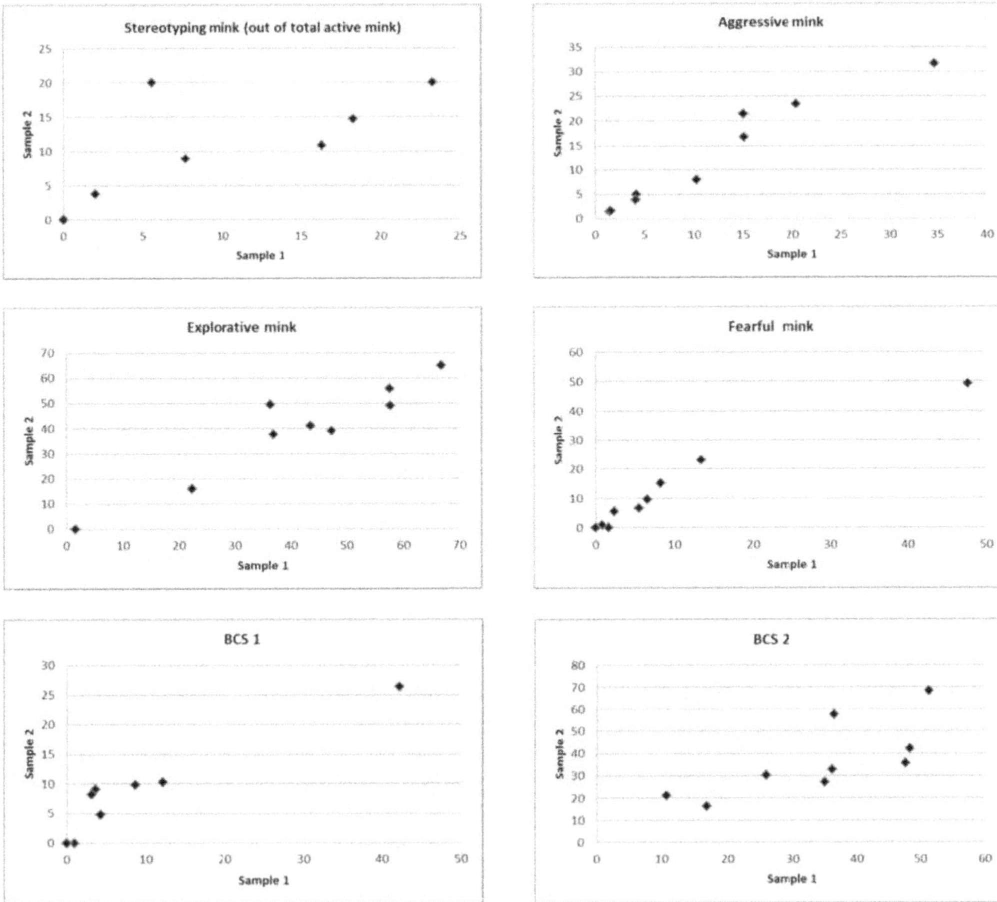

Figure 1. Pair wise herd prevalence of selected animal measures and measure classes for the two stratified random samples.

As illustrated in Figure 1 and Table 2 all measures and classes within measures shoved large variation between herds, and the correlation between samples 1 and 2 were fairly high. The statistical analyses – an unpaired t-tests – showed that there was no significant difference when comparing the two samples.

Discussion

WelFur is a multi-dimensional welfare assessment system based on animal parameters wherever possible. Applicability of measuring methods as well as sample sizes are both crucial elements in developing welfare assessment systems relevant for on-farm use in as well decision support systems, certification and classification systems. The statistical analyses showed that for the animal measures *'body condition', 'stereotypies' and temperament'* there was no significant difference when

Table 2. Mean herd prevalence (min-max) for measures and measure classed and correlations between the two samples.

Measure	Measure class	Sample 1	Sample 2
		Mean (min-max) herd prevalence (%)	
		Correlation btw. herds (P-value,t-test)	
Body condition score	1	8.4 (0-42.2) 0.94 (*P*=0.89)	7.6 (0-26.4)
	2	34.3 (10.7-51.4) 0.74 (*P*=0.75)	36.7 (16.3-68.2)
	3	50.7 (5.5-75.2) 0.88 (*P*=0.98)	50.5 (5.5-79.3)
	4	13.4 (0-20.4) 0.81 (*P*=0.68)	9.6 (0-15.2)
	5	0.2 (0-1.9) na	0 (0-0)
Stereotypies	ratio out of total active	8.1 (0-23.2) 0.78 (*P*=0.89)	8.7 (0-20.0)
Temperament (stick-test)	explorative	41(1.6-66.7) 0.95 (*P*=0.85)	39 (0-65.0)
	fearful	9.6 (0-47.6) 0.97 (*P*=0.72)	12.2 (0-49.2)
	aggressive	11.9 (1.5-34.7) 0.97 (*P*=0.89)	12.6 (1.5-31.6)
	undecided	37.4 (17.5-65.3) 0.97 (*P*=0.85)	36.0 (12.8-63.9)

comparing two samples of 20 sections. Based on this, we accept the hypothesis that a random stratified samples of 20 cage sections (120-140 cages with mink) has a sufficient size to give a valid estimate of the farm. The herd prevalence of the measures *cleanliness of the fur, diarrhoea, sticky kits*, and *obviously sick mink the* where in the 9 mink farms studied were very low, for what reason sample size effect could not be studied.

Conclusion

Based on the comparison of the two samples taken on each of the nine farms we accept the hypothesis that a sample size of 20 cage-sections i.e. 120-140 cages with mink per farm is a robust estimate of the herd level of the animal based measures *'body condition', 'stereotypies' and temperament'*. In this case scenario it means being able to collect all the animal based measures included in the WelFur protocol within a 5 hour visit to the individual farm, also including colleting the resource-based measures.

References

Mononen, J., Møller, S.H., Hansen, S.W. Hovland, A.L., Koistinen, T., Lidfors, L., Malmkvist, J. Vinke, C.M. and Ahola, L., 2012. The development of on-farm welfare assessment protocols for foxes and mink: the WelFur project. Animal Welfare, in press.

Møller, S.H., 1993. Suggestion for a weighing programme for mink production. Annual Academic Report 1992, Danish Fur Breeders Association, pp. 155-161. [In Danish].

Møller, S.H. and Berg, P., 2002. Group size, statistical power and inference in fur animal science. NJF seminar nr. 347. Vuokatti, Finland.

Møller, S.H., Hansen, S.W., Rousing,T. and Malmkvist, J., 2012. WelFur - Mink. Development of on-farm welfare assessment protocols for mink. Scientifur 35, 4, IFASA 2012 Congress proceedings.

Møller, S.H., Rousing,T. and Hansen, S.W., 2012. WelFur - Mink. Inter Observer Reliability of on-farm welfare assessment in the growing season. Scientifur 35, 4, IFASA 2012 Congress proceedings.

Abnormal behaviour in Swedish farm mink during winter

L. Olofsson and L. Lidfors
Department of Animal Environment and Health, Swedish University of Agricultural Sciences, P.O. Box 234, 532 23 Skara, Sweden; lena.lidfors@slu.se

Abstract

The aim was to evaluate the presence of abnormal behaviour in farmed Swedish female mink. Stereotypic behaviour and fur chewing was recorded in 19 farms, including 2843 mink (stereotypies) versus 2803 mink (fur chewing). Behaviour observations were performed during two hours before feeding (starting two hours before the feed machine was started) and two hours after feeding (starting one hour after feeding). One-zero sampling was used with an interval of 10 second and each mink was observed four times each period. Overall, 22% of the mink showed stereotypic behaviour, but there were large differences between farms from three to 76 animals. The presence of stereotypic behaviour did not increase with date of visit. Overall, 36% of the mink showed any form of fur chewing. Severe (> two cm or bald tip of tail) fur chewing was observed in 14% of these. Also, in this case farm differences seemed to be the main explaining factor. No relation to date of visit was found. According to the large differences in management and occurrence of abnormal behaviour future studies should focus on standardised observation procedures and identifying features in farm management that should changed in order to decrease stereotypic behaviour and fur chewing.

Keywords: mink, stereotypies, fur chewing, management

Introduction

Abnormal behaviour such as stereotypic behaviour is frequently found in animals in captivity (Hansen and Damgaard, 2009). Farmed mink are no exception where stereotypies and fur chewing are common abnormal behaviours (Mason, 1994; Mason and Lathamn, 2004). Occurrence of abnormal behaviour is connected to reduced welfare (Mason, 1994; Mason and Lathamn, 2004). Abnormal behaviour is mostly seen during winter when the mink are restrictively fed (Damgaard *et al.*, 2004). Peak levels of stereotypic behaviour are often observed at feeding times and increase prior to feeding (Vinke *et al.*, 2002; Hansen and Damgaard, 2009).

Fur chewing has been shown to be related to poor farm environment such as lack of stimulation or stress (Mason, 1994; Hansen *et al.*, 1998). Hansen *et al.* (2007) found that fur chewing can be decreased through enrichment of the environment. Sensitiveness for development of fur chewing has been shown to be affected by the genetic background of the individual (Zong-yue *et al.*, 2011). Studies have also shown that early separation from the mother can increase the risk of developing fur chewing (Mason, 1994).

The aim of this study was to evaluate the presence of abnormal behaviour such as stereotypic behaviour and fur chewing in farmed Swedish female mink, and to investigate the effect of farm, farm size, age and before or after feeding.

Material and methods

The observations were performed at 19 Swedish mink farms between the 27 January and 16 February 2011. The observations of stereotypic behaviour were based on 2,843 female mink and fur-chewing were based on 2,803 female mink (because some animals were hiding inside the nest during observation). Approximately 150 minks on each farm were observed, divided into 75 standard (brown, n=765; black, n=422) and 75 light colour (silverblue, n=815; sapphire, n=300; hedlund white, n=256; silvercross, n=127; blackcross, n=82; pearl, n=76) mink. Behavioural observations were performed during two periods; the first was done before feeding during two hours (the farmer adjusted feeding time so feed was given when observation stopped) and the second after feeding (starting one hour after feeding). Recorded behaviours are presented in Table 1. One-zero sampling (Martin and Bateson, 2007) was used with an interval of 10 seconds and four recordings were made per mink each period. Observation of stereotypy was done from the opposite shed and all observations were done by one observer.

The age (age 1 born 2010, age 2 born 2009 and age 3 born 2008 or earlier) and colour type was also recorded. Fur chewing was scored based on observation of self-mutilation damage (fur damage in the neck was not included in the results; hence it is not self-inflicted). It was defined from the tip of the tail as category 1 (hair cut 0-2 cm), 2 (hair cut >2 cm), 3 (tail is bald 0-2 cm), 4 (tail is bald >2 cm) and 5 (starting from tip of tail and upwards onto the body or only body). Feed consumption at cage level or as farm average was specified by the farmer.

All farms fulfilled the minimal cage size stated by the Swedish animal welfare legislation (SJVFS, 2010:15; Saknr L100) and the mean size of the bottom area was 0.29 m^2 (min 0.26, max 0.39). The nest's size (according to length, width and height), inside shape (space reduction purpose) and size of entrance varied greatly. Two thirds of the mink was given straw one to four times a month. Wood shavings were commonly used combined with straw or as single bedding material. Most farmers bought feed for the mink from an external source. Estimation of energy consumption was made by the National Veterinary Institute for evaluation against the Nordic Farmer researcher's union NJF production 'norm', i.e. a calculation of the average need for maintenance (Hansen *et al.*, 1991). Difference from 'norm' varied between -41 to +44. Daily amount per animal varied

Table 1. Observed behaviours in female mink and how they were defined.

Behaviour	Definitions
In nest	the mink is in the nest (with more than half of the body inside)
Active in cage	the mink is physically active (standing or moving)
Inactive in cage	lying outside the nest (with more than half of the body outside nest)
Pace	the mink is moving from side to side in cage performing the same momentous movement at least three times with no obvious function
Other stereotypy	the mink is performing a momentous/repetitive movement at least three times with no obvious function

between 125 to ≥210 gram. Farmers reported that they fed the mink once daily from 9-14 á clock, but most often around noon. Temperature varied from +8 to -8 °C.

Statistical analysis

Behavioural data was analyzed with SAS ver. 9.2 (SAS Institute, Cary, USA, 2002-2008). Analyses were done with Wald statistics for Type 3 GEE with the procedure for Generalised Models on behaviours (proc Genmod) and fur chewing (proc Glimmix). One model included behaviours (independent) and the depended variables farm (equal to date of visit as farms were only visited one day), age (1, 2, 3-6 year), period (before and after feeding), colour type (standard and light) and the following interactions: age×period and colour type×period. The other model included farm size (small (1001-3,000), medium (3,001-10,000), large (>10,000)), age, colour type, enrichment and straw. Repeated measure was animal within farm and factor was exchangeable. Pair-wise post hoc comparisons were analyzed with Chi^2-test and confidence values were calculated. Results from colour type, enrichment and straw are not reported here. Spearman rank correlation tested if there was any correlation between stereotypies and fur chewing. Mean values were calculated for the overall behaviours, and Least Square means were provided from the statistical analysis.

Results

There were significant differences between farms regarding number of recordings spent in the nest (χ^2 = 788.21; P<0.0001), being active (χ^2 = 622.34; P<0.0001) and being inactive (χ^2 = 84.49; P<0.0001). Recordings of being in the nest increased with age (LS means 1=0.62±0.037, 2=0.67±0.0692, 3=0.70±0.086). Recordings of being active was also affected by age (P<0.0001, χ^2 = 29.66) where one year old mink were most active (LS means 0.27±0.033) followed by two year old (0.23±0.065) and three years and older were least active (0.19±0.081).

After feeding the female mink were recorded in the nest more often (χ^2 =132.25, P<0.0001) and active out in the cage less often (χ^2 = 91.8, P<0.0001) compared to before feeding (Figure 1).

Stereotypic behaviour

Stereotypies were recorded more often before than after feeding (χ^2 = 91.80; P<0.0001, Figure 1). Recordings of stereotypic behaviour was different between dates of visit (χ^2 = 206.31; P<0.0001, Figure 2), which was the same as different farms. There were differences between the farms regarding pacing (χ^2 = 121.61; P<0.0001) and other stereotypies (χ^2 = 130.42; P<0.0001).

Stereotypies was affected by age (χ^2 = 9.16; P<0.05) and it was higher in young (age 1 and 2) than older (age 3-6) mink. The size of the farm had an effect (χ^2 = 28.67; P<0.0001) on the occurrence of stereotypies, where small farms had the lowest occurrence. However, pacing was more often recorded on large farms and other stereotypic behaviour was more often recorded on middle sized farms.

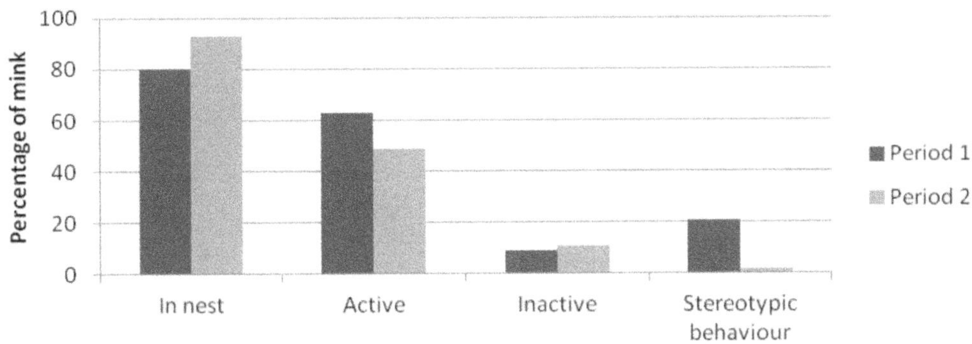

Figure 1. Percentage of female mink observed to perform the behaviour in period 1 (before feeding) and period 2 (after feeding) at least one time during four observations per period and animal on 19 farms (n=2,843 mink).

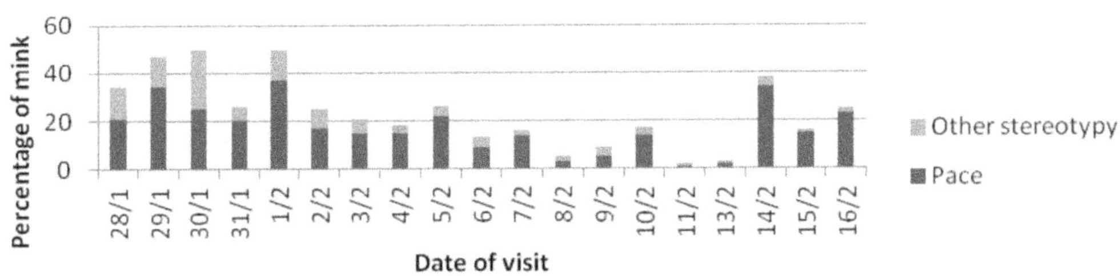

Figure 2. Percentage of female mink that showed stereotypic behaviour on 19 farms during 8 observations (n=150 minks at each farm except 2/2 and 10/2 where n=149).

Fur chewing

Among the observed mink, 36% showed some form of fur chewing (Figure 3). If minor damages (category 1) were removed the number dropped to 14%. Only 6% of the observed mink had a bald tail (category 3 and 4). Less than 1% of the observed mink had fur chewing on the body (category 5).

There were significant differences regarding the presence of fur chewing between farms (F=6.02, P<0.0001). There were significant effects of age (P<0.0075, F=4.90) on presence of fur chewing, where one year old mink hade the highest presence followed by three year and older mink. There was no correlation between number of stereotypies and category of fur chewing (n.s., r= -0.003) (Figure 4), or ambient temperature and performance of stereotypies (n.s., r=-0.098).

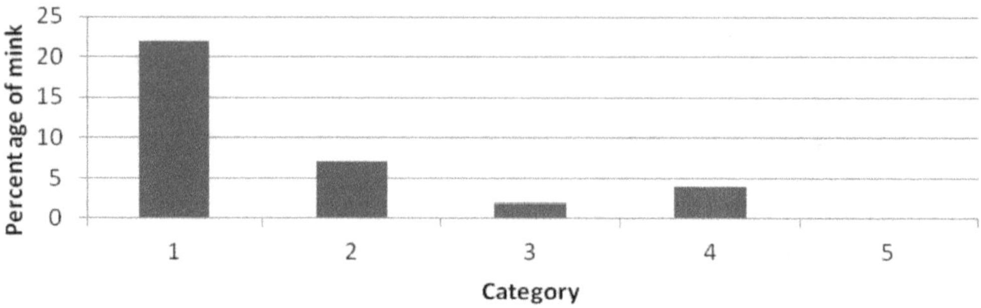

Figure 3. Percentage of female mink with fur chewing from tip of tail (1=hair cut 0-2 cm, 2=hair cut >2 cm, 3=tail bald 0-2 cm, 4=tail bald >2 cm, 5= on body) on 19 farms (n=2,803 animals).

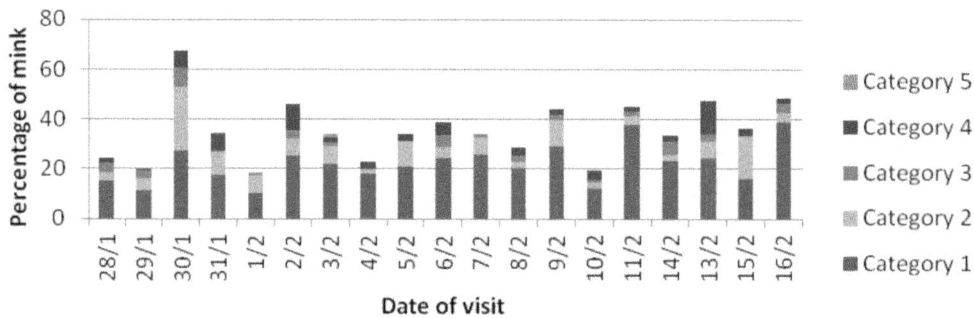

Figure 4. Percentage of mink with fur chewing in different categories based on 19 farms (n=134-150 for each farm).

Discussion

The percentages of mink that showed stereotypies before and after feeding in the present study is much lower compared to findings in previous studies. Mason and Latham (2004) estimated the occurrence of stereotypies in mink to be 80% based on two European studies. An earlier Swedish study by Axelsson *et al.* (2009) used the same method as in the present study but included a larger number of repeated scans in each observation and found that 65% of the mink stereotyped (result based on two farms). Therefore, the large difference in number of mink with stereotypies in the present study can be a consequence of too few scanning-observations in the present study.

The large difference between farms found in the present study should be considered. The conclusion of difference between farms and not date of visit is drawn from the expectation that stereotypic behaviour should increase with date of visit. However, it did not in the present study. The total amount of feed farmers reported to give their mink differed greatly between farms and there are doubts if the numbers are correct. It is therefore difficult to know how amount of feed could be related to the occurrence of stereotypies in the farms visited. Also the large differences in feeding strategy and management between farms can explain some of the difference, as some farms did not feed at the same time every day, two thirds of the farms gave straw of different origin, and several farms provided enrichment (shelf, tube, ball, plastic pipe). Feeding strategies (Hansen and Møller, 2008) and enrichment (Hansen *et al.*, 2007) has been shown to affect the presence of stereotypies. The sensitiveness for developing stereotypies has a genetic component (Hansen *et al.*, 2010; Jeppesen *et al.*, 2004), hence different breeding strategies between farms could also explain some of the difference.

Restrictively fed mink adjust their diurnal rhythm according to time of feeding which affect the presence of stereotypic behaviour (Bildstøe *et al.*, 1991; Hansen and Damgaard, 2009; Vinke *et al.*, 2002; Axelsson *et al.*, 2009). Hence, there is a connection between stereotypic behaviour and increase in activity (Axelsson *et al.*, 2009; Bildstøe *et al.*, 1991). Hansen and Møller (2008) found that mink that were feed *ad libitum* had two peaks of activity and that those that were fed restrictively also had one peak in activity before feeding. Hence, the correlation between feeding and activity can have resulted in a lower presence of stereotypic behaviour than is actually the case on those farms that fed the mink *ad libitum*. Since *ad libitum* fed mink has to be provoked by the feeding procedure to show anticipatory activity (Hansen and Møller 2008). There is also a possibility that the farmers can have affected the presence of stereotypies by changing the feeding routine before the visit. Hence, only one day of visit made it impossible to control for previous feeding routines. By exclusion of the four farms where feeding routine was suspected to have been changed before the visit, the numbers of mink that showed stereotypies increased from 22% to 26%.

Fur chewing was also strongly affected by farm and not dates of visit, since no increase of fur chewing with date of visit was seen. Furthermore, by exclusion of minor damage (category 1) the number dropped to 14%. As the sensitiveness for development of fur chewing is believed to be of genetic origin (Zong-yue *et al.*, 2011) different breeding strategies by farmers might explain some of the large difference between farms. It was expected that mink with no stereotypies might have more fur chewing. However, no correlation between the amount of stereotypies and the categories of fur chewing were found. As the number of females showing stereotypies were quite low this could be one reason for why we did not find any correlations.

In conclusion this study had a considerably lower presence of stereotypies than previous studies, but there were large differences between farms. Management, feeding routines and low number of recordings per observation can have contributed to the lower occurrence of stereotypic behaviour. Hence, the results in this study should be considered with cautions. Future studies should focus on standardised robust observation procedures and identifying which features in the farm management that needs to be changed in order to decrease stereotypic behaviour and fur chewing in mink.

Acknowledgment

To all farmers and Jörgen Martinsson who made this study possible, thank you. For all support during the study, thank you Eva Aldén and Anna-Maria Andersson at SVA. Also we want to thank Jan Hultgren, Helené Axelsson and Gunilla Jacobsson at SLU for helpful advice. The Swedish Fur Breeders Association financed this study.

References

Axelsson, H.M.K., Aldén, E. and Lidfors, L., 2009. Behaviour in female mink housed in enriched standard cages during winter. Applied Animal Behaviour Science 121: 222-229.

Bildstøe M., Heller, K.E. and Jeppesen, L.L., 1991. Effects of immobility stress and food restriction on stereotypies in low and high stereotyping female ranch mink. Behavioural Processes 25: 179-1 89.

Dunstone, N., 1993. The Mink. T and AD Poyser Ltd., London, UK.

Hansen, N.E., Finne, L., Skrede, A. and Tauson, A., 1991. Energiforsyningen hos mink og rave. DSR Forlag. Landbohøjskolen, København.

Hansen, B.K., Jeppesen, L.L. and Berg, P., 2010. Stereotypic behaviour in farm mink (*Neovison vison*) can be reduced by selection. J. Anim. Breed. Genet. 127: 64-73.

Hansen, S. W., Houbak, B. and Malmkvist, J., 1998. Development and possible causes of fur damage in farm mink-significance of social environment. Acta Agriculture Scandinavic Section A, Animal Science 48: 58-64.

Hansen, S.W., Malmkvist, J., Palme, R. and Damgaard, B.M., 2007. Do double cages and access to occupational materials improve the welfare of farmed mink? Animal Welfare 16: 63-76.

Hansen, S.W. and Jeppesen, L.L., 2008. Bidmærker som velfærdsindikator hos mink. Faglig Årsberetning 2007, Pelsdyrerhvervets Forsøgs- og ForskningsCenter, Holstebro, Denmark, pp. 13-23.

Hansen, S.W. and Møller, S.H., 2008. Diurnal activity patterns of farm mink (*Mustela vison*) subjected to different feeding routines. Applied Animal Behaviour Science 111: 146-157.

Hansen, S.W. and Damgaard, B.M., 2009. Running in a running wheel substitutes for stereotypies in mink (*Mustela vison*) but does it improve their welfare? Applied Animal Behaviour Science 118: 76-83.

Jeppesen, L.L, Heller, K.E. and Bildstøe, M. 2004. Stereotypies in female farm mink may be genetically transmitted and associated with higher fertility due to effects on body weight. Applied Animal Behviour Science 86: 137-143.

Martin, P. and Bateson, P., 1993. Measuring Behaviour, an Introductory Guide, 2nd ed. Cambridge University Press, Cambridge, UK.

Mason, G., 1991. Stereotypies: a critical review. Animal Behavior 41: 1015-1037.

Mason, G., 1994. Tail-biting in mink (*Mustela vison*) is influenced by age at removal from the mother. Animal Welfare 3: 305-3011.

Mason, G. and Latham, N.R, 2004. Can't stop, won't stop: is stereotypy a reliable animal welfare indicator? Animal Welfare 13: 57-69.

Vinke, C.M., Eenkhoorn, N.C., Netto, W.J., Fermont, P.C.J. and Spruijt, B.M., 2002. Stereotypic behavior and tail biting in farmed mink (*Mustela vison*) in a new housing system. Animal Welfare 11: 231-245.

Zong-yue, L., Fang-yong, N., Hong-yan, Y., Lai, W. and Xiu-Juan, B., 2011. Rapid detection of self-biting disease of mink by specific sequence-characterized amplified regions. Journal of Forestry Research 22: 123-126.

WelFur – foxes: development of the on-farm welfare assessment scheme for foxes

J. Mononen[1,3], *T. Koistinen*[1], *A.L. Hovland*[2], *H. Huuki*[1,3] *and L. Ahola*[1]
[1]*Department of Biology, University of Eastern Finland, P.O. Box 1627, 70211 Kuopio, Finland;* [2]*Department of Animal and Aquacultural sciences, Norwegian University of Life Sciences, P.O. Box 5003, 1432 Ås, Norway;* [3]*MTT Agrifood Research Finland, Halolantie 31A, 71750 Maaninka, Finland; jaakko.mononen@uef.fi*

Abstract

In 2009, European Fur Breeders' Association made an initiative to develop welfare assurance schemes for farmed foxes and mink. The WelFur welfare assessment protocols are the core of the schemes. The protocols are based on four welfare principles (*good feeding, good housing, good health* and *appropriate behaviour*) with 12 welfare criteria (2-4 per principle) developed originally for cattle, pigs and poultry in the Welfare Quality® project. Here we describe the development of the WelFur fox protocol and scheme step by step. These steps were: (1) Potential welfare measures were identified and evaluated for their validity, reliability and feasibility. (2) The preliminary protocol were tested on farm visits and refined based on the experiences from these visits. (3) The final protocol with descriptions of the 25 chosen measures and detailed instructions on how to carry out an assessment visit was written. (4) The scoring system was developed and will be tested in 2012. (5) The implementation of WelFur on-farm welfare assessment scheme for foxes will be launched in 2012. (6) The scientific documentation of the development work has been initiated. (7) The need to further refine the WelFur fox protocol and scheme will be evaluated continuously.

Keywords: animal welfare, fox farming, welfare measures

Introduction

The European Fur Breeders' Association (EFBA) launched in 2009 the development of WelFur on-farm welfare assessment protocols and scheme for foxes and mink (Mononen *et al.*, 2012). The scheme will be used for certification and advisory purposes, and the ultimate aim is to improve animal welfare on European fur farms. An overview of the fox protocol and the measures included in it are presented in Table 1. In the present paper we summarise (Table 2) and discuss the steps of the development work of the WelFur fox protocol and scheme.

From Welfare Quality® to WelFur

WelFur fox and mink protocols (Mononen *et al.*, 2012) are based on four principles (*Good feeding, Good housing, Good health* and *Appropriate behaviour*) and 12 criteria (2-4 per principle, Table 1) developed originally for cattle, pigs and poultry in Welfare Quality® (WQ) project (Veissier *et al.*, 2009). The WQ welfare principles and criteria are independent of animal species, but valid, reliable and feasible welfare measures for each criterion must be identified separately for each species.

Table 1. The welfare measures included in the WelFur on-farm welfare assessment protocol for foxes. O=output-based and I=input-based measure. V=validity and R=reliability evaluation: 1=low, 2=medium, 3=high. P=period of the measurement: W=winter (breeding animals), S=summer (breeding animals and pre-weaned cubs) and A=autumn (breeding animals and growing cubs).

Principles and criteria	Measures	O/I	V	R	P
1. Good feeding					
Absence of prolonged hunger	body condition score	O	2	3	W,S,A
Absence of prolonged thirst	continuous water availability	I	3	3	W,S,A
2. Good housing					
Comfort around resting	availability of a platform	I	2	3	W,S,A
	cleanliness of the fur	O	2	2	W,S,A
Thermal comfort	protection from exceptional weather conditions	I	2	2	W,S,A
Ease of movement	space available for moving (area and height)	I	2	3	W,S,A
3. Good health					
Absence of injuries	difficulties in moving	O	3	3	A
	skin lesions or other observed injuries to the body	O	3	3	A
Absence of disease	bent feet	O	2	2	A
	ocular inflammation	O	3	3	A
	impaired mouth and teeth health	O	3	3	A
	diarrhoea	O	2	2	A
	urinary tract infection	O	3	2	W
	obviously sick fox	O	3	3	A
	mortality	O	3	3	W,S,A
Absence of pain induced my management procedures	use and type of neck tongs	I	2	3	W,S,A
	killing method	I	3	3	W,S,A
4. Appropriate behaviour					
Expression of social behaviours	social housing	I	3	3	A
Expression of other behaviours	opportunity to use enrichment	I	3	3	W,S,A
	opportunity to observe surroundings	I	3	3	W,S,A
	stereotypic behaviour	O	3	2	W,S,A
	fur chewing	O	2	3	W,A
Good human-animal relationship	feeding test	O	3	3	W
Positive emotional state	temperament test (stick test)	O	1	2	W
	transport of live animals	I	3	3	W,S,A

The development of WelFur fox protocols was initiated with writing of literature reviews (unpublished so far) on each of the 12 criteria, and by listing all potential welfare measures that have been used in fox welfare research. More than 200 measures were identified, but most of them were not feasible (e.g. all measures requiring video-recording or blood sampling). The validity

Table 2. The main steps in the development of WelFur fox protocols and scheme.

Step	Completeness (April 2012)
0. Getting familiarised with Welfare Quality® project and protocols	100%
1. Identifying valid, reliable and feasible welfare measures	100%
2. Testing of preliminary protocols	100%
3. Writing the final protocol	90%
4. Developing the scoring system	90%
5. Implementation	0%
6. Documentation of the development work	10%
7. Refining the protocol and scheme	0%

and reliability of the remaining measures was evaluated and discussed in six meetings of a review group consisting of fur animal scientists, external animal welfare experts (including WQ experts), and EFBA representatives.

Validity and reliability of measures

The validity and reliability of the measures were evaluated on a rough three point scale (Table 1) modified from the European Food Safety Authority (2008: 64):

1 = low certainty: Scarce or no data available; evidence provided in unpublished reports, or based on personal observations or communications; authors' conclusions vary considerably between the reports;

2 = medium certainty: Some or only incomplete data available; evidence provided in small number of references; authors' conclusions vary from one to the other; solid and complete data available from other species which can be extrapolated to the species considered; and/or experts agree to some extent;

3 = high certainty: Solid and complete data available; strong evidence in multiple references with most authors coming to the same conclusion; and/or experts agree fully.

The experts' opinion was emphasised more in the WelFur's validity and reliability evaluation than in the original EFSA evaluation because of the relatively few number of scientific reports on several of the measures. Typically, the experts' opinion gives the benefit of doubt to the animals.

Traditional on-farm welfare assessment protocols comprised predominantly of input-based measures (i.e. resource-based or management-based measures) (e.g. Bartussek 1999; Finnish Fur Breeders' Association, 2011: 147-152) and only few output-based measures (i.e. animal-based measures). The welfare of the animals is affected by numerous internal and external factors in a complicated manner, and output-based welfare measures are inherently more valid than input-based welfare measures. Therefore, and also in line with WQ (Veissier, 2009), output-based measures were preferred over input-based measures in WelFur fox protocol whenever possible. Sixty percent of the measures in the WelFur fox protocol are output-based, and most of them are under the *good health* and *appropriate behaviour* principles (Table 1). Altogether, finding valid measures was easier for the *good health* and *appropriate behaviour* principles than for the *good feeding* or *good housing* principles.

Including *body condition score* (*BCS*) and *continuous water availability* as the two measures for the hunger and thirst criterion, respectively, of the *good feeding* principle (Table 1) was quite evident from an early stage of the development of the protocol. However, the validity of the low BCS, i.e. very lean animals, is not fully clear (validity score = 2) since strong fluctuation in the body weight may be natural particularly for blue foxes (Korhonen 1988). Instead, the validity of *continuous water availability* has been confirmed in a series of studies (see discussion in Mononen *et al.*, 2012).

None of the measures under the *good housing* principle reached the highest validity scores (Table 1). The research results on the relation between the four measures and the welfare of foxes are either contradictory (*availability of platform* and *space available for moving*) or research is lacking (*cleanliness of the fur* and *protection from exceptional weather conditions*) (see discussion in Mononen *et al.*, 2012).

Only little scientific research has been done on foxes' health, and the choice of injury and disease measures under the *good health* of WelFur fox protocol (Table 1) was based more on practical knowledge on current health issues on fur farms than scientific evidence on the effects of injuries and diseases on foxes' welfare. Therefore, the high validity scores for seven out of the nine injury and disease measures (Table 1) are based solely on opinion shared by animal welfare experts. On the other hand, the experts did not fully agree (validity score = 2) on the welfare effects of *bent feet* and *diarrhoea*, which indicates a need for further research at least on these two measures. The medium validity score for *use and type of neck tongs* results from the lack of evidence on whether the use of the neck tongs *per se* is painful for the welfare of the foxes. Instead, it has been shown that rubber or plastic cover on the animal-end of the tongs prevents foxes from assumedly painful tooth and mouth injuries (see discussion in Mononen *et al.*, 2012). The suitability of electrocution as a killing method for farmed foxes has been proved in both practice and scientific research (Korhonen *et al.*, 2009).

A majority of fox welfare research has been done in the field of applied ethology, and the validity estimation for six out of the eight measures under the *appropriate behaviour* principle have high validity scores (Table 1). The relation of *fur chewing* to welfare has not been studied in foxes, and accordingly its validity was scored at medium level only. The validity score for *temperament test* is low (Table 1), which indicates that the true validity of this test as a fox welfare measure remains to be proved. Interestingly, the farm-level results from *temperament test* correlate with the results from both *feeding test* and *stereotypic behaviour*, indicating that these tests are not necessarily independent of each other (Koistinen *et al.*, 2012).

All but one input measures got the highest reliability score, since their assessment does not include any major subjective elements (Table 1). The exception is *protection from exceptional weather conditions*. The effects of farm structures, vegetation and landscape on the protection from heat and cold are many times difficult to judge. The inter-observer reliability of several output-based measures has been confirmed during the WelFur project (Ahola *et al.*, 2011, 2012). The slight uncertainty (score 2) of the reliability of three output measures results either from the subjective nature of judging the seriousness of the health problem (*bent feet* and *diarrhoea*) or the very low incidence of the phenomenon to be measured (dirty fur in the case of *cleanliness of the fur*).

Feasibility of the measures and protocol

As already pointed out above, most of the welfare measures used in scientific studies of farmed foxes were not feasible for an on-farm welfare assessment purpose. Additionally, several factors affect the feasibility of the protocol and an assessment visit as a whole, as well as the implementation of WelFur scheme. The preliminary versions of the protocol have been then tested on altogether 62 farm visits in Finland and Norway. The measure descriptions and the assessment protocol have been refined based on the experiences from these farm visits.

Output-based measurements are typically more laborious to carry out than input-based measurements. The preference for the former has the profound practical effect that the duration of a farm visit is than if only input measures were used. To balance this, attempts were made to simplify the work of the assessors on the farm. For example, the same measures are used for both fox species, blue fox and silver fox, and their hybrids. Some harmonisation of the measures was done also between WelFur fox and mink protocols in cases where similar measures were included in both protocols (e.g. *Continuous water availability*).

It was also decided quite early in the project, that in order to get a holistic picture of welfare of foxes (and mink: Møller 2011) on a farm, welfare assessments have to be accomplished in all three production periods: winter (breeding animals), summer (breeding animals and pre-weaning cubs) and autumn (growing juveniles and breeding animals). Although it is not necessary or possible to take all the measures in each period (Table 1), the need for three visits on each farm is a challenge to the implementation of the WelFur animal welfare scheme.

All measures have been taken in all periods on the preliminary assessment visits to confirm that the decision to make some measurements in one or two periods only is correct. Also the requirement for an assessment on all three periods can be re-evaluated as soon as sufficiently data has been collected from the farms. If any of the three periods does not bring added value from a certification or advisory point of view and/or there are major difficulties in doing the assessment in the period, this period can or must be abandoned.

Experiences until now indicate that the summer period may be the biggest challenge. Parturition and weaning time in silver foxes are weeks earlier than in blue foxes, and since the optimal time window for doing the visit is at the cubs' age of 4-8 weeks, probably two visits are needed on farms with both fox species in this period. Furthermore, the cubs and vixens may spend a majority of their time in nest boxes, which complicates measurements.

Development of the scoring system

Accordingly, the process of choosing the measures was a mixture of validity, reliability and feasibility evaluations. Then detailed descriptions of the chosen measures and how to carry out the measurements on fox farms were written. The next steps were: (1) to transform the results of each individual measure to a score 0-100, (2) to combine these scores to form 12 criteria scores (0-100), (3) to combine the criteria scores to principle scores (0-100), and (4) to combine the principle scores to an overall assessment (not classified, acceptable, enhanced, excellent). Researchers from French National Institute for Agricultural Research (INRA), who had developed the scoring system for WQ, were responsible also for the development of the WelFur scoring system. The steps 3 and 4 (Veissier *et al.*, 2009) were copied as such from WQ. A panel consisting of five fox welfare experts gave the input for the steps 1 and 2, and these steps are illustrated briefly below and in more detail in Botreau *et al.* (2012).

The members of the fox expert panel were asked to score imaginary results given for each measure, e.g. 14 virtual fox farms with the percentage of stereotyping animals ranging from 0 to 50%. The worst situation was scored 0 and the best 100, and all other situations between these two. However, the experts could also set limits for totally unacceptable situation below which all situations are scored 0. For example, all situations with the percentage of stereotyping animals above 40% could be scored 0.

If the measure is to be taken in more than two production periods, scoring was made separately for each period, and the experts were then asked to rank and score (0-100) virtual farms with various score combinations for these periods. Finally, the scores from the measures within each criterion were aggregated by asking the experts to rank and score virtual farms with various score combinations.

Aggregating period scores to measure scores and measure scores to criterion scores was not always an easy job. Researchers from INRA evaluated the experts' rankings and scorings of various combinations to ensure that they were logical, and if not, the experts were asked to revise their scorings. The five experts agreed fairly well on the scorings of most measures, and the results could be averaged. However, in some cases there was more deviation between the experts' scorings. In these cases it was ensured that all experts had understood the measures in the same way, after which the experts revised their scorings accordingly.

Finally, the average scorings were used to create mathematical formulas transforming the original measurement results on the farms to measure scores and then to criterion scores.

Concluding remarks

The first version of the WelFur fox welfare assessment protocol and scoring system is almost complete, and the implementation of WelFur fox scheme can be launched already in 2012 (Table 2). In Finland the implementation will be supported by a project 'Turkistila 2020' ('Fur Farm 2020') in 2012-2013. In that project welfare situation is assessed on approximately 100 fox farms. Although the present protocol has already been tested on 44 fox farms, future improvement of the protocol may be needed if experiences from the implementation phase show that this is necessary. In addition, WelFur scheme has to be integrated into the existing advisory system to ensure that the WelFur assessment data can be utilised on individual farms to improve animal welfare.

The development of WelFur protocols for foxes and mink are based on not only scientific evidence but also fur animal experts' opinion. Therefore, it is of uttermost importance that the development work is documented comprehensively and publicly to expose WelFur on-farm welfare assessment schemes to external scrutiny. This documentation work, a part of which the present paper is, has just started for foxes (Table 2).

Acknowledgements

We are grateful to the numerous animal welfare and fur animal experts who have helped us in our work: scientists, veterinarians, advisors, clerical workers from the national fur farmers' associations and fur farmers. EFBA is acknowledged for funding and co-ordination of the WelFur project.

References

Ahola, L, Koistinen, T. and Mononen, J., 2011. Health and behavioural measures in farmed foxes: inter-observer reliability of farm averages. In: Widowski, T., Lawlis, P. and Sheppard, K. (eds.). Proceedings of the 5th International Conference on the Assessment of Animal Welfare at Farm and Group Level. August 8-11, 2011, Guelph, Canada, p. 51.

Ahola, L.K., Huuki, H., Hovland, A.L., Koistinen, T. and Mononen, J., 2012. WelFur – foxes: the inter-observer reliability of the WelFur health measures, and the prevalence of health disorders on fox farms during the growth period. Scientifur. 35, 4, pp. 6.

Bartussek, H., 1999. A review of the animal needs index (ANI) for the assessment of animals` well-being in the housing systems for Austrian proprietary products and legislation. Livestock Production Science 61: 179-192.

Botreau, R., Gaborit, M. and Veissier, I., 2012. Applying Welfare Quality® strategy to design a welfare assessment tool for foxes and mink farms. Scientifur. 35, 4, pp. 6.

European Food Safety Authority, 2008. Animal welfare aspects of husbandry systems for farmed Trout, Prepared by Working Group on Trout Welfare, Issued on 11 September 2008. Annex I to The EFSA Journal 796: 1-97.

Finnish Fur Breeders' Association, 2011. Toimintakäsikirja. Suomen Turkiseläinten Kasvattajain Liitto, Vantaa, Finland. 119 pp.

Koistinen, T., Huuki, H., Hovland, A.L., Mononen, J. and Ahola, L., 2012. WelFur – Foxes: Do feeding test, temperament test and a measure of stereotypic behaviour differentiate farms? Scientifur. 35, 4, pp. 6.

Korhonen, H., 1988. Regulation of energy economy in raccoon dogs and blue foxes: A process of dynamic interactions. Comparative Biochemistry and Physiology. Part A: Physiology 91: 263-268.

Korhonen, H.T., Cizinauskas, S. and Viitmaa, R., 2009 Evaluation of the traditional way of euthanasia of farmed foxes from an animal welfare point of view. Annals of Animal Science 9: 73-87.

Mononen, J., Møller, S.H., Hansen S. W., Hovland A. L., Koistinen, T., Lidfors, L., Malmkvist, J., Vinke, C. and Ahola, L., 2012. The development of on-farm welfare assessment protocols for foxes and mink: the WelFur project. Animal Welfare, in press.

Møller, S. H. and Hansen, S.W., 2011. Challenges using Welfare Quality® principles for the development of an on-farm welfare assessment system for mink. In: Widowski, T., Lawlis, P. and Sheppard, K. (eds.). Proceedings of the 5th International Conference on the Assessment of Animal Welfare at Farm and Group Level. August 8-11, 2011, Guelph, Canada, p. 62.

Veissier, I., Botreau, R. and Perny, P., 2009 Scoring animal welfare: difficulties and Welfare Quality® solutions. In: Keeling, L. (ed.), Welfare Quality Report No. 12. An overview of the development of the Welfare Quality® project assessment system, Cardiff University, Cardiff, UK, pp. 15-32.

Welfare Quality®, 2009a Welfare Quality® assessment protocol for cattle. Welfare Quality® Consortium: Lelystad, the Netherlands, 182 pp.

Welfare Quality®, 2009b Welfare Quality® assessment protocol for pigs. Welfare Quality® Consortium: Lelystad, the Netherlands, 122 pp.

Welfare Quality®, 2009c Welfare Quality® assessment protocol for poultry. Welfare Quality® Consortium: Lelystad, the Netherlands, 114 pp.

WelFur – foxes: the inter-observer reliability of the WelFur health measures, and the prevalence of health disorders on fox farms during the growth period

L.K. Ahola[1], H. Huuki[1], A.L. Hovland[2], T. Koistinen[1] and J. Mononen[1,3]
[1]Department of Biology, University of Eastern Finland, P.O. Box 1627, 70211 Kuopio, Finland, [2]Department of Animal and Aquacultural Sciences, Norwegian University of Life Sciences, P.O. Box 5003, 1432 Ås, Norway, [3]MTT Agrifood Research Finland, Halolantie 31A, 71500 Maaninka, Finland; leena.ahola@uef.fi

Abstract

The aim of the present study was to assess the prevalence of health disorders on Finnish and Norwegian fox farms, and the inter-observer reliability (IOR) of the health measures included in the present WelFur fox protocol. Ten and five fox farms were visited in Finland and Norway, respectively, in October-December 2011. Three (Finland) and two (Norway) assessors assessed the same animals on each farm. The results revealed that obesity-related problems, i.e. bent feet, moving difficulties and diarrhoea, were the most common problems on the Finnish fox farms with the majority of the foxes being blue foxes. Bent feet was the most common problem also on the Norwegian fox farms where the majority of the assessed foxes were silver foxes. However, in Norway the percentage of foxes with bent feet was much lower than in Finland. The results also show that IORs of all WelFur health measures but one, i.e. bent feet, are good or even excellent indicating the consistency of assessment between different assessors. It is concluded that all health measures included in the WelFur fox protocol are relevant welfare measures that can be measured reliably enough during a one-day farm visit.

Keywords: blue fox, diseases, health, silver fox, welfare

Introduction

The WelFur project was launched by the European Fur Breeders' Association (EFBA) in 2009. The project's main aim is to develop on-farm welfare assessment protocols for farmed foxes (blue fox, silver fox) and mink (see Mononen *et al.*, 2012). The WelFur protocols are constructed according to the four welfare principles, i.e. *good feeding, good housing, good health* and *Appropriate behaviour*, and 12 welfare criteria defined in the Welfare Quality® project (Veissier *et al.*, 2009). The principle *good health* includes three criteria, out of which two, i.e. *absence of injuries* and *absence of diseases*, include animal-based health measures whereas the third criterion, *absence of pain induced by management procedures*, includes management-based measures. In the present study, only the two criteria including animal-based health measures will be discussed. From the two criteria including animal-based measures, the first one states that 'the animals should be free of injuries, e.g. skin damage and locomotory disorders' (see e.g. Welfare Quality®, 2009). The second one states that 'the animals should be free from diseases, i.e. animal unit managers should maintain high standards of hygiene and care'.

At present, the WelFur fox protocol incorporates 25 measures, out of which nine are animal-based health measures included in the principle *good health*. These measures include *difficulties in moving, skin lesions or other observed injuries to the body, bent feet* (referring to foreleg weakness), *ocular inflammation, impaired mouth and teeth health, diarrhoea, urinary tract infection, mortality* and *obviously sick fox* (referring to health problems that cannot be clearly defined within the former categories). All these measures are observed straight from the animals except the measure *mortality* that is obtained from farm records. It is important that the on-farm welfare assessment protocols include animal-based (i.e. output) measures since these measures have, in general, higher validity than resource- and management-based (i.e. input) measures (see Keeling, 2009). In particular, animal-based health measures are relevant indicators of the welfare of animals. To begin with, health problems may cause pain and suffering to animals, just as they do to humans. Additionally, health problems can be seen as outcomes of experienced long-term stress, e.g. due to inappropriate housing conditions or management. According to Moberg (2000), an animal confronted with stressful situations needs to shift its (energy) resources away from normal biological functions to cope with the stress. Energy can be shifted away from basic biological functions such as growth and reproduction, or immunological defence. If the stress experiences do not cease, the animal will finally enter a pathological state, as Moberg (2000) puts it. With regard to the animal's health status, this means that the animal will become more vulnerable to different kinds of hazards to its health and will finally get ill and die. Basically, this means that a high prevalence of any health disorder may reflect not only an acute problem, e.g. a bacterial attack, but also long term problems related to the animals' ability to cope with the housing conditions on the farm.

Besides being valid (here: reflect some aspects of the welfare of farmed foxes) and feasible (here: possible to measure during a one day farm visit), a welfare measure should also be reliable, i.e. it should be precise and robust enough towards random variation in e.g. the time of day or weather conditions. Furthermore, the result from the measure should be more or less the same regardless of the assessor, i.e. the inter-observer reliability (IOR) of the measure should be high. WelFur IOR studies, assessing the extent to which two or more assessors obtain similar results when measuring the same thing (Martin and Bateson, 2007), have already earlier shown that most of the health measures included in the WelFur fox protocol have good or excellent IOR. The data gathered with the preliminary fox protocol in autumn 2010 revealed that IORs of farm level prevalences, measured with Kendall correlation coefficient W, were high ($W \geq 0.65$) for six out of 10 health measures when the three assessors observed the same animals on the farms (Ahola *et al.*, 2011a). Lower reliabilities were found for the measures *toe and paw problems, bent feet, skin lesions* and *obviously sick*. After this pilot study, the measure descriptions of the measures included in the protocol were refined in order to help the assessors to reach a better consensus on their scorings of health measures. The results from the summer 2011 farm visits show that agreement between the assessors was excellent for most of the health measures although lower IORs were still observed for the measures *bent feet* and *diarrhoea* (Ahola *et al.*, 2011b).

The aim of the present study was to continue the assessment of IORs of the WelFur health measures, now with the latest version of the WelFur fox protocol. Additionally, we present the prevalences of different health disorders on Finnish and Norwegian fox farms during the growth period in autumn 2011.

Material and methods

Ten and five fox farms were visited in Finland and Norway, respectively, in October-December 2011. Altogether there were some 73,000 and 5,500 foxes on the farms that were visited in Finland and Norway, respectively. The largest farm in Finland had about 25,000 foxes, the smallest about 2,000 foxes. The corresponding numbers for the Norwegian farms were 1,140 and 600. Out of the foxes on each farm, 2.3±1.6% were assessed for their health in Finland and 6.0±3.3% in Norway. According to the WelFur scheme, the sample of foxes that is assessed on the farm needs to be a representative sample of the population on the farm, taking into account e.g. the species (silver fox and blue fox) and age (breeding animals and juveniles) of the animals. Accordingly, the majority of the assessed foxes in Finland were blue foxes (71%) and cross breeds (23%) whereas in Norway almost all assessed foxes were silver foxes (92%). In Finland, 23% and 77% of the assessed foxes were adult and juvenile foxes, respectively. In Norway, the corresponding percentages were 26% and 74%.

There were three assessors visiting the farms in Finland, and two in Norway. One of the assessors visited all farms in both Finland and Norway. During the farm visits, all measures included in the fox protocol (see Mononen *et al.*, 2012) were assessed. However, in the present study, only the results on the health measures included in the principle *good health* will be shown. These measures include: *difficulties in moving*, *skin lesions or other observed injuries to the body*, *bent feet*, *ocular inflammation*, *impaired mouth and teeth health*, *diarrhoea* and *urinary tract infection*. Also other health problems not individually listed here were assessed with the measure labelled *obviously sick fox*. The number of foxes euthanized and found dead, i.e. the measure *mortality*, is not considered here since the mortality results are obtained from each farmer's farm records, and not assessed by the assessors. Accordingly, there is no similar need for reliability studies with regard to this measure.

During the assessments, all assessors observed the same individual foxes on each farm. The prevalences of different health disorders on Finnish and Norwegian fox farms are presented as mean values from three and two assessors, respectively. The assessment of the same individual animals by three (Finland) or two (Norway) assessors enabled the calculation of the inter-observer reliabilities (IOR) for the assessed health measures. IORs of the health measures were estimated using the index of concordance (IC) (Martin and Bateson, 2007). ICs for each pair of assessors and for each measure was calculated as A / (A + D), where A = the number of agreements and D = the number of disagreements. IC gets values between 0 and 1, 1 meaning a total agreement between the pair of assessors. An IC value >0.8 was considered as good.

Results

The most common health problems found on Finnish fox farms were moving difficulties, bent feet and diarrhoea (Table 1). On Norwegian farms, the percentages of foxes with health disorders were, in general, lower than on Finnish farms. The most common health problems on Norwegian farms were bent feet, moving difficulties and skin lesions.

Table 1. Mean±SD, minimum and maximum percentages of foxes with different health disorders observed on Finnish and Norwegian fox farms during the farm visits in autumn 2011. The results from Finnish and Norwegian farms are mean values from three and two assessors, respectively.

	Finland (10 farms)		Norway (5 farms)	
	Mean±SD	Min-max	Mean±SD	Min-max
Some difficulties in moving	43±17[a]	26-66[a]	1.9±4.3	0-9.6
Major difficulties in moving	2.7±1.4[a]	1.5-4.8[a]	0±0	0
Difficulties in moving: animal does not move	1.0±1.2[a]	0-2.2[a]	7.6±5.3	0-13
Mild skin lesions	1.6±1.4	0-5.3	2.3±2.2	0-4.8
Severe skin lesions	0.037±0.11	0-0.37	1.1±1.1	0-2.3
Slightly bent feet	54±13	22-69	24±9.6	15-38
Severely bent feet	23±10	0-35	2.4±5.4	0-12
Ocular inflammation	1.4±1.4	0-4.4	1.9±1.9	0-4.4
Impaired mouth and teeth health	0.96±1.9	0-5.6	1.4±1.3	0-2.8
Diarrhoea	45±19[b]	9.8-78[b]	0±0	0
Urinary tract infection	0±0[b]	0[b]	0.15±0.34	0-0.75
Obviously sick	0.86±1.2	0-3.6	0.23±0.52	0-1.2

[a] Result from five farms only.

[b] Expressed as% of cages with clear sign of disorder, results from two assessors only.

Inter-observer reliabilities of the health measures, measured with the index of concordance (IC), ranged from 0.67 to 1.00 for the assessor pairs (Table 2). In general, the lowest IC between the assessors was found for the measure *bent feet*, on both Finnish and Norwegian farm visits. In Finland, there was some disagreement also with regard to the assessment of the measures *difficulties in moving* and *diarrhoea*.

Discussion

The present results show that the most common health problems on the Finnish fox farms, where the majority of the assessed foxes were blue foxes, were bent feet (referring to foreleg weakness), moving difficulties and diarrhoea. All these problems seemed to be more or less obesity-related, i.e. it was most often the fattest blue foxes that had moving difficulties and severely bent feet, as well as signs of diarrhoea under their cage (L. Ahola and H. Huuki, personal communications). It has also earlier been shown that poor leg structure (Korhonen *et al.*, 2005; Kempe *et al.*, 2010) and ability to move (Kempe *et al.*, 2010) are linked to obesity in farmed blue foxes. Although the observed problems were common, they are not insurmountable. What is needed, especially on blue fox farms, is goal-orientated management including changes in feeding regime and breeding animal selection in order to decrease the number of too fat foxes, and along with this the number of foxes suffering from obesity-related problems.

Table 2. Inter-observer reliabilities of the health measures assessed on Finnish and Norwegian fox farms during the farm visits in autumn 2011. In parentheses: the number of animals, or cages in regard to diarrhoea and urinary tract infection on the Finnish data, observed by the pair. IC: Index of concordance measured for the pair of assessors (Finland: observers A, B and C; Norway: observers B and D).

	IC: A & B	IC: A & C	IC: B & C	IC: Mean
Finland				
Difficulties in moving	0.87 (455)	0.82 (455)	0.82 (459)	0.84
Skin lesions	0.99 (863)	0.99 (775)	0.99 (791)	0.99
Bent feet	0.77 (849)	0.72 (777)	0.75 (775)	0.75
Ocular inflammation	0.99 (871)	0.98 (783)	0.99 (798)	0.99
Impaired mouth and teeth health	1.00 (871)	0.99 (783)	0.99 (798)	0.99
Diarrhoea	0.87 (491)	0.88 (473)	0.83 (486)	0.86
Urinary tract infection	1.00 (517)	1.00 (473)	1.00 (544)	1.00
Obviously sick	0.98 (873)	0.89 (873)	0.89 (881)	0.92
Norway	IC: B & D			
Difficulties in moving	0.98 (164)			
Skin lesions	0.96 (144)			
Bent feet	0.67 (135)			
Ocular inflammation	0.99 (166)			
Impaired mouth and teeth health	0.98 (166)			
Diarrhoea	1.00 (163)			
Urinary tract infection	0.99 (163)			
Obviously sick	1.00 (242)			

On the Norwegian farms, the most common problems were bent feet and moving difficulties, albeit the percentages of foxes with these problems were lower than on Finnish fox farms. The percentage of foxes that did not move during the assessment was, nevertheless, higher in Norway (7.6%) than in Finland (1.0%). This higher percentage in Norway, however, likely reflects temperament of the foxes rather that actual moving difficulties: typically, silver foxes that did not move during the assessment were animals that refused to leave the wire mesh platform during the assessment (A.L. Hovland and H. Huuki, personal observations). The differences between the two countries in the percentages of foxes with obesity-related problems, and of immobile foxes possibly only relate to the species distribution of the foxes that were assessed in these two countries. Since the aim of the WelFur scheme is to assess a representative sample of the animals on the farms, the majority of the assessed foxes in Finland, where there are about 10 times more blue foxes than silver foxes, were blue foxes whereas in Norway, where majority of all farmed foxes are silver foxes, the majority of the assessed foxes were silver foxes. Therefore, it is not surprising that the percentage of foxes with bent feet was lower and the percentage of foxes that did not move was higher on Norwegian farms since silver foxes, in general, are not that prone to obesity and often respond to human disturbance by jumping onto the platform.

The indices of concordance calculated for the assessor pairs were good or even excellent for all but one health measures included in the WelFur fox protocol. The lowest ICs were observed for the measure *bent feet*, i.e. the assessors disagreed on the scoring of normal feet, mildly bent feet and severely bent feet. Future studies should evaluate potential reasons for these inconsistencies, e.g. to what extent improving the scoring system of *bent feet* would increase the agreement between the assessor, or whether the differences relate to e.g. observer skills and thus the level of training of the assessors.

In conclusion, the present results show that the most common health problems among farmed foxes are bent feet and moving difficulties, as well as diarrhoea which was observed only on the Finnish fox farms. Furthermore, the results show that all but one health measures included in the WelFur fox protocol have good or excellent inter-observer reliability, indicating that the results from assessments are the same regardless of the assessor.

Acknowledgements

This study was financed by the European Fur Breeders' Association through the WelFur project. We wish to thank the whole WelFur project group and WelFur external reviewers for their help in building up the fox protocol. We are most grateful to Mr. Hobo Kukkonen for the practical assistance in the data collection on the farms, and to the farmers that allowed us to visit their farms during the autumn 2011.

References

Ahola, L., Koistinen, T. and Mononen, J., 2011a. Health and behavioral measures in farmed foxes: inter-observer reliability of farm averages. In: Widowski, T., Lawlis, P. and Sheppard K. (eds.) Proceedings of the 5th International Conference on the Assessment of Animal Welfare at Farm and Group Level. August 8-11 2011, Guelph, Canada, p. 51.

Ahola, L., Huuki, H., Hovland, A.L., Koistinen, T. and Mononen, J., 2011b. WelFur – Foxes: State of affairs in autumn 2011. In: NJF seminar, Fur animal research, November 1-3 2011, Knivsta, Sweden.

Keeling, L., 2009. How did we design the welfare measures? In: Butterworth, A., Blockhuis, H., Jones, B. and Veissier, I. (eds.) Delivering animal welfare and quality: transparency in the food production chain. October 8-9 2009, Uppsala, Sweden, pp. 23-26.

Kempe, R., Koskinen, N., Mäntysaari, E. and Stranden, I., 2010. The genetics of body condition and leg weakness in the blue fox (*Alopex lagopus*). Acta Agriculturae Scandinavica, Section A 60: 141-150.

Korhonen, H.T., Happo, M., Rekilä, T., Valaja, J. and Pölönen, I., 2005. Effects of diet calcium:phosphorus ratio and metabolize energy content on development of osteochondrosis, foot bending and performance in blue foxes. Animal Science 80: 325-331.

Martin, P. and Bateson, P., 2007. Measuring behaviour. An introductory guide. Cambridge University Press, Cambridge, UK. 176 pp.

Moberg, G.P., 2000. Biological response to stress: implications for animal welfare. In: Moberg, G.P. and Mench, J.A. (eds.) The biology of animals stress. Basic principles and implications for animal welfare. CABI Publishing, UK, pp. 1-21.

Mononen, J., Møller, S.H., Hansen, S.W., Hovland, A.L., Koistinen, T., Lidfors, L., Malmkvist, J., Vinke, C.M. and Ahola, L., 2012. The development of on-farm welfare assessment protocols for foxes and min: the WelFur project. Animal Welfare, in press.

Veissier, I., Botreau, R. and Perny P., 2009. Scoring animal welfare: difficulties and Welfare Quality® solutions. In: Keeling, L. (ed.) An overview of the development of the Welfare Quality® project assessment system, pp. 15-32.

Welfare Quality®, 2009. Welfare Quality® assessment protocol fox cattle. Welfare Quality® Consortium, Lelystad, the Netherlands, 182 pp.

WelFur – foxes: do feeding test, temperament test and a measure of stereotypic behaviour differentiate between farms?

T. Koistinen[1], H. Huuki[1], A.L. Hovland[2], J. Mononen[1,3] and L. Ahola[1]
[1]Department of Biology, University of Eastern Finland, P.O. Box 1627, 70211 Kuopio, Finland, [2]Department of Animal and Aquacultural sciences, Norwegian University of Life Sciences, P.O. Box 5003, 1432 Ås, Norway, [3]MTT Agrifood Research Finland, Halolantie 31A, 71750 Maaninka, Finland; tarja.koistinen@uef.fi

Abstract

In the present study we evaluate the suitability of the feeding test, temperament test and assessment of stereotypic behaviour for an on-farm welfare assessment protocol. Ten fox farms in Finland and five fox farms in Norway were visited in January or February. On each farm, a representative sample of foxes was chosen for each measure. The percentage of foxes eating in the feeding test, touching a stick in the temperament test and behaving stereotypically varied between 24-73%, 16-60% and 0-13% on the farms, respectively. On farm-level, the percentage of foxes touching the stick in the temperament test correlated positively with the percentage of foxes eating in the feeding test (r_s=0.609, P=0.016), and the percentage of foxes attacking the stick aggressively correlated positively with the percentage of stereotyping foxes (r_s=0.604, P=0.017). Accordingly, all three measures can be used to differentiate farms. Furthermore, the correlations between the results from different measures indicate that the temperament test measures partially the same behavioural features of the stock of the animals on the farm as the feeding test and the measure of stereotypic behaviour.

Keywords: fur farming, human-animal relationship, emotional state, appropriate behaviour

Introduction

The WelFur project develops on-farm welfare assessment protocols for farmed foxes and mink (Mononen *et al.*, 2012). The protocols are based on the same four principles and 12 criteria as the Welfare Quality® protocols developed for pig, cattle and poultry (Welfare Quality® 2009abc). One of the four principles is *appropriate behaviour*, which includes four welfare criteria: *expression of social behaviours, expression of other behaviours, Good human-animal relationship* and *positive emotional state*. In the WelFur protocol for farmed foxes (blue fox, *Vulpes lagopus* and silver fox, *Vulpes vulpes*), the following measures were chosen to represent these four criteria: social housing, opportunity to use enrichment, opportunity to observe surroundings, stereotypic behaviour, fur chewing, feeding test, temperament test and transport of live animals (Mononen *et al.*, 2012).

In the present work, we evaluate the three behavioural measures under the principle *appropriate behaviour*, i.e. the feeding test, temperament test and stereotypic behaviour. According to the WelFur protocol, the feeding test and temperament tests are measured in early winter when there are only breeding animals on the farm. The feeding test is a test of human-animal relationship in farmed foxes validated by Rekilä (1999). Various tests, e.g. novel object test, pencil test and glove test, have been used to test the temperament of farmed foxes (see review in Rekilä, 1999), but none of these tests has been appropriately validated for any of the fox species. Based on the principles of the stick test (or pencil test), which is validated for farmed mink (Kirkden *et al.*, 2010), and utilised also in foxes (e.g. Korhonen and Niemelä, 1996, Korhonen *et al.*, 2000), a new temperament test was developed in the WelFur project. Stereotypic behaviour, in turn, is a good measure of compromised welfare (Mason, 2006) but no feasible method for assessing stereotypic behaviour on commercial fox farms has been developed (see Anttila *et al.*, 2010). Therefore, a new method for measuring the prevalence of stereotypic behaviour in various housing conditions on private fox farms was developed in the WelFur project (Huuki *et al.*, 2011). Stereotypic behaviour is measured three times a year in order to cover all production periods (for more detail information about periods, see Mononen *et al.*, 2012).

The aim of the present work was to assess whether the percentage of foxes eating in the feeding test, the percentage of foxes touching the stick in the temperament test and prevalence of stereotypic behaviour can differentiate commercial farms. The other aim was to measure whether there are relations between the test results on the farm level, i.e. whether the feeding test, temperament test and stereotypic behaviour measure same attributes of the behavioural features of the stock of foxes on the farm.

Material and methods

Ten fox farms in Finland and five fox farms in Norway were visited in late January or early February 2012. Seven out of ten Finnish farms housed mainly, or only, blue foxes whereas four out of five Norwegian farms housed mainly, or only, silver foxes. Similarly as it is described in the WelFur protocol, a representative sample of foxes was separately chosen for each test, i.e. the feeding test, temperament test and assessment of stereotypic behaviour. All of the tested foxes were housed in traditional wire mesh cages.

In the feeding test, the assessor placed a feed portion onto the feeding tray or on the roof of the cage and stood calmly in front of the cage for 30 s (Rekilä, 1999). It was scored whether the fox ate within the 30 s or not. A total of 87-179 foxes were tested with this test on each farm.

In the temperament test, the assessor approached the fox calmly and inserted 30 cm of a 150 cm long stick in to the fox cage through the front wall of the wire mesh cage. The assessor stood still, avoiding eye contact with the fox, in a distance of approximately 1 m from the cage of the fox for 10 s. The fox's reaction towards the stick was scored as either curious, i.e. the fox explored (touched, sniffed, nibbled) the stick in a curious way; aggressive, i.e. the fox attacked (snapped/bit) the stick in an aggressive way; or unresponsive, i.e. the fox did not respond to the stick by touching the stick curiously or aggressively. A total of 40-91 foxes were tested with this test on each farm.

The recordings of stereotypic behaviour were performed from inside the fox house as the assessor stood in the middle of chosen set of fox cages (Huuki *et al.*, 2011). After a habituation time of three minutes, the assessor observed the foxes 'in one direction of the shed' for one minute, then turned around and observed the foxes 'in the opposite direction' for one minute. The occurrence of stereotypic behaviour (yes/no) and the general activity level (active/resting) was recorded for each fox. Stereotypic behaviour was defined as invariant repetitive (at least 3 reiterations) pacing, jumping, head twirling or gnawing/licking of the cage. The prevalence of stereotypic behaviour was calculated as the percentage of foxes behaving stereotypically out of all active foxes. The stereotypic behaviour was assessed in 102-200 foxes on each farm.

The feeding test, temperament test and stereotypic behaviour were measured during the one-day WelFur assessment visits. The feeding test was carried out first in the morning, followed by the assessment of stereotypic behaviour. The behaviour of foxes in the temperament test was assessed along with other animal-based and resource-based measures in the afternoon. In all farms, the feeding test and also the stereotypic behaviour were assessed before feeding of the animals. The temperament test was assessed either before or after the feeding, depending on the schedule on the farm, since it was expected that the result is not affected by feeding.

Four assessors were involved in the farm visits, but only 1-2 assessors were present on the same farm. Thus, each measure was assessed by either one or two assessors on each farm visit, but no one assessor visited all farms. When two assessors collected data on the same farm, the data are presented as a mean of the two assessors.

In order to evaluate the possible relations between these tests, correlations between the results from the tests were calculated on farm means and analysed with Spearman correlation coefficient r_s.

Results

In general, half of the foxes ate in the feeding test (47±16% of the tested foxes, mean±SD). The percentage of foxes eating in the feeding test varied from 24 to 73% among the farms (Figure 1). More foxes touched the stick in a curious way (34±13%, mean±SD) than attacked the stick

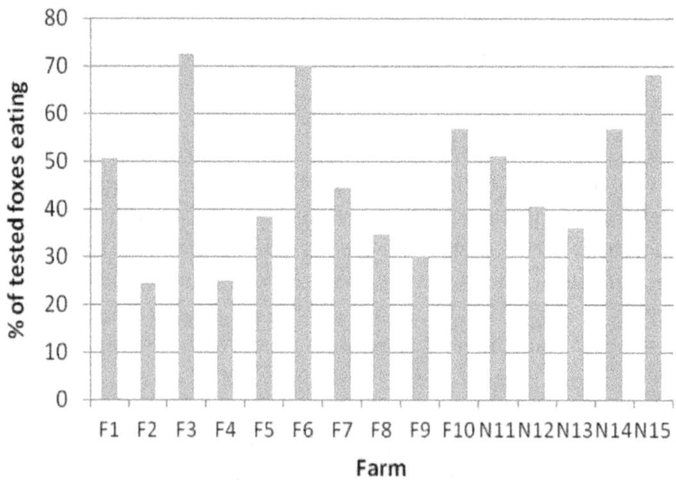

Figure 1. The percentage of foxes eating within the 30 sec test time in the feeding test on the 15 farms. In the farm names, F refers to a Finnish fox farm and N to a Norwegian fox farm.

aggressively (5.9±8.5%). The percentage of foxes touching the stick in a curious way varied from 16 to 60%, and the percentage of foxes attacking the stick varied from 0 to 27% among the farms (Figure 2). Majority of the foxes that were scored unresponsive remained either passive (e.g. laying) during the test or approached the stick quietly without touching it. Obvious fear of the stick was very seldom observed.

The prevalence of stereotypic behaviour was relatively low (Figure 3). Stereotypic behaviour was not observed on two farms. The maximum farm-level percentage of foxes showing some type of stereotypic movement out of all active foxes was 13%.

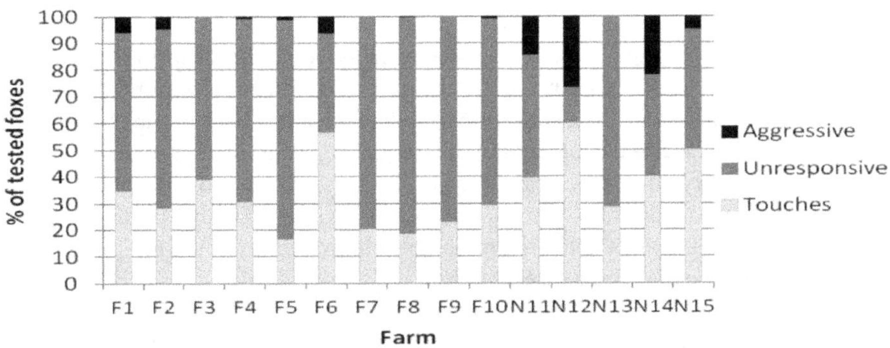

Figure 2. The percentages of the foxes touching the stick, being unresponsive towards the stick and attacking the stick aggressively in the temperament test on the 15 farms. In the farm names, F refers to a Finnish fox farm and N to a Norwegian fox farm.

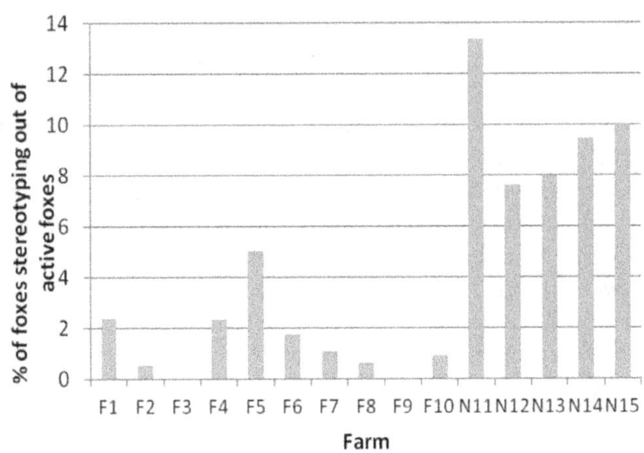

Figure 3. The percentage of foxes behaving stereotypically out of all active foxes in the behavioural observation on the 15 farms. In the farm names, F refers to a Finnish fox farm and N to a Norwegian fox farm.

A farm-level correlation was found between the percentage of foxes touching the stick in the temperament test and the percentage of foxes eating in the feeding test (r_s=0.609, *P*=0.016), and between the percentage of foxes attacking the stick aggressively in the temperament test and the percentage of foxes behaving stereotypically (r_s=0.604, *P*=0.017). No farm-level correlation was found between the percentage of foxes eating in the feeding test and attacking the stick in the temperament test (r_s=0.289, *P*>0.1) or percentage of foxes eating in the feeding test and behaving stereotypically (r_s=0.233, *P*>0.1), or percentage of foxes touching the stick in the temperament test and behaving stereotypically (r_s=0.465, *P*>0.05).

Discussion

With the present relatively small data we demonstrated that the feeding test, temperament test and the measure of stereotypic behaviour are suitable measures in the on-farm welfare assessment protocol for farmed foxes. All measures are feasible to carry out on commercial farms, i.e. they can be performed in various farming conditions and in a relatively high number of foxes within a day. Furthermore, all measures can be used in differentiating farms in relation to the behaviour of the foxes on the farm. Because of the limitations of the present data, no inter-observer reliability could be calculated for any of the measures. The inter-observer reliability of the feeding test has been already proven (Rekilä, 1999), but the inter-observer reliability of the temperament test and stereotypic behaviour remains to be assessed.

No consistent data on the human-animal relationship on commercial farms, assessed as a percentage of foxes eating in the feeding test, has been reported. Also the temperament of foxes on commercial farms is poorly documented. Our data show that there were some variation between the farms in the human-animal relationship and curiosity of farmed foxes. Since it has been shown that confidence of the animals towards humans can be increased through selection (Kenttämies *et al.*, 2002, Trut, 1999), the farmers could possibly use the results of the feeding test and the newly developed temperament test as a selection criterion when selecting future breeding animals to enhance the human-animal relationship and curiosity of the foxes on the farm.

The previous knowledge of the stereotypic behaviour of farmed foxes is sporadic, and is based on studies focusing primarily on other aspects of animal welfare than stereotypic behaviour (Anttila *et al.*, 2010). The knowledge of the prevalence of stereotypic behaviour on commercial farms is limited to one other study carried out in the WelFur project. Consistent with the results from summer, reported by Huuki *et al.* (2011), the percentage of foxes behaving stereotypically showed variation between the visited farms. In our data, more incidents of stereotypic behaviour was observed in Norwegian fox farms, which may be attributable to the species differences, i.e. the greater number of silver foxes in the visited farms in Norway. Huuki *et al.* (2011) observed slightly more stereotypic behaviour in silver foxes than in blue foxes.

On farm-level, the percentage of foxes touching the stick in the temperament test correlated with the percentage of foxes eating in the feeding test on the farm. Thus, these tests measure, at least partially, some of the same aspects of human animal relationship and/or curiosity of the stock of the foxes on the farm. Also the percentage of foxes attacking the stick aggressively in the temperament test correlated with the percentage of foxes stereotyping on the farm. This suggests a possible link between a more active behaviour style (e.g. attacking the stick) and the propensity to develop stereotypies, as discussed by Mason *et al.* (2007). Based on our preliminary results it seems that the feeding test, temperament test and assessment of stereotypic behaviour evaluate some parallel aspects of motivation and behavioural features of the stock of foxes on the farm. In future it is important to properly evaluate the correlations between these tests also at an individual-level.

As a conclusion, the feeding test and the new measures developed for the WelFur protocol, i.e. the temperament test and the measure of stereotypic behaviour, seems to be suitable measures for the on-farm welfare assessment protocol for foxes but further validation and refinement of all measures is needed. The correlations between the feeding test, temperament test and prevalence of stereotypic behaviour should be studied in more detail, e.g. in individual-level, in future studies. Also the possible effect of feeding into the outcome of the temperament test should be studied.

Acknowledgements

This work is part of the WelFur project which is financed by the European Fur Breeders' Association (EFBA). The authors are grateful to Mr. Hobo Kukkonen for the practical assistance in the data collection on the farms and the participated fox farmers.

References

Anttila, R., Koistinen, T., Mononen, J. and Ahola, L., 2010. A review of stereotypic behaviour in farmed foxes. Nordic Association of Agricultural Scientists, Seminar no. 440, Fur Animal Research, Autumn meeting 29.9.-1.10.2010, Oslo, Norway. 5 pp.

Huuki, H., Hovland, A.L., Koistinen, T., Mononen, J. and Ahola, L., 2011. WelFur: Stereotypic behaviour in silver foxes and blue foxes in Norway and Finland during the whelping season. Nordic Association of Agricultural Scientists no 450, Fur animal research, Autumn Meeting, Knivsta, Sweden.

Kenttämies, H., Nordrum, N.V., Brenøe, U.T., Smeds, K., Johannesson, K.R. and Bakken, M., 2002. Selection for more confident foxes in Finland and Norway: heritability and selection response for confident behavior in blue foxes (*Alopex lagopus*). Applied Animal Behaviour Science 78: 67-82.

Kirkden, R.D., Rochlitz, I., Broom, D.M. and Pearce, G.P., 2010. Assessment of on-farm methods to measure confidence in mink and foxes on Norwegian farms. Report prepared for Dyrevernalliansen (Norwegian Animal protection Alliance), Oslo, Norway. Campridge University Animal Welfare Information Centre, Department of Veterinary Medicine, University of Campridge, Campridge, UK, 43 pp.

Korhonen, H. and Niemelä, P., 1996. Temperament and reproductive success in farmbred silver foxes housed with and without platform. Journal of Animal Breeding and Genetics 113: 209-218.

Korhonen, H.T., Niemelä, P., Jauhiainen, L. and Tupasela, T., 2000. Effects of space allowance and earthen floor on welfare-related physiological and behavioural responses in male blue foxes. Physiology and Behaviour 69: 571-580.

Mason, G., 2006. Stereotypic behaviour in captive animals. In: Mason, G. and Rushen, J. (eds.) Animal behaviour: fundamentals and applications to welfare CABI Wallingford, UK, pp. 325-356.

Mason, G., Clubb, R., Latham, N. and Vickery, S., 2007. Why and how should we use environmental enrichment to tackle stereotypic behaviour? Applied Animal Behaviour Science 102: 163-188.

Mononen, J., Møller, S.H., Hansen, S.W., Hovland, A.L., Koistinen, T., Lidfors, L., Malmkvist, J., Vinke, C.M. and Ahola, L., 2012. The development of on-farm welfare assessment protocols for foxes and mink: the WelFur project. Animal Welfare, in press.

Rekilä, T., 1999. Behavioural tests in welfare research of foxes. Kuopio University Publications C. Natural and Environmental Sciences 92. 52 pp.

Trut, L., 1999. Early canid domestication: the farm-fox experiment. American Scientist 87: 160-169.

Welfare Quality®, 2009a. Welfare Quality® assessment protocol for pigs (sows and piglets, growing and finishing pigs). Welfare Quality® consortium, Lelystad, the Netherlands. 122 pp.

Welfare Quality®, 2009b. Welfare Quality® assessment protocol for cattle. Welfare Quality® consortium, Lelystad, the Netherlands. 182 pp.

Welfare Quality®, 2009c. Welfare Quality® assessment protocol for poultry (broilers, laying hens). Welfare Quality® consortium, Lelystad, the Netherlands. 122 pp.

Electrophysiological indices in mink (*Neovison vison*) during CO and CO_2 euthanasia

H.T. Korhonen[1]*, S. Cizinauskas*[2]*, J. Jesernics*[2] *and P. Eskeli*[1]
[1]*MTT Agrifood Research Finland, Animal Production Research, Finland;* [2]*Animal Neurology Clinic AISTI, Vantaa, Finland; hannu.t.korhomen@mtt.fi*

Abstract

The objective of the present study was to find out at what time point standard black male mink (*Neovison vison*) is dying and how quickly changes in the body function can be observed. Experimental groups evaluated were: euthanasia with exhaust gases (petrol engine 4%), carbon dioxide (cylinder 80%), carbon monoxide (cylinder 2%, 4%). Brainstem auditory evoked responses (BAER), electroencephalography (EEG), electrocardiography (ECG) and respiratory rate were measured. Mean time of decline/absent BAER was 112/176, 138/183, 235/390 and 528/833 seconds after gas apply in groups, respectively. Mean time for first changes/absence of EEG was 42/86, 39/75, 55/190 and 176/426 seconds after gas apply in groups, respectively. Mean time for first changes in respiration/absent breathing was 42/217, 28/227, 144/477 and 331/901 seconds after gas application in groups, respectively. Mean time for first changes/absence of ECG was 105/292, 117/220, 215/289 and 481/682 seconds after gas apply in groups, respectively. Gases first affected the brain and brainstem which was seen as loss of EEG and BAER and thereafter respiration and heart rate ceased. A sensitivity to pain is related to consciousness and function of brain. Studied gases can be considered to primary lead to state of non-pain. They are also effective to kill the animal.

Keywords: gas exposure, killing practice, animal welfare, fur farming

Introduction

Several methods and techniques have been considered on the proper euthanasia of farm-raised mink (*Neovison vison*). Such methods are neck breaking, electrocution, carbon monoxide (CO), carbon dioxide (CO_2), nitrogen (N_2), argon (Ar) and various lethal injections (Lamboy *et al.*, 1985, Hansen *et al.*, 1991, EFBA, 1999). Among those, neck breaking is today prohibited. Electrocution has been employed to a certain extent, but it is not a favoured method for today's practice and, therefore, not actually used. Exposure to gases is favoured in the mink today. The predominant killing gas on European fur farms is carbon monoxide (CO), which originates either from a petrol machine (filtered exhaust gases) or, alternatively, from a cylinder (pure source). The other common cylinder gas employed is carbon dioxide (CO_2). Its use seems to be increasing (Korhonen *et al.*, 2011).

Despite the fact that the euthanasia method employed is the best possible presently known, animal euthanasia can cause pain, stress, fear, and other dysfunction. Therefore, special attention must be paid to animal euthanasia in order not to cause them unnecessary suffering. There must be a distinct goal to perform euthanasia as fast and as painlessly as possible. The regulation on euthanasia having been prepared in the EU will be implemented at the beginning of 2013. At present, there is not enough accurate information available on killing methods in farm-raised mink. Therefore, more focused scientific research and background information on relevant gas euthanasia is needed. The euthanasia with filtered exhaust gases, carbon monoxide, and carbon dioxide from a cylinder was compared in this study. These gases are the most commonly used at Europe. The objective was to examine the timeline of changes in respiration and the function of brain and heart. The present study provides new data on the dying process of the mink from an electrophysiological point of view.

Materials and methods

Four groups of standard black male mink were tested in this study. Euthanasia with: (1) filtered exhaust gases (combination of CO, CO_2, HC, O_2); CO concentration in the killing box 4%, n=8 animals; (2) carbon monoxide (CO) from a cylinder, concentration in the killing box 4%, n=9 animals; (3) carbon monoxide (CO) from a cylinder, concentration in the killing box 2%, n=6 animals; (4) carbon dioxide (CO_2) from a cylinder, concentration in the killing box 80%, n=8 animals. Body weights of animals were measured before the start of experiment. They were fed and kept according to normal farming procedures. The health of animals was visually checked daily before study.

Animals were sedated with a combination of Dorbene (0.4 ml i.m.) and Zoletil (0.2 ml i.m.). An otoscopy was performed in order to evaluate the external ear canal and the tympanic membrane. The animal's heart rate and breathing as well as palpebral, corneal and withdrawal reflexes were examined before placing the animal into a wooden-glass chamber (60×30×35 cm; L×W×H), and once before euthanasia with CO and CO_2. The wooden-glass chamber had openings for wires of subcutaneous needle electrodes for recording of brainstem auditory evoked potentials (BAER), electroencephalography (EEG) and electrocardiography (ECG). BAER was recorded on every side twice before euthanasia. Simultaneously EEG and ECG recording of 5 minute duration were performed before euthanasia with CO and CO_2 in order to record the EEG pattern.

The animals were immobilized in the neck and the tail area and gas euthanasia were performed while recording BAER, EEG and ECG continuously. BAER and EEG recordings were continued until no brain activity could be recognised. ECG was performed until the heart arrest (asystole) could be detected. The animal's heart rate, breathing, palperal, corneal and withdrawal reflexes were examined after discontinuation of the BAER, EEG and ECG recordings and after removal of the animal from the chamber.

Results

Basic data are given in Table 1. The range of duration of normal BAER recording during euthanasia was in the same order of magnitude for the exhaust gases group and in CO_2 group. In CO 4% and CO 2% groups, however, it was 2-3 times and 5-6 times longer.

The range of duration of decline of BAER during euthanasia was similar in exhaust gases group, and in CO_2 group, but almost 2 to 4 times longer in CO 4% group and CO 2% groups. The time

Table 1. Summary of EEG, BAER, respiration and ECG results in seconds.

Group name	EEG		BAER	
	First change	**Absent**	**First change**	**Absent**
Exhaust	42±10[a]	86±35[ab]	63±19[ab]	176±37[a]
CO_2	39±10[a]	75±28[a]	45±16[a]	183±19[ab]
CO 4%	55±11[ab]	190±81b[c]	143±70[bc]	390±139[bc]
CO 2%	176±126[b]	426±252[c]	305±148[c]	833±345[c]
P-value$_{group}$	<0.01	<0.0001	<0.0001	<0.0001
Group name	**Respiration**		**ECG**	
	First change	**Absent**	**First change**	**Absent**
Exhaust	42±17[ab]	217±53[a]	105±37[a]	292±130[ab]
CO_2	28±5[a]	227±26[a]	117±31[a]	220±92[a]
CO 4%	144±67[bc]	477±151[b]	219±100[ab]	289±126[ab]
CO 2%	331±191[c]	901±293[b]	481±184[b]	682±333[b]
P-value$_{group}$	<0.0001	<0.0001	<0.001	<0.01

[a-c] Letters mark a statistical difference (P<0.05) between the groups. Statistical analyses by Kruskal-Wallis ANOVA.

point of disappearing of BAER during euthanasia was lowest for exhaust gases group and highest for CO 2% group, respectively.

Typical finding after gas application was burst suppression in EEG, increased respiratory rate and amplitude and decreased amplitude and frequency of ECG recording, arrhythmia or asystole. Changes in EEG, respiration and heart activity appeared to occur earlier in engine exhaust gas (Table 1). Termination of EEG, breathing and ECG also was faster in first two groups (exhaust CO, CO_2) compared to cylinder CO groups (Table 1).

Discussion

The euthanizing agent in the present study was either CO or CO_2 or a mixture of these two gases. The main principle for proper killing here is that any gas that is inhaled must reach a certain concentration in the alveoli before it can be effective. Therefore, euthanasia with gases always takes some time and may last longer than other killing methods.

Consciousness can be considered as the animal's ability to feel emotions and control its voluntary mobility. An animal can be presumed to be unconscious when it loses its natural standing position, is not awake and does not show signs of positive or negative emotions such as excitement or fear.

The onset of unconsciousness is very essential for humane euthanasia because it is an indication that the cerebral cortex has been rendered non-functional. Painful effects can only be experienced when the brain function of the animal is intact (Jepsen *et al.*, 1981; Hansen *et al.*,. 1991; AVMA, 2007). EEG recording shows the activity of the cerebral cortex. Our EEG recording results indicate that the first changes in cortical function occurred in average within the first minute of gas euthanasia in filtered exhaust gas, cylinder CO_2 and cylinder CO 4% gas groups. In contrast the changes in cortical activity of cylinder CO 2% group happened within the third minute. It is impossible to say for sure what the exact time point is when animal does not perceive normally any more (unresponsive), but it most likely correlate with the onset of the first changes in EEG recording. Absent EEG indicate that animal is unconscious (Lamboy *et al.*, 1985; Korhonen and Huuki, 2012).

The present results clearly showed that the four studied gases first affect the brain and brainstem which was seen as loss of EEG and BAER. Thereafter respiration was affected and at last the heart rate. While sensitivity to pain and distress is essentially related to consciousness and function of the brain, the tested gases can be considered primary to lead to a state of non-pain.

In 2009 European Fur Breeders' Association (EFBA) initiated the 'Welfur' project in order to develop a welfare assessment protocol for mink (Mononen *et al.*, 2012). The protocol includes four general welfare principles. Within these principles welfare measures should be taken within 12 welfare criteria. One of these criteria is 'Absence of pain induced by management procedures'. The present study is related to this specific criteria. The main question in protocol here is: 'are there efficient methods of killing mink?'. The results of the present study provide a clear answer to this question. They showed that studies gases are efficient to kill mink.

Acknowledgements

This study was financially supported by the Ministry of Agriculture and Forestry (MMM) and the European Fur Breeders' Association (EFBA). The staff of MTT Fur Animal Group (Kannus) and Turkistila Luova Oy are acknowledged for their valuable help in carrying out this experiment.

References

AVMA, 2007. AVMA Guidelines on Euthanasia. American Veterinary Medicine Association. June 2007. 27 pp.

EFBA, 1999. Code of practice for the care and handling of farmed mink, fitch and fox in Europe. Coordinated and issued by European Fur Breeders' Association. 2nd edition, September 1999. 18 pp.

Jepsen, Ø.R., Poulsen Dirch, J.S. and Jorgensen, G., 1981. A Haematological and Clinical-Chemical Study. Nordisk Veterinaer-Medicin. Supplementum 1. 99 pp.

Hansen,N., Creutzberg, A. and Simonsen, H.B., 1991. Euthanasia of mink (*Mustela vison*) by means of carbon dioxide (CO_2), carbon monoxide (CO) and nitrogen (N_2). British Veterinary Journal 147(2):140-146.

Korhonen, H.T., Sepponen, J. and Eskeli, P., 2011. A questionnaire study on euthanasia in mink. Report No 2. MTT Agrifood Research Finland. 26 pp.

Korhonen, H.T. and Huuki, H., 2012. Code of good practice for CO and CO_2 euthanasia in mink. MTT Agrifood Research Finland. 28 pp.

Lamboy, E., Roelofs, J.A. and Van Voorst, N., 1985. Euthanasia of mink with carbon monoxide. Veterinary Record 116: 416.

Mononen, J., Møller, S.H., Hansen S. W., Hovland A.L., Koistinen, T., Lidfors, L., Malmkvist, J., Vinke, C. and Ahola, L., 2012. The development of on-farm welfare assessment protocols for foxes and mink: the WelFur project. Animal Welfare, in press.

Applying Welfare Quality® strategy to design a welfare assessment tool for foxes and mink farms

R. Botreau[1,2]*, M. Gaborit*[1,2] *and I. Veissier*[1,2]
[1]*INRA, UMR1213 Herbivores, 63122 Saint-Genès-Champanelle, France;* [2]*Clermont Université, VetAgro Sup, UMR Herbivores, BP 10448, 63000, Clermont-Ferrand, France;*
raphaelle.botreau@clermont.inra.fr

Abstract

WelFur project aims to develop on-farm welfare assessment protocols for mink and foxes, using the Welfare Quality® framework. The evaluation relies on a sequential structure, where four welfare principles, subdivided into 12 criteria, are constructed and then aggregated to obtain an overall evaluation at farm level. WelFur partners developed measures to check the compliance of fox and mink farms with all criteria. We applied the multicriteria methods developed in Welfare Quality® to construct criteria, consulting two expert panels to set parameters. To aggregate criteria and then principles to produce an overall assessment, we directly used the Welfare Quality® protocol. The general framework remains the same, but several specificities emerged. First, the whole production cycle, from breeding to slaughter, occurs on the farm, implying to consider three different periods to cover the whole cycle. Depending on the period, number and types of animals, and resources used differ. Second, two different species of foxes and their hybrids are to be considered. They differ in behaviour and needs, impacting the interpretation of some measures. We thus had to run separate consultations for periods, animal types or species, and then aggregate them. This work shows that the rationale followed in Welfare Quality® can be extended to other species once taking into account the specificities linked to species or production systems.

Keywords: animal welfare, overall assessment, multicriteria evaluation, expert opinion, foxes, mink.

Introduction

This study is part of the WelFur project (2010-2012), supported by EFBA in collaboration with seven research institutes and universities. The objective of this project is to introduce a certification program at farm level that will guarantee a good level of welfare on certified farms. WelFur thus aims at applying to fox and mink farms the welfare assessment rationale developed in the Welfare Quality® project (2004-2009), initially developed for cattle, pigs and poultry.

Background information: architecture of the Welfare Quality® assessment tool

Since welfare is a multidimensional concept, covering several aspects such as physical comfort, absence of hunger, thirst, diseases, injuries and stress, and expression of positive behaviour (e.g. the five freedoms, Farm Animal Welfare Council, 1992), the Welfare Quality® project developed a multicriteria evaluation tool. This system is based on a sequential structure, where four principles of welfare (*good feeding, good housing, good health, appropriate behaviour*), subdivided into

12 criteria, are constructed and then aggregated to obtain an overall evaluation at farm level (Table 1). This structure was designed to be common to all animal production systems.

Farms are assessed on each criterion thanks to one to several measures relying on different aspects of the considered criterion. All the measures and the construction process are described in the Welfare Quality® protocols (Welfare Quality® consortium 2009a, b, c).

Specificities of fur farming systems in comparison with Welfare Quality®

The WelFur project aims at providing welfare assessment protocol over the whole production cycle of farmed foxes and mink. Even if the general construction procedure is the same as in Welfare Quality®, several specificities, listed below, have to be taken into consideration in the model construction for fur animals.

Each model built in Welfare Quality® was associated with only one species (e.g. cattle) and one type of animal (for cattle: fattening cattle, veal calves, or dairy cows). On the contrary, for foxes and mink systems the whole production cycle (including breeding period, lactation, weaning, fattening and finally slaughter and pelting) occurs on the same farm. As a consequence, to assess the farm it is necessary to consider all types of animals (adult males, adult females, under-weaning juveniles and growing juveniles). In addition, two different species of foxes and their hybrids are to be considered (commonly reared on the same farms): Blue fox and Silver fox. The different animal types and species differ in behaviour and needs, and thus the interpretation of a given measure can vary. Thus, the construction of some criteria needed to be adjusted according to the different animal types and/or species so that their differences could be taken into account for the interpretation of the related measure.

Table 1. Animal welfare principles and criteria defined in the Welfare Quality® project.

Principles	Criteria	
Good feeding	1	absence of prolonged hunger
	2	absence of prolonged thirst
Good housing	3	comfort around resting
	4	thermal comfort
	5	ease of movement
Good health	6	absence of injuries
	7	absence of diseases
	8	absence of pain due to management procedures
Appropriate behaviour	9	expression of social behaviours
	10	expression of other behaviours
	11	good human-animal relationship
	12	positive emotional state

Moreover, to have an overall view of the whole fur farm, the entire production cycle is to be evaluated. Three periods (pelting to mating (P1) / mating to weaning (P2) / weaning to pelting (P3)) were defined and have to be assessed (Figure 1). Depending on the period, the number and types of animals (adult males, adult females, under-weaning and growing juveniles), and the resources used differ.

The construction of criteria was thus directly impacted by these specificities, e.g. at criterion level the data collected at several periods have to be integrated and this requires specific arrangements for the calculation of scores.

Material and methods

Construction process and methods

The overall welfare assessment is the result of three construction steps: (1) construction of criteria from measures, (2) construction of principles from criteria, and (3) aggregation of the four principles to produce an overall evaluation at farm level. Methods used are described in Botreau *et al.* (2008) for Steps 1 and 2, and in Botreau *et al.* (2009) for Step 3, in addition to Welfare Quality® protocols. Only Step 1 differs from one species to another, and thus was done in WelFur specifically for foxes and mink. Measures, developed by WelFur partners, related to a given criterion have to be interpreted and synthesized to produce one score at criterion level, expressed on a 0-100 scale indicating the degree of compliance of the farm with this specific criterion (from 0 the worst possible situation to 100 the ideal one). Methods used to calculate scores differ from one criterion to another in order to be adapted to the number of measures to be considered, to the type of scale on which the data is observed, and to the relative importance to be assigned to each measure within the criterion. Two main methods of calculation were

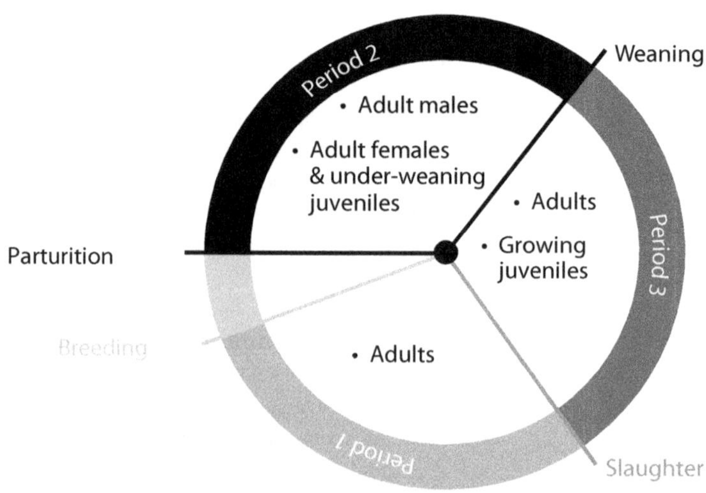

Figure 1. The three periods of the production cycle considered for foxes and mink.

thus used to construct the criteria: construction of a decision-tree, and calculation of weighted sums combined with non-linear transformation (thanks to I-spline functions). While deciding of the type of construction, we and WelFur partners considered the whole range of existing variability across the European production area (e.g. for mink from Finland to Greece). Finally, and specifically to WelFur, when a given measure is assessed at several periods of the production cycle and/or on several animal types (e.g. adult vs. growing juveniles), periods or animal types were aggregated using Choquet integrals (Grabisch and Roubens, 2000).

Consultation process

As in Welfare Quality®, the assessment tool was tuned to fit the average judgement of experts, thus modelling their opinion on (1) the interpretation in terms of welfare of the measures, (2) the relative importance of the different measures, criteria and principles, and (3) the level of compensation to be authorized between welfare issues. To do that with WelFur partners we created two expert panels (5 experts for foxes and 7 experts for mink), composed of animal scientists and vets, chosen for their knowledge on the two species in terms of physiology and behaviour and also on farming practices. We consulted each panel during a one day meeting. When necessary, we pursued the consultation by e-mail exchanges.

Data treatment

Depending on the type of construction, we treated the data using Excel (to calculate means and also to use the solver to set the optimized parameters for the construction), and/or GenStat® 11th edition, using the Biometris set of procedures, to fit I-spline functions (by least-squares minimization).

Results

Thanks to exchanges with WelFur partners in charge of the development of foxes and mink welfare measures, we identified the measures to be looked at for each criterion (Mononen *et al.*, 2012) and we proposed for each criterion a construction method. These choices led to 55 and 80 different questionnaires, for foxes and mink respectively. Only few examples are presented below in order to illustrate the different methods used.

Example 1: interpretation of the use and type of neck tongs in Period 1, as part of Criterion 8 'Absence of pain due to management procedures' in foxes

Two measures are considered to assess Criterion 8, including the handling procedure of animals using neck tongs. In Period 1, when only adult blue and/or silver foxes are present, a decision tree is designed including the frequency of use of the neck tong to handle animals, the type of material that covers the animal-end of the neck tong and the species which is handled with this use and type of neck-tong (since the behaviour of the fox in respect to the handling is different from one species to another). The resulting decision tree thus leads to seven possible situations (Figure 2).

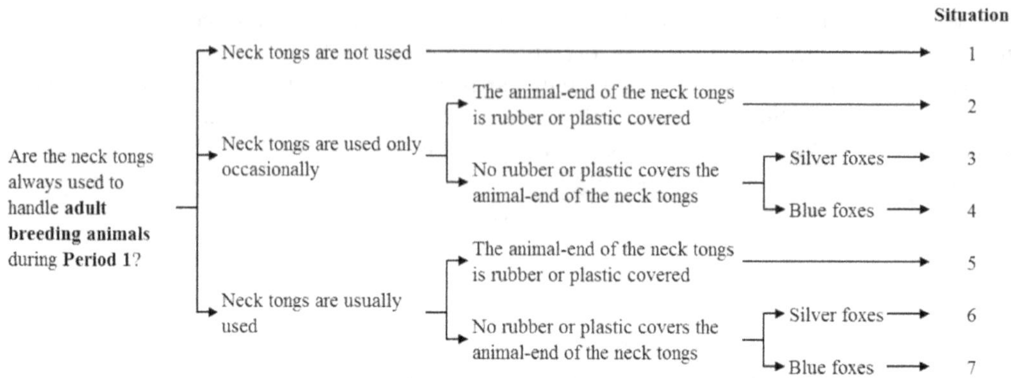

Figure 2. Decision tree used to assess the use and type of neck tongs in period 1 for foxes, as part of Criterion 8.

We asked the experts to assign a score to each of the seven possible situations from 1 to 7 (Table 2). These scores represent their own evaluation using the standardized scale [0,100]. The score to be assigned to each situation thus corresponds to the average score assigned by the experts. Since there may be several types of neck tongs and both species on a given farm, several situations may be observed simultaneously on the farm. Then, the score to be assigned to the farm is the one corresponding to the worst situation observed on at least 15% of the animals.

Example 2: interpretation of quality of the nest box in Period 3, as part of Criterion 3 'Comfort around resting' in mink

Criterion 3 (Comfort around resting) in mink is assessed through two measures observed at individual level during the three periods: the access to a nest box and the quality of the nest box/resting area. At cage level we consider four categories of quality of the nest box/resting,

Table 2. Experts' answers and final score (i.e. mean of experts' scores) assigned to each of this seven possible situations about use of the neck tong in period 1 for foxes, as part of Criterion 8.

Situation	Score assigned by experts					Score (mean)
	Expert 1	Expert 2	Expert 3	Expert 4	Expert 5	
Situation 6	0	0	0	15	0	3
Situation 7	10	0	7	15	0	6
Situation 3	30	17	60	50	0	31
Situation 4	40	17	67	50	0	35
Situation 5	20	17	33	40	100	42
Situation 2	50	100	93	95	100	88
Situation 1	100	100	100	100	100	100

corresponding to different severities of problems: good (Category 0), moderately good (Category 1), moderately bad (Category 2), and bad (Category 3) resting comfort. At farm level, the percentages of mink in each category of nest box quality (called p0, p1, p2 and p3, for categories 0, 1, 2 and 3 respectively) are combined in a weighted sum. This sum is then transformed into an index (*I*) that varies from 0 to 100, and finally this index is computed into a score using I-spline functions. To define the weights to be assigned to the categories and the parameters of the I-spline functions, we asked the experts to rank and then to score a set of virtual farms characterized by their percentages of animals in the four categories (cf. Table 3 for the questionnaire presented to experts, and Table 4 for their answers and the results of our calculations based on their answers).

Weights were defined using the solver of Excel in order to obtain the same ranking of farms with experts' mean score and Index value. Then, we defined the parameters of I-spline functions so that to minimize the sum of square errors between experts' mean scores and the calculated ones.

Example 3: aggregation of the scores obtained for periods 1, 2 and 3 for water provision, as part of Criterion 2 'absence of prolonged hunger' in mink

Table 3. Questionnaire presented to the experts to score quality of the nest box in period 3 for mink, as part of Criterion 3.

Farms	% good resting comfort (p0)	% moderately good resting comfort (p1)	% moderately bad resting comfort (p2)	% bad resting comfort (p3)	Rank	Score
Farm 1	100	0	0	0		
Farm 2	75	25	0	0		
Farm 3	75	0	25	0		
Farm 4	50	50	0	0		
Farm 5	50	0	25	25		
Farm 6	25	75	0	0		
Farm 7	25	50	25	0		
Farm 8	0	100	0	0		
Farm 9	0	75	25	0		
Farm 10	0	50	50	0		
Farm 11	0	50	25	25		
Farm 12	0	25	75	0		
Farm 13	0	0	100	0		
Farm 14	0	0	75	25		
Farm 15	0	0	50	50		
Farm 16	0	0	25	75		
Farm 17	0	0	0	100		

Table 4. Experts' answers, weights, index value and final score assigned to each of the 17 farms composing the questionnaire about quality of the nest box in period 3 for mink, as part of Criterion 3.

	Situation				**Scores given by experts Ei**								**Index**	**Score**
	p0	**p1**	**p2**	**p3**	**E1**	**E2**	**E3**	**E4**	**E5**	**E6**	**E7**	**Mean**		
Weights	**0**	**1**	**2**	**3**										
Farm 17	0	0	0	100	0	0	0	0	0	0	0	0	0	0
Farm 16	0	0	25	75	0	0	0	0	0	0	0	0	8	1
Farm 15	0	0	50	50	10	0	3	10	0	0	0	3	17	3
Farm 14	0	0	75	25	10	0	5	15	10	0	0	6	25	7
Farm 13	0	0	100	0	30	0	10	25	5	10	0	11	33	12
Farm 12	0	25	75	0	40	10	15	25	15	15	0	17	42	18
Farm 11	0	50	25	25	40	20	25	20	15	15	0	19	42	18
Farm 10	0	50	50	0	50	25	30	30	18	20	0	25	50	24
Farm 9	0	75	25	0	50	30	40	35	21	30	0	29	58	31
Farm 5	50	0	25	25	50	35	35	25	18	30	35	33	58	31
Farm 7	25	50	25	0	60	40	45	40	20	50	25	40	67	40
Farm 8	0	100	0	0	60	50	50	40	20	50	20	41	67	40
Farm 6	25	75	0	0	75	60	60	50	19	60	30	51	75	51
Farm 3	75	0	25	0	75	70	65	60	50	70	50	63	83	64
Farm 4	50	50	0	0	90	80	65	70	30	80	40	65	83	64
Farm 2	75	25	0	0	100	90	70	80	75	90	60	81	92	80
Farm 1	100	0	0	0	100	100	100	100	100	100	100	100	100	100

To assess the absence of prolonged thirst on a mink farm, several questions based on resources and management are used, which differ according to the period by taking into account the risk of freezing and the risk of overheating during each period. The Criterion 2 is assessed thanks to three decision trees specific to each period. After calculating the sub-score for each period (similar to Example 1), we need to merge them into a global score covering the three periods. Consulted experts were asked to react on combinations of sub-scores obtained on periods 1, 2 and 3, assigning a score to each of the combinations (the first 4 columns of Table 5 correspond to the questionnaire and the last 9 columns to experts' answers and calculated scores).

We used the Excel solver to find the parameters of the Choquet integral that best match the experts' mean scores by least square minimization. Two indices can be calculated to help the interpretation of the parameters of the Choquet both in terms of relative importance of the three periods and of interaction between sub-scores.

The analysis of results highlights two characteristics in experts' reasoning. First, there is a clear hierarchy between periods: Period 2 is more important than Period 3 which is more important

Table 5. Experts' answers and final criterion-score assigned to each of the 13 farms composing the questionnaire about aggregation of the three periods' sub-scores for water provision for mink, as part of Criterion 2. Grey scales represent general marking of scenarios addressing the relative importance of welfare scores between periods.

Farms	Score for period 1	Score for period 2	Score for period 3	Experts' answers								Criterion-score
				E1	E2	E3	E4	E5	E6	E7	Mean	
Farm 1	25	50	75	25	40	27	35	53	50	30	37	38
Farm 2	25	75	50	30	45	30	40	75	60	30	44	44
Farm 3	40	50	60	40	50	42	50	48	50	40	46	45
Farm 4	40	60	50	50	50	45	50	60	50	40	49	48
Farm 5	50	25	75	25	35	25	30	30	30	25	29	28
Farm 6	50	40	60	40	50	40	45	38	45	40	43	41
Farm 7	50	50	50	50	50	50	50	50	50	50	50	50
Farm 8	50	60	40	50	50	40	50	65	50	45	50	46
Farm 9	50	75	25	30	40	25	40	70	30	25	37	39
Farm 10	60	40	50	40	50	40	45	35	40	40	41	41
Farm 11	60	50	40	40	50	40	40	40	40	45	42	43
Farm 12	75	25	50	25	35	25	30	25	25	20	26	27
Farm 13	75	50	25	25	40	25	35	45	25	25	31	32

than Period 1 in respect to 'absence of prolonged thirst' (with indices of importance of 0.52, 0.30, and 0.18, respectively). This may be explained by the fact that there are much more animals in Periods 2 and 3 than in Period 1 and that in Period 2 a problem around water provision may be dramatic for lactating females and their litters. Second, all interaction indices are positive (between 0.23 and 0.45), showing that the experts want to limit compensations between the periods: to obtain a low score on one period is sufficient for the farm to be considered bad for that measure (i.e. experts are not satisfied when one of the three scores is low). This can be easily seen in Table 6: all farms (with an average score of 50) are scored below 50, even Farm 2 that scores only 44 whereas it would have scored 58.5 using a weighted sum for the calculation, showing that this farm is penalized for the bad score (25) obtained in Period 1.

Discussion

Variability among experts exists. In Welfare Quality® similar differences between experts also appeared. Since we are dealing here with an interpretation process, it is not surprising to find such variability in answers, illustrating the diversity of viewpoints.

One important step was lacking when preparing this paper: the test of the assessment protocols on commercial farms over a whole production cycle. Concerning raw data, all three periods were tested on mink and fox farms in 2011 and early 2012. However score calculations will be possible only when the software derived from foxes and mink welfare assessment protocols will be finalized. This software is under construction and would be ready in a β version for summer 2012. It would be then possible to process the calculations and present the results to experts to check the plausibility of the results. The objective is to make the final protocols and software tool available through the web by the end of 2012.

As a conclusion, this work shows that the rationale developed in Welfare Quality® for cattle, pigs and poultry may be successfully adapted to other species, taking however into consideration the specificities of their production systems. For foxes and mink, these adaptations led to a higher complexity of the construction so that the different animal types and periods of the production-cycle can be considered. One indicator of this complexity is the number of questionnaires to be filled in by experts: 14 for dairy cows on farm in Welfare Quality® vs. 55 and 80 for foxes and mink farms respectively.

Acknowledgements

We thank all WelFur partners and members of the expert panels for fine-tuning the evaluation models. We are especially grateful to the partners in charge of the development of the measures for their great participation to our work, for the fox team: Leena Ahola, Jaakko Mononen, Tarja Koistinen, and Hanna Huuki, and for the mink team: Steen Henrik Møller, Steffen W. Hansen, and Jens Malmkvist. Finally, we thank EFBA for the financial and practical support to the WelFur research project.

References

Botreau, R., Capdeville, J., Perny, P. and Veissier, I., 2008. Multicriteria evaluation of animal welfare at farm level: an application of MCDA methodologies. Foundations of Computing and Decision Sciences 33: 287-316.

Botreau, R., Veissier, I. and Perny, P., 2009. Overall assessment of cow welfare: strategy adopted in Welfare Quality®. Animal Welfare 18: 363-370.

Farm Animal Welfare Council, 1992. FAWC updates the five freedoms. Veterinary Record 17: 357.

Grabisch, M. and Roubens, M., 2000. Application of the Choquet integral in multicriteria decision making. In: Grabisch, M., Murofushi, T. and Sugeno, M. (eds.) Fuzzy Measures and Integrals – Theory and Applications, Studies in Fuzziness and Soft Computing. Physica Verlag: Heidelberg, Germany. pp. 348-374.

Mononen, J., Møller, S.H., Hansen, S.W., Hovland, A.L., Koistinen, T., Lidfors, L., Malmkvist, J., Vinke, C.M. and Ahola, L., 2012. The development of on-farm welfare assessment protocols for foxes and mink: the WelFur project. Animal Welfare, in press.

Welfare Quality® consortium, 2009a. Welfare Quality® Assessment Protocol for Cattle. Lelystad, the Netherlands.

Welfare Quality® consortium, 2009b. Welfare Quality® Assessment Protocol for Pigs. Lelystad, the Netherlands.

Welfare Quality® consortium, 2009c. Welfare Quality® Assessment Protocol for Poultry. Lelystad, the Netherlands.

WelFur – mink: inter-observer reliability of on-farm welfare assessment in the growth season

S.H. Møller, T. Rousing and S.W. Hansen
Department of Animal Sciences, Aarhus University, Blichers Allé 20, P.O. Box 50, 8830 Tjele, Denmark; steenh.moller@agrsci.dk

Abstract

A welfare assessment system should be 'high' in validity, robustness and feasibility – the latter both as regards time and costs. Therefore, observers must be able to perform the on-farm assessment with acceptable validity after some training. Based on empiric data this paper evaluates the consequences of operating with several observers. Animal based measures on 9 Danish mink farms were taken in November 2011. Eight observers individually, but in pairs on herd level, carried out data collection on the measures involving subjective grading, e.g. mink 'activity', 'injuries' and 'fur chewing' on approximately 120 cages with mink per farm. The assessment of the two observers gave similar frequencies of welfare problems and thus similar welfare assessments. The individual problems observed were however, not the same leading to poor or fair, but rarely good inter observer reliability. Despite the skilled assessors, the short training was not sufficient to get highly reliable results. No overall difference was found between the inter observer reliability of cages with ≤2 or ≥3 mink in a cage. More training and better training material and, for some measures, observation procedures are needed in order to increase reliability of the difficult subjective measures.

Keywords: *Neovison vison*, Kappa, index of concordance, welfare

Introduction

The European Fur Breeders' Association (EFBA) launched the project 'WelFur' in 2009 with the aim to develop a protocol for use in welfare assessment of mink and fox farms by the same standards as developed in the EU project Welfare Quality® (Mononen *et al.*, 2012; Møller *et al.*, 2012). A welfare assessment describes the level of welfare on a farm by use of a number of indicators or measures reflecting different aspects of the welfare of the animals. The indicators can be based on the resource provided for the mink (housing system or management) or the animals in terms of health or behaviour (Møller *et al.*, 2012). All the measure included were selected because they were assessed valid (reflect some aspects of the actual welfare), reliable (acceptable inter or intra observer reliability and robustness to external factors), and feasible (possible and practical to apply during a one day visit). Previous inter-observer reliability (IOR) studies, assessing the extent to which two or more assessors obtain similar results when measuring the same thing, have shown that most of the measures taken at animal- or cage-level in the WelFur mink protocol can have good IOR (e.g. Hansen *et al.*, 2009). Despite this validation, some measures are to be applied in other seasons and situations than they have been developed and tested for. Furthermore, the assessors will not be scientists with experience in observing animal health and behavior. The interesting question is therefore not *if* it is possible to obtain good IOR, which is mainly a matter of skills and training. The question is rather the ease and precision with which they can be measured and how much training, what kind of training, and what training material is needed for those that will do the assessments. Training in order to reach reliability is mainly needed for animal-based measures involving elements of subjective grading, e.g. of an underlying continuous variable into two or more categories. In situations where a specific point on a continuous scale is a threshold for pass or fail on a binary welfare scale, e.g. a certain body condition score and where the position and posture of the mink moving in the cage affects the observation, extensive training may be needed. To score the presence or not of a given enrichment in the cage requires no or little training while the grading of body condition or lesions require more intensive training. Finally, few, if any, of the animal-based measures have been tested for IOR in group housing which can be expected to be an extra challenge for a reliable scoring.

As all farms have to be assessed within the same few weeks in each period (Mononen *et al.*, 2012; Møller *et al.*, 2012), a relatively large number of assessors, with sufficient skills and training will be needed. In order to evaluate the training of the assessors performing the tests of the WelFur protocols in the three seasons in Denmark, we designed the test of the WelFur protocol in the growth season 3 in 2011 in order to analyse the agreement between the two observers visiting each farm. This period was chosen for two reasons: (1) It involves many combinations of sex, age and social grouping of mink (single housed adults, pair-housed and group-housed juveniles) for which the measures may require different levels of training and for which the measures may not previously have been tested. (2) It allowed the test assessors to reach some level of familiarity with the measures in the WelFur protocol, as this was the last of the three seasons to be tested. The aim of this paper is to test the hypotheses that: (1) The outcome of the WelFur assessment will be the same for the two assessors. (2) Due to the skilled assessors, the short training session was sufficient to get reliable results. (3) Single or pair housed mink have higher IOR than group housed mink.

Materials and methods

The resource-based measure *quality of the nest box,* and the animal-based measures: *cleanliness of the fur, lesions and injuries, lameness and impaired movement, obviously sick mink, stereotypy*, and *fur-chewing* were tested for IOR in the autumn (season 3) between October 10. and November 14. Measures were taken by 2 assessors at each of nine farms. A total of 8 observers were involved, each visiting one to three farms. The assessors were all skilled farm animal scientists, research technicians and mink production advisors, i.e. they were all familiar with animal and/or mink behaviour. Before each test season an assessment protocol with data collection descriptions, sampling procedures and registration schemes were distributed and all assessors met at a the research farm at Aarhus University and went through the procedures. No formalised education or training programme was followed. The farms are numbered randomly from 1-9 and the observers are numbered 1 and 2 at each farm assessment. In total 1,100 cages with 2,616 mink were observed by two observers.

As there were one to four mink in each cage, the reliability of the observations are based on the number of mink scored with each grade of a condition (e.g. *lesion, stereotypy, fur-chewing*, see Table 1) which is somewhat different from IOR on individually identifiable animals. This allows

Table 1. Categories of each measure tested in the present study. The definition of each category is from the WelFur mink protocol used during the study in the autumn 2011. Some have later been revised in order to increase the precision of the scoring.

Measure	Categories and their definition
Quality of the nest box	0 – dry or 1 – wet 0 – clean or 1 – dirty 0 – whole/undamaged or 1 – damaged (e.g. sharp edges) 0 – free from fleas or 1 – infested with fleas
Cleanliness of the fur	0 – clean: the fur is clean. no faeces, feed or other stains are observable in any part of the animal 1 – moderately dirty: the fur of the mink is dirty in some parts of the body, i.e. there are a few dirty spots or one larger dirty area 2 – severely dirty: the fur of the mink is dirty in larger parts (>5×5 cm) of the body
Lesions and injuries	0 – no lesions 1 – unhealed wound or lesions with a diameter <10 mm or healed lesions e.g. missing an ear or less than half the tail 2 – unhealed wounds or lesions with a diameter <30 mm or healed lesions e.g. missing more than half the tail 3 – unhealed wounds or lesions with a diameter >30 mm or unhealed wound or lesion from e.g. missing more than half the tail
Lameness and impaired movement	0 – no evidence of lameness or impaired movement 1 – evidence of lameness or impaired movement
Obviously sick mink	0 – no signs of poor health or disease 1 – obvious signs of poor health or disease
Stereotypy	0 – not active (moving less than one body length during the 2 minutes of observation) 1 – active (moving more than one body length during the 2 minutes of observation) 2 – stereotypy (three or more repetitions in a row of the same behavioural pattern)
Fur-chewing	0 – no/very little fur-chewing. the outer tip of the tail is not visible (naked) and/or less than 0.5 cm^2 of the tail or body has been chewed 1 – moderate fur-chewing. the outer tip of the tail is visible (naked) and/or fur-chewing of less than 3 cm of the tail or less than 3×3 cm fur-chewing of the body 2 – severe fur-chewing. >3 cm of the tail or >3×3 cm of the body has been chewed 3 – extensive fur-chewing. >10 cm of the tail or >10×10 cm of the body has been chewed

for an evaluation of the effect of number of animals in the cage on the IOR. Potential effects of colour types, cage systems or other sampling strategy factors on the IOR have not been tested.

During the assessments, the two assessors observed the same cages and thus the same mink on each farm. The observation of *stereotypy* was done simultaneously so the two observers could score the same situations while this was not needed for the other measures that are not expected to change during the hours of observation. The prevalence of different health disorders on the farms are presented as mean values from all assessors. The difference at farm level between the two observers was calculated in order to evaluate if they would lead to different welfare assessments. The data enabled the calculation of the inter-observer reliabilities (IOR) for the assessed measures. IORs were estimated using the Kappa value (Coen, 1960) as well as the index of concordance (IC) (Martin and Bateson 2007). ICs for each measure was calculated as A / (A + D), where A = the number of agreements and D = the number of disagreements while Kappa is adjusted for agreements by chance. If there is complete agreement both Kappa and IC is 1. For no agreement IC is 0 while Kappa is below 0 if observed agreement is less than chance agreement (Fleiss *et al.*, 2003). Kappa value interpretations have been suggested as: $k \leq 0.40$ is poor, $0.40 < k \leq 0.60$ is fair, $0.60 < k \leq 0.75$ is good, and $k > 0.75$ is excellent (Kaufman and Rosenthal, 2009), while an IC value >0.8 can be considered as acceptable to good, depending on bias and prevalence of the categories.

Results

Both observed measures for *comfort around resting,* i.e. *quality of the nest box* and *cleanliness of the fur* and the health measures *lameness and impaired movement* and *obviously sick mink* were extremely rare on all farms in this period (Table 2). A few per cent of mink with mainly small lesions were seen while only *active mink* and *moderate fur-chewing* were seen in significant numbers (Table 2).

The difference between the level of problems observed by the two observers at the same farm visit was on average small for all measures, as indicated by the 'observer difference' in Table 2.

Kappa values are generally lower than expected and in the cases of *cleanliness of the fur, lesions and injuries,* and for *fur-chewing* Kappa values were poor, often only one third or half the IC values (Table 3). Some measures had fair IOR in terms of Kappa values while only the observation of *activity* and *stereotypy* had good IOR. The Kappa values for cages with one or two mink are in 4 cases equal to, in 4 cases higher than, and in 6 cases lower than for cages with three or four mink. Kappa is lower for *cleanliness of the fur,* moderate *lesions and injuries,* and for *fur-chewing,* i.e. the measures with poor Kappa values. The scores for small lesions is an illustrative example of the low Kappa values and the large difference between Kappa and IC (Table 4).

The total number of small lesions observed were similar, 52 (27+17+2×3+2×1) for observer 1 and 56 for observer 2, even though the observers did not observe the same small lesions (not in the same cages). In most cases no lesions were observed while most lesions were only observed by one of the two observers. This severely reduces the Kappa value while the IC remains high due to the many agreements on 'no lesions'.

Table 2. Means±SD of measures and difference between the two observers, minimum and maximum percentages of mink in different categories observed by two assessors at 9 Danish mink farms during the autumn 2011.

Measure	Category	Measure mean ±SD	Min-max	Obs. difference mean±SD
Quality of the nest box	wet	0.03±0.14	0.00-0.59	0.07±0.20
	dirty	0.00±0.00	0.00-0.00	0.00±0.00
	damaged	0.09±0.31	0.00-1.31	0.11±0.33
	fleas	0.00±0.00	0.00-0.00	0.00±0.00
Cleanliness of the fur	moderate	0.15±0.25	0.00-0.74	0.07±0.13
	severe	0.00±0.00	0.00-0.00	0.00±0.00
Lesions and injuries	small	2.08±1.01	0.32-4.26	0.74±0.82
	moderate	0.56±0.65	0.00-2.27	0.64±0.61
	severe	0.16±0.26	0.00-0.85	0.22±0.23
Lameness and impaired movement	observed	0.00±0.00	0.00-0.00	0.00±0.00
Obviously sick mink	observed	0.12±0.23	0.00-0.85	0.16±0.29
Stereotypy	active	51.61±23.27	19.76-84.45	4.06±2.34
	stereotypy	1.10±1.32	0.00-4.85	0.45±0.48
	stereotypy/active	1.92±1.70	0.00-5.79	0.55±0.57
Fur-chewing	moderate	11.90±6.27	3.59-27.31	7.10±5.18
	severe	1.73±1.55	0.00-5.46	1.51±1.99
	extensive	0.53±0.46	0.00-1.26	0.38±0.43

Discussion

The total frequency of observed problems are fairly similar for the two observers on most farms and measures and the hypothesis that the outcome of the WelFur assessment will be the same for the two assessors is therefore accepted. In general, very few welfare problems were observed during the autumn (WelFur season 3) and only *small lesions* and *moderate fur-chewing* were seen in average on more than 2% of the mink. This is of course good for the welfare, but problematic for the validation of the observations in terms of IOR. The few observations of problems relative to no problems renders the Kappa analysis of IOR less useful (Byrt *et al.*, 1993) while the IC has the same meaning no matter how skewed the prevalence of the distribution is. The IOR is generally high in terms of IC except for the prevalence of active mink in each cage, while the Kappa value is almost the opposite. This is due to the many agreements on 'nothing to observe' in situations with few 'problems' observed (Byrt *et al.*, 1993), as illustrated by the 1016 agreements of 'no small lesions' and 17 agreements of '1 small lesion' in Table 4 as opposed to the 67 small lesions in one or two animals observed by only one of the two observers. This severely reduces the Kappa value while the IC remains high due to the many agreements on 'no lesions'. The 67 small lesions observed by only one of the two observers indicates a general underestimation of the total number of lesions, as most of these observations must be assumed to be correct (few

Table 3. Inter-observer reliabilities of the each measure tested at 9 Danish mink farms during the autumn. In total 2,616 mink in 1,100 cages were observed. Simple Kappa values and Index of concordance (IC) are calculated for the assessor pairs for all 1,100 cages. Kappa is also calculated separately for cages with 1 or 2 mink and 3 or 4 mink.

Measure	**Category**	**Kappa**	**IC**	**Kappa ≤2 (n=826)**	**Kappa ≥3 (n=270)**
Quality of the nest box	wet	-	0.998	-	1
	dirty	1	1	1	1
	damaged	0.399	0.997	1	0.396
	fleas	1	1	1	1
Cleanliness of the fur	moderate	0.247	0.996	-0.002	0.395
	severe	-	-	-	-
Lesions and injuries	small	0.314	0.939	0.326	-
	moderate	0.132	0.978	0.126	0.133
	severe	0.247	0.995	0.331	-
Lameness and impaired movement	observed	1	1	1	1
Obviously sick mink	observed	0.284	0.995	-	0.497
Stereotypy	active	0.649	0.753	0.703	0.48
	stereotypy	0.62	0.982	0.606	0.657
Fur-chewing	moderate	0.24	0.718	0.23	0.266
	severe	0.195	0.95	0.11	0.37
	extensive	0.472	0.989	0.425	0.567

Table 4. Number of small lesions observed per cage by two observers. Data from cages with (≤2; ≥3) mink in brackets. Data from 9 farms visited by 9 different pairs of observers scoring 2,616 mink in 1,100 cages in total.

Observer 1	**Observer 2**			Total
Lesions observed/cage	**0**	**1**	**2**	
0	1,016 (768;244)	33 (25;8)	3 (3;0)	1,052 (796;252)
1	27 (16;11)	17 (12;5)	0 (0;0)	44 (28;16)
2	3 (1;2)	1 (1;0)	0 (0;0)	4 (2;2)
Total	1,046 (785;257)	51 (38;13)	3 (3;0)	1,100 (826;270)

false positives). It also indicates a need for better definitions and training of observers, probably mainly in ways to make sure that the animals are observed and present themselves in such a way that lesions can be observed.

The Kappa values for cages with one or two mink are not systematically different from cages with three or four mink and therefore the hypothesis that single or pair housed mink have higher IOR than group housed mink must be rejected.

The lack of agreement on the rare cases of *damaged nest box, moderate dirty fur,* and *obviously sick mink* illustrates that the training of observers and/or the definition of these problems should be improved. However, a slight underestimation of IOR may be involved, as cages where one assessor observes one problem in a cage and the other observed two problems. This is in the IOR calculated as a disagreement, while one of the two observations might in fact be in agreement, e.g. the one observation of a small lesion in one or two animals in Table 4.

A clear threshold between the four categories of nest box quality must be illustrated by better descriptions and illustrative photos in order to standardise the observation and qualify the discrimination between 'no problem' and 'problem'. Due the low prevalence of these problems the best way to proceed is to take pictures every time an observer classifies a nest box problem.

For the health measures the challenge is larger, as both the definition of categories and the method of observation are important. The categories are well described and good pictures will facilitate a reliable scoring. However, the chance to actually observe e.g. a lesion or an injury depends on how the mink moves and presents itself to the observer. Bite wounds are most often inflicted at the tip or the base of the tail (Hansen and Møller, 2012) making the observation difficult when the mink are active and orientated towards the observer. Therefore precise and efficient observation methods must be developed and documented.

The low kappa values for *damaged nest box, moderate dirty fur,* and *obviously sick mink* illustrates that the few cases observed by the two observers were not the same. The high IC values illustrate the high agreement on cages without observed problems which is not disturbed by the few un-agreed problems. However, these high IC values do not indicate high IOR, unless the few but critical problems are actually also scored the same by the observers. A large number of IOR tests are available and prevalence-adjusted bias-adjusted kappa (PABAK) (Byrt *et al.*, 1993) could be used. It requires, however, that the same number of categories are scored by the two observers, which is not always the case. The true reliability of e.g. scores of lesions can be investigated against a golden standard, e.g. obtained by post mortem examination of the mink that have been scored would qualify the validation.

Despite the qualifications of the assessors, this was not enough to obtain acceptable IOR with the fairly little amount of training. Therefore better training of assessors, definition (pictures) of the problems and thresholds between categories are needed in order to standardise the observation and qualify the discrimination between 'no problem' and 'problem'. For a larger body of observers without previous qualifications or experience with fur animals, a more intensive on-farm training programme, including documentation of acquired skills, will be needed.

Conclusions

The total frequency of observed problems are fairly similar for the two observers and the hypothesis that the outcome of the WelFur assessment will be the same for the two assessors is accepted. The definite test will be if the difference between observers will in any way affect the final classification of the farms in the four WelFur categories described by Botreau *et al.* (2012). This test will be possible by the end of 2012. The hypotheses that single or pair housed mink have higher IOR than group housed mink is rejected. The hypotheses that a short training session was sufficient to get reliable results due to the skilled assessors must also be rejected as too many measures had poor or fair IOR in terms of Kappa values and only the observation of *activity and stereotypy* had good IOR.

The overall results of the IOR tests are that the agreement for most measures seems to be acceptable in terms of index of concordance but are only poor or fair when Kappa is applied. His is partly due to the high prevalence of 'no problems' while the few problems observed are scored with rather poor reliability.

Better training of assessors, better definitions and pictures of the problems and the threshold between categories in order to standardise the observation and qualify the discrimination between 'no problem'/'problem', and probably most important, ways observe animals thoroughly in the farm situation will be needed.

References

Botreau, R., Gaborit, M., and Veissier, I., 2012. Applying Welfare Quality® strategy to design a welfare assessment tool for foxes and mink farms. Scientifur 35, 4, IFASA 2012 Congress proceedings. pp 6.

Byrth, T., Bishop, J., and Carlin, J.B., 1993. Bias, Prevalence and Kappa. J. Clin. Epidemiol. 46: 423-429.

Cohen, J.A., 1960. A coefficient of agreement for norminal scales. Educational and Psychological Measurement 20: 37-46.

Fleiss, J.L., Levin, B. and Cho Paik, M., 2003. Statistical methods for Rates and Proportions. New York, John Wiley.

Hansen, B.K, Møller, S.H., Berg, P. and. Bækgaard, H., 2009. Validation of a method for body condition scoring. Annual report 2008, Danish Fur Breeders Research Center, Holstebro, Danmark. pp. 129-135 [In Danish].

Hansen, S.W. and Møller, S.H., 2012. Mink's adaptation to group housing in practice. Scientifur 35, 4, IFASA 2012 Congress proceedings.

Kaufman, A.B. and Rosenthal R., 2009. Can you believe my eyes? The importance of interobserver reliability statistics in observations of animal behaviour. Animal Behaviour 78: 1487-1491.

Martin, P. and Bateson, P., 2007. Measuring Behaviour. An Introductory Guide. third ed. Cambridge University Press, Cambridge, UK. 176 pp.

Part 6. Short communications

Lessons learned from not fattening juvenile mink kept as future breeders

S.H. Møller
Aarhus University, Animal Sciences, Blichers Allé 20, P.O. Box 50, 8830 Tjele, Denmark; steenh.moller@agrsci.dk

Abstract

New breeder mink are usually selected from juveniles fed for pelting and thus in very good body condition. As the body size of mink increases this also increases the risk of impaired welfare due to too drastic or too fast weight loss during the winter in order to achieve a suitable body condition for mating. Alternatively the risk of impaired reproduction due to obesity during mating or gestation increases. One solution could be not to feed potential breeders *ad libitum* in the autumn. In order to test this solution, two groups of approximately 400 male+female pairs of juvenile mink were fed individually until mid-September. Thereafter one group was restricted 20% in feed allowance and maintained at a prolonged period of 6 hours without feed before next feeding, while the control group remained on *ad libitum* feeding. The plan was 6 hours without feed, but as the restricted mink ate faster and faster the time without feed had to be increased to 12 hours before a significant difference in feed allowance was achieved. The result of this was that the planned difference in feed intake, weight development and body condition was much smaller than planned, especially in the dams. This means that different feeding management tools than hours without feed has to be applied in order to control the body condition of future breeders by the feeding strategy. As expected the restricted feeding had larger effect on males than females. As only 1 male for every 5 to 6 females are needed as future breeders for next year's production, this is a challenge for the selection of breeders in a male-female pair-wise production.

Effect of dietary protein level during preparative mating period in female minks

Q.K. Jiang[1], Z.Q. Zhang[2], X.H. Gao[2], X.M. Xing[1] and F.H. Yang[1]
[1]Institue of wild animal and plant science, CAAS, 15 Lumingdajie, Changyi District, Jilin, 132109, China; [2]Feed Research Institue, CAAS, 12, Zhongguancunnandajie, Haidian District, Beijing, 100081, China; yangfh@126.com

One hundred and eighty female minks were assigned randomly to six groups with 30 females each. Group1 to Group 4 received diets of fresh material with protein level of 28.59%, 32.31%, 36.21% and 40.35% respectively. Group 5 to Group 6 received diets of dry powder with protein level of 32.66%, 40.47% respectively. The test included a 7-day pre-trial period and a 44-day test period. The results were as follows: There was no significant difference in DM digestibility, CP digestibility, EE digestibility, nitrogen deposition, NPU and BV between groups fed fresh material diet (P>0.05).With the dietary protein level increased, the indicators mentioned above firstly increased and then decreased, reaching the peak when dietary protein level was 36.21%; Contents of TP, GLU and activity of GOT were significantly lower in the 28.59% protein level group (P<0.05), but all the serum biochemical indicators showed no significant difference between groups with dietary protein level higher than 32.31%; nitrogen deposition, NPU, BV, contents of BUN, ALB, GLU in groups fed dry powder diets were not significantly different with those fed fresh material diet of the same dietary protein level (P>0.05), but the DM digestibility, CP digestibility, EE digestibility, activity of GPT and GOT were significantly lower than their dry powder counterparts (P<0.05). The combination of feeding experiment, digestion metabolism experiment, serum biochemical indicators showed that the optimum dietary protein levels in preparative mating period was approximate 32%.

Effect of dietary protein level during gestation period in female minks

Q.K. Jiang[1], Z.Q. Zhang[2], X.H. Gao[2], X.M. Xing[1] and F.H. Yang[1]
[1]Institue of wild animal and plant science, CAAS, 15 Lumingdajie, Changyi District, Jilin, 132109, China; [2]Feed Research Institue, CAAS, 12, Zhongguancunnandajie, Haidian District, Beijing, 100081, China; yangfh@126.com

180 female minks were assigned randomly to 6 groups with 30 females each. Group 1 to Group 4 received diets of fresh material with protein level of 31.79%, 36.53%, 41.39% and 45.39% respectively. Group 5 to Group 6 received diets of dry powder with protein level of 36.55%, 44.47% respectively. The results were as follows: In groups fed fresh material diet, CP digestibility, EE digestibility, nitrogen deposition, NPU and BV increased with dietary protein level(R^2 were 0.9811, 0.9531, 0.9047 and 0.8532, respectively). Indicators mentioned above reached the peak when protein level was 45.39% (P>0.05).No significant difference was observed in DM digestibility and EE digestibility in fresh material diet fed groups. Compared with groups fed fresh material diet of same dietary protein level, the dry powder ones showed significantly lower values in DM digestibility, CP digestibility, EE digestibility, nitrogen deposition, NPU and BV (P<0.05). Different dietary protein levels showed no significant difference in mating rate, parturition rate, litter size, number of born alive, birth survive rate, Individual birth weight among fresh material fed groups, but parturition rate, litter size, number of born alive firstly increased and then decreased with increasing dietary protein level, reaching the peak at 36.53% dietary protein level. Mating rate, parturition rate, litter size, number of born alive, birth survive rate, individual birth weight decreased or significantly decreased in groups fed dry powder diet compared with groups fed flesh material diet of same protein level (P<0.05).The results showed that the optimum dietary protein level in gestation period was approximate 36%.

Effect of dietary protein level during lactation period in female minks

Q.K. Jiang[1], Z.Q. Zhang[2], X.H. Gao[2], X.M. Xing[1] and F.H. Yang[1]
[1]Institue of wild animal and plant science, CAAS, 15 Lumingdajie, Changyi District, Jilin, 132109, China; [2]Feed Research Institue, CAAS, 12, Zhongguancunnandajie, Haidian District, Beijing, 100081, China; yangfh@126.com

Eighty pregnant minks with similar due date were selected and randomly assigned to four groups with 20 minks each. Group 1 to Group 4 received diet of fresh material with protein level of 32.28%, 36.33%, 40.94%, 44.68% respectively. Test ended when the last litter was weaned with a 7-day pre-trial period at the beginning. The results were as follows: dietary protein level significantly affected weaning weight loss and weaning weight loss rate of minks in lactation period ($P<0.05$). Daily weight gain and weaning survival rate of kits increased with the dietary protein level, although no significant difference was observed between the four groups. When the dietary protein was 44.68%, daily weight gain and weaning survival rate reached peak (R^2 were 0.846 and 0.9987, respectively), weaning weight loss and weaning weight loss rate were the lowest. Values of BUN, TP, GLU and GOT were significantly affected by dietary protein level with the tendency of increasing with it, which showed that minks in lactation had high utilization rates of high protein diet. The combination of feeding experiment, serum biochemical indicators and reproductive performance showed that the optimum dietary protein level in lactation period were approximate 44%.

Effects of dietary protein levels and implanted melatonin on growth performance, serum biochemical parameters and nutrient digestibility of minks (*Mustela vison*)

X. Song, H. Xue, H. Wei, G. Li and F. Yang
Institute of Special Wild Economic Animal and Plant Science Chinese Academy of Agricultural and Scie, No 15, Luming street, Zuojia town, Changyi district, Jilin City, Jilin Province, 132109, China; songxingchao_888@126.com

Ninety healthy male mink at the age of fifty-day-old with a similar body weight($P>0.05$) were randomly allotted to six groups, each of which was replicated fifteen times with one mink per replicate. A 3×2(CP×MT) completely random experiment was designed. The minks in 3 groups (1, 2 and 3) which were injected melatonin capsules (10 mg each) on 15 July 2010 were fed experimental diets with protein levels of 32% (CP1), 36% (CP2) and 40% (CP3), respectively, while those in control groups (4 to 6), which were not injected with melatonin were fed the same diets with trial groups. The feeding trial lasted for 64 days. The results showed as follows: (1) For growing minks, injecting melatonin 10 mg merly increased the body weight and average daily gain which were the highest group fed with CP1 level. The body weight of MT groups fed with CP1level was significantly higher than that of controlled group ($P<0.05$). The body weight and average daily gain were not significantly affected by the interaction of the protein levels and exogenous melatonin. (2) The serum ALB concentration of MT treatment groups was significantly lower than that of controlled fed with CP1 and CP3 levels ($P<0.01$). For treatment and controlled minks, the serum TG concentration of CP1 and CP2 levels were significantly higher than those of CP3 level ($P<0.05$). (3) For MT group, the fat digestibility of CP2 level was higher siginificantly than that of CP1 and CP3 level ($P<0.05$). The interaction between protein levels and implanted MT had no influence on the protein, fat digestibility ($P>0.05$).

Effect of different diet protein levels on the digestibility and metabolism of nutritions and growth performance in young male mink

T.T. Zhang and X.H. Gao
Chinese Academy of Agricultural Science, Feed Research Institute, Zhongguancun south street 12, Haidian distriction, Beijing, China, 86100081, China; zhangtietao@yeah.net

The experiment was studied the ratio of digestibility and metabolism in male mink which fed with different protein level diets in the period of growth, and the effect of protein quantity on the growth and development in mink and the feed transformation efficiency. Sixty healthy male sables of 45d old were randomly divided into 6 groups with 10 replicates in which had 1 sable. The sables were fed with protein levels of 28% (group 1), 30% (group 2), 32% (group 3), 34% (group 4), 36% (group 5), 38% (group 6), respectively. The digestibility and metabolic ratio of nutritions were determined in 14d feed with trial diets. At the beginning of the trail, the male minks were weighted and registered in the interval of 2 weeks in order to calculating the average daily gain and the feed transformation efficiency. The results indicated that nitrogen of feed intake was greatly significantly difference among groups (P<0.01). The indexes included that the digestibility of calcium, nitrogen of fence, nitrogen deposition, net protein utilization, biological value of protein were not significantly difference among groups (P>0.05). Compared with the group 1, the group 4, group 5 and 6 group had greatly significantly difference in the average daily gain and feed conversion (P<0.01). In conclusion, when the protein level of diet was lower than 28% protein in the feed, then it had negetive effect on the growth and development of mink. When the protein in the diet met with quantity at the 34% 0f feed, the paremeters that included digestibility of nutrition, average daily gain and feed conversion rate in mink were more optimization than others.

Preliminary determination and analysis of 12 serum biochemical parameters in mink (*Mustela vison*)

X. Song, H. Wei, H. Xue, X. Chen and Y. Yang
Institute of Special Wild Economic Animal and Plant Science, Chinese Academy of Agricultural and Sci, The rearing of special economic animals, No. 15 Luming Street, Zuojia Town, Changyi district, Jilin City, Jilin Province, 132109 China, songxingchao_888@126.com

Blood biochemical parameters are the key index reflecting the physical condition and health of animals. In order to establish the standard reference range of blood biochemical parameters of northeast black mink and explore if a significant difference exist between males and females, we performed the following study. 60 healthy (♂30, ♀30), 8 months of age northeast black mink were selected randomly. The blood of the heart was collected and the serum gained for determination and analysis of the biochemical parameters using automatic biochemical analytic instruments and biochemical reagent kits. The results showed that the concentration of ALP between males and females were significantly different ($P<0.01$). The concentration of UA, GOT, TG between males and females was also different ($P<0.05$). No significant differences were observed in TP, ALB, Cre, BUN, GPT, CK, T-CHO, and GLU between males and females ($P>0.05$). These data could be used for breeding purposes, diagnosing diseases and biological characteristics of mink.

Detection of the Aleutian mink disease virus DNA and anti-viral antibodies in experimentally infected mink 10 days post-inoculation

A.H. Farid, I. Hussain and P.P. Rupasinghe
Nova Scotia Agricultural College, Plant and Animal Sciences, Truro, Nova Scotia, B2N 5E3, Canada; hfarid@nsac.ca

Black (n=44) and sapphire (n=12) mink were injected intraperitoneally with 10% spleen homogenates containing a local strain of the Aleutian mink disease virus (AMDV). Animals were euthanized 10 days post-inoculation (pi) and samples of blood and seven organs were collected. Viral DNA was extracted from blood plasma and cell-free organ homogenates (25% W/V) by magnetic beads and tested by PCR. Anti-viral antibodies in blood plasma were tested by counterimmunoelectrophoresis (CIEP). The virus was present in the spleen, lymph nodes, bone marrow and lung samples of all inoculated mink, but was not detected in some intestine (1.9%), kidney (10.7%) and liver (35.7%) samples. Viral DNA was detected in plasma of all inoculated mink by PCR, although two of the samples showed inconclusive test results. Anti-viral antibodies were detected in plasma of three sapphire (25.0%) and 19 black (43.2%) mink. The relative specificity and sensitivity of the CIEP test on plasma samples were 100% and 39.3%, respectively, implying that low levels of antibodies during early stages of infection resulted in false negative CIEP tests. Histopathology revealed some lesions characteristics of AMDV infection in the liver and/or kidneys of three black and one sapphire mink, but heart and lung tissues had no detectible lesions. The results suggested that PCR was more reliable than CIEP for viral detection in the early period after infection, and that there were considerable differences among mink of each color type for production of detectable levels of anti-viral antibodies and development of disease symptoms. Samples of saliva, rectal swabs and feces collected on day 10 pi did not produce consistent and reliable results for viral detection by PCR.

Aleutian mink disease virus in furbearing mammals in Nova Scotia, Canada

A.H. Farid
Nova Scotia Agricultural College, Plant and Animal Sciences, Truro, Nova Scotia, B2N 5E3, Canada; hfarid@nsac.ca

Presence of Aleutian mink disease virus (AMDV) was tested in 466 wild animals of 12 furbearing species that were trapped in 10 Nova Scotia counties. DNA was extracted from cell-free spleen homogenates by magnetic beads and presence of the virus was tested for by PCR using two primer pairs, each with three DNA volumes. Anti-viral antibodies in spleen homogenates were also tested by counterimmunoelectrophoresis (CIEP) in duplicate. Positive PCR or CIEP samples were detected in 96.9% of 64 mink, 70.5% of 61 weasels, 25.0% of 8 skunks, 18.2% of 11 otters, 10.6% of 85 raccoons and 10.0% of 20 bobcats. Samples of six fishers, 24 coyotes, 25 red foxes, 58 beavers, 45 squirrels and 59 muskrats were negative. Anti-viral antibodies were detected in 16 (25%) of the mink and one of the skunk samples, possibly due to the low antibody titer in spleen homogenates. All CIEP positive animals were also PCR positive, except for three mink and one skunk which were PCR negative. This observation could be the manifestation of persistent antibody production and short-lived viral replication which are signatures of AMDV pathogenesis. The numbers of samples amplified by the two primer pairs were significantly different, suggesting that PCR amplification failure of CIEP positive animals could be the result of differences in the sequence of viruses circulating in different animals. The results imply that the combination of CIEP and PCR using multiple primers would increase the accuracy of detecting AMDV exposure in wild animal species. Infected animals were present in all counties from which samples were collected. The high proportion of animals of so many species across the province carrying AMDV has important epidemiological ramifications and could pose a serious threat to ranched mink as well as susceptible wild animal species.

A real-time polymerase chain reaction assay for the quantification of the Aleutian mink disease virus

A.H. Farid and P.P. Rupasinghe
Nova Scotia Agricultural College, Plant and Animal Sciences, Truro, Nova Scotia, B2N 5E3, Canada; hfarid@nsac.ca

Estimation of the Aleutian mink disease virus (AMDV) copy number in infected animals and environmental samples has a wide range of applications. For instance, while AMDV replication persists in some infected mink for a long period of time, the virus becomes latent in others. Estimate of viral copy number at specific time points in the mink is important for determining the dynamics of AMDV proliferation and designing selection strategies for establishing resistant or tolerant lines of mink. To develop and validate a SYBR Green quantitative real-time PCR assay (RT-qPCR) for estimating the AMDV copy number, 1,973 bp PCR products containing the entire VP2 gene (1,944 bp) of a local strain of the virus were ligated into the pGEM-T Easy plasmid. This 4,989 bp construct was propagated in *E. coli* and linearized with the PstI restriction enzyme. Five pairs of primers were designed for a conserved region of the VP2 gene and were tested in a Bio-Rad iQ5 cycler using nine 10-fold serially diluted samples of the purified plasmis-VP2 construct. One of the primer pairs resulted in a linear standard curve in the range of 11 to 11×106 dsDNA per µl of the sample, corresponding to 22 to 22×106 AMDV copies. The standard curve had high efficiency (98%-102%) and accuracy ($R_2 \geq 0.98$) in multiple runs. Melt curve analysis confirmed specificity of the primer pair. Inter- and intra-assay variability were low when plasma and tissue samples from infected mink were tested in triplicate in multiple runs. This RT-qPCR method is highly sensitive and reproducible for the detection and quantification of AMDV over a 6-log range. Reproducibility of this standard curve was considerably higher than that when 532 bp PCR products of the VP2 gene or the circular plasmid-VP2 construct were used.

Status of Aleutian mink disease virus in Denmark

J. Østergaard and T.H. Jensen
Kopenhagen Fur, Langagervej 74, 2600, Denmark; joe@kopenhagenfur.com

In Denmark Aleutian mink disease (AMD) is a reportable disease with law enforced annual screening of all mink farms. Furthermore, all mink moved between farms have to be tested negative for AMD before moving. If more than 3 seropositive animals are found by counter current immune electrophoresis (CIEP), the farm is regarded as AMD infected according to the legislation. If AMD virus is found in one mink the farm is regarded as AMDV infected. Approximately 3.5 million sera are tested by CIEP annually; PCR is conducted on a small scale (50-100 mink annually). AMD has been confined only to the area north to the Limfjord since 2003. However, in 2011 several outbreaks of AMDV were detected south to the Limfjord in the area of Himmerland, Djursland, Thy/Mors. In total 81 farms were AMD infected by the end of 2011 (the total number of farms in Denmark is about 1,400). The reasons for spreading of AMD is speculated to be caused by various reasons most likely moving between farms of staff, machines etc., aerogen transmission can not be excluded although earlier studies showed it was only over small distances (<1 km). In 2010 an outbreak involving several farms was traced to a pelting facility on one of the farms. Mink with unknown status were pelted on this facility and later it appeared that the mink were AMDV infected at the time of pelting. This highlights the importance of continuous alertness about biosecurity.

The assembly of the mink blood transcriptoms generated by the next generation sequencing using the dog genome

A.H. Farid[1], S.S. Moore[2] and U. Basu[2]
[1]Nova Scotia Agricultural College, Department of Plant and Animal Sciences, Truro. Nova Scotia, B2N 5E3, Canada; [2]University of Alberta, Department of Agriculture, Food and Nutritional Science, Edmonton, Alberta, T6G 2P5, Canada; hfarid@nsac.ca

The next generation sequencing techniques provide a great opportunity for deciphering various aspects of the mink biology. Alignment and mapping of the massive number of short reads produced by these techniques is computationally challenging, specifically for species such as the mink with little genome sequence information. The objectives of this study was to evaluate the merit of the dog genome, the closest species to the mink with complete genome sequence data, as a reference for the assembly of short reads generated by the next generation sequencing. Blood samples from 12 anaesthetized mink were collected by heart puncture into the PAXgene Blood RNA tubes. Libraries were prepared from total RNA using the Illumina TruSeq™ RNA sample preparation kit, and were sequenced using HiScanSQ, producing 75 bp reads from each end. A total of 794,241,780 paired-end reads (59.5×10^9 bp) were generated (ranging from 7.5 million to 182 million reads per sample). The reads were aligned to the dog genome using different stringencies (0 to 8 mismatches/75 bp). Only 5.7% of the reads were aligned within exons of approximately 11,200 known genes on the dog genome with varying coverage depth. Some reads were mapped outside annotated dog transcripts (4.3%), in intronic regions (0.62%), at the exon boundaries (1.6%) or to rRNA and snRNA (0.02%). The results suggested that genomes of other species, such as the cat and ferret, should be explored to indentify the remaining 87.7% of the unmapped mink transcripts. This sequence database is useful for the assembly and annotation of the mink genome.

Simple year-round cage enrichments for mink

G.J. Mason, K. Bahlmann, L. Dawson, J. Ahloy Dallaire, R.K. Meagher, S. Bowyer, M. Diez-Leon, and D. Campbell
University of Guelph, Animal Science, Guelph, Ontario N1G 2W1, Canada; gmason@uoguelph.ca

Subjects were 2,400 kits from 600 families, on 3 commercial farms in Ontario, pair-housed from July-December. 75% were Blacks, 25% Demis. Two male-female pairs were used per family: one was provided with stimuli for interaction; the other housed in standard conditions. The supplemented cages received a golf ball, plastic wiffle ball, and hanging device to tug /chew (hose or plastic chain) (costing under $1.50 a cage). Behaviour, temperament, mortality rates and pelt quality were assessed. Supplemented kits were found to play more than their non-enriched siblings. Social play (defined as 'rough and tumble' interactions without screaming or hard biting) was unaffected, but enhanced object and locomotor play led to significantly higher overall levels in supplemented juveniles. There was no evidence that the added objects increased aggression. The added objects also had no effects on mortality. In stick tests in December, supplemented mink were significantly less likely to be fearful than their standard-housed siblings. When scored for reactions to the live-grading of their fur in November, supplemented mink were significantly less likely to scream than their standard-housed siblings. Across Blacks, the added objects did not seem to affect fur quality, but in Demis, supplemented mink were significantly likely to receive better live-grades than their standard-housed siblings. For pelted subjects, tails of dead animals were inspected for signs of tail-chewing/clipping. Typically 1 cm or less of the tailtip was affected. However, 20 mink across all three farms showed damage to 2 cm or more the tail, and 14 of these were standard-housed rather than supplemented: a trend to significance. Furthermore, pooling all instances of tail-chewing/-biting together, 7.9% supplemented mink had sucked/bitten tails, compared to 11.5 standard-housed: a significant effect. The pelts of all experimental mink were bar-coded for auction, and prices will be obtained in summer 2012. Data during mating showed that supplemented males were more effective maters than standard mink (e.g. showing longer copulations). Subjects selected for breeding have also been evaluated for stereotypic behaviour and faecal cortisol metabolite output, and for whelping success: data still to be analysed. Overall, these findings to date show that simple, long-term, cheap additions to standard cages can act as effective environmental enrichments for farmed mink.

Elevated resting bunks for nursing mink dams

G.J. Mason[1], D. Haley[2] and L. Dawson[2]
[1]University of Guelph, Animal Science, Guelph, Ontario N1G 2W1, Canada; [2]University of Guelph, Population Medicine, Guelph, Ontario N1G 2W1, Canada; gmason@uoguelph.ca

A half-cylinder bunk of wire-mesh was attached to the cage ceiling of 80 dams with litters. Each bunk was 35 cm long, 16.5 cm wide and 10 cm deep, along an adult female to rest in it easily; and placement in the cage was high enough that no kits could climb on or in the bunks before the dam was removed when they were 6 weeks of age (typical weaning age in North America). 80 dams with litters acted as bunk-less controls. Behavioural data were collected daily between kit ages 21 days and weaning (42 days). Kits were weighed at weaning, and post-natal mortalities recorded from their cage records. After weaning, the teats of all dams were inspected for signs of damage or infection. Dams with bunks used them regularly. Consequently, bunk-supplemented dams rested away from their kits more than did nursing dams without bunks. However, compared to bunk-less nursing dams, nursing dams with bunks spent no less time resting with their teats accessible to their kits, and also no less time nursing (a behaviour that was observed in both nest-box and cage). Their time resting away from their kits thus represented enhanced levels of inactivity: compared to standard-housed dams, they spent about 10% more observations resting/sleeping rather than being active. Furthermore, the activity of bunk-supplemented dams also differed qualitatively: when active, bunk-supplemented dams were less likely to be stereotypic. Pacing and head-twirling was thus significantly reduced in bunk dams, both in terms of the proportion of observations devoted to this behaviour, and the proportion of their active time. Furthermore, bunk dams had improved teat health: at weaning, a significantly smaller proportion of enriched nursing females had nipples that were scored as inflamed or crusted with secretions suggestive of mastitis. Despite their mothers' enhanced abilities to rest away from them, the kits of females with bunks did not seem to be adversely affected: along with nursing and teat availability apparently being adversely affected by the provision of bunks, kit growth rates and kit mortality rates were not significantly different between the two groups (although future research should supplement these data with variables better able to detect stress and frustration in kits). Overall, elevated resting bunks enhanced the health and welfare of dams, with no apparent major ill effects to their kits.

Keyword index

W

Y

Z

www.ingramcontent.com/pod-product-compliance
Lightning Source LLC
LaVergne TN
LVHW080117011026
849160LV00013B/1123

* 9 7 8 9 0 8 6 8 6 2 0 5 4 *